CONCORDIA UNIVERSITY CHICAGO
3 4211 00188 4496

Y0-BEB-372

ANNUAL REVIEW OF EARTH AND PLANETARY SCIENCES

ANNUAL REVIEW OF EARTH AND PLANETARY SCIENCES

VOLUME 13, 1985

GEORGE W. WETHERILL, *Editor*
Carnegie Institution of Washington

ARDEN L. ALBEE, *Associate Editor*
California Institute of Technology

FRANCIS G. STEHLI, *Associate Editor*
University of Oklahoma

ANNUAL REVIEWS INC. 4139 EL CAMINO WAY PALO ALTO, CALIFORNIA 94306 USA

ANNUAL REVIEWS INC.
Palo Alto, California, USA

International Standard Serial Number: 0084-6597
International Standard Book Number: 0-8243-2013-1
Library of Congress Catalog Card Number: 72-82137

TYPESET BY A.U.P. TYPESETTERS (GLASGOW) LTD., SCOTLAND
PRINTED AND BOUND IN THE UNITED STATES OF AMERICA

Annual Review of Earth and Planetary Sciences
Volume 13, 1985

CONTENTS

SOME RELATED ARTICLES IN OTHER *ANNUAL REVIEWS*

From the *Annual Review of Astronomy and Astrophysics*, Volume 22 (1984)

Solar Rotation, Robert Howard

Origin and History of the Outer Planets: Theoretical Models and Observational Constraints, James B. Pollack

Helioseismology: Oscillations as a Diagnostic of the Solar Interior, Franz-Ludwig Deubner and Douglas Gough

From the *Annual Review of Biochemistry*, Volume 53 (1984)

Protein–Nucleic Acid Interactions in Transcription: A Molecular Analysis, Peter H. von Hippel, David G. Bear, William D. Morgan, and James A. McSwiggen

Gene Amplification, George R. Stark and Geoffrey M. Wahl

Structure and Function of the Primary Cell Walls of Plants, Michael McNeil, Alan G. Darvill, Stephen C. Fry, and Peter Albersheim

From the *Annual Review of Ecology and Systematics*, Volume 15 (1984)

Geographic Patterns and Environmental Gradients: The Central–Marginal Model in Drosophila *Revisited*, Peter F. Brussard

Ecological Determinants of Genetic Structure in Plant Populations, M. D. Loveless and J. L. Hamrick

Genetic Revolutions in Relation to Speciation Phenomena: The Founding of New Populations, Hampton L. Carson and Alan R. Templeton

Genetic Revolutions, Founder Effects, and Speciation, N. H. Barton and B. Charlesworth

Migration and Genetic Population Structure With Special Reference to Man, E. M. Wijsman and L. L. Cavalli-Sforza

Flow Environments of Aquatic Benthos, A. R. M. Nowell and P. A. Jumars

The Evolution of Food Caching by Birds and Mammals, C. C. Smith and O. J. Reichman

The Role of Disturbance in Natural Communities, Wayne P. Sousa

The Ontogenetic Niche and Species Interactions in Size-Structured Populations, Earl E. Werner and James F. Gilliam

Restitution of r- *and* K-*Selection as a Model of Density-Dependent Natural Selection*, Mark S. Boyce

The Application of Electrophoretic Data in Systematic Studies, Donald G. Buth

Optimal Foraging Theory: A Critical Review, Graham H. Pyke

From the *Annual Review of Fluid Mechanics*, Volume 17 (1985)

Multicomponent Convection, J. S. Turner

Mathematical Models of Dispersion in Rivers and Estuaries, P. C. Chatwin and C. M. Allen

Sound Transmission in the Ocean, Robert C. Spindel

Modeling Equatorial Ocean Circulation, Julian P. McCreary, Jr.

Mantle Convection and Viscoelasticity, W. R. Peltier

From the *Annual Review of Microbiology*, Volume 38 (1984)

Nutrient Transport by Anoxygenic and Oxygenic Photosynthetic Bacteria, Jane Gibson

Alterations in Outer Membrane Permeability, R. E. W. Hancock

Deep-Sea Microbiology, Holger W. Jannasch and Craig D. Taylor

Biology of Iron- and Manganese-Depositing Bacteria, W. C. Ghiorse

Hydrogenase, Electron-Transfer Proteins, and Energy Coupling in the Sulfate-Reducing Bacteria Desulfovibrio, James M. Odom and Harry D. Peck, Jr.

From the *Annual Review of Nuclear and Particle Science*, Volume 34 (1984)

Nucleosynthesis, James W. Truran

From the *Annual Review of Physical Chemistry*, Volume 35 (1984)

Human Effects on the Global Atmosphere, Harold S. Johnston

ANNUAL REVIEWS INC. is a nonprofit scientific publisher established to promote the advancement of the sciences. Beginning in 1932 with the *Annual Review of Biochemistry*, the Company has pursued as its principal function the publication of high quality, reasonably priced *Annual Review* volumes. The volumes are organized by Editors and Editorial Committees who invite qualified authors to contribute critical articles reviewing significant developments within each major discipline. The Editor-in-Chief invites those interested in serving as future Editorial Committee members to communicate directly with him. Annual Reviews Inc. is administered by a Board of Directors, whose members serve without compensation.

ANNUAL REVIEWS OF
Anthropology
Astronomy and Astrophysics
Biochemistry
Biophysics and Biophysical Chemistry
Cell Biology
Earth and Planetary Sciences
Ecology and Systematics
Energy
Entomology
Fluid Mechanics
Genetics
Immunology
Materials Science
Medicine
Microbiology
Neuroscience
Nuclear and Particle Science
Nutrition
Pharmacology and Toxicology
Physical Chemistry
Physiology
Phytopathology
Plant Physiology
Psychology
Public Health
Sociology

SPECIAL PUBLICATIONS
Annual Reviews Reprints:
Cell Membranes, 1975–1977
Cell Membranes, 1978–1980
Immunology, 1977–1979

Excitement and Fascination of Science, Vols. 1 and 2

History of Entomology

Intelligence and Affectivity, by Jean Piaget

Telescopes for the 1980s

A detachable order form/envelope is bound into the back of this volume.

Ann. Rev. Earth Planet. Sci. 1985. 13: 1–4

WITNESSING REVOLUTIONS IN THE EARTH SCIENCES

John Rodgers

Department of Geology and Geophysics, Yale University,
New Haven, Connecticut 06511

The Editorical Committee of the *Annual Review of Earth and Planetary Sciences* has asked me to write a prefatory chapter for this volume, to be based on my experiences in science. A formal biographical account is not in order, but I would like to write, informally and personally if I may, about some of the major developments or "revolutions" that have strongly affected the way I look at and do science. I am reminded of a story told by an old and wise friend on the occasion of his ninetieth birthday. (I still have 20 years to go.) It seems a preacher had been asked to deliver the funeral eulogy for an old man who had been a most unpleasant character in the community—breaking with his family, quarreling with the neighbors, a thorough no-good. The preacher racked his brains for a way to say something good but true about the man, well knowing that all the family and neighbors would be there listening, and he finally found the answer. The old man *had* lived a long time, so the preacher talked not about his life but about all the events and developments the man had witnessed in the course of it. So let me talk about a few of the exciting developments that I have witnessed, that have transformed geology during the years I have been a geologist.

The first subject I shall discuss—it was the first subject, believe it or not, that I taught—is mineralogy. (Please notice the *a* and pronounce the word accordingly—it is not minerology.) When I first studied it, it seemed hopelessly complex and confusing. The simpler compounds attached themselves fairly well to what I knew of chemistry—valence, ionic bonding, etc.—though even here pyrite (FeS_2) was a scandal; the student chemists to whom I taught mineralogy always refused to believe that formula and insisted that our analyses were in error. But the silicates —! They were of course at the core of geology, but how to understand them, or even find

0084–6597/85/0515–0001$02.00

clues for remembering them and their compositions. The oxide formulas then in vogue were impossible to memorize as well as meaningless; I remember also trying to use the concept of silicic acids (ortho, meta, pyro, etc.) and being defeated. And then, during my sophomore year at Cornell, the Chemistry Department invited William Lawrence Bragg (the younger Bragg) to give the George Fisher Baker lectures. I am eternally grateful to my mineralogy professor, J. D. Burfoot, for urging me to attend Bragg's lectures; after the first few talks the chemists stayed away in droves, and Bragg lectured mainly to the physicists and a handful of us geologists. As I listened to him, the scales fell from my eyes, and all the beauty of mineralogy as structural chemistry was revealed. The silicates suddenly became clear, their formulas easy to memorize, and their classification a cornerstone for geological thinking. In his summary lecture, Bragg began with the SiO_4^{4-} tetrahedron and its ways of combining and built up logically to the distinction between continents and ocean basins. This experience led me to recognize at once the truth of King Hubbert's dictum, when I heard it later, that as science advances it becomes not more complex but simpler.

The second transformation I shall discuss was in stratigraphy, my principal subject in the earlier stages of my career. As I grew up in geology, I witnessed and took a small part in the acceptance of the concept of facies in North America—it is difficult now to understand the strength of the then ruling theory of layer-cake stratigraphy and the bitter opposition in some quarters to any deviation from it. Yet it had to be abandoned as we made the transition from studying the stratigraphy of the continental platforms to studying that in the geosynclines, especially in their more interior portions. This change was only a part of the "new wave" in stratigraphy, which ceased to be merely naming and describing stratigraphic units and arguing about their correlation in time from one region to another and became the study of the geologic history they record in terms of sedimentary environments that we can observe in the world today. Not only facies, but deltas, turbidites, and reefs, became key concepts. If we had only listened to A. W. Grabau, we would have moved much faster, but he left the US under a cloud just after the First World War (he had at one point been strongly pro-German), and China, not North America, profited by his broad philosophy. When Carl Dunbar and I tried to summarize the first wave of the "new stratigraphy" in our book published in 1957, we deliberately chose the name of Grabau's great book of 1913 for our own. We hope that we helped to reinstate his thinking into North American geology and to trigger the second wave, which exploded almost immediately afterwards.

My third subject is structural geology, especially of mountain ranges—what is now generally called tectonics. As a graduate student again, I came up against the welter of theories professing to explain orogeny, the series of deformational events that leads to mountain chains, notably the evident shortening of the rocks transverse to the chains. Prof. C. R. Longwell had each member of his class take one such theory and present it (I drew Erich Haarmann, the geotumor man), and the result was about as bewildering as the oxide formulas for the silicates. In general, the theories could be ranged in three major groups, according to the mechanism invoked to explain the shortening, though fundamental differences remained within each group. The classical idea, coming down from the nineteenth century, was contraction, producing compression in the already cooled crust as it contracted around the still cooling interior of the Earth. A major challenger was the gravitational idea, that shortening is a secondary phenomenon produced by gravity sliding off the sides of large, primary, vertical uplifts (the geotumors or undations). A third group of theories called on convection currents within the Earth, though the evident relation to the theory of continental drift, plus the prevailing idea that the Earth's mantle is rigid (as shown by its reaction to seismic waves), made these theories definitely unorthodox. In 1938, I witnessed the memorable experiment on convection currents performed by David Griggs before the Geological Society of America (the paper was published in 1939) to the great scandal of such outstanding senior geologists as Andrew C. Lawson (that year's Penrose medalist) and Bailey Willis, who agreed on this at least, if on little else. My prejudice has been for such ideas ever since, even during the dark days after the Second World War when it seemed proved that the major discontinuities in the mantle were compositional. Whatever my prejudice, however, it was evident that if so many able thinkers about orogeny were entirely unable to agree (there was roughly one theory per theorist), then something must be missing, some essential facts or ideas or both.

My own contribution to the solution of this dilemma was negative; I left North America for a year to study the Alps. As a result, the Yale Geology Department could appoint a visiting professor for that year; we chose S. Warren Carey, and North American geology has never been the same since. He traveled all over the continent, he lectured in his inimitable now-you-see-it-now-you-don't style, he talked to anyone who would listen, and when he was through, no one could laugh off continental drift any more. Then, like St. Paul at Damascus, the archpriest of the geophysical fixists, J. Tuzo Wilson, was converted, and the revolution was on. The geophysicists finally stopped saying that large-scale thrust faulting, continental drift, and convection currents are manifestly impossible, whatever the stupid geologists might adduce as evidence, and began to look at the facts, first their own new evidence from the oceans and then even the old geological evidence from the continents.

Because of my prejudice for convection currents, I was from the first preadapted, to use the paleontological term, to accept the new global tectonics as a replacement for all the older theories, the tectonic equivalents of the old oxide formulas. Others haven't been so fortunate; the contractionists (e.g. Jeffreys, the Meyerhoffs) and the gravitational sliders (e.g. van Bemmelen, Belousov) are still not convinced. Who is right? In the past, some very fashionable hypotheses have come and gone, but others have stayed with us—the glacial theory, organic evolution, the structural chemistry of minerals—because they integrated a large and hitherto inchoate body of facts within a manageable theoretical framework. To which group does the new global tectonics belong? The decision won't be made by the protagonists and antagonists who are now arguing over the hypothesis—before they change their minds they will die off—but by the younger generation, who having watched the debate will make up their minds. (They will then start arguing over other hypotheses, to be decided by the next younger generation.) The one thing we can say with confidence is that *no* hypothesis is strictly correct; it may more or less approach correctness, but in due course it too will be modified or replaced. Already we have witnessed how the first form of the global tectonics, so enthusiastically acclaimed as the truth (eight great rigid plates, only three kinds of boundaries), has been superseded as we add marginal seas, accretionary wedges, soft-plate tectonics; further surprises are doubtless in store. What we should ask of a hypothesis is not that it be true, but that it be fruitful, that it lead to a *simplification* of ideas of the kind that has revolutionized mineralogy, stratigraphy, and tectonics within my lifetime.

Ann. Rev. Earth Planet. Sci. 1985. 13: 5–27

PATTERNS OF ALLUVIAL RIVERS

S. A. Schumm

Department of Earth Resources, Colorado State University, Fort Collins, Colorado 80523

INTRODUCTION

The pattern (planform) of a river can be considered at vastly different scales, depending upon both the size of the river and the part of the fluvial system that is under consideration (Figure 1). For example, in the broadest sense, river patterns comprise a drainage network (dendritic, parallel, trellis, etc; Figure 1*A*). The type of pattern is of interest to geomorphologists and geologists who interpret geologic conditions from aerial photographs.

At another scale a river reach (which in Figure 1*B* is meandering) is of interest to the geomorphologist who is interested in what that pattern reveals about river history and behavior, and to the engineer who is charged with maintaining navigation and preventing major instability. When a single meander is examined (Figure 1*C*), the hydraulics of flow, the sediment transport, and the potential for bank erosion are of concern. In addition, the sedimentologist is interested in the distribution of sediment within the bend, bed forms within the channel (Figure 1*D*), and sedimentary structures (Figure 1*E*), which also establish a component of roughness for the hydraulic engineer. Finally, the individual grains (Figure 1*F*) provide geologic information on the sediment sources, the nature of sediment loads, and the feasibility of dredging for gravel. There is an interaction of hydrology, hydraulics, geology, and geomorphology at all scales, which emphasizes the point that the fluvial system as a whole cannot be ignored, even though only a component of the system is to be studied.

In this review only the patterns or planforms of alluvial rivers are discussed, although it is apparent that the hydrologic and sediment yield characteristics of the drainage basin (Figure 1*A*), as well as its geologic history, cannot be ignored in the explanation of the pattern of any river

0084–6597/85/0515–0005$02.00

reach. This article discusses the great variability and dynamic behavior of river patterns, considers the reasons for the diversity of patterns, and explains why an understanding of patterns is essential for mined-land reclamation, channel modification for flood control and navigation, the identification of areas of active tectonics, and the litigation of boundaries (Elliott 1984).

River patterns provide information on modern river characteristics and behavior. The civil engineer is involved with river patterns primarily because of the pattern changes that may occur at bridges and other sites of

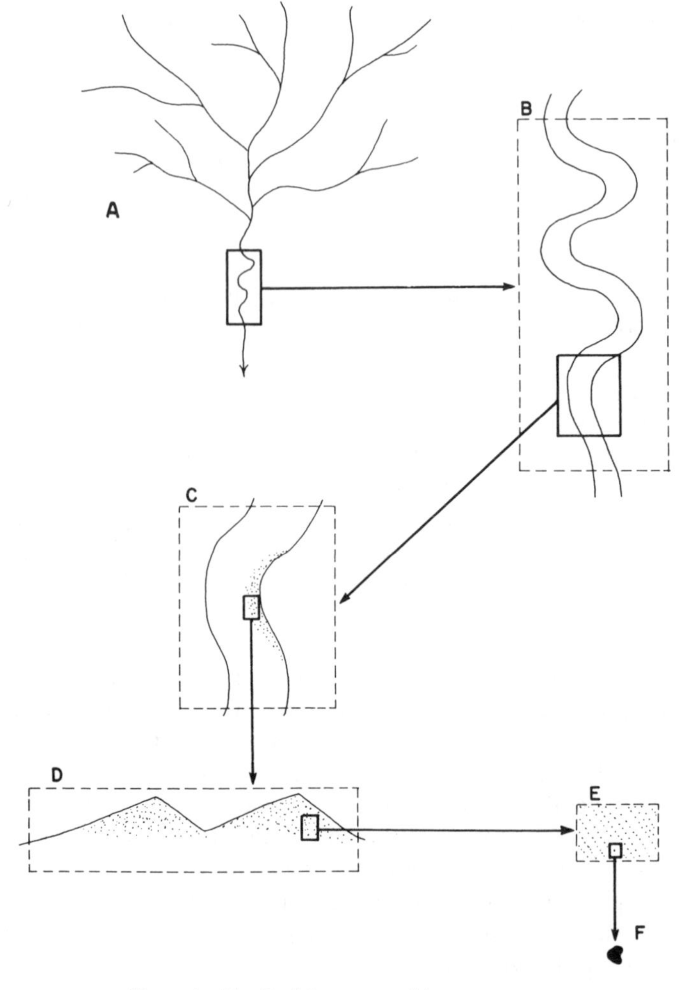

Figure 1 The fluvial system and its components.

construction (Shen et al 1981). Hence, not only the pattern and its characteristics are of interest, but also the dynamics of a particular pattern and its potential for change through time are of practical concern. In addition, an understanding of river patterns provides a basis for understanding ancient fluvial deposits and inferring environments of deposition (Galloway & Hobday 1983), and it also provides an empirical basis for determining past river morphology and paleohydrology (Schumm 1977, Gregory 1983).

RIVER PATTERNS

Depending on the nature of the materials through which a river flows, there are three major categories of stream channels: bedrock, semicontrolled, and alluvial. The bedrock channel is fixed in position, and it is stable over long periods of time. If the bedrock is weak, there can be lateral shift of the channel, but in most cases bedrock control means that the channel is stable. The semicontrolled channel refers to rivers that are controlled only locally by bedrock or resistant alluvium. The pattern may change where the channel encounters more resistant materials, and the channel either can be very stable at that particular locality or can shift away from the bedrock controls. The bed and banks of alluvial channels are composed of sediment transported by the stream. Therefore, the alluvial channel is susceptible to major pattern change and to significant shifts in channel position as the alluvium is eroded, transported, and deposited, and as the sediment load and water discharge change.

Geomorphic and engineering studies have demonstrated that there is a great range of alluvial river types; thus any attempt to classify them based upon pattern characteristics alone is a frustrating task. For example, Brice et al (1978) have illustrated the range of channel patterns (Figure 2). They recognized three basic types of channels that are characterized by degrees of sinuosity (Figures 2*A*,*B*), braiding (Figures 2*C*,*D*), and anabranching (Figures 2*E*,*F*). Sinuosity is the ratio of channel length to valley length (L_c/L_v) or valley slope to channel slope (S_v/S_c). The range of sinuosity is from 1.0 (straight) to about 3.0. Some of the most sinuous channels appear to have one meander pattern superimposed on another (two-phase patterns; see Figure 2*B*, parts 6, 7).

There are also different degrees of braiding, expressed as the percentage of channel length that is divided by islands or bars (Figure 2*C*). There are two types of braided channels: island braided and bar braided (Figure 2*D*). Islands are relatively permanent features that are vegetated, whereas bars are bare sand-and-gravel deposits. Obviously, the stability of the channel is greatly enhanced when vegetation colonizes the bars and the channel becomes island braided.

Another pattern that has been identified is the anabranch channel (Figures 2*E,F*). Anabranching is the division of a river by islands whose width is greater than three times water width at average discharge (Brice et al 1978). The degree of anabranching is the percentage of reach length that is occupied by large islands.

Anastomosing channels are distinct from the anabranched channels, as they are multiple channel systems (Table 1) having major secondary channels that separate and rejoin the main channel to form a network (Schumm 1977, p. 155). The individual branches of anastomosing channels can be meandering, straight, or braided; therefore, they are not considered

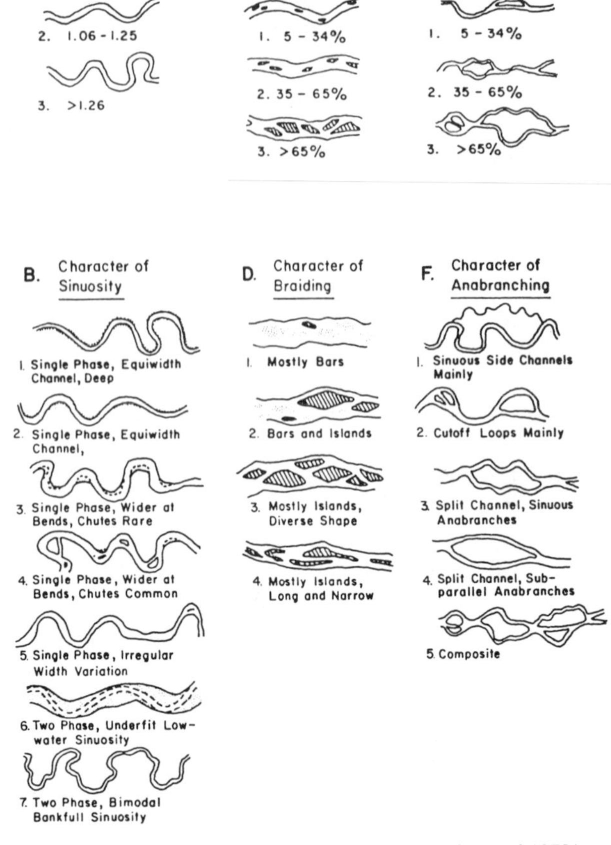

Figure 2 Types of channel patterns. (From Brice et al 1978.)

Table 1 Classification of stable alluvial channels (after Schumm 1977)

Type of channel	Bed load (% of total load)	Type of River: Single channel	Type of River: Multiple channel
Suspended load	<3	Suspended-load channel. Width-depth ratio <10; sinuosity >2.0; gradient relatively gentle.	Anastomosing system
Mixed load	3–11	Mixed-load channel. Width-depth ratio >10, <40; sinuosity <2.0, >1.3; gradient moderate. Can be braided.	Delta distributaries Alluvial plain distributaries
Bed load	>11	Bed-load channel. Width-depth ratio >40; sinuosity, <1.3; gradient relatively steep. Can be braided.	Alluvial fan distributaries

separately from the three basic patterns (meandering, braided, and straight) identified by Leopold & Wolman (1957).

The variety of channel patterns, as illustrated by Figure 2, is a result of the great range of hydrologic conditions, sediment characteristics, and geologic histories of the rivers of the world. Therefore, river patterns provide a key to other river characteristics, both morphologic and dynamic. For example, the pattern of the alluvial channels on the Great Plains of the western United States is related to channel shape. Low width-depth ratio (i.e. relatively narrow and deep) channels are relatively sinuous, whereas wide and shallow channels are relatively straight (Schumm 1977). Therefore, an indication of the morphology and relative stability of a river can be obtained by studying its pattern or planform.

CLASSIFICATION

For simplicity and convenience of discussion, the range of common channel patterns can be grouped into five basic patterns (Figure 3). These five patterns illustrate the overall range of channel patterns to be expected in nature, but of course they do not show the details that can be seen in Mollard's (1973) 17-pattern classification or the 14 patterns described by Schumm (1981). Nevertheless, Figure 3 is more meaningful than a purely descriptive classification of channels because it is based on cause-and-effect relations and illustrates the differences to be expected when the type of sediment load, flow velocity, and stream power differ among rivers. It also explains pattern differences along the same river (Schumm 1977).

A classification of alluvial channels should be based not only on channel pattern but also on the variables that influence channel morphology. This is particularly true if the classification is to provide information on channel stability. Numerous empirical relations demonstrate that channel dimensions are largely due to water discharge, whereas channel shape and pattern are related to the type and amount of sediment load moved through the channel (Table 1). Galloway & Hobday (1983) used this classification as a basis for identifying fluvial clastic depositional systems.

As indicated by Figure 3, when the channel pattern changes from 1 to 5, other morphologic aspects of the channel also change; that is, for a given discharge, both the gradient and the width-depth ratio increase. In addition, peak discharge, sediment size, and sediment load probably increase from pattern 1 to pattern 5. With such geomorphic and hydrologic

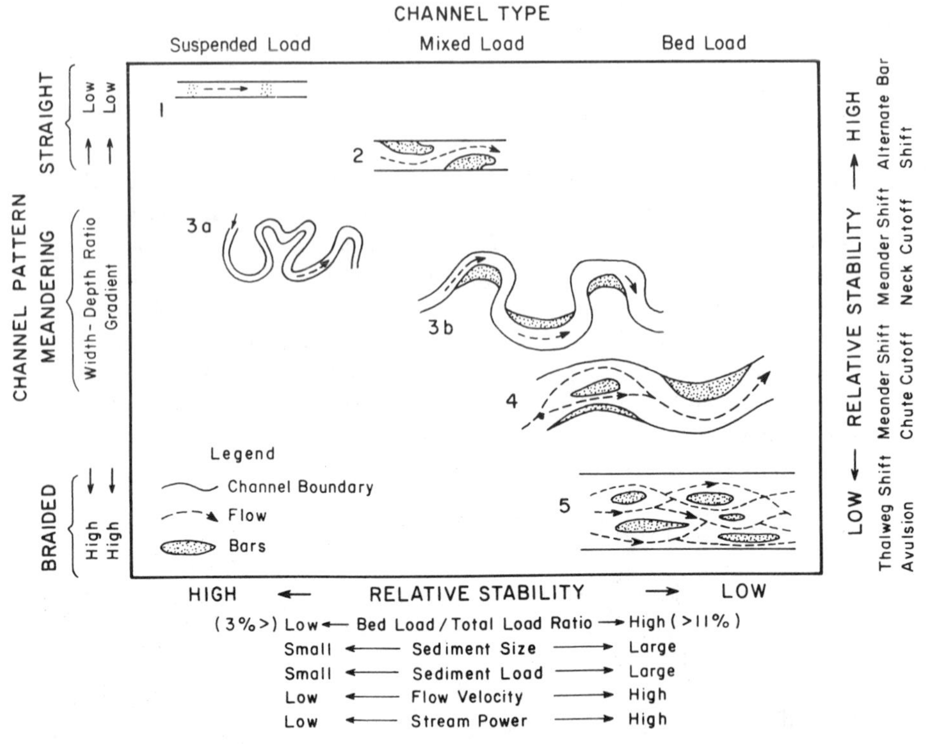

Figure 3 Channel classification based on pattern and type of sediment load, showing types of channels, their relative stability, and some associated variables. (After Schumm 1981.)

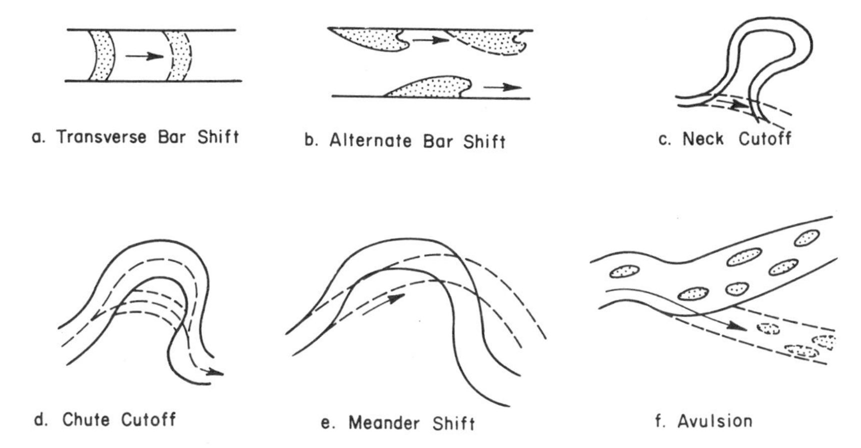

Figure 4 Some channel changes that can be expected along alluvial rivers. Dashed lines indicate future conditions.

changes, hydraulic differences can be expected, and flow velocity, tractive force, and stream power also increase from pattern 1 to 5. Therefore, the stability of a graded stream decreases from pattern 1 to pattern 5, with patterns 4 and 5 being the least stable.

A brief discussion of the five basic patterns is presented in what follows.

PATTERN 1 The suspended-load channel is straight with a relatively uniform width (Figure 3). It carries a very small load of sand and gravel. Gradients are low, and the channel is relatively narrow and deep (low width-depth ratio). The banks are relatively stable because of their high silt-clay content. Therefore, the channel is not characterized by serious bank erosion or channel shift. Bars may migrate through the channel (Figure 4*a*), but such changes should not create undue instability. Pattern 1 channels are rare, but they are stable unless the channel has been artificially straightened and therefore steepened. A naturally straight channel poses few problems, but an artificially straightened channel is subject to degradation and scour, bank erosion, and an increase of sinuosity (Schumm et al 1984).

PATTERN 2 The mixed-load straight channel has a sinuous thalweg (Figure 3). It is relatively stable and carries a small load of coarse sediment, which may move through the channel as alternate bars (Figure 4*b*). As these bars shift through the channel, banks are alternately exposed and protected by the alternate bars. Hence, at any one location the thalweg will shift with time. This means that the apparent deposition or fill at one side of the channel will be replaced by scour as an alternate bar migrates downstream.

Also, at any time, one side of the channel may be filling while the other is scouring.

PATTERN 3 This pattern is represented by two channel patterns, which are only two of a continuum of meandering patterns (Figure 2*B*). Pattern 3*a* shows a suspended-load channel that is very sinuous. It carries a small amount of coarse sediment. The channel width is roughly equal and the banks are stable, but meanders will tend to cut off at their necks (Figure 4*c*). Pattern 3*b* shows a less stable type of meandering stream. Mixed-load channels with high bed loads and banks containing low-cohesion sediment will be less stable than the suspended-load channels. The sediment load is large, and coarse sediment is a significant part of the total load. The channel is wider at bends, and point bars are large. Meander growth and shift (Figure 4*e*) and neck and chute cutoffs are also characteristic (Figure 4*c*). The channel, therefore, is relatively unstable, but the location of the cutoffs and the pattern of meander shift can be predicted. The shifting of the banks and thalweg follows a more or less regular pattern.

The shift of a meander (Figure 4*e*) creates major channel problems, as the flow alignment is drastically altered and bank erosion may become very serious. The rate of a meander shift will vary greatly depending on where in the continuum of meandering patterns the river fits.

PATTERN 4 This pattern represents a meander-braided transition (Figure 3). Sediment loads are large, and sand, gravel, and cobbles are a significant fraction of the sediment load. The channel width is variable but is relatively large compared with the depth (high width-depth ratio), and the gradient is steep. Chute cutoffs, thalweg and meander shift, and bank erosion are all typical of this pattern (Figures 4*d*,*e*). In addition to these problems, which are also characteristic of pattern 3, the development of bars and islands may modify flow alignments and change the location of bank erosion.

PATTERN 5 This bed-load channel is a typical bar-braided stream (Figure 3). The bars and thalweg shift within the unstable channel, and the sediment load and size are large. Braided streams are frequently located on alluvial plains and alluvial fans. Their steep gradients reflect a large bed load. Bank sediments are easily eroded, gravel bars and islands form and migrate through the channel, and avulsion (Figure 4*f*) may be common.

The other type of braided stream is the island-braided stream. This is a much more stable channel, and it would appear to the left of pattern 5 on Figure 3. The Mississippi River above the junction of the Missouri River is of this type. Island formation, erosion, and shift occur in these channels, but at a much slower rate than in a bar-braided channel. The anabranching or anastomosing channels may also be of this type (Figure 2*F*).

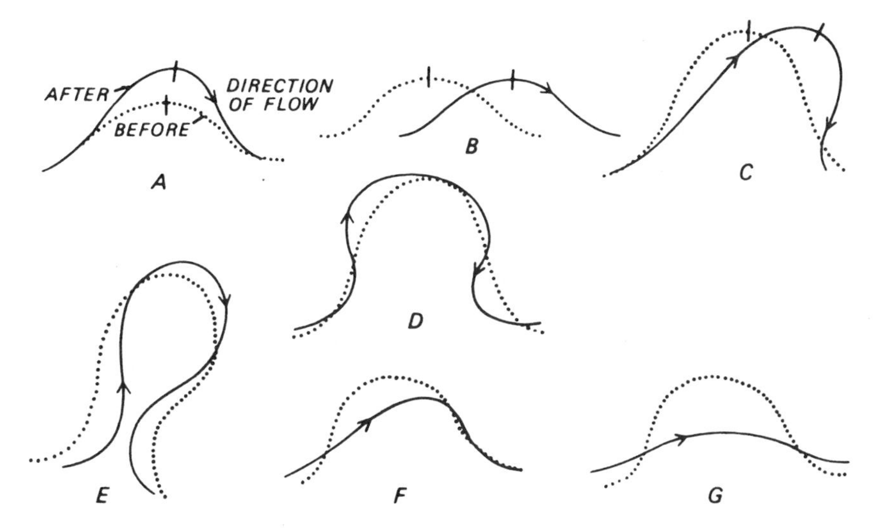

Figure 5 Patterns of meander growth and shift: (*a*) extension, (*b*) translation, (*c*) rotation, (*d*) conversion to a compound meander, (*e*) neck cutoff, (*f*, *g*) chute cutoffs. (From Brice 1974.)

PATTERN VARIABILITY

Historical studies of rivers reveal that rivers are shortened and steepened when meanders cut off, whereas rivers are lengthened and the gradient is decreased when meanders grow. Pattern change is to be expected whenever water and sediment flow through a stream channel. Figure 5 shows various modes of meander-loop behavior as observed by Brice (1974). All of these changes are associated with rivers that could be considered to be unstable. However, there are degrees of stability, and a stable river may have cutoffs and rapid meander shift and growth. From several points of view, this is an important consideration that engineers and geomorphologists should address.

Channel Stability

Because alluvial channels are composed of sediment transported by the stream, the bed and banks are erodible; therefore, no alluvial channel is actually stable in the sense that no change occurs. Rather there are degrees of stability depending upon the rate and type of channel change. For example, aggradation or degradation, caused by changes of baselevel or hydrologic conditions, will change the channel cross-section shape and dimensions. Although this is certainly an instability, the channel pattern may not be affected. On the other hand, local but dramatic changes of pattern such as cutoffs (Figures 4*c,d*) and meander shift (Figures 4*e*, 5) may be considered to be normal river behavior. Therefore, meander growth and shift alone is not a criterion of instability. However, a riparian landowner whose property is being destroyed by the meander change will not be convinced that such a river is relatively stable.

In straight channels, bar shift may greatly alter within-channel patterns and cause major problems with docking facilities and water intakes (Figures 4*a,b*), but the bank line of the channel may nevertheless be generally stable. Therefore, lateral change may be normal river behavior, but vertical change or change of channel size reflects true instability. This is an important distinction, because it is necessary to identify "stable" (graded, equilibrium, regime) channels in order to stratify river data for analysis. In order to develop relations between channel morphology, hydrology, and other characteristics, only data from stable channels should be used.

Thresholds

An important factor relating to pattern variability is the concept of geomorphic thresholds (Schumm 1977, 1979). Thresholds have been recognized in many fields, and perhaps the best known to geologists and engineers are the threshold velocities required to set in motion sediment particles of a given size. With a continuous increase in velocity, threshold velocities are encountered at which movement begins; conversely, with a progressive decrease in velocity, threshold velocities are encountered at which movement ceases. Particularly dramatic also are the changes in bed forms at threshold values of stream power.

In the examples cited, an external variable changes progressively, thereby triggering abrupt changes or failure within the affected system. The response of a system to an external influence occurs at what is referred to as the *extrinsic threshold*; that is, the threshold exists within the system, but it is not crossed and change does not occur without the influence of an external variable.

The other type of threshold is the *intrinsic threshold*, where changes occur without a change in an external variable. In dry regions, sediment storage progressively increases the slope of valley floors and alluvial fans until failure occurs by gullying. This is an intrinsic geomorphic threshold (Schumm 1979).

The variability of sinuosity and the range of pattern changes from meandering to braided provide excellent examples of the effects of both intrinsic and extrinsic threshold conditions. Fisk (1944) and Winkley (1970) showed that the sinuosity and length of the Mississippi River have varied dramatically through time. Its sinuosity decreased to a minimum when an avulsion or a series of cutoffs straightened its channel. Such changes may be related to major changes of sediment load or to an increase of peak discharge, but they may also be due to a progressive increase of sinuosity (with an accompanying reduction of channel gradient) to the point that aggradation and cutoffs or avulsion results. Such a situation appears to exist along the sinuous parts of the Rio Puerco arroyo, New Mexico, where the meander amplitude has increased in some reaches to the point that sediment is being deposited in the upstream limb of each meander and the bends are being cut off (Figure 9-3 of Schumm 1977). These changes reflect an intrinsic control by the channel pattern itself.

The work of Lane (1957) and Leopold & Wolman (1957) indicates that there is a gradient or discharge threshold above which rivers tend to be braided (Figures 6, 7). The experimental work reported by Schumm & Khan (1972) shows that for a given discharge, as valley-floor slope is progressively increased, a straight river becomes sinuous and then eventually braided at high values of stream power and sediment transport (Figure 8). Rivers that are situated close to the meandering-braided threshold should have a history characterized by transitions in morphology from braided to meandering and vice versa.

The suggestion made here is that if one can identify the natural range of patterns along a river, then within that range the most appropriate channel

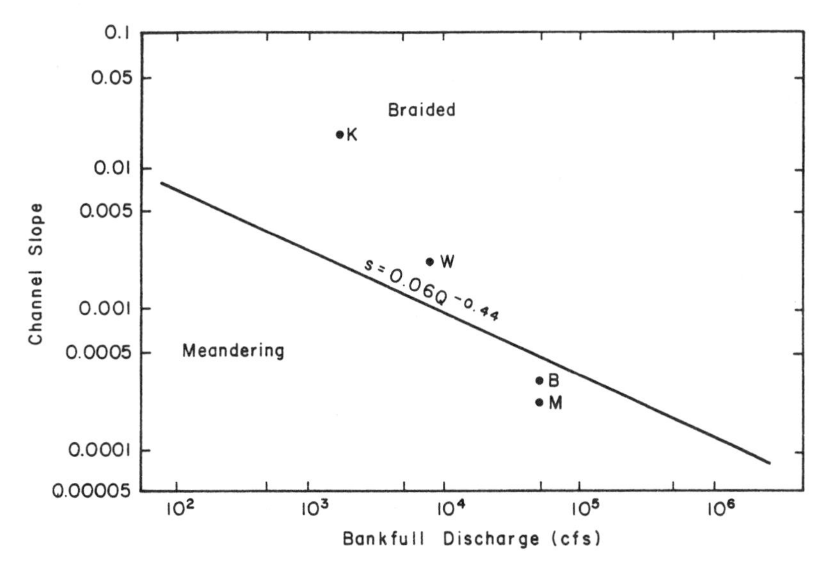

Figure 6 Leopold & Wolman's (1957) relation between channel patterns, channel gradient, and bankfull discharge. The letters B and M identify braided and meandering reaches of the Chippewa River, while the letters K and W refer to the Kowhai and Wairau rivers. (From Schumm & Beathard 1976.)

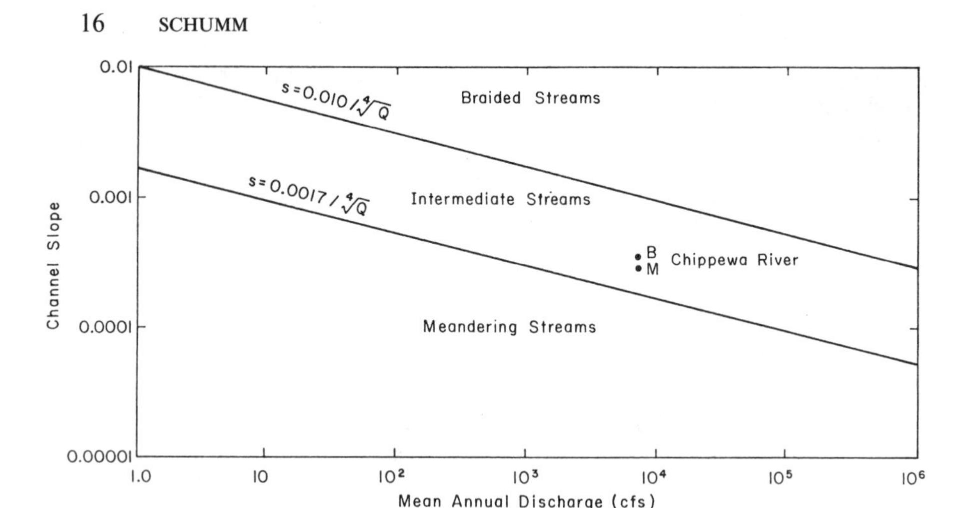

Figure 7 Lane's (1957) relation between channel patterns, channel gradient, and mean discharge. The letters B and M identify the positions of the braided and meandering reaches of the Chippewa River. (From Schumm & Beathard 1976.)

pattern and sinuosity probably can be identified. If so, the engineer can work with the river to produce its most efficient or most stable channel. Obviously a river can be forced into a straight configuration, just as it can also be made more sinuous, but there is a limit to the changes that can be induced; beyond this limit, the channel cannot function without a radical morphologic adjustment, as suggested by Figures 6, 7, and 8. The

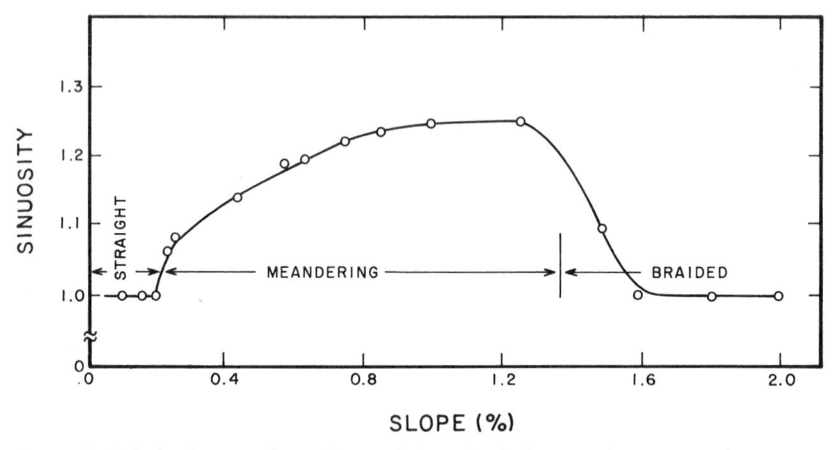

Figure 8 Relation between flume slope and sinuosity during experiments at constant water discharge. Sediment load, stream power, and velocity increase with flume slope, and a similar relation can be developed with these variables. (From Schumm & Khan 1972.)

identification of rivers that are near the pattern threshold would be useful, because a braided river near the threshold might be converted to a more stable single-thalweg stream. On the other hand, a meandering stream near the threshold should be identified in order that steps might be taken to prevent braiding due perhaps to sediment load changes as a result of changes of land use.

In most cases it is difficult to determine if a river is susceptible to the type of treatment discussed in the preceding sections. Perhaps the best qualitative technique for gauging river stability is to compare the morphology of numerous reaches and then determine whether or not there has been a change in the position and morphology of the channel during the last few centuries. Another approach might be to identify the position of the river on the Leopold & Wolman (1957) or Lane (1957) gradient-discharge graphs (Figures 6 and 7). If a braided river plots among the meandering channels, or vice versa, it is unstable and a likely candidate for change.

An example is provided by the Chippewa River of Wisconsin (Schumm & Beathard 1976), a major tributary of the Mississippi River. The Chippewa rises in northern Wisconsin and flows 320 km to the Mississippi, entering it 120 km below Saint Paul. It is the second largest river in Wisconsin, with a drainage basin area of 24,600 km^2.

From its confluence with the Mississippi to the town of Durand, 26.5 km up the valley, the Chippewa is braided. The main channel is characteristically broad and shallow, and it contains shifting sand bars. The bankfull width, as measured from US Geological Survey topographic maps, is 333 m. The sinuosity of this reach is very low (1.06). However, in the 68-km reach upstream from Durand to Eau Claire, the Chippewa River abruptly changes to a meandering configuration, with a bankfull width of 194 m and a sinuosity of 1.49. The braided reach has a channel gradient of 0.00033, whereas the meandering reach has a gradient of 0.00028.

The relations described by Leopold & Wolman (1957) and Lane (1957) provide a means of evaluating the relative stability of the modern channel patterns of the Chippewa River. The bankfull discharge was plotted against channel slope in Figure 6 for both the braided and the meandering reaches of the Chippewa. The value used for the bankfull discharge is 1503 $m^3 s^{-1}$ (53,082 cfs), which is the flood discharge having a return period of 2.33 years. The braided reach plots higher than the meandering reach, but both are well within the meandering zone, as defined by Leopold & Wolman. This suggests that the braided reach is anomalous; that is, according to this relation the lower Chippewa would be expected to display a meandering pattern rather than a braided one. Even when the 25-year flood of 98,416 cfs is used, the braided reach still plots within the meandering region of Figure 6.

When the Chippewa data are plotted on Lane's graph (Figure 7), the same relation exists. The Chippewa falls in the intermediate region, but within the range of scatter about the regression line for meandering streams. Again the braided reach is seen to be anomalous because it should plot much closer to or above the braided-stream regression line. The position of the braided reach, as plotted in both figures, indicates that this reach should be meandering, and historical studies reveal that it had a sinuosity of about 1.3 in the late eighteenth century, before an avulsion changed the position and pattern of the channel.

It appears that the lower Chippewa has not been able to adjust as yet to its new position and steeper gradient, and the resulting bed and bank erosion has supplied large amounts of sediment to the Mississippi. The normal configuration of the lower Chippewa is sinuous, and if it could be induced to assume such a pattern, the high sediment delivery from the Chippewa could be controlled. An appropriate means of channel stabilization and sediment load reduction in this case is the development of a sinuous channel.

More detailed studies of the Chippewa River basin since the above suggestions were made (Schumm & Beathard 1976) indicate that upstream sediment production must be controlled, especially where the upper Chippewa River is cutting into the Pleistocene outwash terraces. If the contribution of sediment from these sources were reduced, the lower Chippewa could resume its sinuous course.

An indication that the pattern conversion of the Chippewa could be successful if the upstream sediment sources were controlled is provided by the Rangitata River of New Zealand (Schumm 1979). The Rangitata is the southernmost of the major rivers that traverse the Canterbury Plain of the South Island. It leaves the mountains through a bedrock gorge. Above the gorge, the valley of the Rangitata is braided, and it appears that the Rangitata should be a braided stream below the gorge, as are all the other rivers crossing the Canterbury Plain. However, below the gorge, the Rangitata is meandering. A few miles farther downstream, the river cuts into high Pleistocene outwash terraces, and it abruptly converts from a meandering to a braided stream. The braided pattern then persists to the sea. If the Rangitata could be isolated from the gravel terraces, it probably could be converted to a single-thalweg sinuous channel, because the Rangitata is a river near the pattern threshold.

Other New Zealand rivers are also near the pattern threshold, and therefore they are susceptible to pattern change. In fact, New Zealand engineers are attempting to accomplish this pattern change on these rivers in order to produce "single-thread" channels that will reduce flood damage and be less likely to acquire large sediment loads from their banks and terraces. For example, the Wairau River, a major braided stream, has been converted from its uncontrolled braided mode to that of a slightly sinuous, single-thalweg channel that is relatively more stable (Pascoe 1976). The increase in sinuosity is only from 1.0 to 1.05, and this was accomplished by the construction of curved training banks. In Figure 6 the Wairau River plots close to the threshold line, and with the reduction of sediment load produced by bank stabilization it appears that the pattern threshold can be crossed successfully.

The Kowhai River of New Zealand is being modified in the same manner as the Wairau (Thomson & MacArthur 1969). Whereas much of the sediment load in the Wairau River is derived from bank and terrace erosion, which can be controlled, high sediment loads are delivered to the Kowhai River directly from steep and unstable mountain slopes. In Figure 6 the Kowhai River plots well above the threshold line, and without a major reduction in upstream sediment, it may be difficult to maintain a single-thalweg channel at this location.

The variability of the Rangitata River pattern indicates that conversions from braided to single-thalweg channels should be possible for the Chippewa and Wairau rivers. However, not all braided rivers can be so readily modified, as such a change depends on their position with regard to the pattern thresholds in Figures 6, 7, and 8.

RIVER METAMORPHOSIS

In addition to the expectable alterations of channel patterns that have been already discussed, major changes of discharge, sediment type, and sediment load can also occur as a result of either man's activities or past climatic changes, and these changes may drastically and totally alter river morphology (Hickin 1983). This transformation has been referred to as river metamorphosis (Schumm 1969, 1977, pp. 159–71). There are six possibilities of metamorphosis as each of the three channel types change to the other two types (that is, a straight channel may become sinuous or braided, a meandering channel may become straight or braided, and a braided channel may become straight or meandering). In each case, the change in the controlling variables differs and the river response is dramatically different.

The South Platte and Arkansas rivers in eastern Colorado are excellent examples of rivers that have undergone dramatic historic changes so extensive that they can be termed a metamorphosis (Schumm 1969, Nadler & Schumm 1981). Measurements and reports by explorers in the early part of the nineteenth century show that both rivers were wide (up to 1.5 km), shallow braided streams. The rivers today exhibit very different channel characteristics (Figure 9).

A

B

Figure 9 *A* Aerial photograph of the South Platte River near Julesburg, Colorado, in July 1977. The channel in the nineteenth century occupied the area now supporting cottonwood tree growth. *B* Aerial photograph of the Arkansas River near Bent's Old Fort (east of La Junta, Colorado) in June 1977. The channel in the nineteenth century was braided.

The South Platte and Arkansas rivers originate in the mountains of central Colorado and flow eastward on valley alluvium of Pleistocene and Holocene age. Both rivers experienced large seasonal fluctuations in discharge due to snow melt, and they decreased in volume as they crossed the semiarid plain, owing to seepage and evaporation losses. Agriculture began in both basins immediately following the first gold rush to Colorado in 1858. By 1895 there were 20 major irrigation diversions on the Arkansas River between Pueblo and the Kansas border, and similar diversions were occurring on the South Platte River.

The hydrologic nature and type of floodplain vegetation of both rivers changed appreciably. As water tables rose, stream flows became perennial, flood peaks decreased, and floodplains were able to sustain denser vegetation. According to early descriptions, woody vegetation was sparse along the rivers. However, the floodplains are now occupied by cottonwoods, and it is apparent that there is more vegetation along the rivers today. This increase in vegetation probably reflects a higher water table, a result of increased irrigation activity. In addition, salt cedars invaded the Arkansas River Valley. These hydrologic and vegetative changes produced major morphologic changes along both rivers.

Figure 10 depicts the manner of the South Platte River metamorphosis, which is characterized by stream narrowing and floodplain construction by vertical accretion. The thalweg did not aggrade, but a floodplain was formed adjacent to the thalweg by island construction (Figure 10*B*) and channel filling. When flood peak decreased, vegetation quickly colonized areas below the mean high water level of the channel (Figure 10*C*). In this way, newly formed bars were stabilized by vegetation and became islands. The channels, which surrounded these islands, no longer shifted, because vegetation had fixed the position of the banks and islands. Channel abandonment and island attachment to the floodplain followed (Figure 10*D*).

The former braided pattern can be seen on aerial photographs (Figure 9*A*). The ages of the largest cottonwood trees on the floodplain indicate that the islands were being colonized by woody vegetation at least 60 years ago. This suggests that the metamorphosis began after 1900.

Figure 11 depicts the manner of river metamorphosis at Bent's Old Fort on the Arkansas River. The important characteristics of this model are point-bar stabilization and meander-loop enlargement. Perennial flows, flood-peak reduction, and especially dense salt-cedar growth were major factors leading to the metamorphosis. Salt cedars colonized the channel below mean high water level and stabilized the point bars during the drought of the 1930s. This process allowed meander loops to enlarge as channel width decreased.

The development of the meandering pattern at this reach of the Arkansas

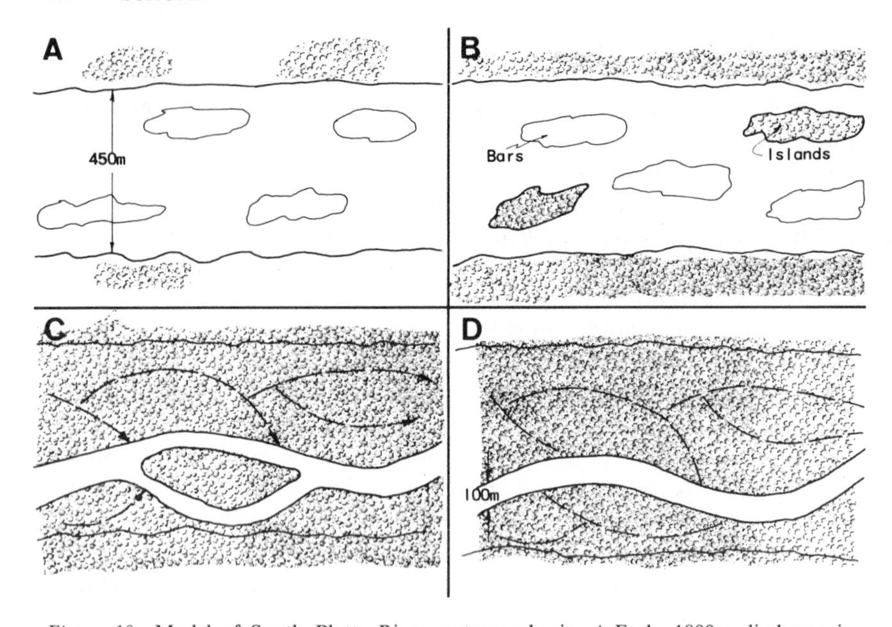

Figure 10 Model of South Platte River metamorphosis. *A* Early 1800s: discharge is intermittent, bars are transient. *B* Late 1800s: discharge is perennial, vegetation is thicker on floodplain and islands. *C* Early 1900s: droughts allow vegetation to establish itself below mean annual high water level, bars become islands, single thalweg is dominant. *D* Modern channel: islands attached to floodplain, braided patterns on floodplain are vestiges of historic channels. (From Nadler & Schumm 1981.)

was the result of an influx of fine sediment (suspended load) from a tributary (Timpas Creek) that incised deeply into the valley fill. The sediment influx converted the Arkansas River in this reach from a braided bed-load channel to a meandering mixed-load channel (Figure 3). Farther downstream, beyond the effects of Timpas Creek, the channel remained straight.

CONTROLS

Alluvial rivers are open channels that clearly are fashioned by the water conveyed by the channel. Therefore, an explanation of channel patterns should be largely hydrologic. However, all types of channels are formed at similar average discharges in the field and in the laboratory by changes of valley slope and sediment supply. The bankfull, average, or mean-annual discharge determines to a large extent channel dimensions such as width, depth, meander amplitude, and meander wavelength, but the quantity of water moving through a channel does not affect the basic pattern. Nevertheless, under otherwise similar conditions a river that has flashy

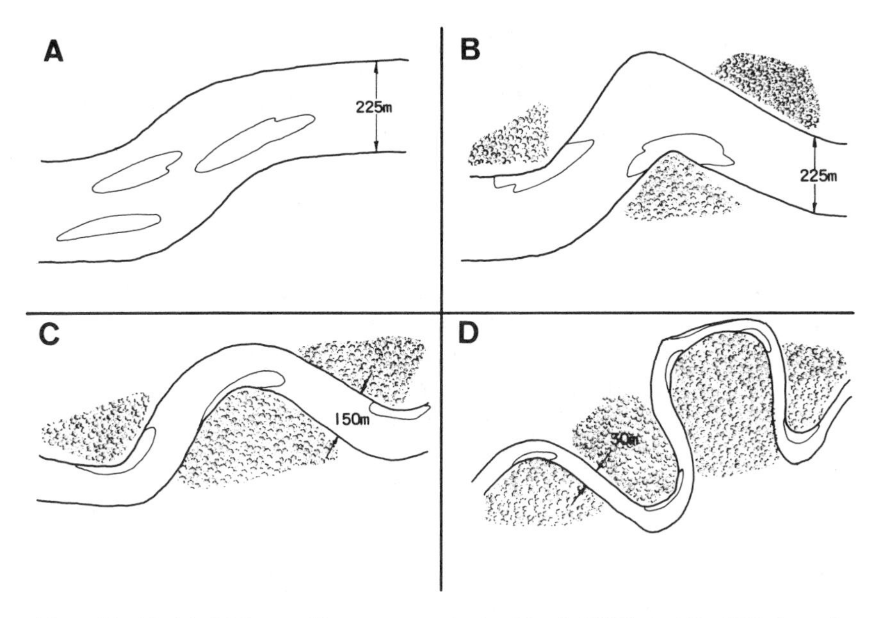

Figure 11 Model of Arkansas River metamorphosis at Bent's Old Fort. *A* Pre-1900 channel. *B* 1926 channel. *C* Channel between 1926 and 1953. *D* Modern channel. (From Nadler & Schumm 1981.)

discharge (high peak discharge to mean discharge ratio) can be braided, in contrast to the channel with a more regular or uniform discharge. Therefore, under some conditions the hydrologic character of the fluvial system is significant (Schumm 1977).

The type of sediment load transported by the river also plays a role (Table 1, Figure 3). Channels that transport relatively low sand-and-gravel (bed load) loads are more likely to be sinuous, and they are not braided. Studies of Great Plains rivers show that channels with a low ratio of bed load to suspended load are relatively narrow, deep, and sinuous, whereas when the ratio is high, the channels are relatively wide, shallow, and straight (Schumm 1977). These differences can occur without a change in the size of the sediment on the channel bed. Therefore, the suggestion that braided rivers are a result of coarse sediment alone is incorrect.

Part of the difficulty in explaining river patterns is that both convergence and divergence are significant. Convergence means that similar landforms (river patterns) can be developed by different causes or processes, and divergence means that different landforms can be developed by similar causes or processes. The fact that a braided river can be caused by high flood peaks, by high bed-load transport, and by aggradation is an example of convergence (Schumm 1984). With these complications, the explanation for the variety of river patterns must necessarily be elusive.

Another problem is that although much has been written concerning river patterns, the emphasis has been upon meandering. Even Albert Einstein (1926) contributed to this literature. The repetitive patterns of curves, and the significant deviation from a straight line, are puzzling. If nature is efficient, why does a meandering river take a long and inefficient course to its destination? The jet stream meanders, the Gulf Stream meanders, and streams on ice and on glass plates meander (Gorycki 1973). The conclusion is that the meandering tendency is inherent in fluid. Hydraulic engineers show that fluid flow patterns develop instability as a result of bank and bed friction and internal shear and turbulence within the flow (Callander 1978); this instability then leads to the formation of alternate bars (Figure 3*b*). The helicoidal flow in bends has been cited as a cause of meanders (Ikeda 1980). However, in other channels these secondary flow patterns do not cause meandering. Experimental studies reveal that major flow deflection caused by the introduction of water into the head of a flume at an angle may produce excellent meander patterns, whereas at the same discharge but under lower energy conditions meanders do not form (Schumm & Khan 1972). These experiments suggest that there is a limited range of hydraulic conditions at which meandering occurs. Therefore, part of the difficulty in explaining meandering is that the full range of river patterns is not normally considered; that is, the question "Why do rivers meander?" can be supplemented by the question "Why are some rivers straight?"

In Schumm & Khan's (1972) experimental study the full range of channel patterns developed at the same discharge (Figure 8). At low slopes and with low sediment loads (i.e. low velocity and low stream power), the experimental channels remained straight, even when the water was introduced into the flume at an angle, which should have forced the channel to meander. It appears that although secondary flow patterns exist under these low-energy conditions, they are not powerful enough to move sediment across the channel to form alternate bars, and thus bank attack is minimal. As the slope of the surface on which the experimental channel flowed was increased, velocity, stream power, and sediment transport all increased. Bank erosion became important, and a sinuous pattern with alternate bars developed. At the highest slopes, with high energy and high sediment load, the channel became braided. The momentum of the flow in the downstream direction prevented the development of alternate bars by cross-channel flow, and increased bank erosion also prevented the formation of alternate-bar patterns.

It is clear that there are ranges of energy or stream power, for a given discharge and for a given size of sediment, at which different river patterns develop (Chang 1979). The explanation for these patterns, therefore, seems to be purely hydraulic. However, flow conditions depend on hydrologic conditions, and indeed it is well known that stream gradient depends on water discharge and sediment load. Therefore, the pattern at a given location depends upon hydrology and sediment supply from the upstream watershed. In addition, any changes of valley-floor slope by active tectonics can significantly affect channel patterns (Burnett & Schumm 1983).

It is apparent that there are multiple explanations for river patterns, and much depends upon local circumstances that influence sediment load and energy. All explanations of channel patterns are correct under certain circumstances. Certain hydraulic conditions and bed instability cause meanders, but these hydraulic conditions may be dependent upon valley-floor slope or upon watershed conditions that determine the amount of water and the quantity and type of sediment delivered to a river reach.

APPLICATIONS

Alluvial channels are dynamic and subject to change, but these changes are of different types and occur at highly variable rates. Therefore, rivers with different patterns behave differently, and their other morphologic characteristics (e.g. channel shape, gradient) are different. Hence, pattern identification can be the first step toward evaluating river stability and identifying potential hazards.

River patterns may be stable at one extreme or in the process of total change as a part of metamorphosis at the other. The change of river planform may be natural or in response to man-induced changes of hydrologic regimen.

The mode of pattern change can be very important when property or political boundaries follow the bank line or thalweg of a river. If a river changes position slowly by bank erosion and lateral accretion, as with meander growth or shift (Figure 4*e*), then the boundary shifts with the river. However, if the channel changes position rapidly by avulsion and meander cutoff (Figures 4*c*,*f*), then the boundary remains fixed. Clearly, an understanding of river patterns and river behavior is an important aspect of boundary law and forensic geomorphology.

The suggestion made here is that if one can identify the range of patterns along a river, then within that range the most appropriate channel pattern and sinuosity can probably be identified. If so, the engineer can work with the river to produce its most efficient or most stable channel. Obviously, a river can be forced into a straight configuration or can be made more sinuous, but there is a limit to the changes that can be induced, beyond which the channel cannot function without a radical morphologic adjustment (as suggested by Figures 3, 5, 6, and 7).

A clear understanding of the relation of channel pattern to river stability is required in the design of channels for mined-land reclamation and in the planning for channel modification. If a designed channel is too sinuous, it will not transport the water and sediment delivered to it and will aggrade. If the designed channel is too straight, it will degrade or erode its banks. The pattern must be adjusted to the gradient required by the discharge and sediment load. The practice of channelization, the replacement of a sinuous channel with a straight one to reduce flood hazards, has in many cases resulted in channel incision and long-term instability (Schumm et al 1984). When the channel gradient is steepened significantly by straightening, the channel will deepen, widen, and eventually evolve to a new condition of stability; however, this may take many years and cause major problems of bank erosion and high sediment transport. Brice (1983) has demonstrated that the response can be large or small depending upon the extent of the pattern alteration and the nature of the channel. It is now recognized that a major pattern alteration can create problems worse than those the alteration was designed to solve, and engineers are now considering alternatives to excessive channelization that will produce a straight channel (Schumm et al 1984, Bhowmik 1981). Again, the success of any design will depend on channel pattern (Figure 3), which in turn is dependent upon hydrology, sediment type, and valley morphology (Richards 1982).

ACKNOWLEDGMENTS

Chester C. Watson provided a useful review of this article, which was prepared while I was receiving support from the National Science Foundation and the US Army Research Office.

Literature Cited

Bhowmik, N. G. 1981. Hydraulic considerations in the alteration and design of diversion channels in and around surface-mined areas. *Symp. Surf. Mining Hydrol., Sedimentol. Reclam., Lexington, Ky.*, pp. 97–104

Brice, J. C. 1974. Evolution of meander loops. *Geol. Soc. Am. Bull.* 85: 581–86

Brice, J. C. 1983. Factors in stability of relocated channels. *J. Hydraul. Div. ASCE* 109: 1298–1313

Brice, J. C., Blodgett, J. C., et al. 1978. Countermeasures for hydraulic problems at bridges. *Fed. Highw. Adm. Rep. FHWA-RD-78-162*, Vols. 1, 2, Washington, DC

Burnett, A. W., Schumm, S. A. 1983. Alluvial-river response to neotectonic deformation in Louisiana and Mississippi. *Science* 222: 49–50

Callander, R. A. 1978. River meandering. *Ann. Rev. Fluid Mech.* 10: 129–58

Chang, H. H. 1979. Minimum stream power and river channel patterns. *J. Hydrol.* 41: 303–27

Einstein, A. 1926. Der ursache der Maanderbildung der Flusslaute und des sogenannter Baerschen Gesetzes. *Naturwissenschaften* 11: 223

Elliott, C. M., ed. 1984. *River Meandering, Proc. Conf. Rivers '83*. New York: Am. Soc. Civ. Eng. 1036 pp.

Fisk, H. N. 1944. Geological investigation of

the alluvial valley of the lower Mississippi River. *Miss. Riv. Comm.*, Vicksburg, Miss. 78 pp.
Galloway, W. E., Hobday, D. K. 1983. *Terrigenous Clastic Depositional Systems*. New York: Springer-Verlag. 422 pp.
Gorycki, M. A. 1973. Hydraulic drag: a meander-initiating mechanism: *Geol. Soc. Am. Bull.* 84:175–86
Gregory, K. J., ed. 1983. *Background to Paleohydrology*. New York: Wiley. 486 pp.
Hickin, E. J. 1983. River channel changes: retrospect and prospect. *Int. Assoc. Sedimentol. Spec. Publ.* 6:61–83
Ikeda, S. 1980. Roles of secondary flow in the formation of channel geometry. In *Application of Stochastic Processes in Sediment Transport*, ed. H. W. Shen, H. Kikkawa, Ch. 12. Littleton, Colo: Water Resour. Publ. 24 pp.
Lane, E. W. 1957. A study of the shape of channels formed by natural streams flowing in erodible material. *US Army Corps Eng., Missouri Riv. Div., Omaha, Nebr., Sediment Ser. 9*. 106 pp.
Leopold, L. B., Wolman, M. G. 1957. River channel patterns: braided meandering and straight. *US Geol. Survey Prof. Pap. 282-B*, pp. 39–84
Mollard, J. D. 1973. Air photo interpretation of fluvial features. In *Fluvial Processes and Sedimentation, Proc. Hydrol. Symp., Univ. Alberta*, pp. 341–80. Ottawa: Natl. Res. Counc. Can.
Nadler, C. T., Schumm, S. A. 1981. Metamorphosis of South Platte and Arkansas Rivers, eastern Colorado. *Phys. Geogr.* 2: 95–115
Pascoe, L. N. 1976. The training of braided single rivers into a single thread channel with particular reference to the middle reach of the Wairau River. Unpubl. Marlborough Catchment Board Rep., Blenheim, N.Z. 8 pp.
Richards, K. 1982. *Rivers, Form and Process in Alluvial Channels*. London: Methuen. 358 pp.
Schumm, S. A. 1969. River metamorphosis. *J. Hydraul. Div. ASCE* 95:255–73
Schumm, S. A. 1977. *The Fluvial System*. New York: Wiley-Interscience. 338 pp.
Schumm, S. A. 1979. Geomorphic thresholds, the concept and applications. *Inst. Br. Geogr. Proc.* 4:485–515
Schumm, S. A. 1981. Evolution and response of the fluvial system, sedimentologic implications. *Soc. Econ. Paleontol. Mineral. Spec. Publ.* 31:19–29
Schumm, S. A. 1984. River morphology and behavior: problems of extrapolation. See Elliott 1984, pp. 16–29
Schumm, S. A., Khan, H. R. 1972. Experimental study of channel patterns. *Geol. Soc. Am. Bull.* 83:1755–70
Schumm, S. A., Beathard, R. M. 1976. Geomorphic thresholds: an approach to river management. In *Rivers '76, Symp. Waterways, Harbors and Coastal Eng. Div. Am. Soc. Civ. Eng., 3rd*, 1:707–24
Schumm, S. A., Harvey, M. D., Watson, C. C. 1984. *Incised Channels: Dynamics, Morphology and Control*. Littleton, Colo: Water Resour. Publ. 200 pp.
Shen, H. W., Schumm, S. A., Nelson, J. D., Doehring, D. O., Skinner, M. M., Smith, G. L. 1981. Methods for assessment of stream-related hazards to highways and bridges. *Fed. Highw. Adm. Rep. FHWA/RD-80/160*, Washington, DC. 241 pp.
Thomson, P. A., MacArthur, R. S. 1968. Major river control, drainage and erosion control scheme for Kaikoura. Unpubl. Marlborough Catchment Board Rep., Blenheim, N.Z. 96 pp.
Winkley, B. R. 1970. Influence of geology on the regimen of a river. *Am. Soc. Civ. Eng. Natl. Water Resour. Meet., Memphis, Preprint 1078*. 35 pp.

Ann. Rev. Earth Planet. Sci. 1985. 13 : 29–47

SOLID-STATE NUCLEAR MAGNETIC RESONANCE SPECTROSCOPY OF MINERALS

R. James Kirkpatrick,[1] *Karen Ann Smith,*[2]
Suzanne Schramm,[2] *Gary Turner,*[2] *and Wang-Hong Yang*[1]

Department of Geology[1] and School of Chemical Sciences,[2] University of Illinois, Urbana, Illinois 61801

INTRODUCTION

Nuclear magnetic resonance (NMR) spectroscopy has been a useful tool for examining the structure of molecules in solution since the 1950s. It has only been since the late 1970s and early 1980s, though, that developments in very high field superconducting magnets and magic-angle sample-spinning (MASS) have made NMR routinely useful for examining the structure of solids. In this paper, we briefly review the theory of MASS NMR spectroscopy and some of the recent experimental results for minerals.

NMR spectroscopy examines the local structural environment of atoms, to at most the third and fourth nearest neighbors. For crystalline silicates, silicon-29 and aluminum-27 MASS NMR have already proven useful in examining the validity of the aluminum avoidance principle and the extent of Al(4)/Si order/disorder, in determining the extent of polymerization, in determining the number of crystallographically distinct silicon sites present, in estimating bond strength sums and Si–O–(Si, Al) bond angles, and in detecting the presence of Al(4) and Al(6) and estimating Al(4)/Si and Al(6)/Al(4) ratios. As the NMR behavior of more nuclides (certainly including boron-11, nitrogen-15, oxygen-17, fluorine-19, sodium-23, magnesium-25, and phosphorus-31) becomes better understood, we feel that this method will become a powerful tool for examining the local structural environment of a wide variety of species in a broad range of

0084–6597/85/0515–0029$02.00

inorganic solids, including minerals. The most significant applications will be to those problems that cannot be addressed using diffraction or TEM methods. These include order/disorder, the structure of amorphous and fine-grained materials, and the local structural environment of species present in small quantities. NMR will likely also become useful in following geochemically important reactions, such as those between minerals and aqueous solutions, in real time.

THEORY

An elementary but complete introduction to NMR theory is given by Davis (1965); more advanced treatments are given by Abragam (1961), Farrar & Becker (1971), Becker (1980), and Akitt (1983). Brief summaries of the theory necessary to understand MASS NMR of solids and of the experimental methods are presented by Lippmaa et al (1980), Müller et al (1981), K. A. Smith et al (1983), and Kirkpatrick et al (1985).

The following summarizes the most important points. (*a*) Atomic nuclei possess a quantized property called spin (I). The application of a static magnetic field H_0 removes the degeneracy of the spin energy levels, giving rise to $2I+1$ such levels. If $I = 1/2$, the nucleus has two energy levels and behaves as a magnetic dipole. If $I \geq 1$, the nucleus has more than two energy levels and behaves as a magnetic quadrupole. (*b*) The NMR experiment measures the frequency of the radio-frequency radiation that has the correct energy to cause transitions from one nuclear spin state to another. (*c*) The frequency causing this transition for a particular set of energy levels for a particular nuclide depends primarily on the magnitude of H_0. This resonance frequency is slightly different for different types of sites, because the electrons in the vicinity of the nucleus shield it different amounts from the applied magnetic field, depending on the local structural environment. The resonance frequencies are reported as chemical shifts, which are parts-per-million differences from an experimentally useful standard. More positive (or less negative) chemical shifts indicate less shielding. (*d*) Nuclides with $I = 1/2$, such as silicon-29, yield narrow peaks with positions at the isotropic value. (*e*) Quadrupolar nuclides ($I \geq 1$, such as oxygen-17 and aluminum-27) often yield peaks that are broadened and split and that are displaced from the isotropic value. The magnitudes of these effects are controlled by the quadrupole coupling constant (e^2qQ/h) and the asymmetry parameter (η). For quadrupolar nuclides, usually only the central ($1/2$, $-1/2$) energy level transition is observed. (*f*) For static solids, a variety of internuclear interactions greatly broaden the peaks. Much narrower peaks can be obtained, however, by spinning the sample (usually 100–500 mg of powder) at kilohertz frequencies about an axis oriented at 54.7° to the H_0 magnetic field [magic-angle sample-spinning (MASS); see Andrew 1971, 1981]. This is because the Hamiltonian for most of these interactions contains terms involving $3\cos^2\theta - 1$, where θ is the angle between H_0 and the axis of rotation, and these terms reduce to zero at 54.7°. MASS does not completely eliminate the second-order quadrupolar effects.

There is presently no precise way of calculating the value of the chemical shift for a particular site. Because chemical shifts cannot be easily calculated, much of the recent work has involved examining the MASS NMR behavior of structurally and chemically well-known phases in order to provide a data base from which to interpret the spectra of less well-understood materials.

Variations in the resonance frequencies of a particular nuclide in solids appear to be due primarily to nearest-neighbor (NN) and next-nearest-neighbor (NNN) interactions. Thus, NMR is a powerful technique for examining the local structural environment of atoms, even in, for instance, Al(4)/Si disordered crystals and glasses. To date, most solid-state MASS NMR work on minerals has been done with aluminum-27 and silicon-29, although other nuclides, including boron-11, oxygen-17, sodium-23, magnesium-25, and phosphorus-31, are becoming increasingly important. In addition, fluorine-19, scandium-45, and a variety of other nuclides of mineralogical interest could be examined.

OBSERVATIONAL NMR OF MINERALS

Silicon-29

To date, most studies of the solid-state NMR behavior of minerals have emphasized silicon-29. In addition to being a critical cation in silicate minerals, silicon-29 can be observed at natural abundance (4.7%) and has spin $I = 1/2$, so that quadrupole effects do not occur.

For silicates with silicon as the only tetrahedral cation and for ordered Al/Si and B/Si alumino- and borosilicates, the silicon-29 peaks are typically quite narrow [1–4 ppm full width at half-height (FWHH)], with one peak for each crystallographically distinct silicon site. Ordered low albite, for instance, produces three peaks, corresponding to the three silicon sites. Figures 1*A–C* show typical spectra of such phases.

Al(4)/Si disordered phases, while often containing only one tetrahedral site, generally produce silicon-29 spectra containing multiple peaks. These peaks arise from silicon with different numbers of Al(4) next-nearest neighbors (Lippmaa et al 1980, 1981, K. A. Smith et al 1983). The peaks for these materials are usually broader than for ordered aluminosilicates, probably because of greater-than-NNN effects. Figures 1*D* and 1*E* show typical spectra for such phases.

In most naturally occurring silicates, silicon is tetrahedrally coordinated by oxygen. Such silicons have chemical shifts in the range of −60 to −120 ppm relative to tetramethylsilane (TMS). In a few high-pressure phases, silicon is octahedrally coordinated by oxygen. The only material with

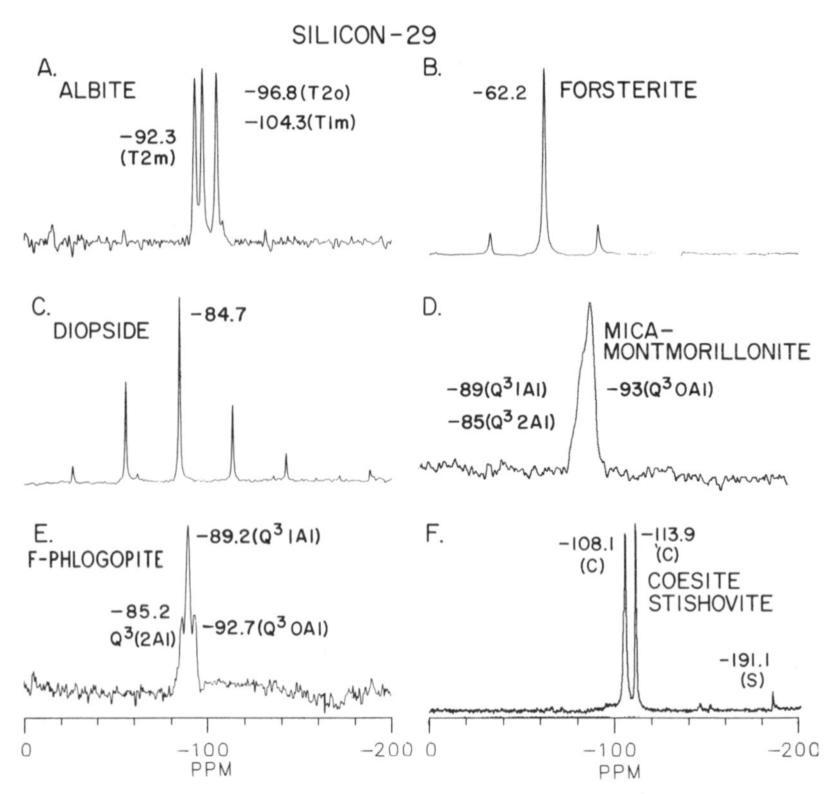

Figure 1 Silicon-29 MASS NMR spectra of silicate minerals obtained at an H_0 field strength of 8.45 T. *A* Ordered low albite (Amelia, Virginia), showing resonances for the three crystallographically different silicon sites. *B* Forsterite (synthetic), showing a resonance for one silicon site. *C* Diopside (synthetic), showing a resonance for one silicon site at −84.7 ppm. The other peaks are spinning sidebands, which are due to MASS. *D* Mica-montmorillonite (synthetic), showing resonances for three silicon sites due to Si/Al disorder. *E* F-phlogopite (synthetic), showing resonances for three silicon sites due to Si/Al disorder. *F* Coesite (C) and stishovite (S) (Meteor Crater, Arizona), with the coesite showing resonances for two crystallographically different silicon sites.

octahedrally coordinated silicon that has been examined so far is stishovite, which has a much more shielded chemical shift of −191.1 ppm (J. V. Smith & Blackwell 1983; Figure 1*F*).

Silicon-29 MASS NMR has been applied to a wide variety of solid silicates to examine the extent of Al/Si order/disorder and the validity of the aluminum avoidance principle (Loewenstein 1954), to determine the NNN Al(4)/Si ratio, to determine the number of bridging oxygens to which the silicon is coordinated (extent of polymerization), and to determine quantitative crystallographic data for the silicon sites. We examine each of these applications in turn.

In this paper the extent of polymerization and the number of NNN Al(4) atoms is described by the usual notation $Q^m(n\text{Al})$, where m is the number of bridging oxygens to which the silicon is coordinated (0–4), and n is the number of Al(4) NNN atoms (0–4 for framework silicates, 0–3 for sheet silicates).

The first studies of the silicon-29 MASS NMR behavior of silicates were by Lippmaa et al (1980, 1981), who examined a variety of synthetic and natural materials. On the basis of these data, they proposed that the chemical shifts for phases of each polymerization state (Q^0–Q^4) with no tetrahedral Al should fall into nearly discrete ranges, and that for framework aluminosilicates [$Q^4(n\text{Al})$] the chemical shifts should fall into discrete ranges depending on the number of Al NNN atoms (0–4). More recent work on a wider variety of silicates (Thomas et al 1983, K. A. Smith et al 1983, Mägi et al 1984) has shown that these correlations do exist, but that there is considerable overlap in the ranges. Figure 2*A* shows the presently known ranges of silicon-29 chemical shifts for silicates with no tetrahedral Al. Much of the overlap of these ranges is due to the effects of octahedrally coordinated Al. Figure 2*B* shows that for silicates containing no Al the ranges are much more distinct. Thus, for Al-free silicates, silicon-29 MASS NMR can be used quite effectively to determine the extent of polymerization of the SiO_4 tetrahedra.

For Q^0 (isolated tetrahedral) to Q^3 (sheet structures), the effect of octahedral Al on the silicon-29 chemical shift decreases as the polymerization of the material increases. Figure 3 plots the differences in silicon-29 chemical shift between phases that contain Al as the octahedral cation and phases that contain Mg as the octahedral cation for Q^0, Q^2, and Q^3 silicates. For Q^0 and Q^2 phases, octahedral Al increases the shielding at silicon relative to Mg, while for Q^3 phases it actually decreases the shielding slightly.

The effect of tetrahedrally coordinated Al is to decrease the shielding at silicon (i.e. to give less negative silicon-29 chemical shifts). Figure 2*C* shows the presently known ranges of chemical shifts for $Q^4(0\text{Al})$ through $Q^4(4\text{Al})$

materials (Thomas et al 1983). This correlation has been extended to sheet silicates (Q^3 structures) by Kinsey et al (1985; Figure 2*D*). These latter ranges will certainly expand as more materials are investigated.

There have also been several attempts to quantitatively correlate the silicon-29 chemical shifts to crystal structure parameters. These parameters include the mean Si–O bond lengths (Higgins & Woessner 1982), the secant of the mean Si–O–*T* bond angle or the mean secant of the Si–O–*T* bond angle (T = Si or Al) (J. V. Smith & Blackwell 1983, J. V. Smith et al 1984), and the sum of the Brown & Shannon (1973) bond strengths to the four oxygens coordinated to the silicon of interest (K. A. Smith et al 1983). Figure 4 shows the silicon-29 chemical shifts plotted versus these three parameters. The results show that the very best correlations are for secant of the mean Si–O–*T* angle for framework (Q^4) silicates, that the mean Si–O bond distance is not a particularly good correlator, and that the best overall correlation is with the sum of the bond strengths to the four oxygens. J. V. Smith & Blackwell (1983) and J. V. Smith et al (1984) have determined least-squares linear fits to these relationships for the various types of silicate

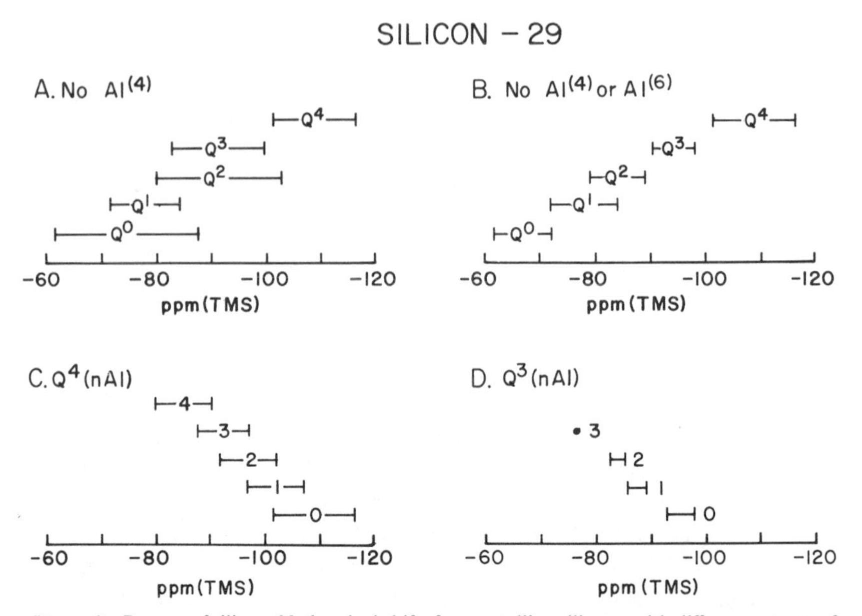

Figure 2 Ranges of silicon-29 chemical shifts for crystalline silicates with different extents of polymerization. *A* Phases with no Al(4). Much of the overlap in ranges is due to Al(6). *B* Phases with no Al(4) or Al(6). Note the well-resolved ranges. *C* Framework silicates with differing numbers (*n*Al) of Al next-nearest neighbors. *D* Sheet silicates with differing numbers of Al(4) (*n*Al) next-nearest neighbors.

structures. Using these relationships, we can now make fair-to-excellent estimates of these structural parameters for some phases directly from silicon-29 NMR data.

Perhaps the most important application of silicon-29 MASS NMR to date has been in the examination of the structure of zeolite catalyst and sorbent materials. These materials are generally too fine grained to examine by single crystal X-ray diffraction, but their economic importance makes an understanding of their structure of considerable significance. Most of this work has been reviewed by Thomas et al (1983), who give an extensive reference list. Although we do not present a detailed discussion here, some of the results are of general interest.

One of the most important results of this work is that silicon-29 MASS NMR can easily distinguish Si atoms with different numbers of NNN Al atoms in Al(4)/Si disordered aluminosilicates and thus that the NMR data can be used to accurately determine the Si/Si + Al(4) ratio in the tetrahedral sites. Figure 5 (after Thomas et al 1983) shows silicon-29 MASS NMR spectra of faujasitic zeolites (zeolites X and Y) with different Si/Si + Al(4) ratios, along with simulations assuming Gaussian peak shapes. All peaks in

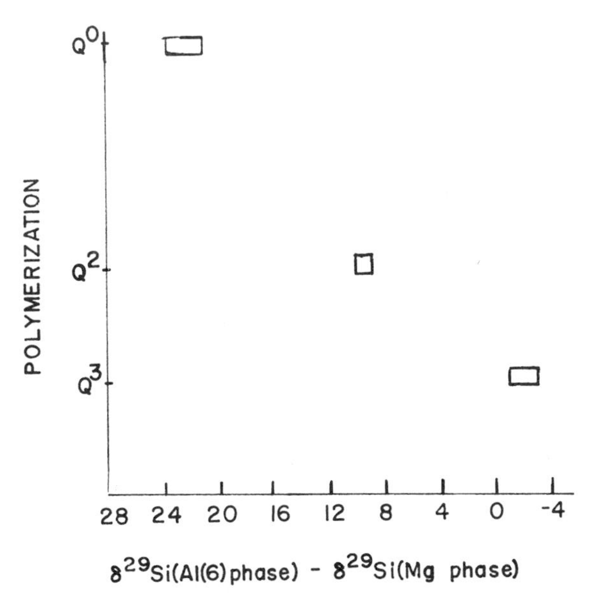

Figure 3 Plot of differences between the silicon-29 chemical shifts for phases with Mg as the octahedral cation and phases with Al as the octahedral cation for Q^0, Q^2, and Q^3 polymerizations. For Q^0 and Q^2 structures, the effect of Al(6) is to increase the shielding. For Q^3 structures, Al(6) decreases the shielding slightly.

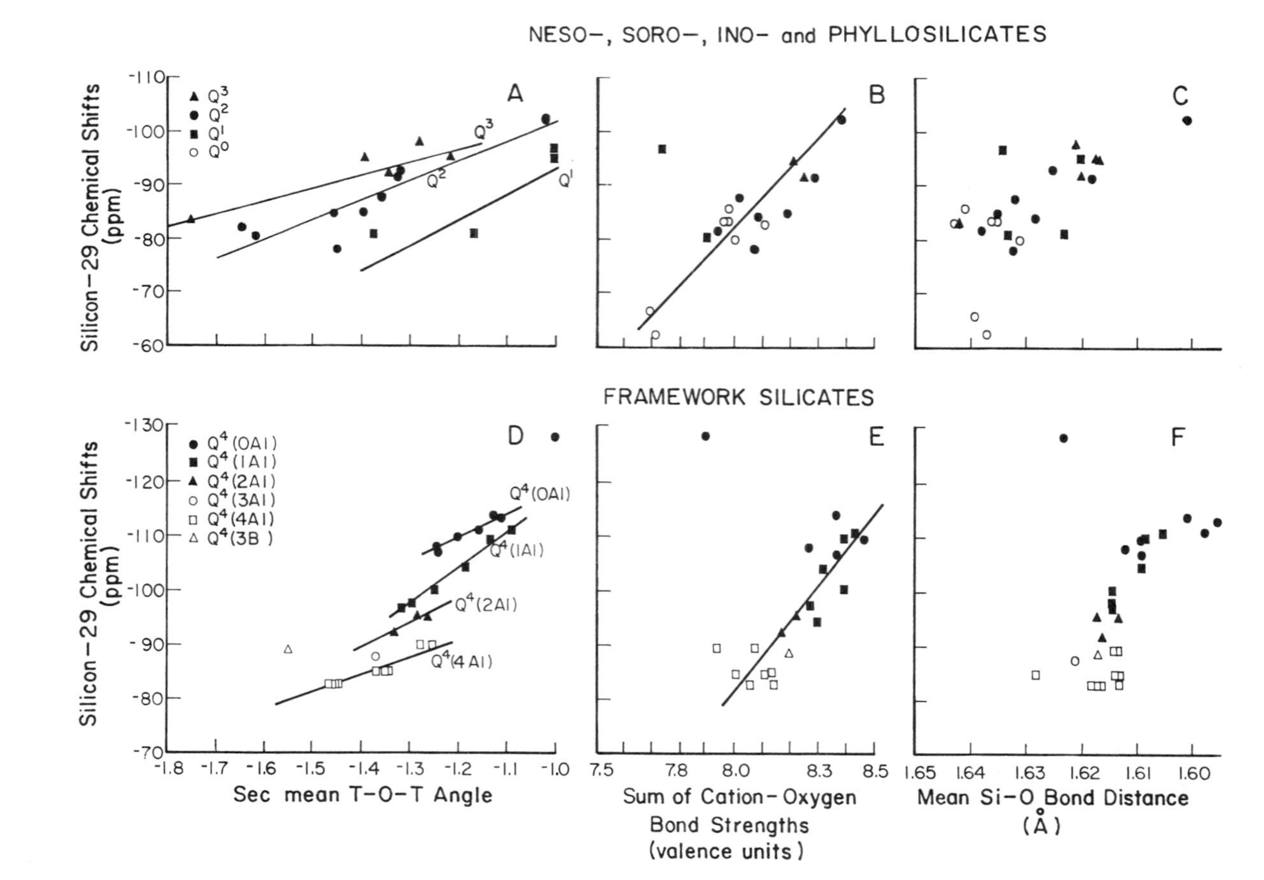

Figure 4 Plots of silicon-29 chemical shifts versus secant of the mean Si–O–T bond angle, the mean Si–O bond distance, and total bond strength sum to the four oxygens coordinated to the silicon of interest. The anomalous point not following any of the trends for framework silicates is the Q^4 site of zunyite, which has 180° Si–O–Si angles.

these spectra represent Q^4 sites, and the numbers next to the peaks in the Si/Si + Al(4) = 0.63 spectrum give the number of NNN Al atoms associated with each peak. It is clear from these spectra that as many as five types of silicon sites [(Q^4(0Al)–Q^4(4Al)] are present in these materials, even though faujasite contains only one tetrahedral site. It is also clear that, as expected, the average number of NNN Al atoms decreases as Si/Si + Al(4) increases and that it is possible to simulate the spectra very accurately. Similar spectra have been obtained by many workers for Al/Si disordered framework silicates. The silicon-29 spectra of synthetic mica-montmorillonite (illite/smectite) and F-phlogopite (Figures 1*D* and 1*E*) show that these minerals are Al/Si disordered. About 46% of the silicon atoms in the F-phlogopite are not in Q^3(1Al) sites, as they would be in a fully ordered sample (Kinsey et al 1985).

An early interpretation of spectra of this sort was that they indicate that

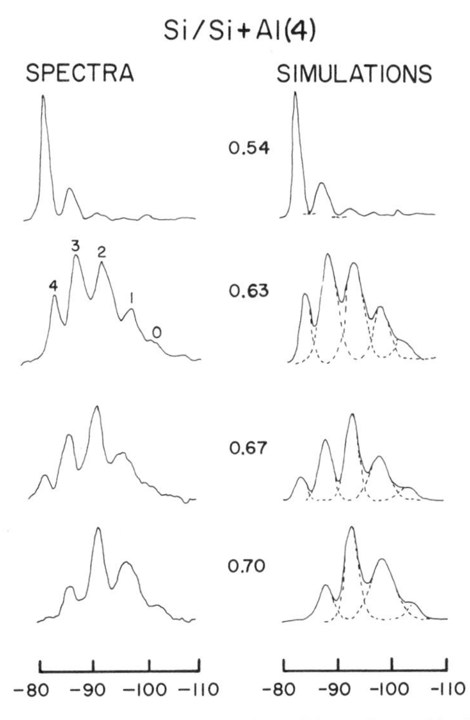

Figure 5 Silicon-29 MASS NMR spectra of faujasitic zeolites with differing Si/Si + Al(4) ratios. Numbers next to peaks in the Si/Si + Al(4) = 0.63 spectrum give the number of Al next-nearest neighbors to Si. Left-hand column gives the observed spectra, right-hand column the simulated spectra. (After Thomas et al 1983.)

the aluminum avoidance principle (Loewenstein 1954) is violated in these materials (Lippmaa et al 1981, Ramdas et al 1981). Recent work, however, has shown that although the Al/Si distributions can be thought of as random, they still obey aluminum avoidance (Klinowsky et al 1982, Melchior et al 1982, Thomas et al 1983).

Another important application of silicon-29 MASS NMR to zeolites has been in resolving the Q^4(0Al) sites in the highly siliceous zeolite catalyst ZSM-5/silicate (Si/Al $\geq$200 to 1000). Fyfe et al (1982) have presented spectra and simulations of this material that indicate that it contains twenty-four distinct tetrahedral sites, all Q^4(0Al). Because of these data, they were able to determine that the correct space group is $Pn2_{1a}$ or $P2_1/n$ rather than Pnma.

The anisotropy of the silicon-29 NMR shielding tensor has received some attention and may become a useful tool for examining the symmetry of silicon sites. Grimmer et al (1981) have shown that for silicon sites with axial symmetry (i.e. three Si–O bonds of almost equal length and one either longer or shorter), the shielding at silicon is less in the long Si–O bond direction and greater in the short Si–O bond direction. K. A. Smith et al (1983) have determined the full shielding tensors for a variety of natural and synthetic silicates. Their results show that $\Delta\sigma\,[=\sigma_{33}-1/2(\sigma_{11}+\sigma_{22})]$ and the asymmetry parameter $\eta\,[=(\sigma_{22}-\sigma_{11})/(\sigma_{33}-\sigma_{\text{i}})]$ are generally largest for soro- and chain silicates and smallest for framework silicates, although there are certainly exceptions. In these expressions σ_{11}, σ_{22}, and σ_{33} are the greatest, intermediate, and smallest elements of the shielding tensor, respectively, and the isotropic chemical shift σ_{i} is one third the trace Tr of the shielding tensor, i.e. $\sigma_{\text{i}} = (1/3)\text{Tr}\,(\sigma_{11}+\sigma_{22}+\sigma_{33})$.

One major limitation of the NMR method is that it cannot tolerate very large concentrations of paramagnetic atoms in the sample, either in solid solution or in a separate phase. In the samples we have examined, iron and manganese appear to have large effects. Little is presently known, however, about the concentrations that can be tolerated or their effects on the observed chemical shift. Oldfield et al (1983) have shown that the addition of powdered magnetite and hematite to sanidine causes static peak broadening and therefore side-band development on MASS, but that it does not significantly effect the chemical shift. Grimmer et al (1983), on the other hand, have examined olivines varying in composition from Fo_{100} to Fo_{91} and find effects on both peak breadth and chemical shift. For Fo_{95} relative to Fo_{100} the static peak breadth is larger, and the center band in the MASS spectra is broader by over a factor of ten. In addition, the chemical shift is at -59.5 ppm instead of the -61.9 ppm value for Fo_{100}. They attribute this effect to dipole-dipole broadening between Si and Fe^{+2}. The samples were made under nitrogen, however, and probably contain significant Fe^{+3}. It is thus possible that the effect is mainly due to paramagnetic effects. This question remains unresolved.

Anecdotal data from our laboratory relevant to the question of how much iron and manganese can be tolerated are that intermediate pyroxenes and amphiboles give no signal, that rhodenite does not even spin when placed in the magnet, and that good spectra can be obtained for illite/smectites with up to as much as 2.7 wt% iron as Fe_2O_3.

Aluminum-27

Aluminum-27 MASS NMR of solids, while not as well investigated as silicon-29, has found significant application in the examination of alumino-silicates. Although spectra of aluminum-27 are relatively easy to obtain, interpretation is complicated by quadrupolar-induced peak broadening and apparent chemical shift changes (Müller et al 1981, Ganapathy et al 1982, Meadows et al 1982, Kirkpatrick et al 1985).

Figure 6 shows aluminum-27 MASS NMR spectra obtained at an H_0 magnetic field strength of 11.7 T for a variety of silicate minerals. Natrolite (Figure 6*A*) has a relatively small quadrupolar coupling constant of 1.66 MHz (Ghose & Tsang 1973) and thus has a relatively narrow peak. Aluminum in petalite (Figure 6*B*) has a larger quadrupole coupling

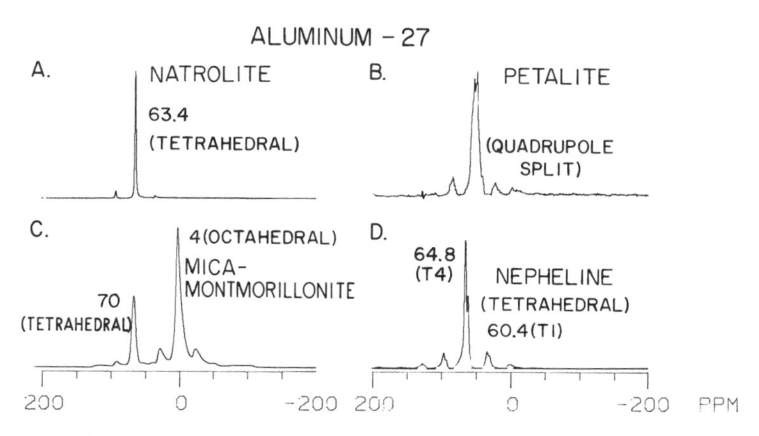

Figure 6 Aluminum-27 MASS NMR spectra of aluminosilicate minerals obtained at 11.7 T. *A* Natrolite (locality unknown), showing a resonance for one aluminum site. *B* Petalite (Varutrasl, Sweden), showing a resonance for one aluminum site. The peak is split and broadened because of large second-order quadrupolar effects ($e^2qQ/h = 4.7$ MHz). *C* Mica-montmorillonite (synthetic), showing resonances for tetrahedral aluminum at 70 ppm and octahedral aluminum at 4 ppm. *D* Nepheline (Havelock, Ontario), showing resonances for two tetrahedral aluminum sites.

constant of 4.7 MHz and thus yields a peak that is broadened and split. Both natrolite and petalite contain only tetrahedral Al and, therefore, have isotropic chemical shifts in the range of +50 to +75 ppm relative to $Al(H_2O)_6^{+3}$ in solution (Müller et al 1981).

Figure 6*C* is an aluminum-27 spectrum of a synthetic illite-smectite. It contains a peak at 70 ppm due to tetrahedral aluminum and a peak at 4 ppm due to octahedral aluminum. Both peaks are broader than typical silicon-29 peaks because of quadrupolar effects. This spectrum is typical of those of many micas and clay minerals that contain both tetrahedral and octahedral aluminum.

Figure 6*D* is an aluminum-27 spectrum of a plutonic nepheline. The smaller peak at 60.4 ppm is from the Tl site, and the larger peak at 64.8 ppm is from the T4 site, which is three times as abundant. Both chemical shifts are in the range for tetrahedral aluminum in framework silicates.

One major difficulty with examining the MASS NMR behavior of aluminum-27 is that quadrupolar effects can be very large. Figure 7 shows static and MASS spectra, along with simulations of the MASS spectra, for low albite (Amelia, Virginia) taken at H_0 magnetic field strengths of 3.5, 8.45, and 11.7 T (Kirkpatrick et al 1985). The spectra taken at 3.5 T are uninterpretable, because MASS cannot sufficiently narrow the peak. The spectra taken at 8.45 T are still relatively broad, and the peaks in the MASS spectrum are at 56 and 53 ppm, significantly displaced from the isotropic value of 62.3 ppm. The spectra taken at 11.7 T are quite narrow, and the peak maximum in the MASS spectrum is at 59.3 ppm, which is closer to the isotropic value. These data and those of Kinsey et al (1985) for sheet silicates show that for aluminum-27 the best resolution is obtained at the highest possible field strengths. These data also demonstrate that great care should be taken in comparing published aluminum-27 chemical shifts, especially for sites with $e^2qQ/h \geq 2$, because the peak position changes significantly with changing field strength, with values closest to the isotropic value being obtained at the largest field strengths.

There have been a number of investigations of synthetic and natural silicates using aluminum-27 MASS NMR. The first was by Müller et al (1981), who showed that the chemical shifts for tetrahedral and octahedral aluminum fall into very distinct ranges. At an H_0 field strength of 11.7 T, tetrahedral aluminum is in the range from about +50 to +80 ppm, while octahedral aluminum is in the range from about −10 to +20 ppm. This easily made distinction has allowed the use of aluminum-27 MASS NMR in following changes in the location of aluminum in reactions involving aluminates, aluminas, and zeolites (Müller et al 1981, Thomas et al 1983).

There have been several attempts to examine the variation in the chemical shift of tetrahedral aluminum with structure and composition,

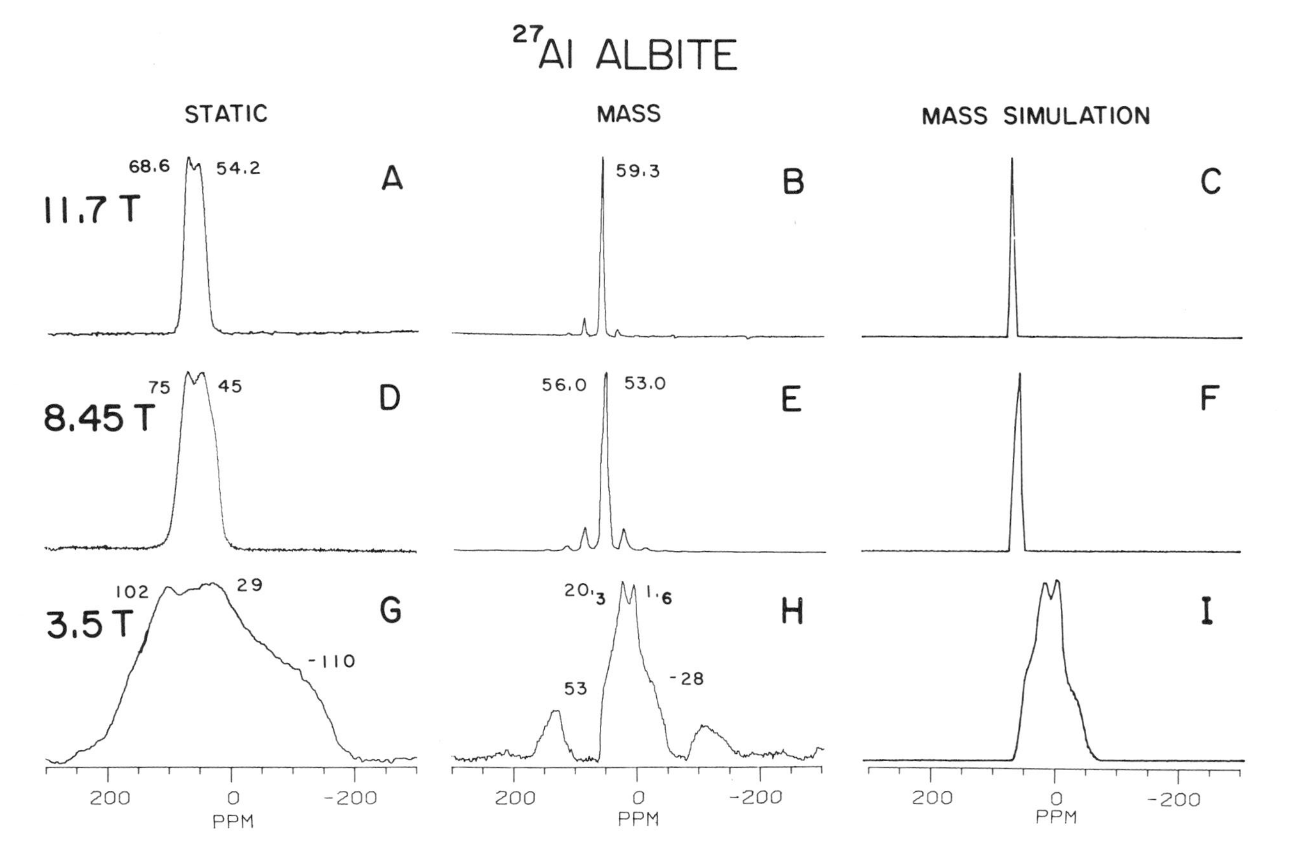

Figure 7 Static, MASS, and simulated aluminum-27 NMR spectra of low albite (Amelia, Virginia) obtained at 3.5, 8.45, and 11.7 T. (After Kirkpatrick et al 1985.)

especially Si/Si + Al(4). Fyfe et al (1982) examined the aluminum-27 MASS NMR behavior of zeolites at an H_0 magnetic field strength of 9.4 T. Their results show that while there is a tendency for the chemical shift to become less shielded (more positive) with decreasing Si/Si + Al(4), there is also considerable scatter in the data. In fact, for faujasitic zeolites (Na–*X* and Na–*Y*) with Si/Si + Al(4) from 0.54 to 0.73, there is essentially no variation in the aluminum-27 chemical shift. This may be due to the presence of aluminum-rich domains with essentially constant Si/Si + Al(4) that become less abundant as the bulk Si/Si + Al(4) increases. For a dealuminated faujasite with Si/Si + Al(4) = 0.98, the chemical shift is much lower (54.8 ppm instead of about 61.2 ppm).

Because the four tetrahedral NNNs to aluminum all appear to be silicons for phases with Si/Si + Al(4) $\geq$ 1 (i.e. aluminum avoidance is obeyed), the variation in the aluminum-27 chemical shift must be primarily due to greater-than-NNN effects, to differences in the nonframework cations, and to variation in the structure (i.e. bond distances and bond angles).

Figure 8 is a plot of all available aluminum-27 chemical shifts

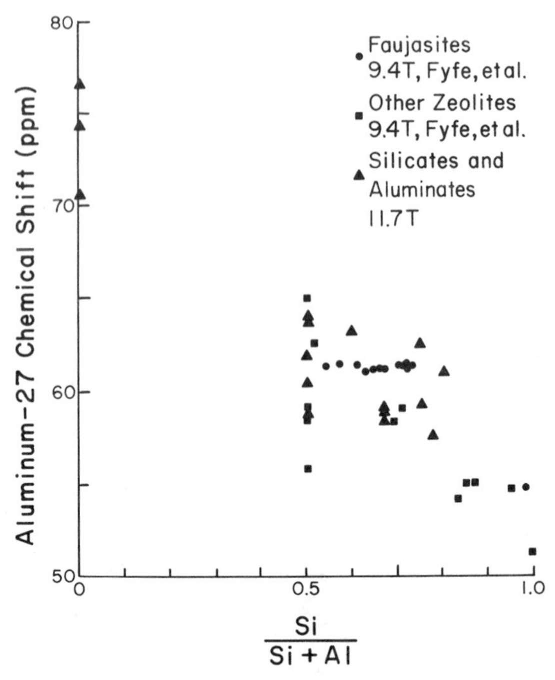

Figure 8 Plot of aluminum-27 chemical shifts versus Si/Si + Al(4) for framework aluminosilicates and aluminates. Scatter is large because of quadrupole effects.

for framework aluminosilicates and aluminates with tetrahedral Al versus Si/Si + Al(4). While it is clear that there is a trend of increasingly positive chemical shift with decreasing Si/Si + Al(4), there is considerable scatter, especially at Si/Si + Al(4) = 0.5. Much of this scatter is probably due to quadrupolar effects, which tend to displace the apparent chemical shifts to less positive values.

For sheet silicates, the correlation of tetrahedral aluminum-27 chemical shift with Si/Si + Al(4) is better (Figure 9; Kinsey et al 1985). There is a trend of increasingly positive chemical shift with decreasing Si/Si + Al(4) for the dioctahedral sheet silicates. The trioctahedral sheet silicates appear to be displaced about 3 ppm to less positive values, although there are few data. The scatter is less than for the framework-structure materials, because all these sheet silicates probably have about the same eq^2Q/h. The presently known range of aluminum-27 chemical shifts for sheet silicates is about +66 to +75 ppm, which is well separated from the +51 to +65 ppm range for framework silicates.

Correlations of octahedral aluminum-27 chemical shift with structure or composition are less well known. It appears, however, that the shielding at octahedral aluminum increases (the chemical shift becomes more positive or less negative) with increasing polymerization. Aluminas and aluminates

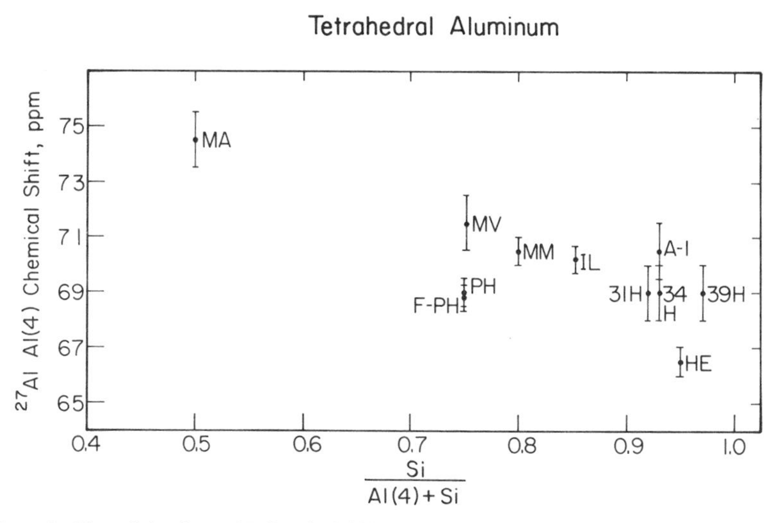

Figure 9 Plot of aluminum-27 chemical shifts at 11.7 T versus Si/Si + Al(4) for sheet silicates. MA = margarite; MV = muscovite; MM = mica-montmorillonite; IL = illite; PH = phlogopite; F-PH = fluoro-phlogopite; HE = hectorite; 31H, 34H, 39H, and A-1 = mixed-layer illite/smectites. (After Kinsey et al 1985.)

have even less shielded (more positive) octahedral aluminum-27 chemical shifts.

One of the potentially useful applications of aluminum-27 MASS NMR is to quantitate the Al(4)/Al(6) ratio in silicates. Kinsey et al (1985) have shown that this is an effective technique for sheet silicates.

OTHER NUCLIDES

Although silicon-29 and aluminum-27 have been the nuclides investigated most by MASS NMR, many other geochemically important elements can also be examined. For many of these a few spectra have been obtained, but there has been no systematic investigation of their behavior.

Potentially the most important of these nuclides is oxygen-17. The difficulty is that oxygen-17 has only 0.04% natural abundance. Except in the

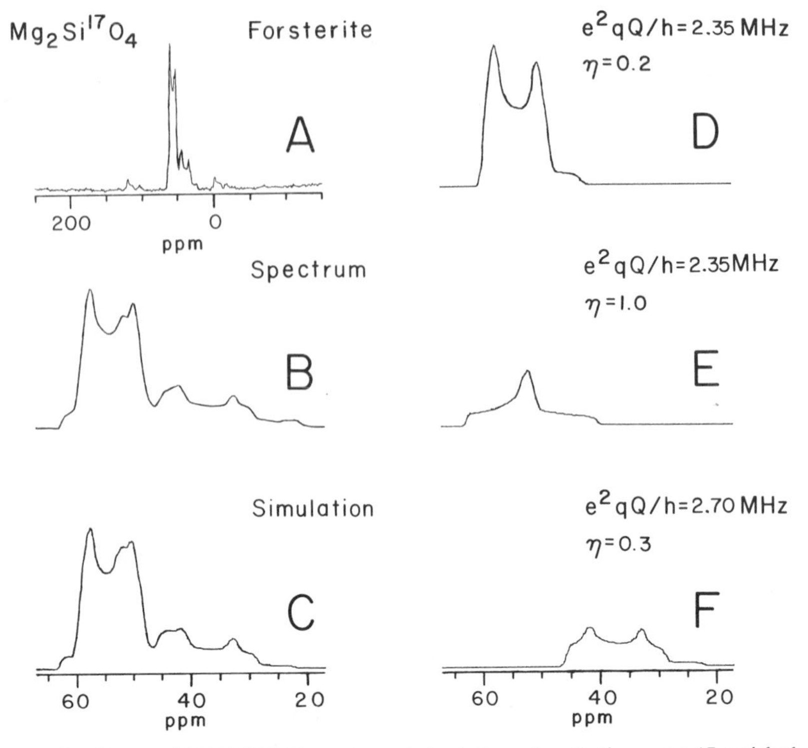

Figure 10 Oxygen-17 MASS NMR spectra and simulations of synthetic oxygen-17-enriched forsterite obtained at 11.7 T. (After Schramm & Oldfield 1984.) *A* Observed spectrum. *B* Expanded, observed spectrum. *C* Simulated spectrum. *D*, *E*, *F* Simulated spectra for the three oxygen sites. The simulated spectrum in *C* is the sum of these three.

most unusual circumstances (pure water, for instance), it must be examined in synthetic, isotopically enriched samples. Figure 10 presents oxygen-17 MASS NMR spectra and simulations of the spectra for synthetic, enriched forsterite obtained by Schramm & Oldfield (1984). These spectra and simulations show that it is possible to resolve the three oxygen sites in forsterite and determine their e^2qQ/h and η values. If future work shows a significant and systematic variation in oxygen-17 chemical shifts, oxygen-17 MASS NMR may become a powerful probe of oxygen sites in mineral phases.

Another potentially useful and easily examined nuclide is sodium-23, which has spin $I = 3/2$. Figure 11*A* shows the sodium-23 MASS NMR spectra of low albite (Amelia, Virginia; Kirkpatrick et al 1985). This spectrum is broadened and split by quadrupolar effects. Simulation yields a single site with $e^2qQ/h = 3.29$ MHz and $\eta = 0.62$. These relatively large values indicate a very distorted site, in agreement with diffraction data (Ribbe 1983).

Another potentially important nuclide is boron-11, which has spin $I = 3/2$. Figures 11*B* and 11*D* show boron-11 spectra of danburite ($CaB_2Si_2O_8$) and colemanite [$CaB_3O_4(OH)_3 \cdot H_2O$]. For danburite, the

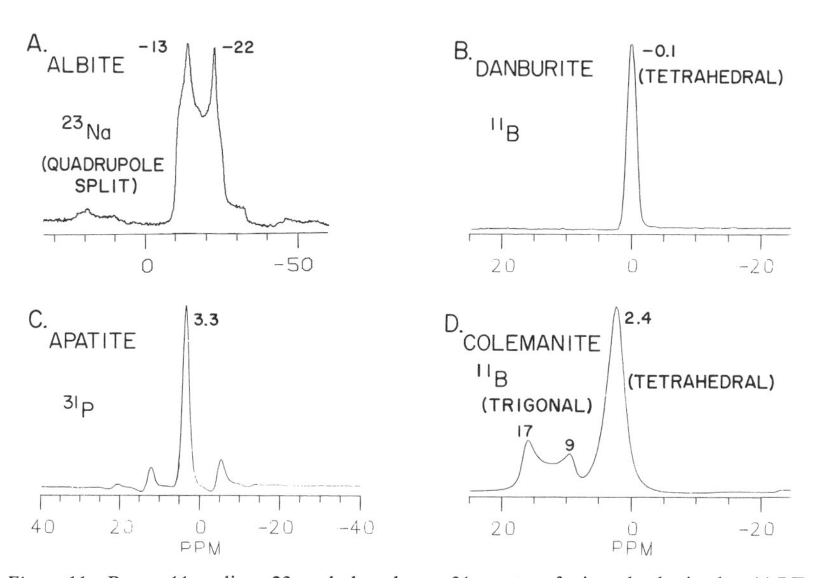

Figure 11 Boron-11, sodium-23, and phosphorus-31 spectra of minerals obtained at 11.7 T. *A* Sodium-23 MASS NMR spectrum of low albite (Amelia, Virginia). *B* Boron-11 MASS NMR spectrum of danburite (Mexico). *C* Phosphorus-31 MASS NMR spectrum of apatite (Cerro de Mercado, Mexico). *D* Boron-11 MASS NMR spectrum of colemanite (Turkey).

narrow peak at −0.1 ppm is from tetrahedral boron. For colemanite, the narrow peak at +2.4 ppm is from tetrahedral boron, and the broad, quadrupole-split peak from about +9 to +17 ppm is from trigonal boron. The value of e^2qQ/h for this site is 2.4 MHz, which is significantly greater than the values of 0.1 MHz for the tetrahedral site of colemanite and ~0.0 MHz for danburite.

Finally, phosphorus-31 is also a potentially useful nuclide. It has spin $I = 1/2$ and 100% natural abundance. Figure 11*C* shows the phosphorus-31 MASS NMR spectrum of apatite. Because the spectra of phosphorus-31 and fluorine-19 are so easy to obtain, the behavior of these nuclides as minor and trace elements should be observable.

ACKNOWLEDGMENTS

We wish to thank Professors Eric Oldfield and Donald M. Henderson and Dr. Robert Kinsey for much useful discussion, and Professor J. M. Thomas for permission to use Figure 5. This research has been supported by NSF Grants EAR 8218741 and EAR 8207260 and has benefited from the NSF-supported Regional NMR facility at the University of Illinois.

Literature Cited

Abragam, A. 1961. *The Principles of Nuclear Magnetism.* Oxford: Clarendon. 599 pp.

Akitt, J. W. 1983. *NMR and Chemistry: An Introduction to the Fourier Transform-Multinuclear Era.* London: Chapman and Hall. 263 pp. 2nd ed.

Andrew, E. R. 1971. The narrowing of NMR spectra of solids by high-speed specimen rotation and the resolution of chemical shift and spin multiplet structures for solids. *Prog. Nucl. Magn. Reson. Spectros.* 8:1–39

Andrew, E. R. 1981. Magic angle spinning. *Int. Rev. Phys. Chem.* 1:195–224

Becker, E. D. 1980. *High Resolution NMR, Theory and Application.* New York: Academic. 354 pp. 2nd ed.

Brown, I. D., Shannon, R. D. 1973. Empirical bond-strength-bond-length curves for oxides. *Acta Crystallogr. Sect. A* 29:266–82

Davis, J. C. 1965. *Advanced Physical Chemistry.* New York: Ronald. 632 pp.

Farrar, T. C., Becker, E. D. 1971. *Pulse and Fourier Transform NMR: Introduction to Theory and Methods.* New York: Academic. 115 pp.

Fyfe, C. A., Gobbi, G. C., Klinowski, J., Thomas, J. M., Ramdas, S. 1982. Resolving crystallographically distinct tetrahedral sites in silicalite and ZSM-5 by solid-state NMR. *Nature* 296:530–36

Ganapathy, S., Schramm, S., Oldfield, E. 1982. Variable-angle sample spinning high resolution NMR of solids. *J. Chem. Phys.* 77:4360–65

Ghose, S., Tsang, T. 1973. Structural dependence of quadrupole coupling constant e^2qQ/h for ^{27}Al and crystal field parameter *D* for Fe^3 in aluminosilicates. *Am. Mineral.* 58:748–55

Grimmer, A.-R., Peter, R., Fechner, E., Molgedey, G. 1981. High-resolution ^{29}Si NMR in solid silicates. Correlations between shielding tensor and Si-O bond length. *Chem. Phys. Lett.* 77:331–35

Grimmer, A.-R., von Lampe, F., Mägi, M., Lippmaa, E. 1983. High-resolution ^{29}Si NMR of solid silicates: influence of Fe^{+2} in olivines. *Z. Chem.* 23:343–44

Higgins, J. B., Woessner, D. E. 1982. ^{29}Si, ^{27}Al, and ^{23}Na spectra of framework silicates. *Eos, Trans. Am. Geophys. Union* 63: 1139 (Abstr.)

Kinsey, R. A., Kirkpatrick, R. J., Hower, J., Smith, K. A., Oldfield, E. 1985. High-resolution aluminum-27 and silicon-29 nuclear magnetic resonance spectroscopy of layer silicates, including clay minerals. *Am. Mineral.* In press

Kirkpatrick, R. J., Kinsey, R. A., Smith, K. A., Henderson, D. M., Oldfield, E. 1985.

High resolution solid state sodium-23, aluminum-27, and silicon-29 nuclear magnetic resonance spectroscopic reconnaissance of alkali and plagioclase feldspars. *Am. Mineral.* In press

Klinowski, J., Ramdas, S., Thomas, J. M., Fyfe, C. A., Hartman, J. S. 1982. A re-examination of Si, Al ordering in zeolites NaX and NaY. *J. Chem. Soc., Faraday Trans. 2* 78: 1025–50

Lippmaa, E., Mägi, M., Samoson, A., Engelhardt, G., Grimmer, A.-R. 1980. Structural studies of silicates by solid-state high-resolution ^{29}Si NMR. *J. Am. Chem. Soc.* 102: 4889–93

Lippmaa, E., Mägi, M., Samoson, A., Tarmak, M., Engelhardt, G. 1981. Investigation of the structure of zeolites by solid-state high-resolution ^{29}Si NMR spectroscopy. *J. Am. Chem. Soc.* 103: 4992–96

Loewenstein, W. 1954. The distribution of aluminum in the tetrahedra of silicates and aluminates. *Am. Mineral.* 39: 92–96

Mägi, M., Lippmaa, E., Samoson, A., Engelhardt, G., Grimmer, A.-R. 1984. Solid-state high-resolution silicon-29 chemical shifts in silicates. *J. Phys. Chem.* 88: 1518–22

Meadows, M. D., Smith, K. A., Kinsey, R. A., Rothgeb, T. M., Skarjune, R. P., Oldfield, E. 1982. High-resolution solid-state NMR of quadrupolar nuclei. *Proc. Natl. Acad. Sci. USA* 79: 1351–55

Melchior, M. T., Vanghan, D. E. W., Jacobson, A. J. 1982. Characterization of the silicon-aluminum distribution in synthetic faujasites by high-resolution solid-state ^{29}Si NMR. *J. Am. Chem. Soc.* 104: 4859–64

Müller, D., Gessner, W., Behrens, H. J., Scheler, G. 1981. Determination of the aluminum coordination in aluminum-oxygen compounds by solid-state high-resolution ^{27}Al NMR. *Chem. Phys. Lett.* 79: 59–62

Oldfield, E., Kinsey, R. A., Smith, K. A., Nichols, J. A., Kirkpatrick, R. J. 1983. High-resolution NMR of inorganic solids. Influence of magnetic centers on magic-angle sample-spinning lineshapes in some natural alumino-silicates. *J. Magn. Reson.* 51: 325–27

Ramdas, S., Thomas, J. M., Klinowski, J., Fyfe, C. A., Hartman, J. S. 1981. Ordering of aluminum and silicon in synthetic faujastites. *Nature* 292: 228–30

Ribbe, P. H., ed. 1983. *Feldspar Mineralogy. Reviews in Mineralogy.* Washington DC: Mineral. Soc. Am. 362 pp. 2nd ed.

Schramm, S. E., Oldfield, E. 1984. High-resolution oxygen-17 NMR of solids. *J. Am. Chem. Soc.* 106: 2502–6

Smith, J. V., Blackwell, C. S. 1983. Nuclear magnetic resonance of silica polymorphs. *Nature* 303: 223–25

Smith, J. V., Blackwell, C. S., Hovis, G. L. 1984. NMR of albite-microcline series. *Nature* 309: 140–42

Smith, K. A., Kirkpatrick, R. J., Oldfield, E., Henderson, D. M. 1983. High-resolution silicon-29 nuclear magnetic resonance spectroscopic study of rock forming silicates. *Am. Mineral.* 68: 1206–15

Thomas, J. M., Klinowsky, J., Ramdas, S., Anderson, M. W., Fyfe, C. A., Gobbi, G. C. 1983. New approaches to the structural characterization of zeolites, magic-angle spinning NMR (MASNMR). In *Intrazeolite Chemistry, Am. Chem. Soc. Symp. Ser. No. 218*, ed. G. D. Stuckey, F. G. Dwyer, pp. 159–80. Washington DC: Am. Chem. Soc.

Ann. Rev. Earth Planet. Sci. 1985. 13: 49–74

EVOLUTION OF THE ARCHEAN CONTINENTAL CRUST

Alfred Kröner

Institut für Geowissenschaften, Johannes Gutenberg-Universität, Postfach 3980, 6500 Mainz, West Germany

INTRODUCTION AND HISTORICAL REVIEW

Some 20 years of plate tectonic theory, combined with new insights into the fine structure of the lithosphere, the application of multielement geochemical and isotopic studies, paleomagnetism, and geophysical modeling of mantle processes, have profoundly influenced present thinking on the origin and evolution of the Earth's early continental crust; previously, our knowledge of the continental crust was based almost exclusively on field geological observations.

Although there is now general agreement on how the Earth worked for the last 200 m.y. because of observable evidence in the oceans and continents (e.g. Bird 1980, Condie 1982), it has proved difficult to extend this history into more ancient times in view of the lost oceanic record and the ambiguity and complexity of the pre-Mesozoic rock relationships in the continents (Dewey 1982). However, preserved characteristic rock assemblages uniquely identifying modern-type Wilson-cycle processes (i.e. opening and closure of oceans underlain by oceanic crust) have now been recognized in continental terranes as old as ~900 m.y., and these assemblages provide strong evidence for the conclusion that the present global tectonic regime has governed the evolution of the lithosphere at least since the late Precambrian (Kröner 1977, 1981a,b; Goodwin 1981). Profound disagreement on the older crustal history, however, prevails to the present day, since typical features characterizing Phanerozoic accretionary terranes (such as obducted ophiolites, blueschists, and Franciscan-type mélanges) have not been found in more ancient regions. Thus, two types of evolutionary models have been developed. One type postulates uniformitarian development back to the earliest Archean (Burke

0084-6597/85/0515-0049$02.00

et al 1976, Burke 1981, Condie 1982, Moorbath 1978, Sleep & Windley 1982), usually by inferring some form of unspecified "primitive" form of plate tectonics and by following the precept that "comparable rocks imply comparable origins" (Dickinson 1981). The other type relates changes in the Earth's internal processes as a result of global cooling to concomitant changes in lithospheric behavior and suggests that tectono-magmatic mechanisms governing crust formation and growth have changed through geologic time (Baer 1981, Goodwin 1981, Kröner 1981a,b).

Worldwide interest in Archean crustal evolution arose in the 1960s from the discovery of komatiites [high-MgO lavas thought to be unique to the early Precambrian (Viljoen & Viljoen 1969)], from the recognition of greenstone belts as tectonic features distinct from younger orogenic belts (Anhaeusser et al 1969), and from reports of rocks with very old ages up to 3.7 b.y. (McGregor 1973) that suddenly extended the geologic record by several hundred million years. This surge of interest coincided with the development of analytical methods for routine but precise and rapid determination of major and trace elements, the application of isotope geochemistry in petrology, and an interest by planetologists in the early evolution of the Earth as compared with other bodies of the solar system (e.g. Lowman 1972, Ringwood 1966).

The wealth of geochemical data thus obtained led to frequent comparisons of Archean rock types with those of modern environments, and remarkable similarities were reported. Greenstone belt basalts were seen as analogues to modern mid-oceanic ridge basalts (MORBs) (Hart et al 1970, Glikson 1971, Jahn & Sun 1979), while the tonalite-trondhjemite igneous suite, found in batholiths surrounding the greenstone belts and in so-called grey gneiss terranes, was compared to various crustal levels of modern calc-alkaline arc complexes (Glikson 1972; see also papers in Windley 1976). This comparison led to the first plate tectonic models for Archean crustal evolution (Condie 1972, Glikson 1972, Anhaeusser 1973, Talbot 1973) that proposed ensimatic (i.e. intraoceanic) crustal growth in "primitive" island-arc systems. These concepts were supported by Sr isotopic data, which were interpreted as suggesting massive juvenile crust formation through greenstone and granitoid magmatism on the basis of very low $^{87}Sr/^{86}Sr$ initial ratios (Moorbath 1978). Gravity-driven vertical tectonics was seen as the dominant force both in forming the deep, synclinal greenstone basins with apparent stratigraphic thicknesses exceeding 20 km in places and in emplacing the numerous and areally extensive dome-shaped granitoid batholiths (Anhaeusser 1973, Wilson 1973).

The preceding paragraphs describe the "state of the art" as reviewed by Anhaeusser (1975) in Volume 3 of this series and as reflected in several conference proceedings and review volumes (e.g. Glover 1971, McCall 1977, Sutton & Windley 1973, Windley 1976). Alternative views suggesting ensialic (i.e. intracontinental) greenstone belt evolution and granitoid formation through multistage processes including anatexis of older sialic crust (e.g. Fyfe 1974, McGregor 1973, Condie & Hunter 1976, Kröner 1977) were clearly in the minority, and a suggestion by Green (1972) that Archean greenstone belts represent terrestrial equivalents of lunar maria was never really taken seriously.

Most of the above models were based on rock types and their chemistry, rather than on rock associations and detailed field relationships; subsequent structural work, together with further chemical data, isotopic constraints, and first detailed seismic profiles through Archean terranes, established new criteria that were incompatible with some of the earlier assumptions. Greenstone belt volcanics, including komatiites, were found to rest unconformably on older continental crust (for references, see Kröner 1981a,b), and in many cases where field relationships are ambiguous, precise dating showed granitoid gneisses to be older than the neighboring greenstones (e.g. Gee et al 1981, Kröner et al 1981, Grachev & Federovsky 1981, Percival & Card 1983). Furthermore, neither the tholeiitic nor the calc-alkaline suites in greenstone belts are strictly identical to modern ocean-floor and island-arc volcanics (Hawkesworth & O'Nions 1977, Glikson 1979, Condie 1981, Grachev & Fedorovsky 1981), and a pronounced bimodality unlike that in oceanic arcs was recognized in many Archean volcanic successions (Condie 1981, Ayres 1983, Thurston & Fryer 1983). There is also increasing evidence for crustal contamination in greenstone volcanics and other mantle-derived igneous rocks (Hegner et al 1984, Chauvel et al 1984).

The earlier proposal of deriving the Archean tonalite-trondhjemite-granodiorite (TTG) suite from anatexis of downsagging mafic greenstone volcanics became untenable in view of REE and isotopic systematics (Hanson 1981, Condie 1981, Martin et al 1983), and extremely precise dating in the Abitibi belt of Canada demonstrated that major TTG plutons are closely related in time (and probably genesis) to the neighboring volcanic rocks (Krogh et al 1982). Furthermore, seismic profiling in Western Australia showed greenstone belts to pinch out at relatively shallow depth and to be apparently underlain by high-grade older siliceous gneisses (Archibald et al 1981). Lastly, early recumbent folds and thrusts were discovered in greenstone piles that have caused stratigraphic repetition, requiring a revision of many published thicknesses (Gorman et al 1978, Williams & Furnell 1979, Platt 1980) and causing the predominance of vertical tectonics in Archean domains to be questioned. Thus, horizontal deformation, as demonstrated for many gneiss terranes (Myers 1976, Park 1981), undoubtedly also played an important role in greenstone belt formation and was not restricted to the Archean high-grade crust.

These new results were partly incorporated into revised crustal evolution models that see Archean granite-greenstone terranes as analogues to modern (Andean-type) arc–back arc or continental rift settings (for summaries, see Windley 1977, Kröner 1981a, Platt 1980, Condie 1981, Archibald et al 1981), while the extensive high-grade gneiss terranes are now variously interpreted as ancient accretionary thrust stacks adjacent to arc complexes (Tarney & Windley 1977) or as segments of largely pregreenstone lower continental crust that were brought to the surface along listric thrusts (Park 1981, Kröner 1984b, Coward 1976).

There is as yet no generally accepted model for the evolution of the Archean continental crust, but the debate vigorously continues and currently receives much impetus from isotope geochemistry (in particular Sm-Nd data), from comparative planetology (in particular, crust formation on Venus and the Moon), and from a growing interest by geophysicists in the early thermal evolution of the Earth. This review summarizes and discusses some of the major problems inherent in currently popular evolutionary models and tries to demonstrate that the dogma "comparable rocks imply comparable origins" may not necessarily apply to the early history of the continental crust.

THERMAL EVOLUTION AND CRUSTAL GROWTH

It is now generally accepted that growth of the continental crust was significantly faster in the Archean than in later times, and that 70% or more of the present crustal material had already separated from the mantle by about 2.5 b.y. ago (Figure 1; for references, see Reymer & Schubert 1984). Later growth was much slower, and post-Archean crustal evolution was apparently dominated by reworking and cannibalistic recycling (Veizer & Jansen 1979, Allègre 1982). This early rapid growth has been explained in terms of much higher arc accretion rates, said to have resulted from faster plate motion and correspondingly faster oceanic plate turnover (Bickle 1978, Dewey & Windley 1981). If the Archean terrestrial heat flow was 2–3 times the present rate (Lambert 1981), upper-mantle temperatures were likely to have been up to 150°C higher than today (Smith 1981), as shown by the occurrence of komatiites in Archean terranes that are commonly ascribed to high degrees of partial melting in the mantle (Green 1975). Hager & O'Connell (1980) calculated that a temperature increase of 100°C in the mantle would result in a viscosity decrease by a factor of 10, which, in turn, would lead to a threefold increase in plate velocities. This would allow for about six times the current oceanic crust formation rate of $\sim 3\ km^2\ yr^{-1}$ by 2.8 b.y. ago (Sleep & Windley 1982) and a correspondingly higher arc accretion rate.

Inherent in these calculations is the assumption that the temperature dependence of the mantle viscosity implies the internal temperature to be strongly coupled with convective heat loss (e.g. Schubert 1979). However, recent numerical work indicates that variable rather than constant viscosity must be expected during mantle convection, and thus the dependence of heat flow on the mantle temperature may be much weaker than expected, implying that plate velocities have not changed much through geologic time (Christensen 1985). This is supported by presently available paleomagnetic data for the Archean that indicate minimum average velocities of 1–4 cm yr^{-1} with respect to the pole for the Superior province of Canada, the Pilbara block of Western Australia, and the Kaapvaal craton of southern Africa (Kröner et al 1984, and unpublished data). Although the data are still limited, they suggest different apparent polar wander paths and individual velocity signatures for each of the above cratons, thus implying Archean continental drift but indicating no difference from the overall mean minimum velocity of 2.8 cm yr^{-1} for the continents in post-Archean times (Ullrich & Van der Voo 1981).

Geological arguments, such as the lack of obducted ophiolites and blueschist assemblages in the Archean, also argue against fast plate motion during that time (Kröner 1984a). Consequently, if plate velocities were broadly constant, arc accretion rates in the Archean could not have been

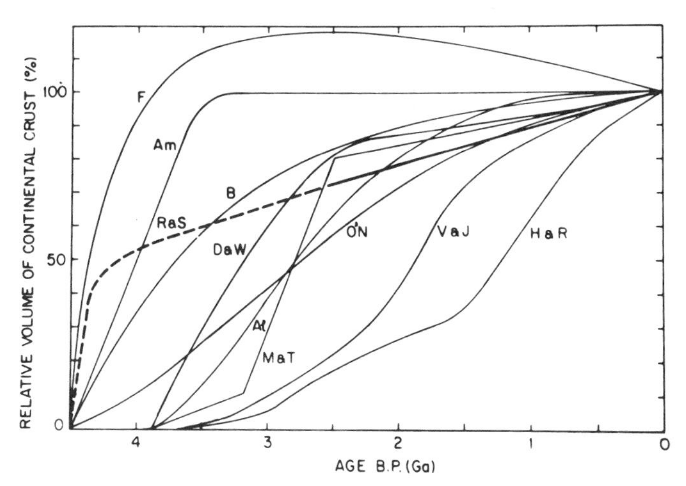

Figure 1 Crustal growth curves proposed in the literature, showing rapid development of continental crust up to $\sim$2.5 b.y. ago in most models. (From Reymer & Schubert 1984; see also the references cited therein.)

substantially higher than today, and additional crust formation processes such as hotspot magmatism may have operated (Lambert 1981, Kröner 1981a, Reymer & Schubert 1984).

The relative importance of hotspot-induced crust formation by vertical accretion (i.e. under- or overplating) and by subduction-related horizontal arc accretion may have changed through the Archean and younger times (Reymer & Schubert 1984), but it seems unlikely that plate tectonics involving rigid lithospheric slabs operated during the formation of the earliest crust. Morgan & Phillips (1984) have presented a model for hotspot crustal generation on Venus that they also suggest for the early Earth, while Kröner (1982) drew an analogy between the growth of present-day Iceland and the formation of earliest Archean continental crust. Condie (1980) proposed a plume model for the time prior to 4 b.y. and suggested rapid recycling of the earliest basaltic crust in "sinks" (Figure 2; see also Condie 1984).

The following scenario may be compatible with such proposals: Following rapid cooling of the originally melted surface of the Earth after core formation (Ringwood 1979), a very thin crust composed chiefly of ultramafic rocks (Condie 1980, Nisbet & Walker 1982) forms small but coherent rigid segments like the crust on modern lava lakes. These segments frequently break up again and are recycled back into the underlying melted mantle because of their high density and viscous drag at their bases resulting from convection below. The breakup and recycling is further enhanced through heavy meteorite bombardment.

As cooling progresses, the early crustal segments thicken and komatiitic-to-basaltic volcanism becomes widespread (Condie 1980, Nisbet & Walker 1982). An Archean heat flow of 200–250 mW m^{-2} implies volcanic activity over virtually the entire surface of the early Earth (Smith 1981). It is unlikely

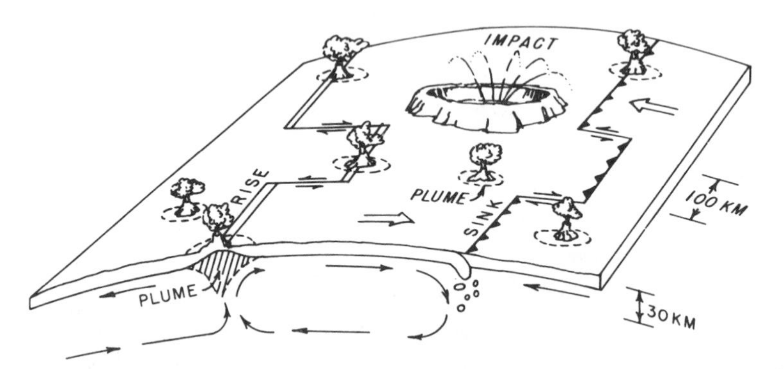

Figure 2 Sketch of early Archean (~4 b.y.) tholeiitic crust generated and recycled in "primitive" plate tectonic system. (From Condie 1980.)

that this volcanism occurred along linear zones comparable to present oceanic ridge systems. Rather, vigorous secondary convection in the partial melt zone under the evolving crust (Figure 3) would suggest a system of closely spaced thermal plumes (Lambert 1981, Smith 1981).

The early ultramafic crust with overlying komatiites and basalts was still too dense to survive and was completely recycled back into the hot mantle at comparatively short time rates. At about 4.3 b.y. ago, mantle temperatures and heat flow had decreased sufficiently to permit komatiite-dominated crust that now had a longer survival rate. Continued volcanism then had the chance to build volcanic centers that grew rapidly in height. The growth of the volcanic pile through partial melting of the subcrustal mantle immediately below resulted in basalt depletion and cooling of the depleted region so as to form a mini-tectosphere (Jordan 1979) that further stabilized the overlying crust and contributed to its continued survival. Regions of the early crust that did not experience extensive and continued volcanism are likely to have maintained relatively thin, cooling lithospheres and were easily dragged back down into the mantle, while the evolving and thickening continental segments over plumes grew rapidly and tended to survive longer (Figure 3). The oldest depleted mantle reservoirs may date

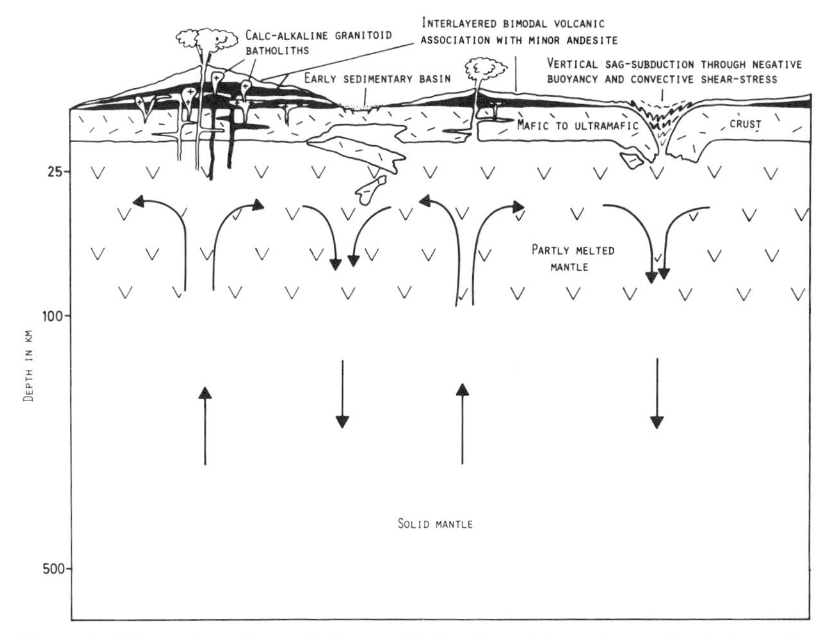

Figure 3 Schematic and speculative model for the origin and growth of the earliest continental crust. (Based on Figure 2 of Condie 1980.)

back to this time (McCulloch 1982, Jagoutz et al 1983), indicating that irreversible differentiation of crust and mantle was now possible. Abbott & Hoffman (1984) have shown that relatively slow plate motion as advocated above would tend to enhance the formation of local, thick hotspot-produced crust, while fast motion would inhibit it.

The evolution of Iceland may be an appropriate scenario for the next stage of continental differentiation (Kröner 1982), since the present heat flow there is about 200 mW m^{-2} (Smith 1981). Continued melting in the mantle produces further komatiitic-to-tholeiitic magmas, but a large proportion of this dense melt no longer reaches the surface of the growing crust; instead, it accumulates at the base of the crust and forms the reservoir for bimodal volcanic suites and the first rising tonalite-trondhjemite plutons (Figure 3). As the volcanic pile thickens, its lower parts and the underlying ultramafic protocrust melt and form a vast reservoir for further differentiated magmas. The dense residue remaining from this process may become decoupled from the crust through viscous drag to allow recycling into the convecting mantle. Continued growth of this kind gradually reduces the overall density of the evolving crust, which is now less than ~3 g cm^{-3}, and, through partial melting at its base, establishes a two-layer structure with the lower part transformed into high-grade assemblages and the upper part consisting of both primitive and differentiated volcanics and granitoid intrusives.

Suitable crustal segments could have evolved to this stage by about 4.1 b.y. ago, since the oldest known zircons of that age (Froude et al 1983) are likely to have been derived from a differentiated source. Eventually a situation may have been reached where a mixture of granitoid intrusives and differentiated volcanics constituted the surface of this now predominantly sialic early upper crust. If it emerged above sea level, erosion would have ensued, and the first sediments reflecting the composition of their sources would have been deposited. The ~3.8 b.y. old Isua sequence of West Greenland (Hamilton et al 1983) and the equally old Sand River gneisses of the Limpopo belt in southern Africa (Barton et al 1983) could be examples of this type.

The evolutionary model postulated above guarantees survival of these early continental nuclei, which were now too buoyant to be recycled and whose subscrustal roots gradually thickened and thus "shielded" the overlying crust against extensive disruption. Neighboring lithospheric segments, where mantle melting and volcanism were insufficient to form sialic crust, remained comparatively thin and dense and were sites of drag-induced "sag-subduction" (Goodwin 1981, Kröner 1981a; see Figure 3). None of this early "oceanic" crust has survived, while the sialic differentiates are considered to form the oldest preserved crustal remnants.

Although the volume of presently known continental crust older than ~3.5 b.y. is relatively small, there is increasing evidence for significant mantle depletion at this time, as shown by positive ε_{Nd} values in the most ancient mantle-derived rocks (Jagoutz et al 1983, Fletcher et al 1982, Macdougall et al 1983); more early Archean sialic crust can therefore be expected.

EVOLUTION OF ARCHEAN GREENSTONE BELTS

Currently popular views on the evolution of granite-greenstone terranes are all compatible with the concept of moving plates in the Archean, but they differ substantially in tectonic development and in the generation and emplacement of magmatic rocks. These models have been described in detail elsewhere (e.g. Condie 1981, Glover & Groves 1981, Kröner 1981b) and can be grouped into two categories: (*a*) subduction-related magmatic arc or marginal basin settings, and (*b*) intracontinental rift settings.

Models of the first category consider processes in modern intraoceanic or Andean-type magmatic arcs and genetically related marginal basins to be the closest analogue to Archean crust-forming mechanisms and are based largely on geochemical arguments. Condie (1980) suggested that the first segments of surviving continental crust evolved entirely from oceanic andesite arc systems at convergent plate boundaries between 4 and 3.5 b.y. ago, while Andean-type and/or marginal basin evolution should have followed between 3.5 and 2.7 b.y. ago. The assumption of an early oceanic development is based on the belief that komatiite-basalt sequences in the lower parts of the oldest known greenstone belts (~3.5 b.y.) that have no recognized basement represent primeval oceanic crust (e.g. Anhaeusser 1973, Condie 1981). For the younger greenstone successions of the late Archean that show abundant evidence for the presence of sialic material or that were found resting on continental crust, the continental margin or marginal basin model appeared more appropriate (Tarney et al 1976, Windley 1977, Condie 1981; see Figure 4).

Several objections have been raised to these views that require short

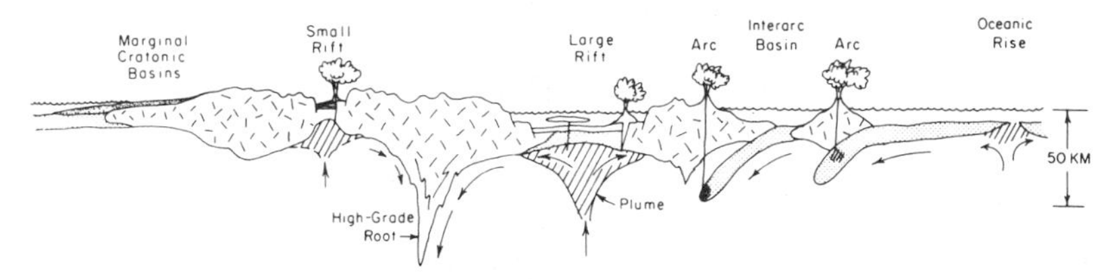

Figure 4 Diagrammatic cross-section combining proposed tectonic settings for greenstone belt evolution between 3.5 and 2.7 b.y. ago. (From Condie 1980.)

discussion. First, the oldest known greenstone suites are all dominated by bimodal volcanic associations and contain no or only a small proportion of true andesites. Second, there is growing evidence that virtually all the early greenstone sequences were deposited in shallow water, as shown by silicified evaporites, stromatolite-bearing carbonates, banded iron formations, and subaerial volcanics (for reviews, see Bickle & Eriksson 1982, Groves & Batt 1984); the depositional basins are therefore interpreted to have evolved on older sialic crust and were marginal to major oceans (Eriksson 1981, De Wit et al 1983, Hickman 1984).

Third, some of the most surprising aspects of early greenstone belts are the excellent state of preservation, the generally low degree of metamorphism and, locally, the lack of significant penetrative deformation in the volcanic piles that have reported thicknesses of 5–15 km (e.g. Viljoen & Viljoen 1969, Hickman 1984). If these rocks indeed represent ancient oceanic crust, it is difficult to understand why such crust—given much higher Archean mantle and ocean-water temperatures than those of today (Fyfe 1978)—should be so well preserved while modern ocean crust has been profoundly altered by superheated aqueous solutions to depths of more than 5 km (East Pacific Rise Study Group 1981). Also, the present oceanic thermal regime is consistent with upper amphibolite facies metamorphism in layer 3 between ~5 km and ~12 km depth (Oskarsson et al 1982). It is feasible, therefore, to assume that the greenstone successions were laid down on cool crust that effectively isolated the depositional environment from the hotter mantle. The lack of layered crust with sheeted dikes in Archean greenstone suites further supports this contention. Lastly, the growing evidence for contamination of at least some greenstone volcanics by significantly older sialic continental crust, resulting in anomalously high isotopic ages as a result of mixing processes (Chauvel et al 1984), provides most convincing proof against an ensimatic environment of deposition.

Crustal contamination and low-temperature alteration may also affect the major and trace element geochemistry of greenstones, thereby obliterating primary magmatic trends, and there are examples where such processes have imposed an apparent calc-alkaline character on originally tholeiitic rocks (for a review, see Kröner 1982). Caution should therefore be exercised in using popular geochemical discrimination diagrams to infer Archean greenstone tectonic settings.

Although it is commonly believed that calc-alkaline magmas are produced in the mantle wedge overlying subduction zones (Gill 1981), there are examples of anorogenic basalt-andesite-dacite-rhyolite associations from the Tertiary (Wilkinson & Binns 1977) to the Archean (Hegner et al 1981); these rocks apparently occur in extensional environments where crustal melting is required to account for the observed chemical variations and isotopic data. Several recent investigations have shown that andesitic-to-rhyolitic rock types in extensional settings may be produced by varying degrees of mixing of lower continental crust and mantle material [for summaries, see *Eos, Trans. Am. Geophys. Union* 61:67–68 (1980)], and Gélinas et al (1984) have proposed the following model for the voluminous andesites of the Abitibi belt in Canada that may have general validity for Archean greenstone generation: "Mafic magmas nourished the central reservoir where melting of the sialic crust took place. The rhyolitic magma occupying the upper part of this reservoir mixed with basaltic magma, producing calc-alkaline andesites."

Inherent to these models is a mafic protolith ultimately derived from the mantle, and underplating of dense picritic-to-gabbroic magmas at the base of the crust has been suggested for Phanerozoic settings (Cox 1980, Ewart et al 1980; see Figure 5). In the Archean such underplating must have occurred on a much more extensive scale, since Hanson (1981) and Phinney et al (1981) have pointed out that mantle-derived melts high in MgO and FeO, such as komatiites, have densities higher than most crustal rocks, and many such melts would therefore not reach upper crustal levels. Given the generally accepted high degrees of partial melting in komatiite source regions and melt densities in excess of 3 g cm^{-3} (Nisbet 1982), much of this material would be trapped near the base of the crust (Sun 1984). Hanson (1981) further suggested that such melts would tend to form immiscible

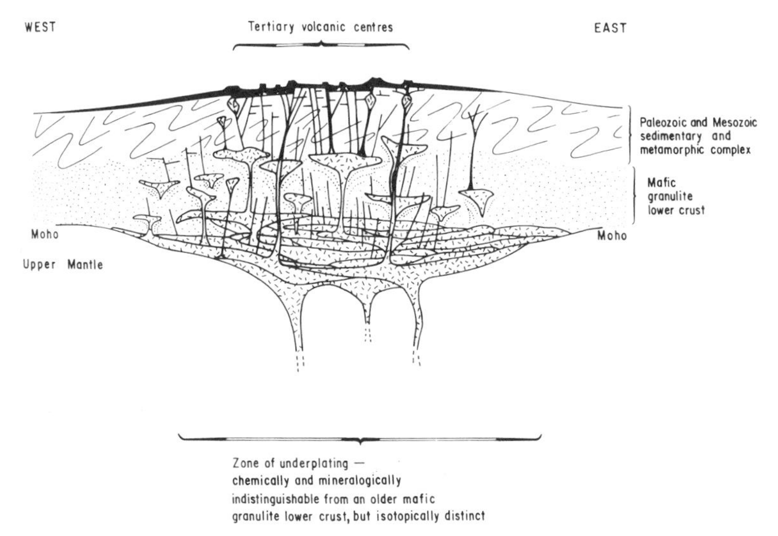

Figure 5 Model of magmatic underplating in continental rift environment providing reservoirs for bimodal volcanic associations. (From Ewart et al 1980.)

liquids, to separate gravitationally, and then to ascend into the upper crust to form either basalt and/or mafic tonalite or to solidify as dacite/rhyolite and/or granodiorite.

Exposed high-grade Archean terranes interpreted as lower crust (Drury et al 1984) and geophysical models (Drummond et al 1981) generally indicate that there is not enough mafic-to-ultramafic residue to account for the observed quantity of felsic igneous rocks in granite-greenstone complexes. Sun (1984) has therefore suggested that the underplated residues convert to eclogite after cooling and may delaminate and sink back into the mantle. Such material may also maintain its primary isotopic signature and remain a distinct, long-lived, depleted source for further melting.

Early models of Archean evolution assumed that greenstone belts follow a pattern from ultramafic-to-mafic-to-siliceous volcanics, succeeded by a suite of granitoids evolving from Na-rich to K-rich types and involving large-scale anatexis of older greenstone material (e.g. Anhaeusser 1975). This process should reflect "maturing" of intraoceanic arcs to microcontinents (Glikson 1972). However, the discoveries of felsic volcanic rocks that occur in the lower parts of greenstone suites and that may be overlain by or intercalated with komatiites, as well as the close spatial relationship (i.e. interlayering) of these rocks with andesites in individual belts (for details, see Kröner 1982), cast strong doubt on the "growing arc" model. The broad contemporaneity of both extrusive and granitoid intrusive rocks (Barton 1981, Krogh et al 1982) furthermore implies that the latter are unlikely to represent melting products of the former.

The marginal basin model for greenstone belt evolution (Figure 6) is also not without its problems, and several authors (for references, see Kröner 1982) have listed objections that are mainly based on field relationships. The model does not seem to account for the apparently large size (several hundred to more than 1000 km) of greenstone basins, the interlayering of tholeiitic and calc-alkaline lavas, and the long evolutionary history of most greenstone belts, extending over ~200 m.y. or more; in contrast, recent marginal basins, on average, have relatively short lifetimes of 40 m.y. or less (Talwani & Langseth 1981).

The preferred occurrence of MgO-rich lavas in the lower parts of greenstone assemblages [also found in Phanerozoic rift environments (Hawkesworth & O'Nions 1977)], the pronounced bimodality of volcanic suites, and the close spatial relationship of greenstones with granitoid crust all favor a rift-related setting for the initial stage of greenstone belt formation. Continental rift models have been discussed at length in recent papers (for reviews, see Condie 1981, Glover & Groves 1981, Kröner 1981b, Bickle & Eriksson 1982), and the suggested scenarios range from grabens where thickened and stabilized older continental lithosphere inhibited

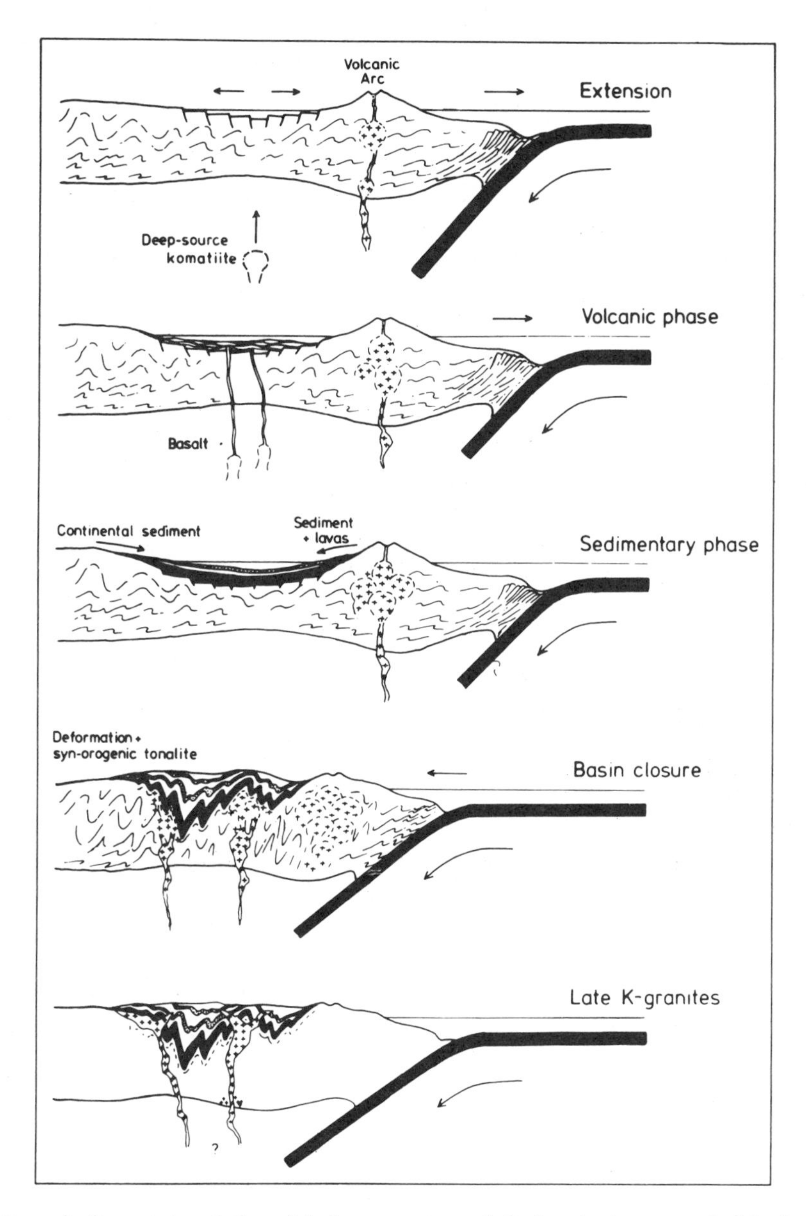

Figure 6 Suggested evolution of Archean greenstone belts in a back-arc marginal basin setting. (From Tarney et al 1976.)

proper greenstone basin evolution through pull-apart basins, deep rifts, and small ocean basins of Red Sea type. Bickle & Eriksson (1982) have related the evolution of several such Archean basins to the two-stage subsidence model of McKenzie (1978) and show that subsidence in greenstone basins cannot be due to loading by the dense volcano-sedimentary pile, since virtually all rocks were deposited in shallow water. Instead, it is likely that crustal thinning produced listric fault patterns in the upper crust, similar to those observed in the Basin and Range province, while the lower crust yielded by laminar flow. Such basin-widening mechanisms could explain the apparent great thickness in some greenstone successions.

Much of the early deformation now recognized in greenstone belts is probably not due to diapiric rise of tonalitic plutons alone but relates to considerable horizontal translation, including nappe formation (De Wit et al 1983; see also papers in Kröner & Greiling 1984). Many of these originally flat structures were later modified and frequently steepened by listric faulting or during the emplacement of granitoid batholiths.

Figure 7 depicts several hypothetical stages in the evolution of a greenstone belt based on the rift setting discussed above. The fundamental difference between the rift and magmatic arc models from the point of view of crustal growth is that the former requires massive vertical accretion as well as crustal reworking, while the latter implies horizontal accretion through arc magmatism and subsequent tectonism.

ORIGIN OF ARCHEAN BIMODAL GNEISS AND GRANITOID SUITES

All presently known Archean terranes consist of varying proportions of greenstone belts, granitoid intrusives, and a variety of banded medium- to high-grade "grey gneisses" that frequently contain metasedimentary remnants and that are either older or formed broadly contemporaneously with the greenstone belt rocks (Windley 1977, Glover & Groves 1981). Taken together, an Archean crustal column composed of these units would have broadly andesitic composition (Taylor 1967), while the individual rock suites are remarkably andesite-free (Barker & Arth 1976). The volumetrically most important granitoid rocks of Archean terranes are tonalites, trondhjemites, and granodiorites (TTG suite of Jahn et al 1981) and their tectonized gneissic equivalents. Much controversy still exists as to the origin of these rocks (Condie 1981, Jahn et al 1981), which typically form large composite batholiths and which often contain xenoliths of greenstone belt material. The banded gneisses and migmatites have been interpreted as older sialic basement or as products of interaction between greenstones and granitoid melts.

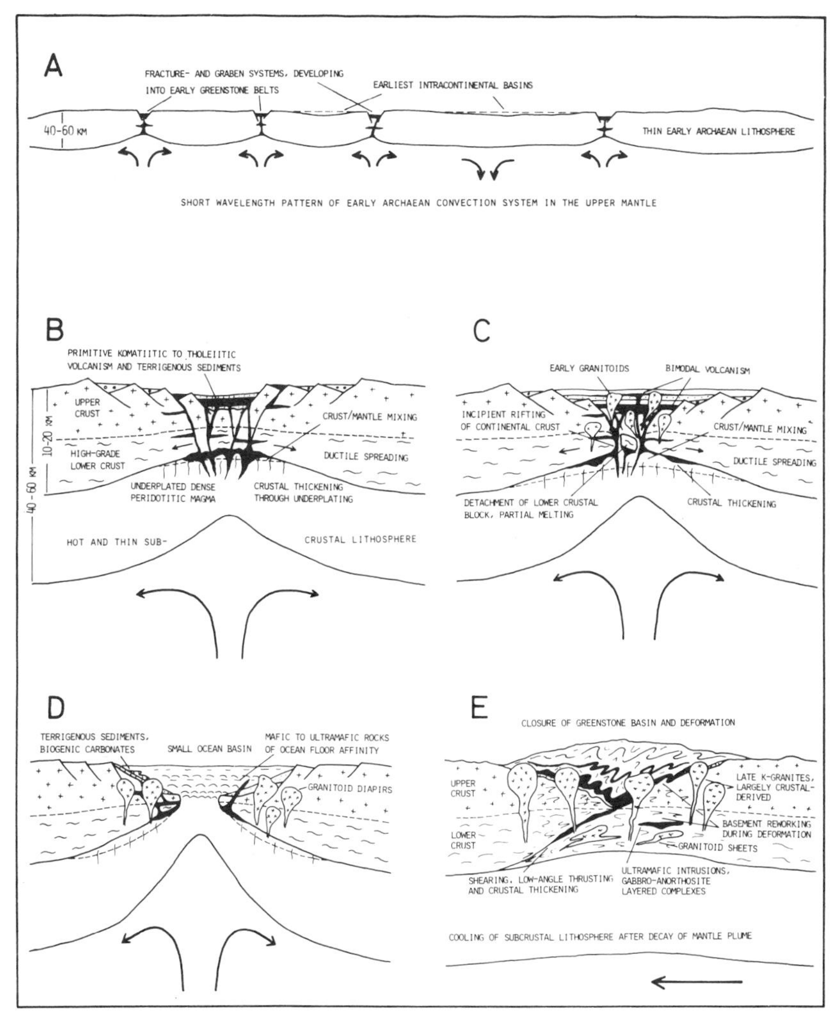

Figure 7 Schematic cross-sections showing suggested evolution of early Archean lithosphere in response to small-scale convection pattern in the upper mantle (*A*) and various stages in the development of a rift-induced greenstone belt. (From Kröner 1982.)

Anatexis of greenstone volcanics or of primitive mafic-ultramafic oceanic crust has been suggested as the origin for the TTG suite (e.g. Anhaeusser 1975), but this is unlikely as shown earlier (pp. 51, 60), and petrogenetic considerations also do not favor this model (Condie & Hunter 1976, Hanson 1981). An alternative origin is proposed by most uniformitarian models, which see similarities between Archean TTG and modern calc-alkaline batholiths and their gneissic roots that characterize Andean-type belts (Windley 1979). In this view, the TTG rocks are generated from mantle-derived juvenile melts that originate above subduction zones and either intrude into the evolving arc or continental margin or accrete and differentiate at or near the base of the crust, where they solidify and develop into granulite-gneiss belts. These processes require no or only short crustal residence times for the granitoid precursors, a feature supported by many isotopic data (Moorbath 1978). Following this line of reasoning, the great volumes of TTG produced in the Archean are conveniently ascribed to the rapid recycling of ocean crust and the frequent collision of numerous, small, fast-moving plates (Dewey & Windley 1981); this scenario, however, neither takes account of the spatial and temporal distribution of these rocks nor considers their cyclic generation during well-defined crustal accretion events (Moorbath 1978). The conflict of this hypothesis with the available paleomagnetic data and crustal growth models has already been discussed.

A further origin for the TTG suite was proposed on the basis of petrogenetic and geochemical modeling. These models show that TTG rocks may be derived from mafic-ultramafic sources through processes of partial melting of amphibolite or eclogite, or through anatexis of older tonalite (for reviews, see Condie 1981, Hanson 1981, Jahn et al 1984). Partial melting of 10–20% amphibolite of basaltic-to-komatiitic composition at or near the base of the crust produces tonalite-trondhjemite melts. In the lower crust these magmas intrude as sheets, and together with their host rocks, they are involved in plastic flow by lateral extension and form high-grade gneisses with generally flat foliations (Gastil 1979). At higher levels these melts may still intrude as sheets and form complex structural relationships with the deformed roots of greenstone belts, including marginal zones of migmatites, but most of the magma rises diapirically to form large batholiths that often engulf the greenstones and obliterate their relationships with older basement.

The existence of trondhjemitic intrusives in modern extensional environments such as Iceland (Sigurdsson & Sparks 1981), which were derived from basaltic sources, supports the above model and suggests that Archean hotspot activity may have been an important cause for bimodal magmatic suites (Condie 1981, Lambert 1981).

The TTG suites with "primitive" isotope characteristics also satisfy the above model, but they do not preclude the possibility that their source regions had very low Rb/Sr, so that considerable crustal residence times of up to several hundred million years are possible for the TTG precursors (e.g. Barton 1981, Hamilton et al 1979). Jahn et al (1984), for example, demonstrated that ~3.1-b.y.-old tonalitic gneisses from northern Finland have Sm-Nd systematics and REE patterns that suggest separation of the basaltic gneiss protolith from the mantle ~3.4–3.5 b.y. ago. This mafic material resided in the lower crust for some 300–400 m.y. before partially melting to produce the tonalitic magma (Figure 8). Likewise, Collerson (1983) demonstrated on the basis of ion microprobe zircon dating that the tonalitic Uivak gneisses of Labrador, emplaced at ~3.7 b.y. ago and equated with the Amitsoq gneiss of West Greenland, were probably derived from older crustal material dating back to ~3.9 b.y. ago.

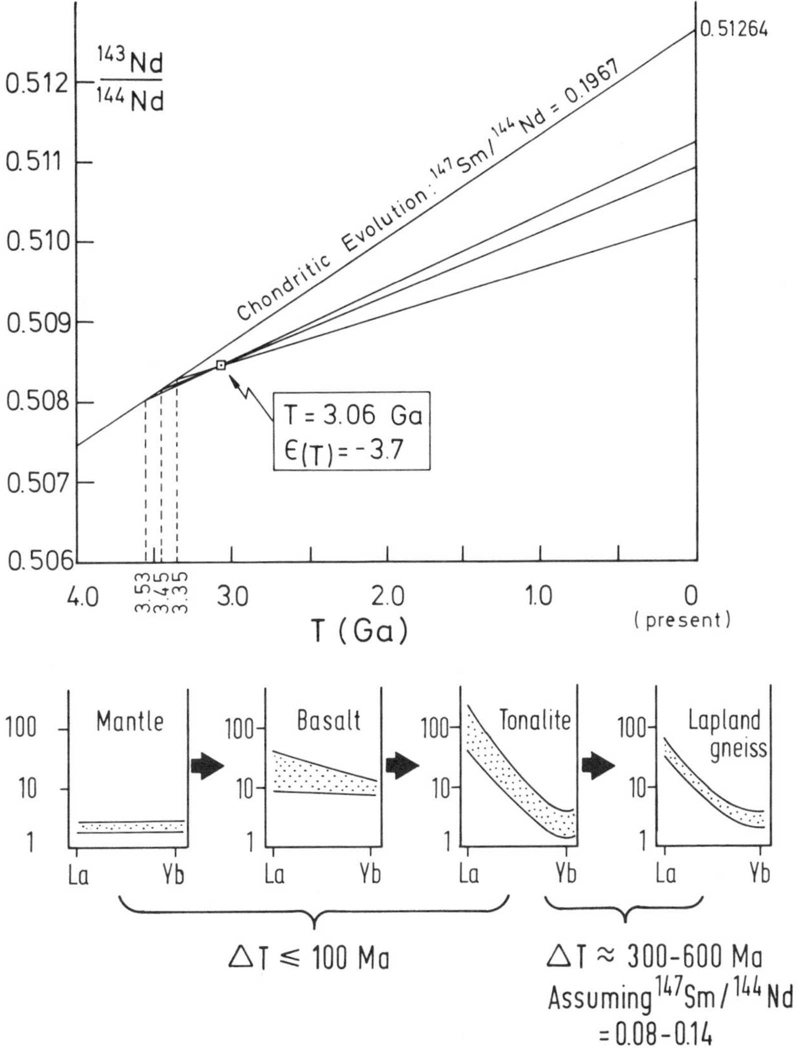

Figure 8 Nd isotopic evolution diagram (*top*) and hypothetical three-stage REE evolution model (*bottom*) for tonalitic gneisses of northern Finland. (From Jahn et al 1984.)

In summary, current models for the origin of Archean TTG suites and compositionally similar grey gneisses either favor Andean-type subduction-related magmatic arc settings or propose partial melting of underplated reservoirs and remelting of older continental crust. These models must account for the episodic nature of granitoid magmatism and must also provide a mechanism to initiate and terminate massive volcano-plutonic cycles over vast distances (Krogh et al 1982). The long duration of such cycles [up to 500 m.y. (Barton 1981)] and the unlikely high crust production rate if the Archean granitoids and gneisses are predominantly derived from juvenile lateral accretion (Reymer & Schubert 1984) cast doubt on simple uniformitarian plate tectonic models, and scenarios involving episodic hotspot-related vertical accretion may also be possible.

ORIGIN OF ARCHEAN HIGH-GRADE GNEISSES

High-grade gneiss terranes are widespread in several ancient shields and are compositionally distinct from the lower-grade greenstone belts with which they are often associated. Well-known and extensive regions include southern India and Ceylon, Western Australia, much of Enderby Land in Antarctica, the Limpopo belt of southern Africa, the West Nile Complex of central Africa, and parts of West Greenland. Minor occurrences are found in all Archean cratons. These regions consist predominantly of granitoid TTG assemblages, bimodal (i.e. amphibolite–leucocratic gneiss) suites and metasedimentary/metavolcanic successions, all in upper amphibolite to granulite metamorphic grade, tectonically interlayered, and displaying a complex structural history (Windley 1977). They also support the concept of average thermal gradients and an average thickness of continental crust between 30 and 40 km as early as 3.5 b.y. ago (Grambling 1981). Of particular importance are the worldwide observations that the supracrustal suites are predominantly of shallow-water origin (e.g. original sandstone-shale-limestone-evaporite sequences) that were deposited on older continental crust, and that high-grade equivalents of greenstones are rare or absent (Gee et al 1981).

There have been suggestions that high-grade gneiss terranes constitute the basement to Archean greenstone belts and may underlie virtually all cratons (for discussions, see Kröner 1981a, 1984b); however, preserved field relationships are equivocal, and in many cases the low- and high-grade domains were tectonically juxtaposed. However, geophysical models show that dense, granulitic gneisses extend beneath the granite-greenstone layer (e.g. Archibald et al 1981). Isobaric or prograde pressure regimes, together with prograde temperatures, have probably been the cause of most high-grade gneisses in the lower crust; these regimes have been ascribed either to magmatic crustal accretion mechanisms (Wells 1981), to tectonic thickening through crustal interstacking (Bridgwater et al 1974, Myers 1976), or to continental collision (Dewey & Burke 1973, Hansen et al 1984).

The first model is based on the concept of massive accretion through juvenile, mantle-derived magmas (Moorbath 1978) at Andean-type continental margins and is discussed in detail by Windley (1979) and Wells (1981). These authors suggest that Archean gneisses are the equivalent of modern accretionary prisms that were thrust under the active continental margin during the subduction process. This would explain the close association of high-grade sedimentary rocks with the TTG suite, and apparently analogous situations are cited from the southern Andes of Chile and from the Himalayas in Pakistan (Windley 1979, Dewey & Windley 1981).

This process, if applicable to the Archean, would have produced linear gneiss belts whose internal primary structure would have been completely destroyed during the accretion event, as demonstrated by modern accretionary thrust stacks and mélanges (Dickinson & Seely 1979) and by older accretion complexes such as the Cretaceous Shimantu belt in Japan (Ogawa & Horiuchi 1978) that display chaotic structures containing tectonically emplaced exotic fragments. Such complexes, however, are contrary to field observations in many Archean gneiss terranes, where metamorphic equivalents of Franciscan-type assemblages have not been identified. Instead, shallow-water supracrustal rocks predominate and frequently display superposed fold patterns without extensive disruption of the original lithologies. In West Greenland, for example, the high-grade Malene supracrustals often define fold limbs that can be traced for several tens of kilometers (Bridgwater et al 1976). Furthermore, the nonlinear nature of many ancient gneiss terranes (e.g. southern India) also renders the horizontal accretion model unlikely for the origin of all Archean granulite complexes.

Continental collision is an alternative and plausible mechanism for the generation of these rocks; presently it converts much of the old Indian upper crust that has disappeared beneath Asia over the last 30 m.y. into high-grade assemblages that may reappear at the surface after substantial erosion following crustal thickening and isostatic equilibration. The process, in the present Wilson-cycle regime, is the final product of orogeny. Present plate tectonic theory holds that continental collision is preceded by ocean opening and closure and the generation of at least one active continental margin. The operation of this process has so far not been demonstrated in the Archean, although it should have occurred if plates moved independently of each other at that time, as suggested earlier. In contrast to modern "orogenic" depositional environments, the supracrustal rocks in Archean high-grade terranes suggest deposition in shallow, restricted basins on older crust, subsequent deformation and metamorphism during their transfer to the lower crust, and eventual reemergence at or near the surface at some later time. For most of these gneiss terranes, this history may have extended over several hundred million years (e.g. Barton 1981) and appears incompatible with the simple collision model.

Ancient granulite terranes are frequently separated from the (overlying) lower-grade granite-greenstone assemblage by transitional zones over which quartzofeldspathic gneisses commonly become migmatized and give way to charnockites, while other lithologies develop a granulite-facies mineralogy (Newton & Hansen 1983). This may suggest that the gneissic basement was transferred, almost vertically, to the lower crust while granitoid intrusions considerably thickened the upper crust. However, the ubiquitous evidence for horizontal shortening in gneiss terranes clearly shows that vertical motion alone was not responsible for the transfer to the lower crust.

The mounting evidence for deep fault and thrust zones in and on the margins of high-grade terranes suggests that plane-strain thrust restacking of previously thinned or mechanically weakened continental crust may have been the predominant mechanism in the Archean for the production of high-grade rocks. Such faults may extend to the base of the crust, where they gradually flatten out (Bally 1981, Oliver et al 1983). The thrusting mechanism, coupled with intense deformation and granitoid intrusions, may have transferred the once shallow-level sediments to deep crustal regions, where they recrystallized and experienced chemical modification. This often led to pronounced depletion in light-ion lithophile elements, perhaps related to CO_2 streaming that is now seen as the major agency of metamorphic dehydration (charnockitization) in the lower crust (Newton et al 1980, Hansen et al 1984, Allen & Condie 1984).

Katz (1981) ascribes the generation of Archean granulite terranes to large-scale intracratonic transform motion along zones of weakness demarcated by sediment-filled aulacogens, while Jahn & Zhang (1984) relate the close association of TTG rocks with metavolcanics and metasediments in granulite complexes to crustal underplating combined with intracrustal thrusting and stacking of crustal slices. The marked structural linearity that commonly defines the narrow Precambrian "mobile belts" may also be due to such thrusting, whereby large segments of the crust are placed "on edge," as suggested for the Archean Limpopo belt of southern Africa (Coward 1976) and the Proterozoic Nagssugtoqidian belt of Greenland (Bak et al 1975). Figure 9 illustrates this model for a hypothetical Archean crustal section and implies that the high-grade gneisses underlie the greenstone belts and only reach the surface after crustal tilting and thickening through thrusting, followed by erosion.

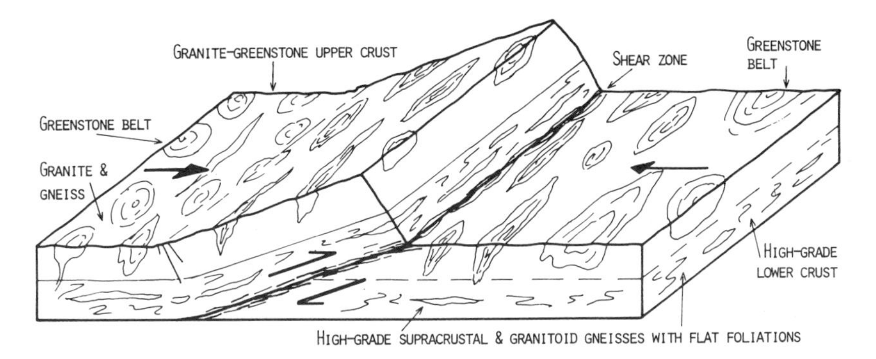

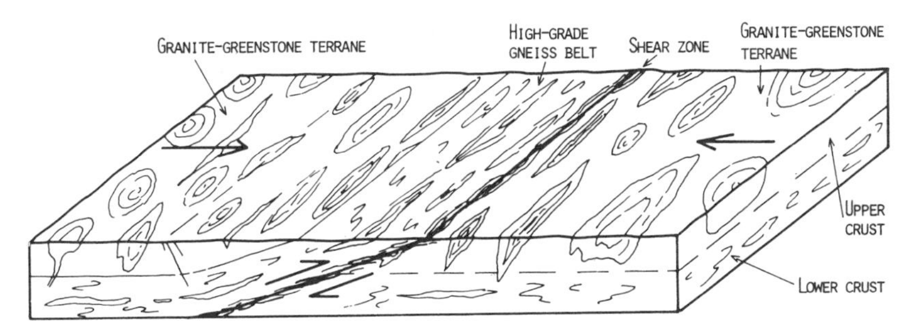

Figure 9 Schematic and idealized section through the Archean continental crust to show suggested relationship between upper crustal granite-greenstone terranes and lower crustal high-grade gneiss terranes. The upper section displays crustal interstacking by low-angle thrust that results in tectonic juxtaposition of lower crustal gneisses with upper crustal granite-greenstone terrane. A wide shear zone develops along the thrust. The lower section shows the same situation after removal of the overthrust block by erosion. The tilting of the crust now results in a transition from upper to lower crust on the exposed erosion surface. The model is not to scale and is drawn to explain the possibile origin of Archean high-grade gneiss belts such as the Limpopo in southern Africa. Note that this model assumes the continuation of high-grade gneisses under the granite-greenstone terrane. (From Kröner 1984b.)

CONCLUSIONS

Although there is general agreement about rapid and massive growth of the continental crust in the Archean, the mechanism for this growth is still not understood. Most presently popular models relate continent formation to voluminous arc accretion resulting from fast plate turnover, but the calculations of Christensen (1985) and the paleomagnetic data do not favor such situations, which also result in unrealistic growth rates (Reymer & Schubert 1984). If Venus has no plate tectonics, as is currently argued (Morgan & Phillips 1984), then this planet demonstrates that alternative heat loss mechanisms are possible, and students of the Earth should consider these, at least for the earliest history of our planet. Although some form of plate tectonics appears to have operated since at least ~3.5 b.y. ago, we have no record of the style of plate interaction other than a predominance of horizontal deformation both in the upper and lower Archean continental crust.

One of the major uncertainties in interpreting Archean tectonic settings remains the source of most igneous rock assemblages; this review has attempted to demonstrate that underplated plume-induced magma reservoirs may be as capable of generating the observed granite-greenstone assemblages as are settings involving contemporary-type subduction processes. Perhaps both mechanisms of crust formation operated in the Archean.

Major new insights can be expected from isotope geochemistry, from precise ion microprobe dating of zircons from very old rocks, and from high-resolution seismic data that should help to elucidate the relationship between granite-greenstone and high-grade gneiss terranes. We should continue to explore whether Archean crustal evolution is compatible with processes in the modern Earth, but we should not necessarily assume that the present is a key to the past.

ACKNOWLEDGMENTS

This research was largely funded by the German Research Council (DFG) and is a contribution to the International Lithosphere Programme.

Literature Cited

Abbott, D. H., Hoffman, S. E. 1984. Archaean plate tectonics revisited. Part 1. Heat flow, spreading rate, and the age of subducting oceanic lithosphere and their effects on the origin and evolution of continents. *Tectonics* 3:429–48

Allègre, C. J. 1982. Chemical geodynamics. *Tectonophysics* 81:109–32

Anhaeusser, C. R. 1973. The evolution of the early Precambrian crust of southern Africa. *Philos. Trans. R. Soc. London Ser. A* 273:359–88

Anhaeusser, C. R. 1975. Precambrian tectonic environments. *Ann. Rev. Earth Planet. Sci.* 3:31–53

Anhaeusser, C. R., Mason, R., Viljoen, M. J., Viljoen, R. P. 1969. A reappraisal of some aspects of Precambrian shield geology. *Geol. Soc. Am. Bull.* 80:2175–2200

Archibald, N. J., Bettenay, L. F., Bickle, M. J., Groves, D. I. 1981. Evolution of Archaean crust in the eastern Goldfields Province of the Yilgarn Block, Western Australia. See Glover & Groves 1981, pp. 491–504

Ayres, L. D. 1983. Bimodal volcanism in Archean greenstone belts exemplified by greywacke composition, Lake Superior Park, Ontario. *Can. J. Earth Sci.* 20:1168–94

Baer, A. J. 1981. Geotherms, evolution of the lithosphere and plate tectonics. *Tectonophysics* 72:203–27

Bak, J., Sørensen, K., Gnocott, J., Korstgard, J. A., Nash, D., Watterson, J. 1975. Tectonic implications of Precambrian shear belts in western Greenland. *Nature* 254:566–69

Bally, A. W. 1981. Thoughts on the tectonics of folded belts. In *Thrust and Nappe Tectonics*, ed. K. McClay, N. J. Price, pp. 13–32. *Geol. Soc. London Spec. Publ. No. 9*

Barker, F., Arth, J. G. 1976. Generation of trondhjemitic-tonalitic liquids and Archean bimodal trondhjemite-basalt suites. *Geology* 4:596–600

Barton, J. M. Jr. 1981. The pattern of Archaean crustal evolution in southern Africa as deduced from the evolution of the Limpopo mobile belt and the Barberton granite-greenstone terrain. See Glover & Groves 1981, pp. 21–32

Barton, J. M. Jr., Ryan, B., Fripp, R. E. P.

1983. Rb-Sr and U-Th-P isotopic studies of the Sand River Gneisses, central zone, Limpopo mobile belt. In *The Limpopo Belt*, ed. W. J. van Biljon, J. H. Legg, pp. 9–18. *Geol. Soc. S. Afr. Spec. Publ. 8*
Bickle, M. J. 1978. Heat loss from the Earth: constraints on Archaean tectonics from the relation between geothermal gradients and the rate of plate production. *Earth Planet. Sci. Lett.* 40:301–15
Bickle, M. J., Eriksson, K. A. 1982. Evolution and subsidence of early Precambrian sedimentary basins. *Philos. Trans. R. Soc. London Ser. A* 305:225–47
Bird, J. M., ed. 1980. *Plate Tectonics*. Washington DC: Am. Geophys. Union. 992 pp.
Bridgwater, D., McGregor, V. R., Myers, J. S. 1974. A horizontal tectonic regime in the Archaean of Greenland and its implications for early crustal thickening. *Precambrian Res.* 1:179–97
Bridgwater, D., Keto, L., McGregor, V. R., Myers, J. S. 1976. Archaean gneiss complex of Greenland. In *Geology of Greenland*, ed. A. Escher, W. S. Watt, pp. 19–75. Copenhagen: Geol. Surv. Greenland
Burke, K. 1981. Did the Wilson cycle dominate Precambrian time? *Eos, Trans. Am. Geophys. Union* 62:418 (Abstr.)
Burke, K., Dewey, J. F., Kidd, W. S. F. 1976. Dominance of horizontal movements, arc and microcontinent collision during the late permobile regime. See Windley 1976, pp. 113–29
Chauvel, C., Dupré, B., Jenner, G. A. 1984. Kambalda greenstone belt: 2700 or 3200 Ma old? *Nature*. In press
Christensen, U. R. 1985. Thermal evolution models for the Earth. *J. Geophys. Res.* In press
Collerson, K. D. 1983. Ion microprobe zircon geochronology of the Uivak gneisses: implications for the evolution of the early terrestrial crust in the North Atlantic Craton. In *Field Guide and Abstracts, Archean Geochem. Early Crustal Genesis Field Workshop, Ottawa*, pp. 14–18
Condie, K. C. 1972. A plate tectonics evolutionary model of the South Pass Archean greenstone belt, southwestern Wyoming. *Int. Geol. Congr., 24th, Montreal, Can.* 1: 104–12
Condie, K. C. 1980. Origin and early development of the Earth's crust. *Precambrian Res.* 11:183–97
Condie, K. C. 1981. *Archaean Greenstone Belts*. Amsterdam: Elsevier. 434 pp.
Condie, K. C. 1982. *Plate Tectonics and Crustal Evolution*. New York: Pergamon. 310 pp.
Condie, K. C. 1984. Archean mantle fractionation. *Geophys. Res. Lett.* 11:283–86
Condie, K. C., Allen, P. 1984. Origin of Archaean charnockites from southern India. In *Archaean Geochemistry*, ed. A. Kröner, G. N. Hanson, A. M. Goodwin, pp. 182–203. Berlin/Heidelberg: Springer-Verlag
Condie, K. C., Hunter, D. R. 1976. Trace element geochemistry of Archaean granitic rocks from the Barberton region, South Africa. *Earth Planet. Sci. Lett.* 29:389–400
Coward, M. P. 1976. Archaean deformation patterns in southern Africa. *Philos. Trans. R. Soc. London Ser. A* 283:313–31
Cox, K. G. 1980. A model for flood basalt volcanism. *J. Petrol.* 21:629–50
Dewey, J. F. 1982. Plate tectonics and the evolution of the British Isles. *J. Geol. Soc. London* 139:371–412
Dewey, J. F., Burke, K. 1973. Tibetan, Variscan and Precambrian basement reactivation, products of continental collision. *J. Geol.* 81:683–92
Dewey, J. F., Windley, B. F. 1981. Growth and differentiation of the continental crust. *Philos. Trans. R. Soc. London Ser. A* 301:189–206
De Wit, M. J., Fripp, R. E. P., Stanistreet, I. G. 1983. Tectonic and stratigraphic implications of new field observations along the southern part of the Barberton greenstone belt. In *Contributions to the Geology of the Barberton Mountain Land*, ed. C. R. Anhaeusser, pp. 21–30. *Geol. Soc. S. Afr. Spec. Publ. No. 9*
Dickinson, W. R. 1981. Plate tectonics through geologic time. *Philos. Trans. R. Soc. London Ser. A* 301:207–15
Dickinson, W. R., Seely, D. R. 1979. Structure and stratigraphy of forearc regions. *Am. Assoc. Pet. Geol. Bull.* 63:2–31
Drummond, B. J., Smith, R. E., Horwitz, R. C. 1981. Crustal structure in the Pilbara and northern Yilgarn Blocks from deep seismic sounding. See Glover & Groves 1981, pp. 33–42
Drury, S. A., Harris, N. B. W., Holt, R. W., Reeves-Smith, G. J., Wightman, R. T. 1984. Precambrian tectonics and crustal evolution in South India. *J. Geol.* 92:3–20
East Pacific Rise Study Group. 1981. Crustal processes of the mid-ocean ridge. *Science* 213:31–40
Eriksson, K. A. 1982. Sedimentation patterns in the Barberton Mountain Land, South Africa, and the Pilbara Block, Australia: evidence for Archean rifted continental margins. *Tectonophysics* 81:179–93
Ewart, A., Baxter, K., Ross, J. A. 1980. The petrology and petrogenesis of the Tertiary anorogenic mafic lavas of southern and central Queensland, Australia—possible implications for crustal thickening. *Contrib. Mineral. Petrol.* 75:129–52
Fletcher, I. R., Rosman, K., Trendall, A., de Laeter, J. R. 1982. Variability of ε^{i}_{Nd} on greenstone belts in the Archaean of

Western Australia. *Abstr. Vol. Int. Conf. Geochron., Cosmochron. Isotope Geol., 5th, Nikko, Jpn.* pp. 100–1

Froude, D. O., Ireland, T. R., Kinny, P. D., Williams, I. S., Compston, W., et al. 1983. Ion microprobe identification of 4,100–4,200 Myr-old terrestrial zircons. *Nature* 304:616–18

Fyfe, W. S. 1974. Archaean tectonics. *Nature* 249:338

Fyfe, W. S. 1978. The evolution of the Earth's crust: modern plate tectonics to ancient hot spot tectonics? *Chem. Geol.* 23:89–114

Gastil, R. G. 1979. A conceptual hypothesis for the relation of differing tectonic terranes to plutonic emplacement. *Geology* 7:542–44

Gee, R. D., Baxter, J. L., Wilde, S. A., Williams, I. R. 1981. Crustal development in the Archaean Yilgarn Block, Western Australia. See Glover & Groves 1981, pp. 43–56

Gélinas, L., Trudel, P., Hubert, C. 1984. Chemostratigraphic division of the Blake River Group, Rouyn-Noranda area, Abitibi, Quebec. *Can. J. Earth Sci.* 21:220–31

Gill, J. 1981. *Orogenic Andesites and Plate Tectonics.* Berlin: Springer-Verlag. 390 pp.

Glikson, A. Y. 1971. Primitive Archaean element distribution patterns: chemical evidence and geotectonic significance. *Earth Planet. Sci. Lett.* 12:309–20

Glikson, A. Y. 1972. Early Precambrian evidence of a primitive ocean crust and island nuclei of sodic granite. *Geol. Soc. Am. Bull.* 83:3323–44

Glikson, A. Y. 1979. Early Precambrian tonalite-trondhjemite sialic nuclei. *Earth Sci. Rev.* 15:1–73

Glover, J. E., ed. 1971. *Symposium on Archaean Rocks Held at Perth, 23–26 May 1970, Geol. Soc. Aust. Spec. Publ. No. 3.* 469 pp.

Glover, J. E., Groves, D. I., eds. 1981. *Archaean Geology, Geol. Soc. Aust. Spec. Publ. No. 7.* 515 pp.

Goodwin, A. M. 1981. Precambrian perspectives. *Science* 213:55–61

Gorman, B. E., Pearce, T. H., Birkett, T. C. 1978. On the structure of Archean greenstone belts. *Precambrian Res.* 6:23–41

Grachev, A. F., Fedorovsky, V. S. 1981. On the nature of greenstone belts in the Precambrian. *Tectonophysics* 73:195–212

Grambling, J. A. 1981. Pressures and temperatures in Precambrian metamorphic rocks. *Earth Planet. Sci. Lett.* 53:63–68

Green, D. H. 1972. Archean greenstone belts may include terrestrial equivalents of lunar maria? *Earth Planet. Sci. Lett.* 15:263–70

Green, D. H. 1975. Genesis of Archaean peridotitic magmas and constraints on Archaean geothermal gradients and tectonics. *Geology* 3:15–18

Groves, D. I., Batt, W. D. 1984. Spatial and temporal variations of Archaean metallogenetic associations in terms of evolution of granitoid-greenstone terrains with particular emphasis on the Western Australian shield. In *Archaean Geochemistry*, ed. A. Kröner, G. N. Hanson, A. M. Goodwin, pp. 73–100. Berlin/Heidelberg: Springer-Verlag

Hager, B. H., O'Connell, R. J. 1980. Lithosphere thickening and subduction, plate motions and mantle convection. In *Physics of the Earth's Interior*, pp. 464–92. Amsterdam: North-Holland

Hamilton, P. J., Evensen, N. M., O'Nions, R. K. 1979. Sm-Nd systematics of Lewisian gneisses: implications for the origin of granulites. *Nature* 277:25–28

Hamilton, P. J., O'Nions, R. K., Bridgwater, D., Nutman, A. 1983. Sm-Nd studies of Archaean metasediments and metavolcanics from West Greenland and their implications for the Earth's early history. *Earth Planet. Sci. Lett.* 62:263–72

Hansen, E. C., Newton, R. C., Janardhan, A. S. 1984. Pressures, temperatures and metamorphic fluids across an unbroken amphibolite facies to granulite facies transition in southern Karnataka, India. In *Archaean Geochemistry*, ed. A. Kröner, G. N. Hanson, A. M. Goodwin, pp. 161–81. Berlin/Heidelberg: Springer-Verlag

Hanson, G. N. 1981. Geochemical constraints on the evolution of the early continental crust. *Philos. Trans. R. Soc. London Ser. A* 301:423–42

Hart, S. R., Brooks, C., Krogh, T. E., Davis, G. L., Nava, D. 1970. Ancient and modern volcanic rocks: a trace element model. *Earth Planet. Sci. Lett.* 10:17–28

Hawkesworth, C. J., O'Nions, R. K. 1977. The petrogenesis of some Archaean volcanic rocks from southern Africa. *J. Petrol.* 18:487–520

Hegner, E., Tegtmeyer, A., Kröner, A. 1981. Geochemie und Petrogenese archaischer Vulkanite der Pongola Gruppe in Natal, Südafrika. *Chem. Erde* 40:23–57

Hegner, E., Kröner, A., Hofmann, A. W. 1984. Age and isotope geochemistry of the Archaean Pongola and Usushwana suites, Swaziland, southern Africa: a case for crustal contamination of mantle derived magma. *Earth Planet. Sci. Lett.* In press

Hickman, A. H. 1984. Archaean diapirism in the Pilbara Block, Western Australia. See Kröner & Greiling 1984, pp. 113–27

Jagoutz, E., Dawson, B., Spettel, B., Wänke, H. 1983. Identification of early differentiation processes on the Earth. *Meteoritics* 18:319–20 (Abstr.)

Jahn, B.-M., Sun, S.-S. 1979. Trace element distribution and isotopic composition of

Archean greenstones. In *Origin and Distribution of the Elements*, ed. L. H. Ahrens, pp. 597–618. Oxford: Pergamon

Jahn, B.-M., Zhang, Z.-Q. 1984. Archean granulite gneisses from eastern Hebei Province, China: rare earth geochemistry and tectonic implication. *Contrib. Mineral. Petrol.* 85: 224–43

Jahn, B.-M., Glikson, A. Y., Peucat, J. J., Hickman, A. H. 1981. REE geochemistry and isotopic data of Archean silicic volcanics and granitoids from the Pilbara Block, Western Australia: implications for the early crustal evolution. *Geochim. Cosmochim. Acta* 45: 1633–52

Jahn, B.-M., Vidal, P., Kröner, A. 1984. Multi-chronometric ages and origin of Archaean tonalitic gneisses in Finnish Lapland: a case for long crustal residence time. *Contrib. Mineral. Petrol.* 86: 398–408

Jordan, T. H. 1979. The deep structure of continents. *Sci. Am.* 240: 70–82

Katz, M. B. 1981. A shear-mobile transform belt in the Precambrian Gondwanaland of Africa–South America. *Geol. Rundsch.* 70: 1012–19

Krogh, T. K., Davis, D. W., Corfu, F. 1982. Archean evolution from precise U-Pb isotopic dating. *Abstr. Vol. Int. Conf. Geochron., Cosmochron. Isotope Geol., 5th, Nikko, Jpn.*, p. 192

Kröner, A. 1977. Precambrian mobile belts of southern and eastern Africa—ancient sutures or sites of ensialic mobility? A case for crustal evolution towards plate tectonics. *Tectonophysics* 40: 101–35

Kröner, A. 1981a. Precambrian plate tectonics. See Kröner 1981b, pp. 57–90

Kröner, A., ed. 1981b. *Precambrian Plate Tectonics.* Amsterdam: Elsevier. 781 pp.

Kröner, A. 1982. Archaean to early Proterozoic tectonics and crustal evolution: a review. *Rev. Bras. Geociên.* 12: 15–31

Kröner, A. 1984a. Evolution, growth and stabilization of the Precambrian lithosphere. *Phys. Chem. Earth* 15: 69–106

Kröner, A. 1984b. Fold belts and plate tectonics in the Precambrian. *Proc. Int. Geol. Congr., 27th, Moscow, USSR*, 5: 247–80. Utrecht: VNU Sci. Press

Kröner, A., Greiling, R., eds. 1984. *Precambrian Tectonics Illustrated.* Stuttgart: E. Schweitzerbart'sche Verlagsbuchhandlung. 419 pp.

Kröner, A., Puustinen, K., Hickman, M. 1981. Geochronology of an Archaean tonalitic gneiss dome in northern Finland and its relation with an unusual overlying volcanic conglomerate and komatiitic greenstone. *Contrib. Mineral. Petrol.* 76: 33–41

Kröner, A., Layer, P. W., McWilliams, M. O. 1984. Archaean palaeomagnetism: evidence for continental drift and the existence of a dipolar magnetic field since ca. 3.5 billion years ago. *Terra Cognita* 4: 78 (Abstr.)

Lambert, R. St. J. 1981. Earth tectonics and thermal history: review and a hot-spot model for the Archaean. See Kröner 1981b, pp. 57–90

Lowman, P. D. Jr. 1972. The geologic evolution of the Moon. *J. Geol.* 80: 125–66

Macdougall, J. D., Gopalan, K., Lugmair, G. W., Roy, A. B. 1983. An ancient depleted mantle source for Archean crust in Rajasthan, India. In *Field Guide and Abstracts, Archean Geochem. Early Crustal Genesis Field Workshop, Ottawa*, pp. 40–41 (Abstr.)

Martin, H., Chauvel, C., Jahn, B.-M. 1983. Major and trace element geochemistry and crustal evolution of Archean granodioritic rocks from eastern Finland. *Precambrian Res.* 21: 159–80

McCall, G. J. H., ed. 1977. *The Archean: Search for a Beginning.* Stroudsburg, Pa: Dowden, Hutchinson & Ross. 505 pp.

McCulloch, M. T. 1982. Identification of Earth's earliest differentiates. *Abstr. Vol. Int. Conf. Geochron., Cosmochron. Isotope Geol., 5th, Nikko, Jpn.*, pp. 244–45

McGregor, V. R. 1973. The early Precambrian gneisses of the Godthåb district, West Greenland. *Philos. Trans. R. Soc. London Ser. A* 273: 343–58

McKenzie, D. 1978. Some remarks on the development of sedimentary basins. *Earth Planet. Sci. Lett.* 40: 25–32

Moorbath, S. 1978. Age and isotope evidence for the evolution of the continental crust. *Philos. Trans. R. Soc. London Ser. A* 288: 401–13

Morgan, P., Phillips, R. J. 1984. Hot spot heat transfer: its application to Venus and implications to Venus and Earth. *J. Geophys. Res.* 88: 8305–17

Myers, J. S. 1976. Granitoid sheets, thrusting and Archaean crustal thickening in West Greenland. *Geology* 4: 265–68

Newton, R. C., Hansen, E. C. 1983. The origin of Proterozoic and late Archean charnockites—evidence from field relations and experimental petrology. In *Proterozoic Geology*, ed. L. G. Medaris Jr., C. W. Byers, D. M. Mickelson, W. C. Shanks. *Geol. Soc. Am. Mem.* 161: 167–78

Newton, R. C., Smith, J. V., Windley, B. F. 1980. Carbonic metamorphism, granulites and crustal growth. *Nature* 288: 45–50

Nisbet, E. G. 1982. The tectonic setting and petrogenesis of komatiites. In *Komatiites*, ed. N. T. Arndt, E. G. Nisbet, pp. 501–20. London: George Allen & Unwin

Nisbet, E. G., Walker, D. 1982. Komatiites and the structure of the Archaean mantle. *Earth Planet. Sci. Lett.* 60: 105–13

Ogawa, Y., Horiuchi, K. 1978. Two types of accretionary fold belts in central Japan. *J. Phys. Earth* 26 (Suppl.): S231–36
Oliver, J., Cook, I., Brown, L. 1983. COCORP and the continental crust. *J. Geophys. Res.* 88: 3329–47
Oskarsson, N., Sigvaldason, G. E., Steinthórsson, S. 1982. A dynamic model of rift zone petrogenesis and the regional petrology of Iceland. *J. Petrol.* 23: 28–74
Park, R. G. 1981. Origin of horizontal structure in high-grade Archaean terrains. See Glover & Groves 1981, pp. 481–90
Percival, J. A., Card, K. D. 1983. Archean crust as revealed in the Kapuskasing uplift, Superior province, Canada. *Geology* 11: 323–26
Phinney, W. C., Morrison, D. A., Ashawal, L. D. 1981. Implications of Archean anorthosites for crust-mantle evolution. *Lunar Planet. Sci. XII*, pp. 830–32 (Abstr.)
Platt, J. P. 1980. Archaean greenstone belts: a structural test of tectonic hypotheses. *Tectonophysics* 65: 127–50
Reymer, A., Schubert, G. 1984. Phanerozoic addition rates to the continental crust and crustal growth. *Tectonics* 3: 63–78
Ringwood, A. E. 1966. Chemical evolution of terrestrial planets. *Geochim. Cosmochim. Acta* 30: 41–104
Ringwood, A. E. 1979. *Origin of the Earth and Moon.* New York: Springer. 295 pp.
Schubert, G. 1979. Subsolidus convection in the mantles of terrestrial planets. *Ann. Rev. Earth Planet Sci.* 7: 289–342
Sigurdsson, H., Sparks, R. S. J. 1981. Petrology of rhyolitic and mixed magma ejecta from the 1875 eruption of Askja, Iceland. *J. Petrol.* 22: 41–84
Sleep, N. H., Windley, B. F. 1982. Archean plate tectonics: constraints and inferences. *J. Geol.* 90: 363–80
Smith, J. V. 1981. The first 800 million years of Earth's history. *Philos. Trans. R. Soc. London Ser. A* 301: 401–22
Sun, S.-S. 1984. Geochemical characteristics of Archaean ultramafic and mafic volcanic rocks: implications for mantle composition and evolution. In *Archaean Geochemistry*, ed. A. Kröner, G. N. Hanson, A. M. Goodwin, pp. 25–47. Berlin/Heidelberg: Springer-Verlag
Sutton, J., Windley, B. F., ed. 1973. A discussion on the evolution of the Precambrian crust. *Philos. Trans. R. Soc. London Ser. A* 273: 315–581
Talbot, C. J. 1973. A plate tectonic model for the Archaean crust. *Philos. Trans. R. Soc. London Ser. A* 273: 413–28
Talwani, M., Langseth, M. 1981. Oceanic crustal dynamics. *Science* 213: 22–31
Tarney, J., Windley, B. F. 1977. Chemistry, thermal gradients and evolution of the lower continental crust. *J. Geol. Soc. London* 134: 153–72
Tarney, J., Dalziel, I. W. D., de Wit, M. J. 1976. Marginal basin "Rocas verdes" complex from S. Chile: a model for Archaean greenstone belt formation See Windley 1976, pp. 131–46
Taylor, S. R. 1967. The origin and growth of continents. *Tectonophysics* 4: 17–34
Thurston, P. C., Fryer, B. J. 1983. The geochemistry of repetitive cyclical volcanism from basalt through rhyolite in the Uchi-Confederation greenstone belt, Canada. *Contrib. Mineral. Petrol.* 83: 204–26
Ullrich, L., Van der Voo, R. 1981. Minimum continental velocities with respect to the pole since the Archean. *Tectonophysics* 74: 17–27
Veizer, J., Jansen, S. L. 1979. Basement and sedimentary recycling and continental evolution. *J. Geol.* 87: 341–70
Viljoen, M. J., Viljoen, R. P. 1969. The geology and geochemistry of the lower ultramafic unit of the Onverwacht Group and a proposed new class of igneous rocks. *Geol. Soc. S. Afr. Spec. Publ.* 2: 55–86
Wells, P. R. A. 1981. Accretion of continental crust: thermal and geochemical consequences. *Philos. Trans. R. Soc. London Ser. A* 301: 347–57
Wilkinson, J. F. G., Binns, R. A. 1977. Relatively iron-rich lherzolite xenoliths of the Cr-diopside suite: a guide to the primary nature of anorogenic tholeiitic andesite magmas. *Contrib. Mineral. Petrol* 65: 199–212
Williams, D. A. C., Furnell, R. G. 1979. Reassessment of part of the Barberton type area. *Precambrian Res.* 9: 325–47
Wilson, J. F. 1973. The Rhodesian Archaean craton—an essay in cratonic evolution. *Philos. Trans. R. Soc. London Ser. A* 273: 389–411
Windley, B. F., ed. 1976. *The Early History of the Earth.* London: Wiley. 619 pp.
Windley, B. F. 1977. *The Evolving Continents.* London: Wiley. 385 pp.
Windley, B. F. 1979. Tectonic evolution of continents in the Precambrian. *Episodes 1979* 4: 12–16

Ann. Rev. Earth Planet. Sci. 1985. 13: 75–95

OXIDATION STATUS OF THE MANTLE: PAST AND PRESENT

Richard J. Arculus

Department of Geological Sciences, University of Michigan, Ann Arbor, Michigan 48109

INTRODUCTION

It is important for a number of reasons that we obtain a thorough understanding of the present oxidation status of the mantle. Sub- and supersolidus phase relationships, the nature of degassed volatile species, electrical conductivity, diffusivity, and mechanical behavior are some of the properties that are a function of this parameter. Of further interest is the fact that the present oxidation state must in some way reflect the complex processes that have occurred since the formation of the proto-mantle some 4.5 Gyr ago. These processes include the redox characteristics of accreted material, heterogeneous equilibria prevailing during the growth of the Earth (such as those involved in core formation), convective cycling in the mantle, subduction of oxidized surface layers of the Earth and mixing in the mantle, incorporation of impacted planetesimals, infiltration by melts and fluids, and selective volatile losses and gains.

It would be fair to characterize the current range of opinions on the oxidation status of the upper mantle as diverse, partly as a result of discrepancies between theoretical and experimental approaches to the problem, and partly because of differences in philosophical viewpoints concerning the major redox controls. Granted this divergence of opinion for the upper mantle, it must also be recognized that knowledge of the redox state(s) of the greater part of the mantle to the boundary with the core is rudimentary. There does appear, however, to be some intriguing evidence not only that the present upper mantle is heterogeneous with respect to oxidation state, but also that a secular trend toward oxidation of the subcontinental mantle portion of the lithosphere has taken place.

0084–6597/85/0515–0075$02.00

In this article, the evidence for the oxidation status of the upper mantle from natural mineral assemblages, experimental studies, and less direct approaches such as the nature of degassed volatiles is reviewed. The apparent discrepancies between some aspects of these studies are discussed, and a synthesis of our current understanding of the redox evolution of the upper mantle is attempted.

MINERALOGICAL CONSTRAINTS

Studies of the mineral assemblages characteristic of cognate and xenolithic inclusion suites in kimberlites (Dawson 1980) have provided some direct evidence for a heterogeneous redox state of the upper mantle traversed by kimberlite magmas. For example, several occurrences of Fe-Ni metallic alloys enclosed by diamond have been reported in kimberlites from the Soviet Union (Bulanova et al 1979, Sobolev et al 1981), which, together with the unusually high Cr content (and plausibly significant Cr^{2+}) of olivine hosted by the same phase (Hervig et al 1980), are prime indicators of a reduced environment in the vicinity of the synthetic iron-wüstite (IW) buffer prevailing during diamond growth. A further intriguing aspect of some inclusion-host diamond relationships is that radiometric dating gives Archean ages for the genesis of the inclusions under barometric/thermometric conditions comparable with those assigned to a modern, subcontinental lithosphere geotherm (Kramers 1979, Boyd & Gurney 1982, S. Richardson, personal communication, 1984).

The discovery of ilmenite-spinel intergrowths among inclusion suites of kimberlites from West Africa permitted Haggerty & Tompkins (1983) to apply the relevant oxygen barometer-thermometer for coexisting Fe-Ti oxides (Spencer & Lindsley 1981); in contrast with inclusions in diamonds, the results are close to the synthetic fayalite-magnetite-quartz (FMQ) oxygen buffer when calculated at 1-bar conditions (Figure 1). These oxygen fugacity (f_{O_2}) values for the equilibrium between spinel and ilmenite phases are about 4 log orders more oxidized than the IW oxygen buffer.

Comparable, if slightly more reduced, data were obtained by Eggler (1983) for inclusion suites in kimberlites with assemblages of olivine, orthopyroxene, and ilmenite. The f_{O_2} defined by these assemblages can be written as

$$\underset{\text{in ilmenite}}{2Fe_2O_3} + \underset{\text{in orthopyroxene}}{4FeSiO_3} \rightleftharpoons \underset{\text{in olivine}}{8FeSi_{0.5}O_2} + O_2. \quad (1)$$

The results calculated by Eggler (1983) have a higher degree of uncertainty than those of Haggerty & Tompkins (1983), in that one or the other of the

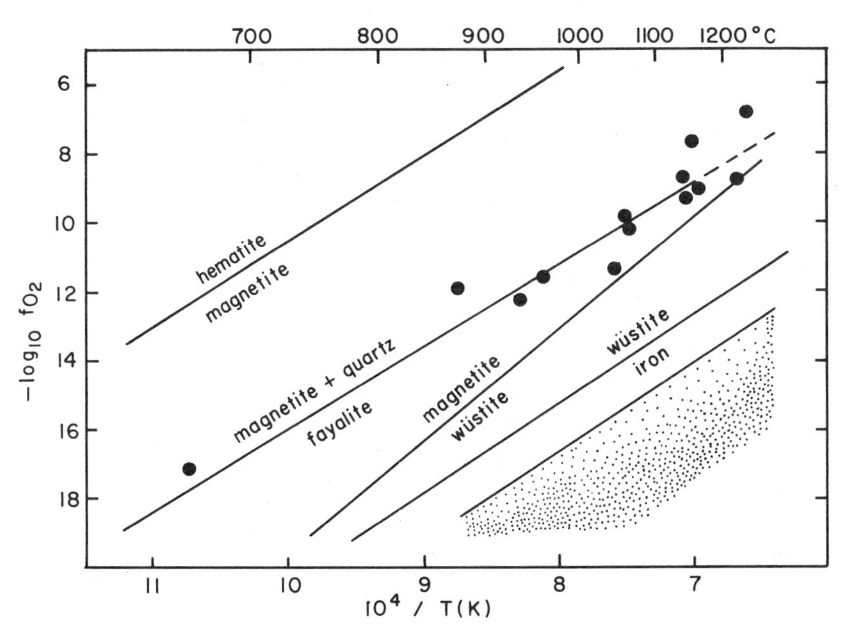

Figure 1 Temperature-f_{O_2} relations for megacryst ilmenite-spinel intergrowths from West African kimberlites with representative f_{O_2} buffers. (After Haggerty & Tompkins 1983.) The dotted field represents peridotite saturated with Fe-rich metal, such as is found in diamonds (Holmes & Arculus 1982). Relations calculated at 1-bar conditions.

silicate phases involved in reaction (1) is typically absent and has to be inferred. Nevertheless, the f_{O_2}'s range from the FMQ buffer when calculated at relatively low pressures (25 kbar) to slightly below the magnetite-wüstite (MW) buffer at pressures around 60 kbar (Figure 2). Interestingly, these data bracket the f_{O_2} defined by the assemblages enstatite-magnesite-olivine-graphite/diamond (EMOG–EMOD of Eggler et al 1980), which is regarded by Eggler (1983) as the most plausible redox buffer in the subcontinental lithosphere and asthenosphere.

The presence of graphite or diamond in samples from the mantle provides an upper limit on the prevailing f_{O_2}, dependent on the relative abundances of C, H, and S in any coexisting fluid, but the extreme pressure dependence of the solid carbon-gas equilibria with respect to solid f_{O_2} buffers such as IW and FMQ means that at pressures greater than 10 kbar, the carbon-gas surface is as oxidized as the FMQ buffer. At pressures greater than about 15 kbar, CO_2 reacts with silicates in the mantle to form carbonates (Eggler 1978, Wyllie 1978), and there is no doubt that carbonate-graphite/diamond-bearing assemblages are a potent f_{O_2} buffer. Despite intensive petrographic observations, carbonates appear to be

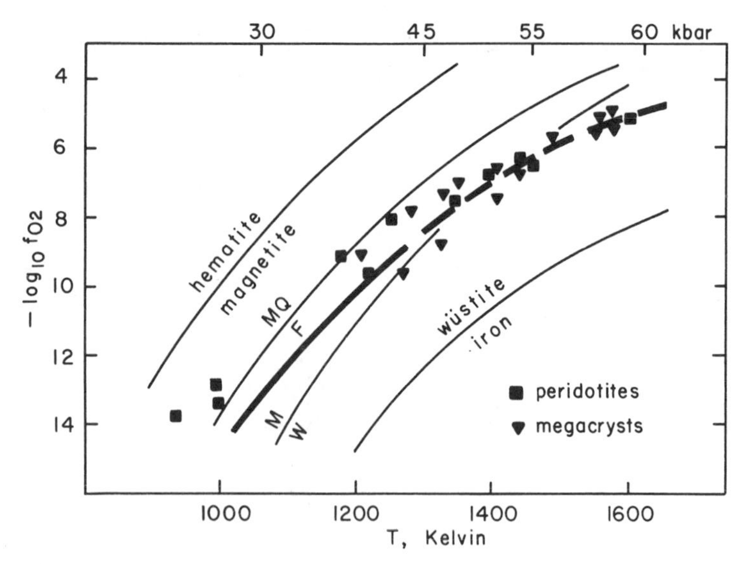

Figure 2 Temperature-f_{O_2} relations for megacrysts and peridotites calculated with reaction (1). (After Eggler 1983.) The pressure-temperature relations are for the Lesotho geotherm. The heavy black line is for EMOG (full) and EMOD (dashed).

exceedingly rare in samples of the upper mantle (Boettcher 1984), and the invocation of EMOG/D f_{O_2} buffers must also appeal to the rapid decomposition of carbonate as a function of sample transport processes in order to account for the absence of carbonate. Nevertheless, in the stability field of carbonate in the mantle, the EMOG/D buffer constitutes an upper limit to f_{O_2}, given the local presence of C–O–H-bearing fluid components and the existence of graphite and/or diamond.

More general equilibria constraining the f_{O_2} prevailing in upper mantle peridotite assemblages may be written as

$$\underset{\text{in olivine}}{6Fe_2SiO_4} + O_2 \rightleftharpoons \underset{\text{in orthopyroxene}}{3Fe_2Si_2O_6} + \underset{\text{in spinel}}{2Fe_3O_4}. \tag{2}$$

Given acceptable thermodynamic data for the pure phases and crystalline solution models, we can calculate the f_{O_2}'s prevailing in the spinel peridotite facies of the upper mantle. Such calculations show a range of values, from IW to more oxidized than FMQ (O'Neill et al 1982), but a note of caution is necessary in that the assumptions of solid-solution characteristics (which require, for example, ideal stoichiometry in the spinel phase) may be invalid, as is discussed further below.

VOLATILES AND BASALTS AS f_{O_2} PROBES OF THE MANTLE

All of the mineralogical constraints described above are based on samples from the subcontinental upper mantle and accordingly represent a biased sample set. A more widely spread sampling approach is to examine the redox state of basaltic rocks and associated volatile species from both continental and oceanic environments.

In a historical sense, the relative abundance of volatiles emitted by volcanoes was one of the major constraints that led Ringwood (1979), among others, to infer a relatively oxidized state (and by implication, one close to FMQ) for the upper mantle source region of the associated magmas. It is well known from a number of studies that CO_2–H_2O–SO_2 are the major gas species associated with modern volcanic activity (Gerlach 1980a,b,c), by far outweighing the abundances of reduced species such as H_2 and CO. At low pressures and high temperatures, these latter gases should predominate if the volatiles were unchanged in terms of atomic ratios from equilibrium at f_{O_2} conditions equivalent to IW in the upper mantle. However, selective leakage of gaseous components from magmas at shallow levels in the Earth's crust can result in dramatic changes in both the redox state of the remaining magma-volatile system and the proportions of gas species emitted at any given time from a volcanic vent (Sato 1978, Mathez 1984).

Numerous authors have shown that under pressure and temperature conditions appropriate for the upper mantle, the dominant volatiles in the system C–O–H with f_{O_2}'s close to IW are CH_4 and H_2, whereas CO_2 and H_2O predominate near the FMQ buffer (e.g. Deines 1980). There are limited data that suggest significant proportions of CH_4 are present in fluids trapped in diamonds (Melton & Giardini 1974, 1981), but in view of the possible leakage of components from the trapped fluids by processes such as C deposition or H_2 diffusion, a more convincing proof of the presence of CH_4 during formation of the diamonds is the fact that ratios of H_2/O_2 in these trapped fluids are in excess of 2 (Deines 1980). Furthermore, part of the spread of C isotopic values observed for diamonds is consistent with fractionation between the diamonds and CH_4-bearing fluid (Deines 1980).

In general, fluid inclusions preserved within phases of mantle-derived peridotites have proved difficult to interpret with respect to the redox conditions prevailing in the upper mantle (Roedder 1965, Roedder & Bodnar 1980). For example, the inclusions rarely show evidence of confining pressures greater than about 10 kbar, presumably as a result of the failure of the host phases to contain higher pressure differentials. Consequently, most inclusions have to be regarded as entrapments during healing of the failed crystal, and during this failure-healing cycle there is a possibility of selective volatile escape (in particular of H_2) and consequent change of volatile abundance ratios. Some inclusions are observed to contain graphite precipitates, so that again the present fluid compositions may not be representative of those in existence in the upper mantle (Roedder 1965, Mathez & Delaney 1981). Recently, the presence of significant quantities of CO as well as CO_2 has been reported in fluid inclusions in mantle-derived peridotites (Bergman & Dubessy 1984), but the rarity of H_2O in these fluid inclusions is most obviously at odds with the relative proportions of gases emitted by volcanoes. A major exception is the presence of 0–30 vol% H_2O in CO_2-dominated fluid inclusions from the Ichinomegata locality in northwestern Honshu, Japan (Trial et al 1984).

There are a number of factors that have to be considered when attempting to invert the volatile abundance ratios produced at 1 bar by volcanoes to those that might be dissolved in the melt or exist as a separate fluid during melting of the upper mantle. These are (*a*) the possibility of selective volatile loss by reaction with the mantle to form amphibole (H_2O-loss) at pressures <25 kbar and carbonate (CO_2-loss) at pressures >15 kbar; (*b*) the contrasts in solubility behavior of CO_2, H_2O, and other volatiles as a function of pressure and temperature (e.g. Wyllie 1978); and (*c*) the rates of degassing at low pressures (<10 kbar) within the volcanic edifice and crust.

Despite all of these difficulties, the sum of the evidence is indicative of the dominance of H_2O and CO_2 in gases associated with most active volcanism, and it is suggestive of oxidized ($\simeq$FMQ) mantle source regions. There do appear, however, to be examples of present-day volcanic activity associated with reduced volatile species, including CO and various hydrocarbons (Karzhavin & Vendillo 1970, Konnerup-Madsen et al 1979, Byers et al 1983); thus the existence of a heterogeneity of volatile distributions in the upper mantle is a sensible conclusion to draw, especially given the evidence of reduced fluids in diamonds. The relative volumetric significance of these potential volatile reservoirs is not so clear.

An alternative and perhaps more direct approach to the problem of the oxidation status of the upper mantle than the nature of fugitive components has been the study of the f_{O_2}–T-sensitive mineralogy of basaltic rocks (Haggerty 1978). The major assumptions are that no redox alteration as a function of volatile loss (especially H_2) has occurred from mantle source to surface (cf Sato 1978), and that the mineralogy has not been affected by subsolidus oxidation (cf Hammond & Taylor 1982). This approach suggests that the great majority of basaltic rocks have equilibrated in the vicinity of the FMQ buffer, and it further implies that the source regions of basalts in the upper mantle are at a similar redox state (Haggerty 1978, Basaltic Volcanism Study Project 1981). Comparable inferences have been made from the study of the pressure-dependent changes of partial molar volumes of FeO and Fe_2O_3 in basaltic melts and the probable redox conditions of their sources (Mo et al 1982).

EXPERIMENTAL STUDIES

The pioneering efforts of Sato (1965, 1972) brought to the attention of the geological community the powerful applicability of oxygen-specific solid electrolytes to the study of redox problems in the Earth sciences. A number of independent workers have subsequently employed stabilized-ZrO_2 electrolytes in the direct measurement of the oxidation status of samples from the upper mantle. For cells arranged as

$$\mathrm{M}|\underset{\text{electrode A}}{\text{sample of unknown } f_{O_2}}|\underset{\text{stabilized } ZrO_2}{Y_2O_3\text{- or CaO-}}|\underset{\text{electrode B}}{\text{reference } f_{O_2}}|\mathrm{M}, \quad (3)$$

where M is a conducting lead, the unknown f_{O_2} of a geological single or polyphase sample is given by the Nernst equation

$$\ln f_{O_2}{}^{A} = \frac{4FE}{RT} + \ln f_{O_2}{}^{B}, \quad (4)$$

where R is the universal gas constant, T is the absolute temperature, F is the Faraday constant, and E is the potential difference across the cell.

It should be noted that the universal presence of point defects ("non-stoichiometry") in minerals means that an intrinsic oxygen fugacity of a single phase, of potentially variable composition, is a measurable thermodynamic property for a specific pressure, temperature, and composition of the phase. Fayalite and magnetite are two well-studied examples of components of widely distributed phases in the upper mantle that are stable over large ranges (4–6 log orders) of f_{O_2} with concurrently small, but highly significant, changes in cation/anion ratios (Nitsan 1974, Stocker & Smyth 1978, Flood & Hill 1957, Aragón & McCallister 1982; see Figure 3).

In general, the individual phases that have been most intensively studied with O^{2-}-specific electrochemical cells are olivine and spinel separated from upper mantle–derived peridotites (Koseluk et al 1979, Ulmer et al 1980, Arculus & Delano 1981a). In the case of separates from relatively Mg–Cr-rich type 1 peridotites (nomenclature of the Basaltic Volcanism Study Project 1981), the intrinsic f_{O_2}'s are all close to the IW buffer at 1 bar (Figure 4). The provenance of this type of peridotite is the subcontinental mantle portion of the lithosphere.

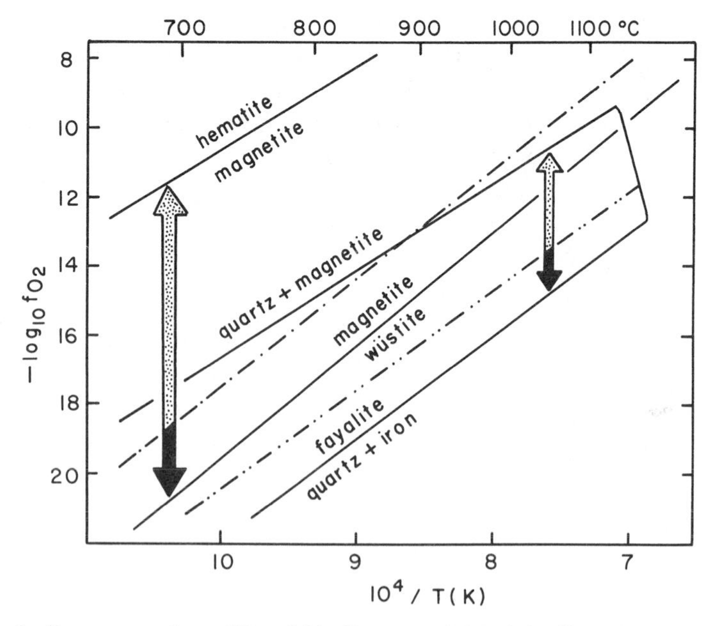

Figure 3 Temperature-f_{O_2} stability fields for magnetite and fayalite. The dashed-dot (magnetite) and dashed–double dot (fayalite) lines are the schematic stoichiometric compositions dividing cation-excessive (full arrow) from cation-deficient (dotted arrow) compositions. (After Nitsan 1974, Stocker & Smyth 1978, and Aragón & McCallister 1982.) Relations calculated at 1-bar conditions.

If independent calculations of the f_{O_2} prevailing in the peridotite assemblages are made via equilibrium (2), then values more oxidized than IW are generally obtained (see also Bohlen et al 1983), but these values depend critically on the solution models employed. For example, the activity of the magnetite component in spinel ($a_{Fe_3O_4}^{spinel}$) can be formulated in different ways (O'Neill & Navrotsky 1984); in general, however, no provision is made with these models for the point defect character of the spinel. The classic formalism (Flood & Hill 1957) relating the intrinsic f_{O_2} of a spinel to the defect character is given by

$$2Fe^{2+} + \tfrac{1}{4}O_2 \rightleftharpoons 2Fe^{3+} + O^{2-} + \tfrac{6}{8}V \qquad (5)$$

for cation-deficient spinels and

$$Fe^{2+} + \tfrac{1}{4}O_2 + \tfrac{3}{8}I \rightleftharpoons Fe^{3+} + \tfrac{1}{2}O^{2-} \qquad (6)$$

for cation-excessive structures, where V refers to a vacant lattice site, I to an interstitial cation site, and O^{2-} is an oxygen ion in the lattice. The proportion of vacancies and interstitials relative to oxygen anions stems from the Fe_3O_4 structural formula.

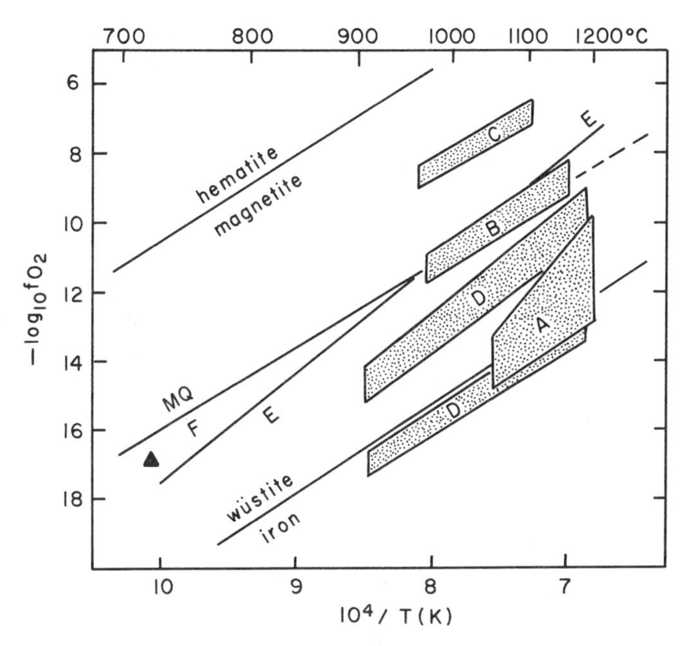

Figure 4 Temperature-f_{O_2} relations for type 1 spinel peridotites (*A*), type 2 peridotites and ilmenite megacrysts (*B*), cumulus magnetite associated with alkali basalts (*C*; from R. J. Arculus & S. Y. O'Reilly, in preparation), horizons in the Bushveld and Skaergaard intrusions (*D*), the Premier kimberlite (*E*), and the Oka carbonatite (▲). References cited in the text. Relations calculated at 1-bar conditions.

In order to account for the reduced character of the intrinsic f_{O_2}'s compared with calculated values, the spinels must be cation excessive to some degree, but the concentration of point defects can be remarkably small ($10^{-5} \simeq 10^{-6}$) despite a shift of f_{O_2} of more than 1–2 log orders from a stoichiometric composition (Aragón & McAllister 1982). In order to calculate $a^{\text{spinel}}_{Fe_3O_4}$, the activity coefficient must include a term that accounts for the degree of shift from the stoichiometric composition, in addition to standard types of site interaction and occupation considerations. A similar formalism relevant to the presence of point defects in olivine has been presented by Stocker & Smyth (1978). It is clear, however, that further studies are required of the relationships of point defects and intrinsic f_{O_2} of spinel and other peridotite phases, such as olivine.

An additional concern is the relevance of 1-bar intrinsic f_{O_2} data for peridotite phases formed at higher pressures. To date, it has been assumed that the appropriate isothermal pressure correction term

$$\ln\left(\frac{f_{O_2}''}{f_{O_2}'}\right) = \frac{-\Delta V_s}{R}\left(\frac{P''-P'}{T}\right), \tag{7}$$

where f_{O_2}'' is the f_{O_2} at some pressure $P'' > P'$ and ΔV_s is the molar volume change of the solid species defining the f_{O_2}, is of the same order as that for a solid-state buffer like FMQ or IW. Consequently, the relative position of the 1-bar intrinsic f_{O_2} data with respect to the FMQ or IW buffers is not expected to change as a function of increased pressure, but this assumption also deserves further study. For example, Virgo et al (1976) pointed out that it is impossible to measure at 1 bar the intrinsic f_{O_2} of a high-pressure wüstite, given the likelihood that depressurizing the phase would result in an increase in the nonstoichiometry of the phase via the separation of metallic Fe. If the defect structures indirectly monitored in 1-bar f_{O_2} studies of olivine and spinel are a function of pressure rather than f_{O_2}, then the 1-bar results cannot be easily extrapolated to higher pressures without further direct experimentation on the magnitude of the nonstoichiometric effects.

An experimental difficulty, which was first encountered by Sato (1979) during electrochemical studies of lunar basalts, is that the presence of metastable carbon in the sample can cause a kinetically controlled "autoreduction" at temperatures in excess of about 1120°C. Conversely, the release of CO_2 from high-pressure fluid inclusions in phases like olivine during heating at 1 bar in a cell can result in thermal decomposition to graphite in the cool parts of the cell and "autooxidation" of the sample by the remaining O_2 (Koseluk et al 1979, Arculus et al 1984). There is no doubt that a small quantity of C is a highly effective reducing agent at 1 bar, where the C–(CO_2 + CO) equilibrium is at f_{O_2}'s well below IW. For example, 100 ppmw (parts per million by weight) of C is capable of totally reducing 482 ppmw of Fe_3O_4 via reactions such as

$$Fe_3O_4 + 4C \rightleftharpoons 4CO + 3Fe, \tag{8}$$

and it can reduce even more Fe_3O_4 if CO_2 is formed instead. Such reactions may be highly significant, given the sensitivity of intrinsic f_{O_2}'s to point defect concentrations, the potentially stable solubility of C in mantle peridotite (Freund et al 1980), and the prolonged heating during decompressional transport to the surface. However, it is worth noting that even with remaining C concentrations in peridotite of the order of 50 ppmw (Mathez et al 1984), no autoreduction effects have been observed either in peridotites of reduced (IW) type 1 character or in the more oxidized type 2 peridotites (see below), which occur in the same host lava and contain comparable C abundances (Arculus et al 1984). Thus, despite all of these considerations, there does appear to be consistent evidence that the relatively refractory (in the sense of a lack of basaltic components) type 1 peridotites from the continental lithosphere are characterized by oxidation states close to IW.

Some support for this concept comes from the close correspondence of experimentally determined intrinsic f_{O_2}'s for ilmenite megacrysts in kimberlite (Arculus et al 1984) and the lower limit of original f_{O_2} equilibration determined by Haggerty & Tompkins (1983) for ilmenite-spinel intergrowths of a similar high-pressure paragenesis. In a sense, this consistency constitutes a natural test of the pressure dependence of measured, 1-bar intrinsic f_{O_2}'s and of concerns over the preservation of the point defect structures from high to low pressure.

There are other electrochemical cell data indicative of oxidized facies of spinel peridotite (the Al–Ti–Fe-rich type 2 peridotites of the Basaltic Volcanism Study Project 1981) that are close to FMQ (Arculus et al 1984), as well as comparably oxidized carbonatite and kimberlite magmas (Friel & Ulmer 1974, Ulmer et al 1976; see Figure 4). All of these oxidized kimberlite-carbonatite-megacryst associations and type 2 peridotites are derived from the subcontinental mantle, but limited data for megacryst species from one oceanic plateau (Ontong Java) are identical (Arculus et al 1984).

Where structural relationships are observed, type 2 peridotites are invariably present as cross-cutting veins in older type 1 peridotites (Irving 1980). Although the individual phases of type 1 peridotites are typically in Nd isotopic equilibrium (Jagoutz et al 1980), there are some indications of disequilibrium and age information (Stueber & Ikramuddin 1974), and formation ages based on model Sm/Nd calculations are typically in excess of 1 Gyr (Menzies et al 1984). It has been suggested by several workers that the type 2 peridotites either are the necessary precursor melts and metasomatic activity for alkaline magmatism or are the result of infiltration of alkaline melts into a more refractory peridotitic lithosphere (Irving 1980, Boettcher & O'Neil 1980).

Intrinsic f_{O_2} measurements have also been made on phenocrysts in basalts (Sato 1972) and bulk andesites from the Cascades (Ulmer et al 1976) and on the major, layered igneous intrusions derived from basaltic magmas, such as the Stillwater, Skaergaard, and Bushveld intrusions (Sato 1972, Sato & Valenza 1980, Elliott et al 1982). Interestingly, there is fair agreement between the electrochemical cell methods, the calculations based on coexisting Fe–Ti oxides, and the reproduction of observed Fe^{3+}/Fe^{2+} ratios via gas equilibration (Fudali 1965) in the case of the andesites, which are slightly more oxidized than FMQ. Significant disagreement between the first two of these approaches exists for the Skaergaard intrusion. In particular, samples from this intrusion are characterized by autoreduction during experimental study due to the inferred presence of metastable graphite (Sato & Valenza 1980), but they appear mostly to have formed between the MW and IW buffers. The Bushveld intrusion spans similar values and extends to f_{O_2}'s below IW. For the Skaergaard intrusion, Sato &

Valenza (1980) suggest that the magmatic sequence was buffered by the C–(CO_2+CO) equilibrium at pressures of the order of 1 kbar, where this equilibrium lies between the MW and IW buffers.

These conclusions have been contested by Morse et al (1980) based on the higher f_{O_2}'s (> MW) calculated from coexisting Fe–Ti oxide compositions, as well as on the mass balance of carbon with respect to total di- and trivalent Fe in the parental magma. If the parental magmas of the major layered intrusions were buffered by the C–(CO_2+CO) equilibrium all the way from the upper mantle source regions to the site of intrusion, then oxidation states approaching FMQ at pressures greater than 10 kbar can be inferred for the source. However, it is not clear that this is a reasonable inference, given the lack of evidence for the presence of a separate, vapor-rich fluid phase during the crystallization of the lower parts of the Skaergaard (Morse et al 1980). To this extent, the introductory remarks concerning philosphical disagreements over the significance of potential redox buffers are encapsulated in the variation of f_{O_2} in the classic natural crucible of the Skaergaard. Resolution of the discrepancy between calculated and experimentally determined f_{O_2}'s of the Skaergaard sequences requires further study of the stability of the Fe–Ti oxides to subsolidus alteration (Hammond & Taylor 1982), of the solution models assumed for the Fe–Ti oxides, and of the relationship of carbon and point defects in the minerals to the measured f_{O_2}'s.

The parental magmas of the major layered intrusions obviously constitute an important sampling agent of the upper mantle, but there is evidence that significant contamination of the magmas within the continental crust has occurred (Allègre & Luck 1980). Consequently, it is not altogether certain how representative the measured f_{O_2}'s are of the upper mantle source regions.

In an attempt to summarize the status of this ongoing debate over the oxidation status of the upper mantle, it seems to be a reasonable assertion that there is a consistent body of data showing that a range of oxidation states are present in the lithosphere beneath the continental crust. There are obviously many problems associated with the interpretation of the experimental data that are amenable to further study, such as the effect of pressure on point defect–structured oxides and silicates, the relationship of point defects and intrinsic f_{O_2}, and the role of carbon in crystalline and melt structures. Even if the extreme position is adopted of discarding all of the experimental evidence for the reduced state of type 1 peridotites, there is still the natural mineral evidence for reduced portions of the upper mantle manifest not only as inclusions within diamonds, but also as larger mineral grains and fragments (e.g. low-Ca, high-Cr garnets) of the same paragenesis associated with kimberlites (Sobolev 1977).

DISCUSSION AND SYNTHESIS

Heterogeneous Oxidation States

If the hypothesis that the mantle portion of the continental lithosphere is characterized by a range of oxidation states from somewhere in the vicinity of IW to FMQ is accepted, and if the further tentative evidence of an age relationship is sound, then we face the fundamental questions as to the causes and timing of the development of this heterogeneity. There is indirect data supporting the notion that the mantle source regions of basalts and komatiites were more reduced in the Archean than at present. For example, the persistently higher Cr contents of Archean melts do not appear to be simply the result of greater degrees of melting of an upper mantle source (Basaltic Volcanism Study Project 1981), and they may be related to the presence of Cr^{2+} as well as Cr^{3+} in the source regions. More reduced basaltic melts become saturated with sulfide at lower f_{S_2}'s (Naldrett 1969), so the temporal variation of the oxidation status of mantle-derived melts has relevance both to the process of sulfur transport from the mantle and to the formation of mineral deposits in the Archean.

There are a number of possible explanations for the development of a heterogeneous oxidation status of the upper mantle, including one or more of the following: (*a*) the partial preservation and intermixing of a primary (i.e. accretion-generated) stratification of the mantle with a relatively oxidized, volatile-rich interior and a more reduced, volatile-poor exterior (Ringwood 1979); (*b*) the result of progressive transfer of volatiles and melts from deeper zones of the mantle with enhanced Fe^{3+}/Fe^{2+} ratios as the result of Fe^{2+} disproportionation reactions, requiring the sequestration of Fe^0 in the lower mantle or core (Mao & Bell 1977); (*c*) infiltration of reduced mantle, relict from equilibration with a core-forming metal phase, by melts and volatiles released from relatively oxidized, subducted lithosphere (Turekian 1968); (*d*) the tendency toward a gravitationally stable, chemical stratification of the mantle (Sato 1980).

With respect to the possible zonation of the mantle during primary accretion, it is of interest that Smith (1982) has proposed the inverse of model (*a*) on the basis of a heterogeneous accretion process in which the lower portion of the mantle is composed of material similar to enstatite chondrites. The rationale for the selection of such a highly reduced component, with Fe-free silicates and Si in solution in Fe–Ni metal, is the identity of enstatite chondrites with the Earth and Moon on a $\delta^{18}O$–$\delta^{17}O$ basis, unlike that observed for any other meteorite type (Clayton et al 1976).

Note that any one of the processes outlined above is capable of rendering the upper mantle heterogeneous with respect to oxidation state, and that there are others that have not been included. It is perhaps no real surprise to find that samples in hand reveal this heterogeneity, even within the restricted region of the mantle accessible to us directly via the products of explosive volcanism. Obviously we are far from a complete resolution of these problems, but it does seem clear that present-day volcanicity and degassing processes of the Earth are dominated by the relatively oxidized signature of the upper mantle, with oxidation status in the vicinity of the FMQ–EMOG/D buffer equivalents.

Core-Mantle Equilibrium

The question of core-mantle equilibrium is of fundamental concern to geochemists, and it is a topic of recurrent debate in the literature. If such equilibrium occurred, then the mantle should bear the redox signature of the event. The issue is complicated of course by our lack of understanding of the heterogeneous equilibria involved at high pressures in a body the size of the Earth. Ringwood (1966) made the seminal observation that the abundances of siderophile trace elements in the upper mantle (in what would now be termed type 1 peridotites) exceed by an order of magnitude those expected if the silicates and oxides of the upper mantle had once been in equilibrium with a metal phase; this result prompted the development of a variety of Earth-forming models aimed at keeping the metal core apart from the mantle either by heterogeneous accretion or by some sort of disequilibrium during homogeneous accretion.

Brett (1971) argued that the effects of high pressure on the distribution of siderophile elements between metal and silicate/oxide might result in elevated abundances in the latter, and if convective overturn followed high-pressure equilibration, then an apparent overabundance might be preserved. The key question is whether the ΔV terms in any exchange reaction involving a given siderophile element between silicate/oxide and metal are of the right sign and magnitude, and the required data to answer this question are not available. With some foresight, Brett (1971) also claimed that the mantle could be considerably more reduced than was previously supposed by workers such as Ringwood (1966).

A curious feature of the type 1 peridotites is the chondritic abundance ratio of Ni/Co of about 20. This fact led Murthy (1976) to propose that the overabundance of siderophile trace elements in the upper mantle results from the intimate mixing of impacted meteoritic debris following core separation. However, the chondrite-normalized abundances of Ni and Co exceed by an order of magnitude those of the precious metals such as Pd, Pt, Ru, Rh, Ir, and Re, so that some other process must be involved to account for the nonchondritic Ni or Co/precious metal ratios. Within the precious metal group itself, unfractionated relative abundances are observed, and Chou (1978) suggested that these metals are present at concentration levels (about 10^{-2} times chondritic) expected of the probable volume of terrestrially accumulated, impacted solar system debris in 500 Myr following the formation of the Earth. Subsequent confirmation of the unfractionated nature of these precious metals has been demonstrated by the discovery of chondritic $^{187}Os/^{186}Os$ ratios in the upper mantle (Allègre & Luck 1980, Luck & Allègre 1983). Given the expected fractionation of the precious metals during melt-solid equilibrium (Hertogen et al 1980) and the apparent residual nature of type 1 peridotites as the result of melt extraction, it is curious that chondritic ratios are preserved, and these ratios may indicate mixing of meteorite contaminant after melt extraction.

The homogenization of a late-stage veneer of planetesimals with the rest of the mantle might account for the relative abundances of the precious metals, but if core-mantle equilibrium did take place, then we must account for the awkwardly overabundant siderophiles such as Ni, Co, W, Mo, and Sn. Arculus & Delano (1981b) argued that this group of siderophiles share a chalcophilic character and suggested that partial retention of a Fe–S-dominated metal phase in the mantle during core formation might explain the overabundances. Jones & Drake (1982) have advanced similar arguments based on more substantive partitioning data, and they suggest that the precious metal abundances might be explained by indigeneous processes without appealing to extraterrestrial intervention.

Holmes & Arculus (1982), using thermodynamic data (obtained with O^{2-}-specific solid electrolyte cells) for the free energies of formation of Ni_2SiO_4, Co_2SiO_4, and Fe_2SiO_4 olivine structures, reasonable solution models for olivine and metal, and published and estimated abundances of Ni, Co, and Fe in these phases in the mantle and core, showed that under no condition could metal-olivine equilibrium simultaneously satisfy observed f_{O_2}'s and metal abundances in type 1 peridotites. Furthermore, a single theoretical f_{O_2} of equilibration could not be determined. That these results are not a failure of the available thermodynamic data, solution models, and siderophile abundance data was shown by the consistency of equilibrium f_{O_2} and element abundances in coexisting metal and olivine in pallasitic meteorites (Holmes & Arculus 1982). Either the silicate phase or the nature of the metal phase assumed to take part in the equilibrium are incorrect, or else (more speculatively) the solution models chosen for the olivine are in error.

Some indication that the identification of the metallic phase is critical has recently been presented by Brett (1984), who gave partitioning data for the siderophile elements between an Fe–S–O phase and silicate. In general, a striking correspondence between calculated and observed abundances of these elements in type 1 peridotites is found, with initial chondritic abundances partitioned between the masses of the mantle and core.

Most of these geochemical arguments, however, are devoid of significant physical constraints of core-forming processes. The most recent review of this topic (Stevenson 1981) has shown that the extent of silicate-metal equilibrium must be a function of the size of the metal aggregations and the rate at which these pools of metal pass through the mantle. It would appear that if core-mantle equilibrium can be shown to have occurred, then inversion of the geochemical data might constrain the physical processes.

If we ignore the problem of the trace siderophile abundances, then the concentration of Fe in olivine of type 1 peridotites is compatible with equilibrium with a Fe-dominated (mole fraction $\simeq 0.9$), core-forming metal phase at f_{O_2}'s about 1 log order below IW. Some experimentally determined intrinsic f_{O_2}'s are close to this value, a result supporting the concept that the oxidation status of type 1 peridotite material could be relict from equilibrium with the core (Arculus & Delano 1981b). However, a simple single-stage process at low pressures has not been substantiated, and further progress on this topic is in critical need of high-pressure partitioning information. That portions of the upper mantle have been at Fe-metal saturation is of course confirmed by the presence of metal inclusions inside diamonds. Whether this was a more generalized redox condition for the whole of the upper mantle at some point in the past is a question that deserves further study.

Oxidation Controls and Profiles for the Mantle

There are currently at least three major schools of thought concerning the redox controls in the upper mantle. These are (*a*) that Fe as the most abundant heterovalent element is the major buffer through $Fe^{2+} \rightleftharpoons Fe^{3+}$ exchange, (*b*) that silicate-carbon-carbonate equilibria at pressures above about 15 kbar are critical, and (*c*) that solid C–gaseous (CO_2+CO) is the major control.

Eggler (1983) has pointed out with respect to melting of the upper mantle that it is most unlikely, given plausible concentrations of potentially volatile gas species and the solubility of these species in silicate melts, that a separate, volatile-rich fluid phase exists. At pressures greater than about 15 kbar, considerable quantities of CO_2 are required to saturate a peridotite with carbonate, and it is doubtful if these conditions are reached subsolidus (except locally). Olafsson & Eggler (1983) have shown that fluids can exist in a phlogopite-carbonate-bearing peridotite ($P > 22$ kbar) only if H_2O is present in excess of that required for phlogopite saturation, and then the fluid will be H_2O-rich because of the ability of the peridotite to react with the available CO_2. In general, therefore, it seems safe to assume that the C–(CO_2+CO) equilibrium cannot be applied to upper mantle conditions, although under anhydrous conditions at low pressures (<10 kbar), evolution of a CO_2+CO-rich vapor may be important in magmatic systems.

The internal consistency of experimentally determined phase relations of peridotite in the presence of CO_2 and H_2O, the compositions of alkaline melts generated therefrom, and the overall similarities of measured and calculated f_{O_2}'s for products of such magmatism all point to the local importance of silicate-carbon-carbonate buffering in the upper mantle (Eggler 1983). Certainly the enrichment of CO_2 and carbonate in highly potassic rocks like lamproites and kimberlites is indicative of the enrichment of these species in the mantle source regions. However, it should be emphasized that analyses of the abundances of carbon (in any form) in chemically diverse samples of the upper mantle show that these abundances rarely exceed concentrations greater than 50–100 ppm (Mathez et al 1984). Even though such quantities are significant in redox terms at low pressures through equilibrium such as (8), it seems unreasonable to ignore the significantly greater abundances of Fe^{2+} and Fe^{3+} in the upper mantle, as well as the potential buffering capacity of heterogeneous, Fe-bearing phase equilibria. This buffering capacity can vary, it should be recalled, as a function of pressure, temperature, and phase composition, including the significant role of defect structures. Furthermore, it does not seem sensible to regard the mantle as a closed system, so that reequilibration between Fe-bearing phases and fluid components can result in heterogeneities of redox states. It seems unlikely that the precipitation of Fe-rich metal in the upper mantle now encapsulated in some diamonds was controlled by a silicate-carbonate-carbon buffer. Carbonate is not present in the inclusion assemblage, and at the pressure range inferred for the inclusions [$\simeq 50$ kbar (F. R. Boyd, personal communication, 1984)], the EMOD buffer is more oxidized than IW and the f_{O_2} levels required for metal saturation (Ryabchikov et al 1981).

Major characterizations of oxidation state profiles of the mantle have recently been attempted (e.g. Sato 1980, Haggerty & Tompkins 1983). There is of course a considerable degree of speculation involved, but the concept of a relatively reduced, old mantle lithosphere beneath continents, and a more oxidized asthenosphere beneath spreading ridges and oceanic and continental lithosphere, seems plausible. The logic of an oxidized peridotitic wedge overlying a subducted lithospheric slab also seems consistent with the relatively oxidized character of island-arc volcanics (f_{O_2}'s generally greater than FMQ), and the subduction process may be one of the major causes of secular oxidation of the mantle.

Sato (1980) has pursued the more abstract concept of an oxidation profile through the mantle corresponding to an equigravitational f_{O_2} distribution. In this approach, equilibrium between oxide and metal is assumed at the core-mantle boundary; the variation of f_{O_2} as a function of gravitational potential is then calculated, resulting in equilibrium f_{O_2}'s considerably more oxidized than the IW buffer at low pressures. Such models (see also Mao & Bell 1977) are hard to evaluate given the assumptions involved, the potential disturbances to the system, and data in conflict with the generalized result, but the overall approach appears to offer first-order insight into possible redox controls on a large scale.

CONCLUSIONS

The whole topic of the oxidation status of the mantle is one of active research and debate. Considerable efforts need to be devoted to the reconciliation of experimental and theoretical models, but such efforts are feasible and practical. There does appear to be evidence for a heterogeneous distribution of redox conditions in the upper mantle, and there are hints that a secular trend toward progressive oxidation is taking place. However, it is clear that we are far from a thorough grasp of the variation of oxidation status either locally in the upper mantle, more generally through the whole of the mantle, or as a function of time and place. Given the interest in the general topic and its relevance for branches of solid-state geophysics, it is apparent that rapid progress can be expected in these fields.

ACKNOWLEDGMENTS

Discussions with Robin Brett, John Delano, David Eggler, Eric Essene, Stephen Haggerty, Richard Holmes, Hugh O'Neill, Roger Powell, and Gene Ulmer have been influential in the preparation of this article, even if the views expressed will appear recalcitrant to some of them.

Literature Cited

Allègre, C. J., Luck, J.-M. 1980. Osmium isotopes as petrogenetic and geological tracers. *Earth Planet. Sci. Lett.* 48: 148–54

Aragón, R., McCallister, R. H. 1982. Phase and point defect equilibria in the titano-magnetite solid solution. *Phys. Chem. Miner.* 8: 112–20

Arculus, R. J., Delano, J. W. 1981a. Intrinsic oxygen fugacity measurements: techniques and results for spinels from upper mantle peridotites and megacryst assemblages. *Geochim. Cosmochim. Acta* 45: 899–913

Arculus, R. J., Delano, J. W. 1981b. Siderophile element abundances in the upper mantle: evidence for a sulfide signature and equilibrium with the core. *Geochim. Cosmochim. Acta* 45: 1331–43

Arculus, R. J., Dawson, J. B., Mitchell, R. H., Gust, D. A., Holmes, R. D. 1984. Oxidation states of the upper mantle recorded by megacryst ilmenite in kimberlite and type A and B spinel lherzolites. *Contrib. Mineral. Petrol.* 85: 85–94

Basaltic Volcanism Study Project. 1981. *Basaltic Volcanism on the Terrestrial Planets*. New York: Pergamon. 1286 pp.

Bergman, S. C., Dubessy, J. 1984. CO_2–CO fluid inclusions in a composite peridotite xenolith: implications for upper mantle oxygen fugacity. *Contrib. Mineral. Petrol.* 85: 1–13

Boettcher, A. L. 1984. The source regions of alkaline volcanoes. In *Explosive Volcan-*

ism: *Inception, Evolution and Hazards*, pp. 13–22. Washington DC: Natl. Acad. Press

Boettcher, A. L., O'Neil, J. R. 1980. The origin of alkalic basalts and kimberlites: isotopic, chemical and petrographic evidence. *Am. J. Sci.* 280A: 594–621

Bohlen, S. R., Metz, G. W., Essene, E. J., Anovitz, L. M., Westrum, E. F., Wall, V. J. 1983. Thermodynamics and phase equilibrium of ferrosilite: potential oxygen barometer in mantle rocks. *Eos, Trans. Am. Geophys. Union* 64: 350 (Abstr.)

Boyd, F. R., Gurney, J. J. 1982. Low-calcium garnets: keys to craton structure and diamond crystallization. *Carnegie Inst. Washington Yearb.* 81: 261–67

Brett, R. 1971. The earth's core: speculations on its chemical equilibrium with the mantle. *Geochim. Cosmochim. Acta* 35: 203–21

Brett, R. 1984. Chemical equilibration of the Earth's core and upper mantle. *Geochim. Cosmochim. Acta* 48: 1183–88

Bulanova, G. P., Varshavskiy, A. V., Leskova, N. V., Nikishova, L. V. 1979. Central inclusions in natural diamonds. *Dokl. Akad. Nauk SSSR* 244: 704–6

Byers, C. D., Muenow, D. W., Garcia, M. O. 1983. Volatiles in basalts and andesites from the Galapagos Spreading Center, 85° to 86°W. *Geochim. Cosmochim. Acta* 47: 1551–58

Chou, C.-L. 1978. Fractionation of siderophile elements in the earth's upper mantle. *Proc. Lunar Planet. Sci. Conf., 9th*, pp. 219–30

Clayton, R. N., Onuma, N., Mayeda, T. K. 1976. A classification of meteorites based on oxygen isotopes. *Earth Planet. Sci. Lett.* 30: 10–18

Dawson, J. B. 1980. *Kimberlites and Their Xenoliths*. New York: Springer-Verlag. 252 pp.

Deines, P. 1980. The carbon isotopic composition of the diamonds: relationship to diamond shape, color, occurrence and vapor composition. *Geochim. Cosmochim. Acta* 44: 943–61

Eggler, D. H. 1978. The effect of CO_2 upon partial melting of peridotite in the system Na_2O–CaO–Al_2O_3–MgO–SiO_2–CO_2 to 35 kb, with an analysis of melting in a peridotite–H_2O–CO_2 system. *Am. J. Sci.* 278: 305–43

Eggler, D. H. 1983. Upper mantle oxidation state: evidence from olivine-orthopyroxene-ilmenite assemblages. *Geophys. Res. Lett.* 10: 365–68

Eggler, D. H., Baker, D. R., Wendlandt, R. F. 1980. f_{O_2} of the assemblage graphite-enstatite-forsterite-magnesite: experiment and application to mantle f_{O_2} and diamond formation. *Geol. Soc. Am. Abstr. with Programs* 12: 420

Elliott, W. C., Grandstaff, D. E., Ulmer, G. C., Buntin, T., Gold, D. P. 1982. An intrinsic oxygen fugacity study of platinum-carbon association in layered intrusions. *Econ. Geol.* 77: 1493–1510

Flood, H., Hill, D. C. 1957. The redox equilibrium in iron oxide spinels and related systems. *Z. Elektrochem.* 61: 18–24

Freund, F., Kathrein, H., Wengler, H., Knobel, R., Heinen, H. J. 1980. Carbon in solid solution in forsterite—a key to the untractable nature of reduced carbon in terrestrial and cosmogenic rocks. *Geochim. Cosmochim. Acta* 44: 1319–33

Friel, J. J., Ulmer, G. C. 1974. Oxygen fugacity thermometry of the Oka carbonatite. *Am. Mineral.* 59: 314–18

Fudali, R. F. 1965. Oxygen fugacities of basaltic and andesitic magmas. *Geochim. Cosmochim. Acta* 29: 1063–75

Gerlach, T. M. 1980a. Evaluation of volcanic gas analyses from Kilauea volcano. *J. Volcanol. Geotherm. Res.* 7: 295–317

Gerlach, T. M. 1980b. Investigation of volcanic gas analyses and magma outgassing from Erta'ale lava lake, Afar, Ethiopia. *J. Volcanol. Geotherm. Res.* 7: 415–41

Gerlach, T. M. 1980c. Volcanic gases from Nyiragongo lava lake. *J. Volcanol. Geotherm. Res.* 8: 177–89

Haggerty, S. E. 1978. The redox state of planetary basalts. *Geophys. Res. Lett.* 5: 443–46

Haggerty, S. E., Tompkins, L. A. 1983. Redox state of Earth's upper mantle from kimberlitic ilmenites. *Nature* 303: 295–300

Hammond, P. A., Taylor, L. A. 1982. The ilmenite/titano-magnetite assemblage: kinetics of re-equilibration. *Earth Planet. Sci. Lett.* 61: 143–50

Hertogen, J., Janssens, M.-J., Palme, H., Anders, E. 1980. Trace elements in ocean ridge basalt glasses: implications for fractionations during mantle evolution and petrogenesis. *Geochim. Cosmochim. Acta* 44: 2125–43

Hervig, R. L., Smith, J. V., Steele, I. M., Gurney, J. J., Meyer, H. O. A., Harris, J. W. 1980. Diamonds: minor elements in silicate inclusions: pressure-temperature implications. *J. Geophys. Res.* 85: 6919–29

Holmes, R. D., Arculus, R. J. 1982. Metal-silicate redox reactions: implications for core-mantle equilibrium and the oxidation state of the upper mantle. *Extended Abstr., Conf. Planet. Volatiles*, pp. 45–46. Houston: Lunar Planet. Inst.

Irving, A. J. 1980. Petrology and geochemistry of composite ultramafic xenoliths in alkalic basalts and implications for magmatic processes within the mantle. *Am. J. Sci.* 280A: 389–426

Jagoutz, E., Carlson, R. W., Lugmair, G. 1980. Equilibrated Nd–unequilibrated Sr isotopes in mantle xenoliths. *Nature* 286: 708–10

Jones, J. H., Drake, M. J. 1982. An experimental approach to early planetary differentiation. *Lunar Planet. Sci. XIII*, pp. 369–70 (Abstr.)

Karzhavin, V. K., Vendillo, V. P. 1970. Thermodynamic equilibrium and conditions for existence of hydrocarbon gases in a magmatic process. *Geokhimiya* 1970(10): 797–803

Konnerup-Madsen, J., Larsen, E., Rose-Hansen, J. 1979. Hydrocarbon-rich fluid inclusions in minerals from the alkaline Ilimaussaq intrusion, south Greenland. *Bull. Minéral.* 102: 642–53

Koseluk, R. A., Elliott, W. C., Ulmer, G. C. 1979. Gas inclusions and f_{O_2}-T data for olivines from San Carlos, Az. *Eos, Trans. Am. Geophys. Union* 60: 419 (Abstr.)

Kramers, J. D. 1979. Lead, uranium, strontium, potassium and rubidium in inclusion-bearing diamonds and mantle-derived xenoliths from southern Africa. *Earth Planet. Sci. Lett.* 42: 58–70

Luck, J.-M., Allègre, C. J. 1983. ^{187}Re–^{187}Os systematics in meteorites and cosmochemical consequences. *Nature* 302: 130–32

Mao, H. K., Bell, P. M. 1977. Disproportionation equilibrium in iron-bearing systems at pressures above 100 kbar with applications to chemistry in the earth's mantle. In *Energetics of Geological Processes*, ed. S. K. Saxena, S. Bhattacharji, pp. 237–49. New York: Springer-Verlag

Mathez, E. A. 1984. Influence of degassing on oxidation states of basaltic magmas. *Nature* 310: 371–74

Mathez, E. A., Delaney, J. R. 1981. The nature and distribution of carbon in submarine basalts and peridotite nodules. *Earth Planet. Sci. Lett.* 56: 217–32

Mathez, E. A., Dietrich, V. J., Irving, A. J. 1984. The geochemistry of carbon in mantle peridotites. *Geochim. Cosmochim. Acta.* In press

Melton, C. E., Giardini, A. A. 1974. The composition and significance of gas released from natural diamonds from Africa and Brazil. *Am. Mineral.* 58: 775–82

Melton, C. E., Giardini, A. A. 1981. The nature and significance of occluded fluids in three Indian diamonds. *Am. Mineral.* 66: 746–50

Menzies, M., Kempton, P., Dungan, M. 1984. Multiple enrichment events in residual MORB-like mantle below the Geronimo volcanic field, Arizona, U.S.A. *J. Petrol.* Submitted for publication

Mo, S., Carmichael, I. S. E., Rivers, M., Stebbins, J. 1982. The partial molar volume of Fe_2O_3 in multicomponent silicate liquids and the pressure dependence of oxygen fugacity in magmas. *Mineral. Mag.* 45: 237–45

Morse, S. A., Lindsley, D. H., Williams, R. J. 1980. Concerning intensive parameters in the Skaergaard intrusion. *Am. J. Sci.* 280A: 159–70

Murthy, V. R. 1976. Composition of the core and the early chemical history of the earth. In *The Early History of the Earth*, ed. B. F. Windley, pp. 21–31. New York: Wiley

Naldrett, A. J. 1969. A portion of the system Fe–S–O between 900 and 1080°C and its application to ore magmas. *J. Petrol.* 10: 171–201

Nitsan, V. 1974. Stability field of olivine with respect to oxidation and reduction. *J. Geophys. Res.* 79: 706–11

Olafsson, M., Eggler, D. H. 1983. Phase relations of amphibole, amphibole-carbonate, and phlogopite-carbonate peridotite: petrologic constraints on the asthenosphere. *Earth Planet. Sci. Lett.* 64: 305–15

O'Neill, H. St. C., Navrotsky, A. 1984. Cation distributions and thermodynamic properties of binary spinel solid solutions. *Am. Mineral.* 69: 733–53

O'Neill, H. St. C., Ortez, N., Arculus, R. J., Wall, V. J., Green, D. H. 1982. Oxygen fugacities from the assemblage olivine-orthopyroxene-spinel. *Terra Cognita* 2: 228

Ringwood, A. E. 1966. The chemical composition and origin of the earth. In *Advances in Earth Sciences*, ed. P. M. Hurlye, pp. 287–356. Cambridge, Mass: MIT Press

Ringwood, A. E. 1979. *Origin of the Earth and Moon.* New York: Springer-Verlag. 295 pp.

Roedder, E. 1965. Liquid CO_2 inclusions in olivine-bearing nodules and phenocrysts from basalts. *Am. Mineral.* 50: 1746–82

Roedder, E., Bodnar, R. J. 1980. Geologic pressure determinations from fluid inclusion studies. *Ann. Rev. Earth Planet. Sci.* 8: 263–301

Ryabchikov, I. D., Green, D. H., Wall, V. J., Brey, G. 1981. The oxidation state of carbon in the environment of the low velocity zone. *Geokhimiya* 1981(2): 221–32

Sato, M. 1965. Electrochemical thermometer: a possible new method of geothermometry with electroconductive minerals. *Econ. Geol.* 60: 812–18

Sato, M. 1972. Intrinsic oxygen fugacities of iron-bearing oxide and silicate minerals under low total pressure. *Geol. Soc. Am. Mem.* 135:289–307

Sato, M. 1978. Oxygen fugacity of basaltic magmas and the role of gas-forming elements. *Geophys. Res. Lett.* 5:447–49

Sato, M. 1979. The driving mechanism of lunar pyroclastic eruptions inferred from the oxygen fugacity behavior of Apollo 17 orange glass. *Proc. Lunar Planet. Sci. Conf., 10th*, pp. 311–25

Sato, M. 1980. Gas fugacities in planetary interiors and their bearing on the origin of metallic cores in the inner planets. *Lunar Planet. Sci. XI*, pp. 974–76 (Abstr.)

Sato, M., Valenza, M. 1980. Oxygen fugacities of the layered series of the Skaergaard intrusion, east Greenland. *Am. J. Sci.* 280A:134–58

Smith, J. V. 1982. Heterogeneous growth of meteorites and planets, especially the Earth and Moon. *J. Geol.* 90:1–125

Sobolev, N. V. 1977. *Deep-Seated Inclusions in Kimberlites and the Problem of the Composition of the Upper Mantle*, ed. F. R. Boyd. Washington DC: Am. Geophys. Union. 279 pp. (Engl. Transl. by D. A. Brown)

Sobolev, N. V., Efimova, E. S., Pospelova, L. N. 1981. Native iron in diamonds of Yakutiya and its paragenesis. *Geol. Geofiz.* 22:25–28

Spencer, K. J., Lindsley, D. H. 1981. A solution model for coexisting iron-titanium oxides. *Am. Mineral.* 66:1189–1201

Stevenson, D. J. 1981. Models of the earth's core. *Science* 214:611–19

Stocker, R. L., Smyth, D. M. 1978. Effect of enstatite activity and oxygen partial pressure on the point-defect chemistry of olivine. *Phys. Earth Planet. Inter.* 16:145–56

Stueber, A. M., Ikramuddin, M. 1974. Rubidium, strontium and the isotopic composition of strontium in ultramafic nodule minerals and host basalts. *Geochim. Cosmochim. Acta* 38:207–16

Trial, A. F., Rudnick, R. L., Ashwal, L. D., Henry, D. J., Bergman, S. C. 1984. Fluid inclusions in mantle xenoliths from Ichinomegata, Japan: evidence for subducted H_2O? *Eos, Trans. Am. Geophys. Union* 65:306 (Abstr.)

Turekian, K. K. 1968. The composition of the crust. In *Origin and Distribution of the Elements*, ed. L. H. Ahrens, pp. 549–57. Oxford: Pergamon

Ulmer, G. C., Rosenhauer, M., Woermann, E., Ginder, J., Drory-Wolff, A., Wasilewski, P. 1976. Applicability of electrochemical oxygen fugacity measurements to geothermometry. *Am. Mineral.* 61:653–60

Ulmer, G. C., Rosenhauer, M., Woermann, E. 1980. Glimpses of mantle redox conditions? *Eos, Trans. Am. Geophys. Union* 61:415 (Abstr.)

Virgo, D., Huggins, F. E., Rosenhauer, M. 1976. Petrologic implications of intrinsic oxygen fugacity measurements on titanium-containing garnets. *Carnegie Inst. Washington Yearb.* 75:730–35

Wyllie, P. J. 1978. Mantle fluid and compositions buffered in peridotite–CO_2–H_2O by carbonates, amphibole and phlogopite. *J. Geol.* 86:687–713

Ann. Rev. Earth Planet. Sci. 1985. 13: 97–117

THE GEOLOGICAL SIGNIFICANCE OF THE GEOID

Clement G. Chase

Department of Geosciences, University of Arizona, Tucson, Arizona 85721

INTRODUCTION

Global gravity has come into its own with the advent of satellite radar altimetry and satellite measurements of the geopotential. The volume of published work mining this rich data set has burgeoned from just a few papers per year in the early 1970s to one or two a week at the present writing. Satellite gravity data are being used to study terrestrial phenomena at scales from tens of thousands of kilometers down to hundreds or even tens of kilometers. This paper focuses on the interpretation of global gravity in the form of the geoid and concentrates on what has already been learned from it about the structure of the Earth at long-to-intermediate wavelengths.

The longest wavelengths of the gravity field are dominated by density distributions that must be supported by very large-scale convective processes. I attempt here to document the claim that these density anomalies are both deep and old. At slightly shorter length scales, the geoidal anomalies give very useful constraints on possible subduction processes and lead to the conclusion that density contrasts in the subducting slabs must be regionally partially compensated and supported from below. At yet shorter length scales, satellite altimetry results give insight into the thermal and mechanical structure of the oceanic lithosphere, seeming to favor plate over thermal boundary layer models of cooling. At the shortest length scales, the location and mode of compensation of individual seamounts can be studied.

0084–6597/85/0515–0097$02.00

THE GEOID

The fundamental data set collected by satellite geodesy is usable in a number of forms. For analysis and modeling of longer wavelength features, a spherical harmonic expansion of the gravitational potential is most convenient. These harmonics are calculated from orbital information on many satellites and may include satellite radar altimetry data and surface gravity observations. For smaller features, individual satellite altimeter passes from *Geos-3* and *Seasat* or regional maps compiled from these data can be used. In both of these cases, the form in which the data have proved to be most useful is geoidal anomalies (mappings of the departure of the sea-level equipotential surface from an equilibrium shape for the Earth) rather than the classically emphasized gravity anomalies. Figure 1 displays a geoid calculated from spherical harmonics with approximately 100-km spatial resolution (Rapp 1981).

To examine the differences between geoidal and gravity anomalies and understand their characteristics, it is useful to present a few of the relevant equations. The spherical harmonic expansion of the Earth's gravitational potential U, leaving out the purely rotational contribution, is given by

$$U(r, \theta, \lambda) = \frac{GM_e}{r} \times \left[1 + \sum_{n=2}^{\infty} \sum_{m=0}^{n} \left[\frac{a}{r}\right]^n [\bar{J}_n^m \cos(m\lambda) + \bar{K}_n^m \sin(m\lambda)] \bar{P}_n^m(\cos\theta) \right], \quad (1)$$

where r is distance from the center of mass, θ is colatitude, λ is longitude, G is Newton's universal gravitational constant, M_e is the mass of the Earth and a its equatorial radius, $\bar{J}_n^m$ and $\bar{K}_n^m$ are the spherical harmonic coefficients that describe the actual gravitational potential, and $\bar{P}_n^m(\cos\theta)$ are fully normalized associated Legendre polynomials of degree n and order m. The spatial resolution of a harmonic expansion up to degree $n_{\max}$ for the Earth goes roughly as 2×10^5 km $n_{\max}^{-1}$, and the number of coefficients required goes as $(n_{\max}+1)^2$. Spherical harmonics are not an efficient descriptor of small features.

The frequency content of the geopotential can be described by the amplitude spectrum σ_n:

$$\sigma_n = \left[\frac{1}{2n+1} \sum_{m=0}^{n} [(\bar{J}_n^m)^2 + (\bar{K}_n^m)^2] \right]^{1/2}. \quad (2)$$

One of the notable characteristics of the Earth's gravity field is the extent to which the low-degree, long-wavelength harmonics dominate. According to Kaula's rule of thumb, σ_n is approximately $10^{-5}\, n^{-2}$ for $n \leq 15$ (Kaula 1967). Another way of looking at the importance of the low-degree

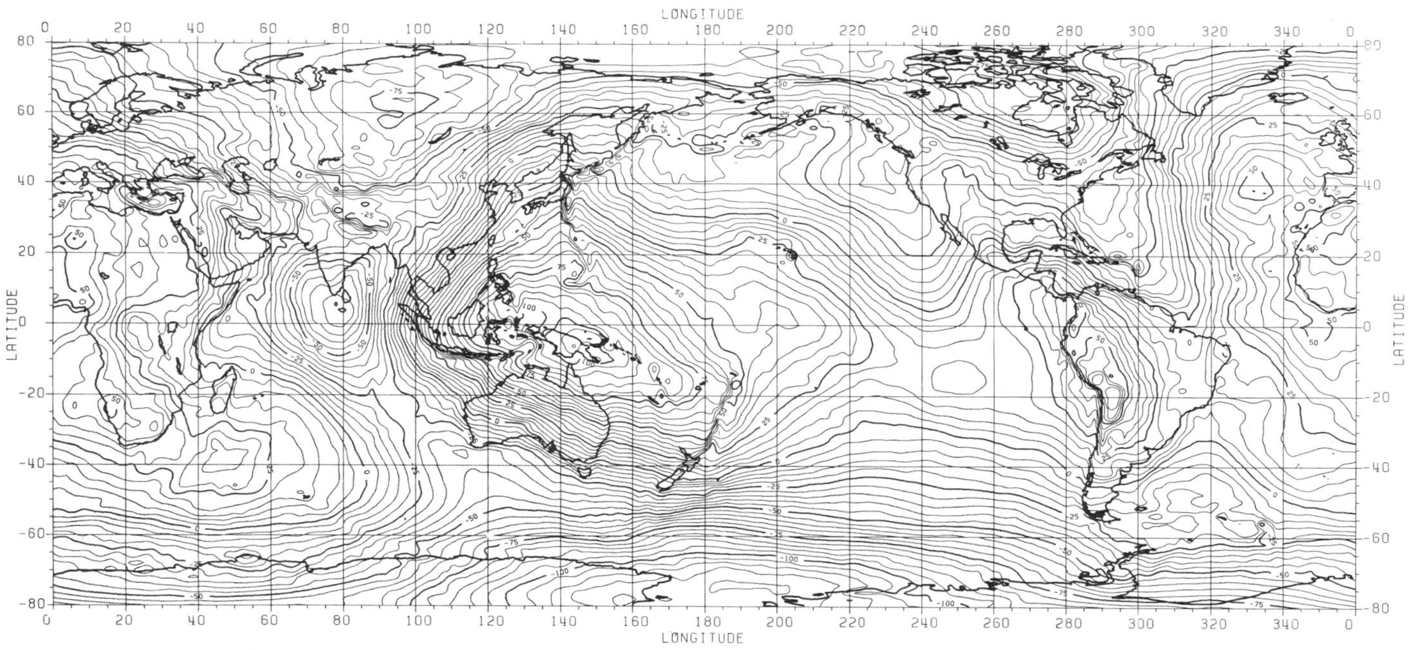

Figure 1 Geoid relative to hydrostatic reference figure, complete to degree and order 180, with coefficients from Rapp (1981). The contour interval is 5 m, and the map is displayed in a "cylindrical equidistant" representation. Figure calculated and kindly provided by R. Rapp.

harmonics is through the geoidal anomalies. Figure 1 is calculated to degree and order 180, and it has a total amplitude range of about 230 m. Of this, the second-degree terms contribute 140 m of relief, and the second- and third-degree terms together contribute 190 m.

The geoid itself is the surface of constant potential U_0 that corresponds to mean sea level. This surface is deformed from a spherical shape by more than 20 km of rotationally supported equatorial bulge, which would thoroughly mask the 230 m of relief that is caused by internal mass heterogeneities. In order to see this geologically interesting relief, it is necessary to refer the geoid to a base level, a reference figure that contains the purely rotational effects. Choosing an ellipsoid of revolution that has the same mass and a "normal" potential W with $W_0 = U_0$ on its surface, we have the anomalous potential $T = U - W$.

The ellipticity, or flattening, of the reference figure must be chosen. For geodetic surveys a best fit to the actual shape is best, but for tectonic purposes the proper choice is a figure in hydrostatic equilibrium (O'Keefe & Kaula 1963, Chase 1979), as all departures from such equilibrium must be supported by internal stresses. The source of these stresses is geologically of great interest. The hydrostatic reference figure for Figure 1, from Nakiboglu (1982), has a flattening of 1/299.638.

To go from the potential to geoidal anomalies N is straightforward:

$$N(\theta, \lambda) = \frac{T(\theta, \lambda)}{\gamma(\theta)}, \tag{3}$$

where $\gamma(\theta)$ is the "normal" gravity calculated on the reference surface from the normal potential W. This varies by only 0.5% from pole to equator, so the geoidal anomalies are directly related to the anomalous potential. The acceleration of gravity is given, to a sufficient approximation, by $\partial U/\partial r$, and the gravity anomaly that results is

$$\Delta g(\theta, \lambda) = \frac{1}{R} \sum_{n=2}^{\infty} (n-1) T_n(\theta, \lambda), \tag{4}$$

where T_n represents all the n-degree harmonics of T, and R is the average radius of the Earth. A comparison of Equations (3) and (4) shows that the gravity anomaly magnifies the higher degree harmonics by the factor $(n-1)$. It is just these higher harmonics that are most uncertain in the satellite solutions, so the geoidal anomalies are more stable representations of the data.

One reason that the geoidal representation has become popular is its stability. In addition, geoidal anomalies resemble the potential in having $1/r$ dependence with distance from their source, in contrast to the $1/r^2$

dependence of gravity. The result is that the geoid is more sensitive to deep mass distributions than is gravity. In a flat-Earth approximation, the geoid will show a step offset across the edge of isostatically compensated features such as continent/ocean margins, while the gravity signature is an edge effect and averages to zero (Crough 1979a) and is thereby more difficult to distinguish from noise. As a final advantage, the radar altimetry passes can be cast as geoidal undulations with little processing.

LONG-WAVELENGTH ANOMALIES AND MANTLE CONVECTION

Perhaps the earliest effort to interpret geoidal anomalies in the spatial domain was a proposal by Runcorn (1964, 1967) that the geopotential anomalies could be converted into maps of sublithospheric stress by assuming Newtonian laminar flow and treating the lithosphere as a rigid shell. There are considerable difficulties with this approach. The simplifications necessary to make the problem tractable, such as near-critical Rayleigh number, tend to be unreasonable, and any three-dimensional component to the convection invalidates the method (Phillips & Ivins 1979). Furthermore, the geoidal anomalies come from many different sources and cannot all be attributed to a simple convective pattern, as we shall see.

Kaula (1972) compared known tectonic features with gravity anomalies calculated from satellite data for harmonics $6 \leq n \leq 16$; he concluded that positive anomalies marked trench and island arc areas, as well as oceanic rises, while negative gravity was typical of both oceanic and continental basins, especially those recently glaciated. Figure 1 shows that the correlation of positive geoid anomalies with areas of subduction certainly holds up, but that oceanic spreading centers do not have a very conspicuous signature.

Statistical and Spectral Approaches

A number of attempts have been made to localize the sources of the longer wavelength geopotential anomalies by statistical analyses in the frequency domain. These techniques rely on the slope of the gravitational spectrum to give a maximum depth estimate for the source of the anomalies. Hide & Horai (1968) attributed the lowest degree harmonics to undulations in the core/mantle boundary. By assuming that a single density interface gave rise to the geopotential anomalies, several workers found a best depth for that interface of less than 1000 km (Higbie & Stacey 1970, 1971, Bott 1971). Further analyses used more density interfaces (McQueen & Stacey 1976) or random source distributions with no spatial coherence (Khan 1977,

Lambeck 1976) to conclude that the sources were predominantly shallow and showed a tendency to anticorrelate or compensate at different levels.

All these approaches suffer from the same problem: The anomalies are generated and supported in the Earth at a variety of depths, and the inherent nonuniqueness of potential interpretation does not permit these variations to be sorted out a priori. They also display the weakness of spectral techniques in that the geography of known tectonic features cannot guide the analysis, because spatial information is lost in forming the spectrum [Equation (2)].

Geoidal Anomalies and Subduction Zones

The very clear significance of subduction zone geoidal anomalies (Figure 1) has inspired considerable attention to modeling the geoidal signatures of global or individual convergent plate boundaries. Three main conclusions emerge: (*a*) the subducting slabs are regionally compensated, (*b*) they must be partially supported by stresses exerted at their deeper ends, and (*c*) they are by no means the only important feature causing the long-wavelength geoidal undulations.

Several studies treated the subduction problem as a static, or rigid-Earth, one, finding a density distribution that would satisfy the observed anomalies without accounting for the loads imposed by the density differences. One could term this approach an "equivalent mass" technique, in that viscous responses are ignored, but the net result of anomalous density and consequent deformation is treated as an equivalent density distribution.

The end result of these "equivalent mass" approaches is that the excess masses of subducting slabs must be regionally partially compensated in some way. A global study that started with thermal models of subducting lithosphere and then adjusted the model densities of individual subducting slabs to match the observed geoidal anomalies for harmonics of $2 \leq n \leq 20$ (Chase 1979) found that the thermal models overestimated the amount of excess density detectable in the geoid by up to a factor of four. Crough & Jurdy (1980) used presubduction thermal subsidence to estimate the excess density of the slabs globally and obtained a reasonable fit to observed anomalies for $n \leq 12$; however, they tapered the density contrast linearly to zero at the depth of deepest earthquakes and thereby used less mass than conductive models of the slab, such as that of McKenzie (1969), would suggest. McAdoo (1981) included lithospheric structure around the Tonga-Kermadec Trench and found densities lower than those from thermal models. Chapman & Talwani (1982), in a shorter wavelength study of the same subduction zone combining gravity and geoid data, could hardly detect the slab at all. Earlier studies with marine gravity data from the northwest Pacific (Watts & Talwani, 1974, 1975) had the same difficulty, though a significant slab effect was reported from the Peru-Chile Trench (Grow & Bowin 1975).

The thermal models do not predict that the slabs will be completely reequilibrated within the time needed to reach 700 km (McKenzie 1969). There is seismic evidence that the slabs may persist to greater depth (Creager & Jordan 1984, Jordan 1977). In addition, conduction of heat in the Earth is so slow that even the conductive models for warming the slabs should *underestimate* the mass excess: As the slab is warmed, the surrounding mantle is cooled, and the resulting cold and dense region is very hard to disperse. The consistent findings of less than predicted net density contrasts indicate that more than just passive insertion of slabs is going on, and some sort of regional compensation must also be taking place.

One major form of compensation is the natural result of a convective process like subduction in a viscous Earth. It has long been known that deformation of density interfaces due to transmission of viscous stresses can be as important in generating gravity anomalies as the density contrasts driving the convection (Pekeris 1935). For example, the weight of a subducting slab will cause viscous depression of the Earth's surface in a wide strip surrounding the subduction zone. The actual gravity or geoid anomaly that results will depend on a delicate balance between the driving density contrasts and the warping of density interfaces (Morgan 1965, McKenzie 1977, Parsons & Daly 1983). The anomaly can be either positive or negative and is a sensitive function of the boundary conditions. Davies (1981) and McAdoo (1982) have investigated this effect with a half-space approximation and find that surface depression, given Newtonian viscosity, would generate a negative geoid anomaly sufficient to render the geoid signature of a subducting slab negative; they appeal to support of the slab from above or below and non-Newtonian flow, respectively, to explain the fact that geoid anomalies over the convergent zones are positive. Obviously, this form of compensation is sufficiently potent to explain the reduction in amplitude compared with a static thermal model.

Using an elegant theory for the effect of internal loads in a Newtonian viscous self-gravitating planet (Richards & Hager 1984), Hager (1984) has calculated geoidal anomalies for global subduction zones and finds the best correlation with the observed geoid for $4 \leq n \leq 9$ (Figure 2). In order to maintain positive geoid anomalies over the slabs, he found that a viscosity increase with depth was required to support the lower end of the slab and prevent surface deformation from turning the geoidal anomaly negative. More excess mass than that contained in the seismically active part of the slabs is also indicated. Given the likely persistence of the thermal anomalies

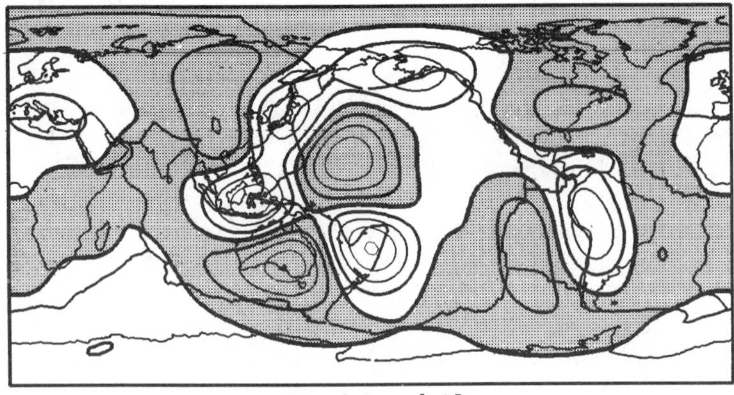

Figure 2 Slab geoid of Hager (1984), using harmonics of degree 4 through 9. The contour interval is 10 m, and negative geoidal anomalies are stippled. Reproduced with permission from the American Geophysical Union.

caused by subduction of cold lithosphere, this result should be expected. The model does not include the lithosphere as a mechanical entity, but because of the long wavelengths used, the conclusion of a downward increase in resisting forces is probably robust.

Figure 3 displays a degree 20 geoid with an older, less sophisticated slab model (Chase 1979) subtracted. From this figure it is obvious that although the slabs are important, contributing a maximum amplitude variation of about 70 m (Figure 2), they are by no means the most important phenomenon determining the long-wavelength geoid (Chase 1979, Crough & Jurdy 1980). In fact, this can be seen without modeling by inspection of Figure 1. The largest positive anomaly stretches right across the equatorial Pacific between the trenches at either end. The second largest positive anomaly, that extending from the southern Indian Ocean through Africa to near Greenland, only peripherally touches any subduction zone. A discontinuous negative anomaly girdles the globe from pole to pole, crossing the equator south of India. The possible causes of this remaining and dominant part of the long-wavelength geoid is the subject of the next section.

Deep Convection

Perhaps the most remarkable thing about the Earth's gravity field is that the largest and highest amplitude anomalies bear so little visible relationship to the convection expressed at the surface in plate tectonics. As we shall

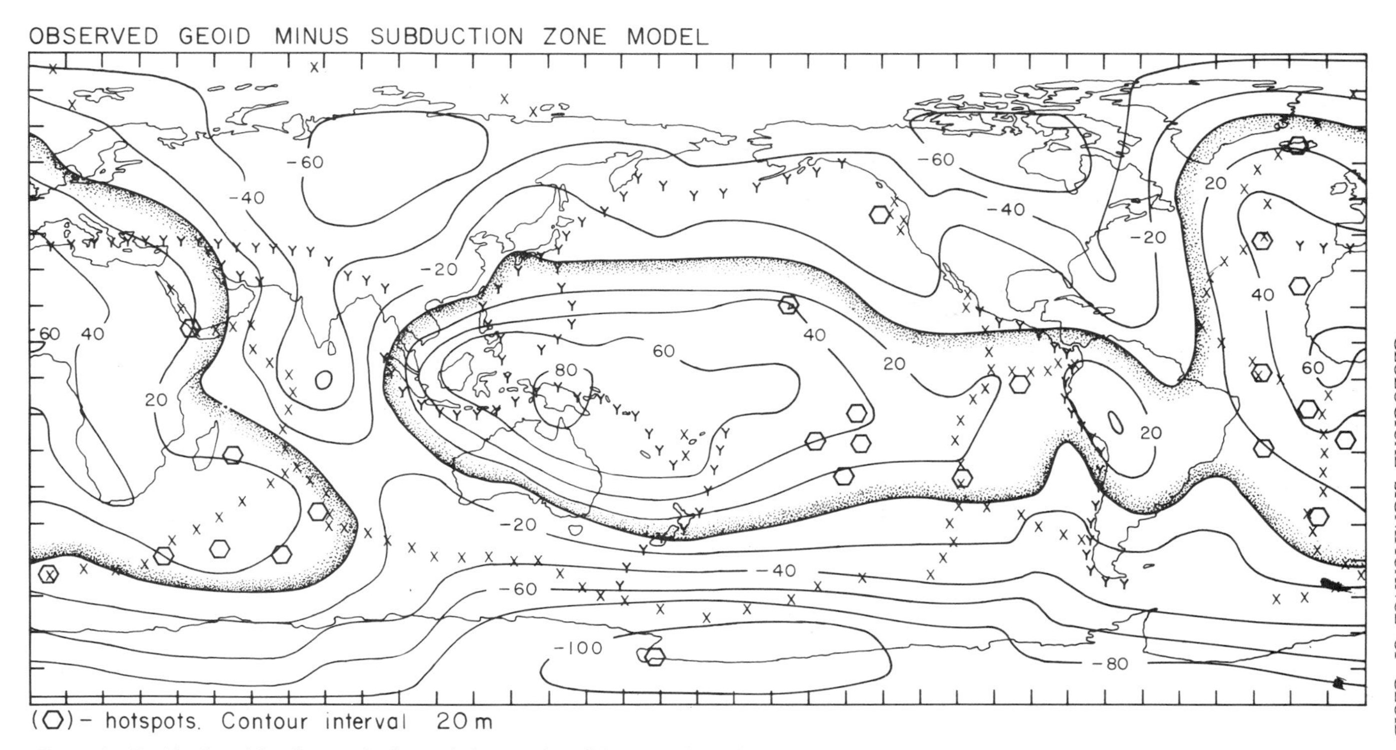

Figure 3 Residual geoid, using nonhydrostatic harmonics of degree 1 through 20, and with the model slab geoid of Chase (1979) subtracted. The contour interval is 20 m, and the positive side of the zero contour is stippled. The hexagons mark the location of large oceanic hotspots, the ×'s are spreading center locations, and the Y's are subduction zone locations. Reproduced by permission of *Nature*. Copyright © 1979 by Macmillan Journals Ltd.

see later, the oceanic spreading centers contribute only 10-m anomalies. The large geoid anomalies residual to the effects of the subducting slabs (Figure 3) are much simpler in pattern than the plate boundaries, and they seem to cross ridges and transform faults and all ages of lithosphere from Archean to Recent at will. Whether one treats the positive anomalies as upwellings or downwellings (Morgan 1965, McKenzie 1977), the major residual anomalies are poorly positioned to explain either return flow beneath the moving plates or intraplate stresses (Chase 1979). It is clear, to me at least, that some alternative explanation, other than convection closely coupled to surficial plate tectonics, must be sought for the large anomalies that represent most of the amplitude in the Earth's gravitational spectrum. That the cause is some form of convection is also quite clear (see reviews by Phillips & Ivins 1979, Phillips & Lambeck 1980).

There are tectonic features that do correlate much better than plates with the long-wavelength geoid: the oceanic hotspots (Chase 1979, Crough & Jurdy 1980). The hotspots show a very strong and statistically significant tendency to occur in regions of positive geoidal anomaly (Figure 3). If the apparent fixity of the hotspots relative to each other (Morgan 1972, 1981) is real, then the sources of the hotspots must be deeper than the region stirred by plate motions. Correlation of the hotspot locations with the large geoidal anomalies then implies that these anomalies are also of deep origin. The integrity of the large anomalies across plate and lithospheric boundaries leads to the same conclusion (Chase 1979).

Seismic velocity analyses of the lower mantle support the deep origin of the long-wavelength geopotential anomalies, especially those of degree 2 and 3. Dziewonski et al (1977) found a negative correlation between velocity anomalies deeper than 1100 km and $n = 2$ and 3 gravity, as might be expected if the velocities represent thermally induced density differences driving convection whose deformation of density boundaries dominates the gravity signal. Recent, more detailed seismic velocity distributions for the lower mantle (Clayton & Dziewonski 1984, Dziewonski 1984) give a good fit to the degree 2 and 3 gravity potential (Hager & Richards 1984) when assigned a velocity/density relationship and passed through the viscous response filters of Richards & Hager (1984).

Size, geographic distribution, and the seismic connection make the case for a deep convective origin of the longest wavelength geoid anomalies seem very sound. It is not at all clear why the lower mantle should have a seemingly simple convective pattern despite its almost certainly very high Rayleigh number. What is clear, as discussed below, is that the $n = 2$ harmonics of the Earth's gravity field are intimately associated with its rotational history, and therefore this gravest part of the geopotential must be relatively stable with time.

True Polar Wander and the "Age" of the Geoid

The close relationship of the geoid with rotation arises because the same integrals over the density distribution of the Earth that determine its second-degree gravity harmonics also determine the moments and products of inertia that go to make up its inertia tensor. The Earth rotates about the principal axis of maximum moment of the inertia tensor. This maximum moment of inertia reflects the equatorial bulge, itself mostly a result of the rotation. Goldreich & Toomre (1969) showed that the axis of maximum moment of the *nonhydrostatic* inertia figure, corresponding to the nonhydrostatic geoids we have been examining, controls the location of the spin axis, and thereby the location of the axis of the hydrostatic figure as well. This is very much a situation of the tail wagging the dog, with the 140-m degree 2 nonhydrostatic bulge determining the location of the North and South Poles and the >20-km hydrostatic bulge.

In a slowly deforming planet such as the Earth, if the nonhydrostatic figure of inertia evolves as a result of convection or other causes, the planet will move relative to its rotational axis (fixed in space) to keep the maximum nonhydrostatic moment aligned with the pole. It is thereby more accurate to say that the equator lies on the large Pacific Ocean geoid anomaly of Figure 1 than to say that the anomaly lies on the equator. Movement of the Earth relative to its rotational axis can be described as true polar wander. Thus the problem of assigning a time scale to long-term, long-wavelength changes in the geoid becomes the problem of measuring true polar wander.

One way of estimating true polar wander is to compare the motions of plates in a paleomagnetic reference frame to their motions relative to presumed fixed hotspots (Jurdy 1981, Morgan 1981). The necessary assumptions are that the average position of the paleomagnetic pole is the same as the spin axis, and that the hotspots in some way represent the mean position of the bulk of the mantle (Morgan 1972). A number of analyses along these lines find possibly 10° of difference between the hotspot and paleomagnetic reconstructions at about 60 Myr ago (Jurdy 1981, Morgan 1981, Gordon 1982). At least for the Pacific plate, the assumption that the paleomagnetic axis is the same as the rotational axis is confirmed by the positions of equatorial sedimentary facies (Gordon & Cape 1981). There are indications of possible rapid polar wander in the late Cretaceous (Gordon 1982, 1983), but the limited amount of wander in the Cenozoic indicates that the degree 2 part of the geoid has not changed very quickly. Thus the underlying convection must be reasonably stable in configuration.

The consequences of large changes in the geoid could be drastic. The current pattern is composed of two large highs (African and Pacific) and a narrower negative (Figures 1, 3). If the African high were to become larger than the Pacific high through evolution of the convective system causing them both, fairly rapid polar wander of some 35° could occur (Chase 1979). Jurdy (1983) has calculated separate inertia tensors for the subduction zones and the hotspots (equated to the rest of the $n = 2$ geoid) and finds that the early Tertiary subduction zone positions could cause the observed 10° difference between hotspot and paleomagnetic reference frames without requiring any change in the hotspot component. This analysis relies on an empirical and somewhat arbitrary scaling of the relative inertial effects of hotspots and subducted lithosphere.

The positive geoidal anomalies may correlate somewhat with subduction zones and quite well with the locations of hotspots, but the negative anomalies show no clear tectonic affinities at all. Curiously enough, both positive and negative long-wavelength anomalies seem to correspond to ancient plate positions. The African high lies over the Mesozoic position of Pangea (Anderson 1982), and the globe-encircling system of geoidal lows is located near the reconstructued 125-Myr-old positions of subduction zones in Morgan's (1981) hotspot reference frame (Chase & Sprowl 1983; see Figure 4). Anderson (1982) suggested that the African high reflects upper-mantle overheating caused by thermal insulation by the Pangean continental assemblage, and that the Pacific high might also be related to since-dispersed continental fragments. It may rather be that the Pacific anomaly is in some way genetically linked to the shallow and volcano-studded Darwin Rise (Menard 1984). Stifling of asthenospheric return flow under the Pangean continent by the circumferential arrangement of subduction zones is as likely a culprit for the thermal event there as continental insulation.

The similarity between Cretaceous subduction zone positions and geoidal lows (Figure 4) is as suggestive as the continental mass/geoid high association. This could represent either a piling up of cold, dead lithospheric slabs in the middle mantle (Hager 1984) or a closer thermal and/or mechanical coupling of plate motions to lower-mantle convection in the Cretaceous, perhaps abetted by the relative stability of plate boundary positions in the late Mesozoic (Chase & Sprowl 1983). In any case, the conclusion is inescapable that if the convective pattern determining the longest wavelength geoidal anomalies does respond to the surface conditions represented by plate tectonics, then there is some 100 Myr of time delay in the response.

THE STRUCTURE OF THE LITHOSPHERE

At shorter wavelengths than we have been hitherto discussing, geoidal anomalies are useful in constraining many aspects of lithospheric structure.

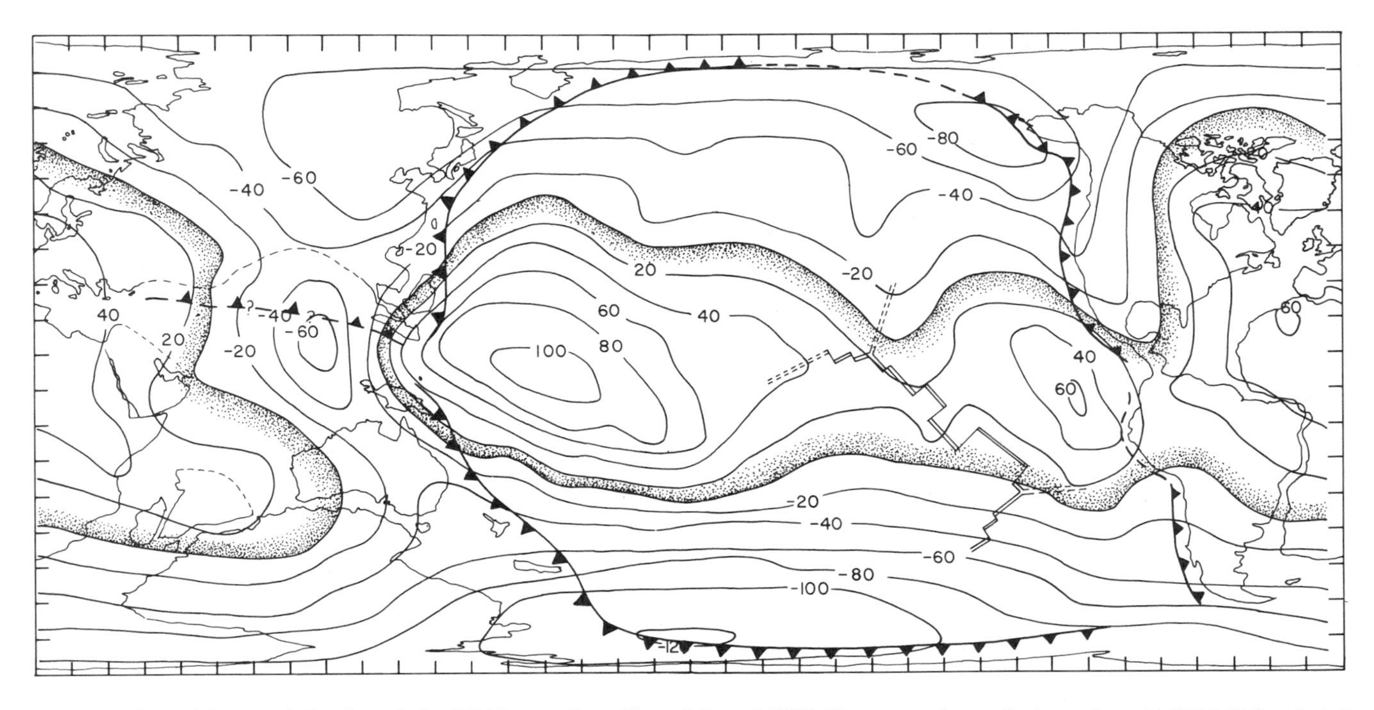

Figure 4 The geoid now and plate boundaries 125 Myr ago (from Chase & Sprowl 1983). The present-day nonhydrostatic geoid GEM-10 (Lerch et al 1979), complete to degree and order 20, is plotted on hotspot-relocated continental positions in the early Cretaceous (Morgan 1981), with subduction zone positions shown by toothed lines. The contour interval is 20 m. Reproduced with permission of Elsevier Science Publishers B.V.

Isostatic differences between continents and oceans need not extend very deep, thermal structure and net driving forces at mid-ocean ridges are measurable from the geoid, and the question of whether heat is supplied convectively or only conductively to the base of the aging oceanic lithosphere can be resolved.

Continent/Ocean Differences and Isostasy

Geoidal anomalies would seem to be a natural for study of the isostatic compensation of continents relative to oceans, but as is frequently the case with potential theory and other problems (Kipling 1970), nonuniqueness complicates the interpretation. Even the definition of isostasy appropriate to a spherical Earth is uncertain, and seemingly equally valid forms can give rise to a factor of 2 difference in predicted amplitudes (Dahlen 1982).

In the face of the large-amplitude geoidal undulations arising elsewhere, it has proved difficult to extract the anomalies distinctive to isostatic compensation of continental crust and lithosphere relative to oceanic. Turcotte & McAdoo (1979) attempted an inverse approach, isolating the continent/ocean geoidal differences by regional averaging, and found a mean value near zero. From this they concluded that thermal differences between continental and oceanic lithosphere extended no deeper than 200 km. However, the averaging process is not particularly reliable given the high amplitudes of the longer wavelengths.

In places where the contours controlled by longer wavelength anomalies of deeper origin are favorably oriented, such as eastern Australia, East Antarctica, and Greenland (Figure 1), it is apparent that old continental crust is marked by a positive geoidal relief of about 10 m relative to the adjacent ocean basins. Forward modeling of an altimeter profile over the east coast of North America (Haxby & Turcotte 1978) and of global topography (Chase & McNutt 1982) suggests that Airy compensation of the continent/ocean differences at a depth of around 40 km will satisfy the 10-m anomaly without appealing to deeper density differences. This conclusion persists independent of the particular model of the thermal structure of the oceanic lithosphere that is used (Hager 1983).

A number of studies have used geoidal data to constrain the depth of isostatic compensation of aseismic ridges on the oceanic plate (Haxby & Turcotte 1978, Crough & Jarrard 1981, Crough 1982, Soriau 1984); these studies find depths of compensation of less than 100 km, i.e. within the lithosphere. For short wavelengths (less than about 2000 km), isostasy should be dominated by flexural response of the lithosphere (see the review by Watts & Daly 1981), and geoidal anomalies are now being employed in studies of flexural isostasy (Watts 1979, Cazenave et al 1980, Lambeck

1981). It is not clear yet that conclusions reached using geoidal data sets are any different than those previously arrived at with marine gravity and bathymetric data.

Mid-Ocean Ridges

A pioneering study by Haxby & Turcotte (1978), based on approximate solutions for thin layers derived by Ockendon & Turcotte (1977), showed that for a simple isostatic thermal model of the elevation of mid-ocean ridge crests, the geoid anomaly should be a linear function of age of the crust. The predicted slope of -0.16 m Myr^{-1} for geoidal slope with age matched a *Geos-3* altimetry profile across the Mid-Atlantic Ridge for crust younger than 50 Myr old. A rather arbitrary trend removal was necessary to filter out the ridge signature of about 10 m from independent, longer wavelength and higher amplitude variations.

Sandwell & Schubert (1980) averaged geoidal heights as a function of crustal age to remove the interfering larger anomalies, and they found roughly constant geoid/age slopes for the North and South Atlantic and southeast Indian Oceans over crust less than 80 Myr old. The slope for each sample was different, however, and separation of the ridge crest anomaly was not successful for the South Pacific. These problems indicate either that contamination by other anomalies is very hard to avoid, or that the thermal parameters of cooling oceanic lithosphere vary regionally (Cazenave et al 1983), or both. This would be an unfortunate characteristic, as stable estimates of the geoidal anomalies can be used to invert for upper-mantle thermal properties (Lister 1982). Another useful property is that for spreading faster than 10 mm yr^{-1}, the driving force resulting from the outward topographic slope of the ridge is directly proportional to the geoidal anomaly (Parsons & Richter 1980). If the ridge anomalies can be reliably distinguished, it will be possible to observationally quantify this important component of plate tectonic driving forces.

Another feature that Sandwell & Schubert (1980) noted leads us into the next topic to be discussed. Instead of continuing to decrease with age, the geoidal anomalies seem to level off over crust older than 80 Myr. This observation has a direct and crucial bearing on models of the aging process in oceanic lithosphere.

Old Oceanic Lithosphere: Cooling Models

There have been two competing conceptual models for the thermal structure of old oceanic lithosphere. The thermal boundary layer model proposed by Turcotte & Oxburgh (1967) envisions the old lithosphere as an effective half-space that only experiences conductive heat transfer. Such an oceanic lithosphere would continue its thickening and isostatic subsidence linearly with the square root of its age (Parker & Oldenburg 1973). Heat flow would continue to decrease indefinitely. The plate model (archetypally McKenzie 1967, following Langseth et al 1966) treats the lithosphere as a layer of bounded thickness and imposes either a constant temperature or constant heat flux condition on its bottom boundary. This model requires that heat be supplied convectively to the bottom of the plate to maintain the constant thickness and heat flow that are attained with age.

A compilation of bathymetry and heat flow data by Parsons & Sclater (1977) showed that, in general, older seafloor departed from the linear depth/square root of time relationship, asymptotically approaching constant depth, and heat flow curves also seemed to flatten. These observations support the plate model as more applicable than the thermal boundary layer approximation. Heestand & Crough (1981) countered in favor of the half-space model by suggesting that flattening of the seafloor at greater age was due to contamination of the depth-age relationship by hotspot-caused bathymetric swells representing later thermal reactivation of the lithosphere (Crough 1978, 1979b, Detrick & Crough 1978). Actually, this position is already fairly close to the plate model philosophically, as it requires convective heat flux into the bottom of the plate to raise the hotspot swells. The heat flux is just grainier than originally conceived for the plate model.

A more sensitive and robust test of whether the oceanic lithosphere continues to cool indefinitely can be obtained from the geoidal signature of fracture zones of large offset (Detrick 1981). Initially, there will be a step offset in the geoid across the fracture zone proportional to the age difference across the fracture zone (Crough 1979a). This step, which is a short-wavelength feature with a simple form, is relatively easy to distinguish from interfering anomalies of other origin (Sandwell & Schubert 1982). The thermal boundary layer model predicts that the size of the step will not decay as the crust enclosing the fracture zone ages, because both sides cool and become denser in parallel. The plate model reaches a constant temperature structure independent of age on both sides if enough time has elapsed, so the initial geoid offset across the fracture zone should decay with age. Detrick (1981) observed just such a decay for the Mendocino fracture zone; this was later confirmed by Sandwell & Schubert (1982) and extended to other northeast Pacific fracture zones by Cazenave et al (1982). Contamination by hotspot tracks is not a serious possibility to explain these short-wavelength observations.

Evidence from the geoid thus seems to weigh heavily in favor of the plate thermal model for the oceanic lithosphere. Gravity observations also are being used to constrain the manner in which convective heat can be delivered to the base of the lithosphere.

Mid-Plate Convection

If indeed hotspot upwellings are not sufficient by themselves to supply the extra heat at the bottom of the oceanic lithosphere, a small-scale convection of some kind is required in the upper mantle (Richter 1973, McKenzie & Weiss 1975). Richter & Parsons (1975) suggested that this convection might take the form of rolls in the direction of plate motion over the lower mantle, stabilized by the shear involved, although the convection should initially be three dimensional and might not have sufficient time to become organized into rolls under any but the fastest-moving plates (Parsons & McKenzie 1978). The question then follows, Is there direct evidence for the nature of small-scale convection under the plates?

Marsh & Marsh (1976) claimed to have detected free-air gravity anomalies in the central Pacific that were elongated in the direction of motion of the Pacific plate relative to the Hawaiian hotspot and might therefore represent the second scale of convection. This interpretation was challenged by Phillips & Ivins (1979) on the basis of insufficient reliability of the satellite harmonic solution GEM-8 (Wagner et al 1977), although satellite-to-satellite tracking results seemed to show the anomalies also (Marsh et al 1981).

Gravity and geoid anomalies alone are not enough to solve the problem of the small-scale convection, however. McKenzie et al (1980) pointed out that bathymetry in the form of residual depth anomalies must also be considered. They found a correlation of bathymetry and geoid anomalies in the central Pacific that suggests irregular rolls, more three dimensional than two dimensional in form, but somewhat elongated in the direction of Pacific/hotspot motion. The transverse wavelength of these features is about 1500 km. As they expected a spacing of roughly twice the thickness of the convecting layer on theoretical grounds, their conclusion was that the small-scale convection does not extend throughout the entire mantle.

CONCLUSIONS

Besides the central conclusion that the geoidal representation of the Earth's gravity field is proving a most fruitful tool for investigating a variety of geologic problems, several statements that seem to me to be highly probable stand out:

1. The excess densities of the subducting slabs are in general regionally compensated, and it is most likely that this compensation involves support from below in the form of resistance, not necessarily successful, to their further penetration into the mantle.
2. The Earth's gravity field is dominated by a large, deep, slowly changing

convective pattern that is not closely linked to present-day upper-mantle convection in the form of plate tectonics, but that is related to the source of hotspots and possibly to past plate boundary configurations.

3. Use of the geoid has not settled many of the long-standing problems in crustal isostasy, but it promises to help do so in the future.
4. Geoidal anomalies across fracture zones confirm the validity of plate models for the thermal evolution of oceanic lithosphere and require small-scale convection in the upper mantle to provide heat to the base of the plates.

Future directions for research using the magnificently detailed global geoids measured by satellites will be many. "Pseudotopography" studies that map geoidal undulations into equivalent topographic features (Haxby et al 1983, Sandwell 1984) will fill in poorly studied areas and suggest new possibilities. To date, the resolution of the marine geoid enabled by satellite radar altimetry of the sea surface has guaranteed an emphasis on oceanic problems. Similar detail and resolution must be attained over continental areas to develop fully the potential of geoidal studies.

ACKNOWLEDGMENTS

I thank R. Rapp and B. Hager for generating and providing illustrations.

Literature Cited

Anderson, D. L. 1982. Hotspots, polar wander, Mesozoic convection, and the geoid. *Nature* 297: 391–93

Bott, M. H. P. 1971. The mantle transition zone as possible source of global gravity anomalies. *Earth Planet. Sci. Lett.* 11: 28–34

Cazenave, A., Lago, B., Dominh, K., Lambeck, K. 1980. On the response of the ocean lithosphere to sea-mount loads from *Geos-3* satellite radar altimeter observations. *Geophys. J. R. Astron. Soc.* 63: 233–52

Cazenave, A., Lago, B., Dominh, K. 1982. Geoid anomalies over the northeast Pacific fracture zones from satellite altimeter data. *Geophys. J. R. Astron. Soc.* 69: 15–31

Cavenave, A., Lago, B., Dominh, K. 1983. Thermal parameters of the oceanic lithosphere estimated from geoid height data. *J. Geophys. Res.* 88: 1105–18

Chapman, M. E., Talwani, M. 1982. Geoid anomalies over deep sea trenches. *Geophys. J. R. Astron. Soc.* 68: 349–69

Chase, C. G. 1979. Subduction, the geoid, and lower mantle convection. *Nature* 282: 464–68

Chase, C. G., McNutt, M. K. 1982. The geoid: effect of compensated topography and uncompensated oceanic trenches. *Geophys. Res. Lett.* 9: 29–32

Chase, C. G., Sprowl, D. R. 1983. The modern geoid and ancient plate boundaries. *Earth Planet. Sci. Lett.* 62: 314–20

Clayton, R. W., Dziewonski, A. M. 1984. Lateral variations in lower mantle velocities determined from travel time data. *Eos, Trans. Am. Geophys. Union* 65: 271 (Abstr.)

Creager, K. C., Jordan, T. H. 1984. Slab penetration into the lower mantle. *J. Geophys. Res.* 89: 3031–50

Crough, S. T. 1978. Thermal origin of mid-plate hot-spot swells. *Geophys. J. R. Astron. Soc.* 55: 451–69

Crough, S. T. 1979a. Geoid anomalies across fracture zones and the thickness of the lithosphere. *Earth Planet. Sci. Lett.* 44: 224–30

Crough, S. T. 1979b. Hotspot epeirogeny. *Tectonophysics* 61: 321–33

Crough, S. T. 1982. Geoid height anomalies over the Cape Verde Rise. *Mar. Geophys. Res.* 5:263–71
Crough, S. T., Jarrard, R. D. 1981. The Marquesas-Line swell. *J. Geophys. Res.* 86:11763–71
Crough, S. T., Jurdy, D. M. 1980. Subducted lithosphere, hotspots, and the geoid. *Earth Planet. Sci. Lett.* 48:15–22
Dahlen, F. A. 1982. Isostatic geoid anomalies on a sphere. *J. Geophys. Res.* 87:3943–47
Davies, G. F. 1981. Regional compensation of subducted lithosphere: effects on geoid, gravity, and topography from a preliminary model. *Earth Planet. Sci. Lett.* 54: 431–41
Detrick, R. S. 1981. An analysis of geoid anomalies across the Mendocino fracture zone: implications for thermal models of the lithosphere. *J. Geophys. Res.* 86: 11751–62
Detrick, R. S., Crough, S. T. 1978. Island subsidence, hot spots, and lithospheric thinning. *J. Geophys. Res.* 83:1236–44
Dziewonski, A. 1984. Mapping the lower mantle: determination of lateral heterogeneity in P velocity up to degree and order 6. *J. Geophys. Res.* 89:5929–52
Dziewonski, A., Hager, B. H., O'Connell, R. J. 1977. Large scale heterogeneities in the lower mantle. *J. Geophys. Res.* 82:239–55
Goldreich, P., Toomre, A. 1969. Some remarks on polar wandering. *J. Geophys. Res.* 74:2555–67
Gordon, R. G. 1982. The late Maastrichtian paleomagnetic pole of the Pacific plate. *Geophys. J. R. Astron. Soc.* 70:129–40
Gordon, R. G. 1983. Late Cretaceous apparent polar wander of the Pacific plate: evidence for rapid shift of the Pacific hotspots with respect to the spin axis. *Geophys. Res. Lett.* 10:709–12
Gordon, R. G., Cape, C. D. 1981. Cenozoic latitudinal shift of the Hawaiian hotspot and its implications for true polar wander. *Earth Planet. Sci. Lett.* 81:37–47
Grow, J. A., Bowin, C. O. 1975. Evidence for high density crust and mantle beneath the Chile trench due to the descending lithosphere. *J. Geophys. Res.* 80:1449–58
Hager, B. H. 1983. Global isostatic geoid anomalies for plate and boundary layer models of the lithosphere. *Earth Planet. Sci. Lett.* 63:97–109
Hager, B. H. 1984. Subducted slabs and the geoid: constraints on mantle rheology and flow. *J. Geophys. Res.* 89:6003–16
Hager, B. H., Richards, M. A. 1984. The source of long wavelength gravity anomalies. *Eos, Trans. Am. Geophys. Union* 65:271 (Abstr.)
Haxby, W. F., Turcotte, D. L. 1978. On isostatic geoid anomalies. *J. Geophys. Res.* 83:5473–78
Haxby, W. F., Karner, G. D., LaBrecque, J. L., Weissel, J. K. 1983. Digital images of combined oceanic and continental data sets and their use in tectonic studies. *Eos, Trans. Am. Geophys. Union* 64:995–1004
Heestand, R. L., Crough, S. T. 1981. The effect of hot spots on the oceanic age-depth relation. *J. Geophys. Res.* 86:6107–14
Hide, R., Horai, K.-I. 1968. On the topography of the core-mantle interface. *Phys. Earth Planet. Inter.* 1:305–8
Higbie, J., Stacey, F. D. 1970. Depth of density variations responsible for features of the satellite geoid. *Phys. Earth Planet. Inter.* 4:145–48
Higbie, J. W., Stacey, F. D. 1971. Interpretation of global gravity anomalies. *Nature Phys. Sci.* 234:130–32
Jordan, T. H. 1977. Lithospheric slab penetration into the lower mantle beneath the Sea of Okhotsk. *J. Geophys.* 43:473–96
Jurdy, D. M. 1981. True polar wander. *Tectonophysics* 74:1–16
Jurdy, D. M. 1983. Early Tertiary subduction zones and hot spots. *J. Geophys. Res.* 88:6395–402
Kaula, W. M. 1967. Geophysical implications of satellite determinations of the Earth's gravitational field. *Space Sci. Rev.* 7:769–94
Kaula, W. M. 1972. Global gravity and tectonics. In *The Nature of the Solid Earth*, ed. E. C. Robertson, pp. 385–405. New York: McGraw-Hill. 677 pp.
Khan, M. A. 1977. Depth of sources of gravity anomalies. *Geophys. J. R. Astron. Soc.* 48:197–209
Kipling, R. 1970. In the Neolithic age. In *The Collected Works of Rudyard Kipling*, 26: 86–87. New York: AMS Press
Lambeck, K. 1976. Lateral density anomalies in the upper mantle. *J. Geophys. Res.* 81: 6333–40
Lambeck, K. 1981. Lithospheric response to volcanic loading in the Southern Cook Islands. *Earth Planet. Sci. Lett.* 55:482–96
Langseth, M. G., Le Pichon, X., Ewing, M. 1966. Crustal structure of mid-ocean ridges, 5. Heat flow through the Atlantic Ocean floor and convection currents. *J. Geophys. Res.* 71:5321–55
Lerch, F. J., Klosko, S. M., Laubscher, R. E., Wagner, C. A. 1979. Gravity model improvement using *Geos-3* (GEM 9 and GEM 10). *J. Geophys. Res.* 84:3897–3916
Lister, C. R. B. 1982. Geoid anomalies over cooling lithosphere: source of a third kernel of upper mantle thermal parameters and thus an inversion. *Geophys. J. R. Astron. Soc.* 68:219–40

Marsh, B. D., Marsh, J. G. 1976. On global gravity anomalies and two-scale mantle convection. *J. Geophys. Res.* 81:5267–80
Marsh, J. G., Marsh, B. D., Williamson, R. G., Wells, W. T. 1981. The gravity field in the central Pacific from satellite-to-satellite tracking. *J. Geophys. Res.* 86: 3979–97
McAdoo, D. C. 1981. Geoid anomalies in the vicinity of subduction zones. *J. Geophys. Res.* 86:6073–90
McAdoo, D. C. 1982. On the compensation of geoid anomalies due to subducting slabs. *J. Geophys. Res.* 87:8684–92
McKenzie, D. P. 1967. Some remarks on heat flow and gravity anomalies. *J. Geophys. Res.* 72:6261–73
McKenzie, D. P. 1969. Speculations on the consequences and causes of plate motions. *Geophys. J. R. Astron. Soc.* 18:1–32
McKenzie, D. P. 1977. Surface deformation, gravity anomalies, and convection. *Geophys. J. R. Astron. Soc.* 48:211–38
McKenzie, D. P., Weiss, N. 1975. Speculations on the thermal and tectonic history of the Earth. *Geophys. J. R. Astron. Soc.* 42:131–74
McKenzie, D. P., Watts, A. B., Parsons, B., Roufosse, M. 1980. Planform of mantle convection beneath the Pacific Ocean. *Nature* 288:442–46
McQueen, H. W. S., Stacey, F. D. 1976. Interpretation of low-degree components of gravitational potential in terms of undulations of mantle phase boundaries. *Tectonophysics* 34:1–8
Menard, H. W. 1984. Darwin reprise. *J. Geophys. Res.* In press
Morgan, W. J. 1965. Gravity anomalies and convection currents 1. A sphere and cylinder sinking beneath the surface of a viscous fluid. *J. Geophys. Res.* 70:6175–85
Morgan, W. J. 1972. Deep mantle convection plumes and plate motions. *Am. Assoc. Pet. Geol. Bull.* 56:203–13
Morgan, W. J. 1981. Hotspot tracks and the opening of the Atlantic and Indian Oceans. In *The Sea*, ed. C. Emiliani, 7:443–87. New York: Wiley
Nakiboglu, S. M. 1982. Hydrostatic theory of the Earth and its mechanical implications. *Phys. Earth Planet. Inter.* 28:302–11
Ockendon, J. R., Turcotte, D. L. 1977. On the gravitational potential and field anomalies due to thin mass layers. *Geophys. J. R. Astron. Soc.* 48:479–92
O'Keefe, J. A., Kaula, W. M. 1963. Stress differences and the reference ellipsoid. *Science* 142:382
Parker, R. L., Oldenburg, D. W. 1973. Thermal model of ocean ridges. *Nature Phys. Sci.* 242:137–39
Parsons, B., Daly, S. 1983. The relationship between surface topography, gravity anomalies, and temperature structure of convection. *J. Geophys. Res.* 88:1129–44
Parsons, B., McKenzie, D. P. 1978. Mantle convection and the thermal structure of plates. *J. Geophys. Res.* 83:4485–96
Parsons, B., Richter, F. M. 1980. A relation between the driving force and geoid anomaly associated with mid-ocean ridges. *Earth Planet. Sci. Lett.* 51:445–50
Parsons, B., Sclater, J. G. 1977. An analysis of the variation of ocean floor bathymetry and heat flow with age. *J. Geophys. Res.* 82:803–27
Pekeris, C. L. 1935. Thermal convection in the interior of the Earth. *Mon. Not. R. Astron. Soc. Geophys. Suppl.* 3:343–67
Phillips, R. J., Ivins, E. R. 1979. Geophysical observations pertaining to solid-state convection in the terrestrial planets. *Phys. Earth Planet. Inter.* 19:107–48
Phillips, R. J., Lambeck, K. 1980. Gravity fields of the terrestrial planets: long-wavelength anomalies and tectonics. *Rev. Geophys. Space Phys.* 18:27–76
Rapp, R. H. 1981. The Earth's gravity field to degree and order 180 using *Seasat* altimeter data, terrestrial gravity data, and other data. *Ohio St. Univ. Dept. Geod. Sci. Surv. Rep. 322*
Richards, M. A., Hager, B. H. 1984. Geoid anomalies in a dynamic Earth. *J. Geophys. Res.* 89:5987–6002
Richter, F. M. 1973. Convection and large-scale circulation of the mantle. *J. Geophys. Res.* 78:8735–45
Richter, F. M., Parsons, B. 1975. On the interaction of two scales of convection in the mantle. *J. Geophys. Res.* 80:2529–41
Runcorn, S. K. 1964. Satellite gravity measurements and a laminar viscous flow model of the Earth's mantle. *J. Geophys. Res.* 69:4389–94
Runcorn, S. K. 1967. Flow in the mantle inferred from the low degree harmonics of the geopotential. *Geophys. J. R. Astron. Soc.* 14:375–84
Sandwell, D. T. 1984. A detailed view of the South Pacific geoid from satellite altimetry. *J. Geophys. Res.* 89:1089–1104
Sandwell, D. T., Schubert, G. 1980. Geoid height versus age for symmetric spreading ridges. *J. Geophys. Res.* 85:7235–41
Sandwell, D. T., Schubert, G. 1982. Geoid height-age relation from *Seasat* altimeter profiles across the Mendocino fracture zone. *J. Geophys. Res.* 87:3949–58
Soriau, A. 1984. Geoid anomalies over Gorringe Ridge, North Atlantic Ocean. *Earth Planet. Sci. Lett.* 68:101–14
Turcotte, D. L., McAdoo, D. C. 1979. Geoid

anomalies and the thickness of the lithosphere. *J. Geophys. Res.* 84:2381–87

Turcotte, D. L., Oxburgh, E. R. 1967. Finite amplitude convective cells and continental drift. *J. Fluid Mech.* 28:29–42

Wagner, C. A., Lerch, F. J., Brownd, J. E., Richardson, J. A. 1977. Improvement in the geopotential derived from satellite and surface data (GEM 7 and GEM 8). *J. Geophys. Res.* 82:901–14

Watts, A. B. 1979. On geoid heights derived from *Geos-3* altimeter data along the Hawaiian-Emperor seamount chain. *J. Geophys. Res.* 84:3817–26

Watts, A. B., Daly, S. F. 1981. Long wavelength gravity and topography anomalies. *Ann. Rev. Earth Planet. Sci.* 9:415–48

Watts, A. B., Talwani, M. 1974. Gravity anomalies seaward of deep-sea trenches and their tectonic implications. *Geophys. J. R. Astron. Soc.* 36:57–90

Watts, A. B., Talwani, M. 1975. Gravity effect of downgoing lithospheric slabs beneath island arcs. *Geol. Soc. Am. Bull.* 86:1–4

Ann. Rev. Earth Planet. Sci. 1985. 13 : 119–46

DIRECT TEM IMAGING OF COMPLEX STRUCTURES AND DEFECTS IN SILICATES

David R. Veblen

Department of Earth and Planetary Sciences, The Johns Hopkins University, Baltimore, Maryland 21218

INTRODUCTION

Seventy years ago, the X-ray diffraction determination of the NaCl crystal structure by W. H. and W. L. Bragg ushered in a new age in the science of mineralogy. Over the following 50 years, the structures of most minerals were determined, and the ideal crystal structures of most of the important rock-forming minerals have now been refined to high levels of precision. It is perhaps only in the last 20 years or so that the importance of the nonideal, nonperiodic aspects of mineral structures has been recognized. These nonperiodic features, or defects, in many cases can control important properties of a mineral (such as color, mechanical properties, and intracrystalline diffusion rates), and they may play important roles in solid-state reaction processes. In the present paper, I limit the discussion to one- and two-dimensional defects, which are sometimes called *extended defects.*

While X-ray and neutron diffraction are ideal tools for the determination of periodic structures with relatively small unit cells, their application in studies of defects is limited to the determination of average defect densities in highly disordered materials. One powerful method that can be used to investigate the structures of extended defects in minerals is transmission electron microscopy (TEM). Although the transmission electron microscope was developed during the 1930s, the intensive application of TEM to the study of defects in minerals did not occur until the early 1970s, when a number of important investigations, many of them on lunar minerals, were performed.

0084–6597/85/0515–0119$02.00

Just as with defect studies, structural studies of minerals with extremely large unit cells and minerals that typically occur finely intergrown with other structures can be difficult with X-ray diffraction methods. Electron microscopy and electron diffraction again can provide powerful means for studying such structures that are intractable with more traditional diffraction techniques.

One TEM technique that has been very successful in the elucidation of both extended defect structures and the structures of long-period materials is the high-resolution TEM method (also called lattice imaging, structure imaging, and multiple-beam bright-field imaging), which was first applied to minerals about ten years ago (see, for example, Iijima et al 1973, Buseck & Iijima 1974). The HRTEM method allows imaging of real features in crystals (e.g. atomic clusters or columns of heavy atoms) at a resolution of about 0.3 nm for many modern instruments and better than 0.2 nm for special high-voltage instruments that are dedicated to high-resolution studies.

This paper reviews some of the contributions that HRTEM studies have made to silicate mineralogy. I first discuss briefly the principles of HRTEM. Different types of defects and long-period ordered structures are then described, followed by a survey of HRTEM results on specific silicate groups.

THEORY OF HRTEM

In this section, a brief qualitative discussion of diffraction in the electron microscope and how HRTEM works is presented. For more detailed discussions of the theoretical basis of HRTEM, the reader is referred to Cowley (1975, 1979), Cowley & Iijima (1976), and Spence (1981).

The most important lens of the TEM is the objective lens, a magnetic lens that is optically analogous to the objective lens of an ordinary light microscope or a simple hand lens as used by a field geologist. Just as in a petrographic microscope, the illuminating radiation passes through a thin specimen and through the objective lens. This optical geometry is shown schematically in Figure 1. (There are differences, however, between light and electron microscopes. For HRTEM, specimen thickness may be less than 10 nm, whereas a petrographic thin section is typically 3000 times thicker; and a TEM specimen generally lies within the magnetic field of the objective lens, rather than entirely before the objective.)

Just as in the case of coherent light optics, a diffraction pattern of the specimen is formed in the back focal plane of the objective lens, and an inverted two-dimensional image of the object is formed in the first image plane. By varying the excitations of the objective lens and several other

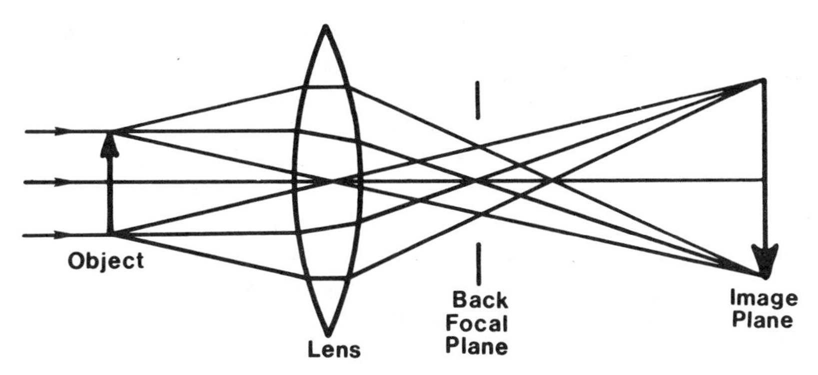

Figure 1 Ideal imaging system. Radiation passes through the specimen ("object") and through the objective lens. A diffraction pattern forms in the back focal plane, and an inverted image of the specimen forms in the image plane.

magnetic lenses that follow the objective, it is possible to project either the diffraction pattern or the image onto a fluorescent viewing screen or sheet of film and to vary the magnifications of either the diffraction pattern or the image. The ultimate magnifications of modern high-resolution microscopes may exceed 1,000,000 times, although most high-resolution work is still performed at lesser magnifications.

Diffraction Modes

There are a number of different types of diffraction patterns that can be formed in a modern TEM, which depend primarily on the geometry of the illuminating electron beam. Of these, I mention three: selected-area electron diffraction (SAED), convergent-beam electron diffraction (CBED), and electron microdiffraction. In SAED, an aperture is placed in the first image plane of the objective lens. This aperture allows only information from that part of the specimen within the aperture to pass through to the viewing screen, so that the diffraction pattern is effectively that of the area selected by the microscopist. Roughly parallel radiation is used to illuminate the specimen, and the SAED patterns can be used to obtain lattice parameter and Laue group symmetry information. Diffraction information can be obtained on areas as small as approximately 0.5 μm.

In CBED, the specimen is illuminated with a solid cone of converging electrons, which is focused on an area that typically is a few tens of nanometers in diameter. The details of the CBED technique are reviewed by Steeds (1979) and are not discussed here. However, CBED, like SAED, can be a useful adjunct to HRTEM studies, in that CBED can detect minor changes in orientation over very small specimen areas and can also be used under favorable circumstances to detect space-group symmetry elements (screws, glides), including centers of inversion. Like CBED, electron microdiffraction can be used to obtain information on very small specimen areas, i.e. a few nanometers in diameter or smaller. The technique employs roughly parallel illumination that is focused into an extremely fine probe.

The above diffraction modes primarily utilize the geometry of diffraction patterns, rather than the diffracted intensities. For geologists and crystallographers who are accustomed to using X-ray intensity information in a quantitative fashion, it must be stressed that electron diffraction intensities are affected strongly by multiple scattering effects. In general, it is not possible to use electron diffraction intensities as one would use X-ray intensities (e.g. for crystal structure refinement). Furthermore, except in special cases, it is not possible to use the presence of electron diffractions to infer violations of screw axes or glide planes (Gjonnes & Moodie 1965). The mineralogical literature contains at least several examples of space-group assignments that are probably incorrect as a result of inappropriate use of electron diffraction data.

High-Resolution Imaging (HRTEM)

In a perfect optical system with no lens aberrations, as in Figure 1, a Fraunhofer diffraction pattern, which is the two-dimensional Fourier transform of the specimen, appears in the back focal plane. Likewise, the image that forms in the first image plane corresponds to the Fourier transform of the diffraction pattern and is thus a faithful representation of the original specimen. In a perfect electron microscope, with a vanishingly thin specimen, an image recorded under appropriate instrumental conditions could indeed reliably show the structure of a crystalline specimen. There are, however, two major factors that prevent such ideal imaging.

First, TEM specimens always have finite thickness. Unlike the case for X-ray diffraction, where single scattering generally is a good assumption (kinematical diffraction), the interaction of electrons with matter is strong enough so that electrons commonly are multiply scattered, even in relatively thin specimens (dynamical diffraction). The multiple scattering of electrons in a specimen perturbs both the phases and the amplitudes of diffracted beams, so that the diffraction pattern that forms in the back focal plane is not simply the Fourier transform of the specimen. Consequently, an image formed by Fourier transformation of this diffraction pattern will not necessarily be a simple representation of the specimen structure.

The other barrier to obtaining perfect images of the specimen in HRTEM is defects in the electron microscope, especially spherical aberration of the objective lens. This lens introduces an additional phase shift to the diffracted beams. The phase shift alters the amplitude distribution in the back focal plane by a factor $\exp(i\chi)$, the phase factor χ being given by

$$\chi = \pi\Delta f\lambda(u^2+v^2)+\tfrac{1}{2}\pi C_s\lambda^3(u^2+v^2)^2, \tag{1}$$

where π is the angle 180°, λ is the electron wavelength, Δf is the amount that the microscope focus setting differs from perfect (Gaussian) focus, C_s is a measure of the spherical aberration of the objective lens, and u, v are the Cartesian coordinates of points in the back focal plane (i.e. in the diffraction pattern). In addition to focus defect and spherical aberration of the objective lens, images are affected by imperfect coherency of the electron beam and smearing due to chromatic aberration and effects such as mechanical vibrations, instabilities in the imaging lens currents, and the fact that the specimen is never illuminated with parallel radiation, but rather by a somewhat convergent beam.

With so many factors affecting HRTEM images, it is clear that this type of microscopy should not be thought of simply as taking pictures of the crystal structure of the specimen. Indeed, by varying specimen thickness and microscope parameters such as defect of focus Δf, it is possible to obtain a tremendous range of different images from the same structure! Luckily for high-resolution microscopists, there is a narrow range of instrumental

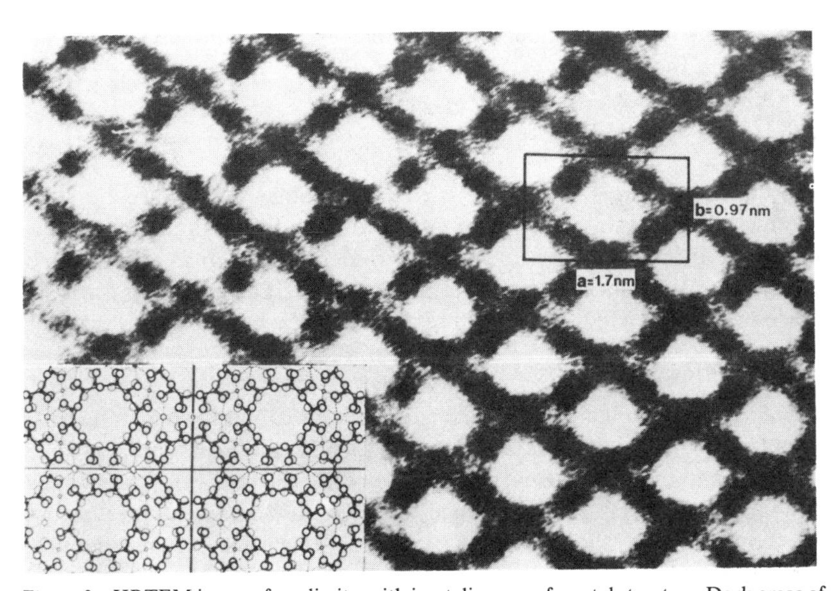

Figure 2 HRTEM image of cordierite, with inset diagram of crystal structure. Dark areas of the image correspond to regions of high atomic density, while light areas correspond to tunnels with low atomic density. The unit cell is outlined (Buseck & Iijima 1974; reproduced with permission from *The American Mineralogist*).

settings over which one can obtain HRTEM images that may look somewhat like the projected atomic density of the crystal. (These are sometimes referred to as structure images.) Specifically, from the phase factor χ (Equation 1) it is clear that a negative value of the defocus setting Δf will partially offset the effects of the spherical aberration coefficient C_s, and there is a focus called the optimum defocus or Scherzer focus (Scherzer 1949) at which differences in phase shift for different diffracted beams are minimized. Images obtained at optimum defocus may be interpretable intuitively in terms of the projected structure of the specimen, as in the image of a cordierite crystal shown in Figure 2 (Buseck & Iijima 1974).

HRTEM images of crystals clearly are dependent on a wide variety of factors, and their interpretation is not trivial. It is vitally important for correct image interpretation that all experimental parameters be controlled, and this requires great attention to experimental technique. If, however, all of the relevant variables, such as lens aberration coefficients, defocus, specimen thickness, etc, are known, then it is possible to calculate images from model structures using computer programs that mimic the imaging process. Such simulated images have become an important tool in the interpretation of HRTEM images, since the ability to reproduce an observed image from a structural model shows that the interpretation is consistent with the experiment. An example of a computer-simulated image is shown in Figure 7.

DEFECT TYPES

Defects are features that violate the long-range periodicity or other symmetry of a crystal. They commonly are classified by dimensionality as point (zero-dimensional), linear (one-dimensional), and planar (two-dimensional) defects. Defects are important in the Earth sciences for many reasons, some of which have been reviewed briefly by Veblen (1985a).

Point Defects

Point defects, such as vacancies (sites where atoms are missing in a structure), can be important determinants of physical and transport properties in minerals. However, with only minor exceptions (see Veblen 1985a), point defects are not imaged with HRTEM and hence are beyond the scope of this article. For reviews on point defects, see Lasaga (1981) and Kröger (1985).

Linear Defects

Linear defects, including various types of dislocations, involve displacements of a crystal structure and resulting distortion localized about a line;

they can play a crucial role in deformation and crystal growth processes. While dislocations can be observed with HRTEM techniques, they have been much more extensively studied with conventional TEM methods and therefore are not covered further in this review. It should be noted, however, that HRTEM holds great promise for the determination of the detailed atomic positions in and near the cores of dislocations, as evidenced by recent work on nonmineral materials. For a thorough discussion of dislocations, the reader is referred to Hirth & Lothe (1982).

Planar Defects

There are a number of different types of planar defects, all of which involve displacements or other violations of crystal symmetry along a two-dimensional surface. Some two-dimensional defects commonly describe complex surfaces; though not strictly planar, these can be lumped with truly planar defects for simplicity. Types of planar defects include stacking faults, Wadsley defects, twin surfaces, and antiphase boundaries. Of all defect types, HRTEM has made its greatest contributions to the understanding of planar defects, and this article reviews much of this work as applied to silicate minerals.

STACKING FAULTS Defects that are truly planar and can be described by a displacement of the structure that lies within the fault plane are called stacking faults. They are common in, though by no means restricted to, structures that can be described in terms of closest packing. Stacking faults can be formed during crystal growth, during deformation, or during polymorphic transformations. Among rock-forming minerals, they are especially important in the chain and sheet silicates.

WADSLEY DEFECTS Faults that are truly planar and can be described by a displacement vector that lies out of the fault plane are called Wadsley defects, intercalation defects, or polysomatic defects. These defects are of particular interest because the structure in the immedate vicinity of the fault plane is chemically distinct from the bulk structure. The presence of Wadsley defects therefore results in nonstoichiometry of the host crystal. These defects form during crystal growth or reactions involving chemical transport.

TWIN PLANES Twin planes (and more complex twinning surfaces) are boundaries across which there is an error in a point symmetry element of the crystal. This error can be in a rotational symmetry element, a mirror, or an inversion. There is a strict geometrical relationship between the structure on the two sides of the defect. Twin planes form during growth, deformation, or phase transformation.

ANTIPHASE BOUNDARIES A surface across which there is an error in the translational symmetry of the crystal is called an antiphase boundary (APB), and the parts of the crystal separated by the APB are called antiphase domains (APDs). The displacement vector between the two APDs is generally half that of a normal lattice translation. (Similar boundaries with other displacements can be called out-of-phase boundaries.) APBs can form during crystal growth but are probably most frequently formed during phase transformations in which a translational symmetry element is lost.

LONG-PERIOD AND DISORDERED STRUCTURES

A superstructure has a unit cell that can be described as a multiple of the unit cell of a more basic structure. When the superstructure unit cell (supercell) is an integral multiple of the subcell, the superstructure is called commensurate. In cases where a nonintegral relationship exists between the superstructure and the substructure, the superstructure is called incommensurate. HRTEM and electron diffraction have been used to study both types of superstructures, as well as other long-period structures and disordered structures lacking periodicity in one dimension.

Commensurate Long-Period Structures

Commensurate superstructures commonly form in a variety of minerals by ordering of cations. In addition, long-period commensurate structures can form in polytypic minerals (those that are formed by different stackings of roughly identical layers). For example, long-period polytypes in micas (Ross et al 1966) are long-period commensurate structures.

Incommensurate (Modulated) Structures

Incommensurate, or modulated, structures are those in which a basic structure is perturbed in some way so that a superstructure forms that is not integrally related to the unit cell of the basic structure. Types of perturbations include, but are not restricted to, static atomic displacements, modulations of conduction-band electron density, time-variant thermal displacements of atoms, and imperfect long-range atomic ordering in which the average supercell is not an integral multiple of the subcell. Applications of HRTEM to modulated structures, plus a few long-period commensurate structures, have been reviewed by Buseck & Cowley (1983), and the theoretical basis of modulated structures is discussed in several papers in Cowley et al (1979) and by McConnell (1983).

Structurally Disordered and Finely Intergrown Minerals

Many crystal structures can be thought of as consisting of two or more structurally distinct slabs that intergrow in a regular, ordered fashion. Such structures have been called *polysomatic* by Thompson (1978). While most polysomatic structures are usually well ordered, it is also possible to combine the slabs in a less-ordered, or even random, sequence. Such structurally disordered minerals are especially well suited for HRTEM study, since this method can in many cases determine the exact disordered structure, which is not accessible to X-ray methods. Similarly, some ordered structures, especially polysomatic ones, sometimes occur finely intergrown with other structures. HRTEM again provides a useful tool for determining the exact structures of such intergrowths.

SURVEY OF HRTEM STUDIES ON SILICATE MINERALS

This section briefly reviews the contributions of direct-imaging studies to silicate mineralogy. While it is not intended to be exhaustive, this survey does reference a relatively large number of the HRTEM studies on silicates that have been published, emphasizing more recent and review contributions that in turn refer to earlier work. The discussion is organized according to structural and chemical type. Other reviews emphasizing the diverse contributions of HRTEM to mineralogy include those by Hutchison et al (1977), Thomas et al (1979), Buseck & Veblen (1981), and Buseck (1984). In addition, there are a number of review papers on specific structure types: Thomas (1984; sheet silicates), Buseck et al (1980; pyroxenes), Veblen (1981; amphiboles and other hydrous pyriboles), and Chisholm (1983; asbestos minerals).

The reader will note that for some minerals and mineral groups, such as some of the island silicates, there has been little or no HRTEM work published. On the other hand, there have been numerous studies on some of the other groups, such as the chain and sheet silicates. This difference is not only a reflection of research interests (rock-forming minerals tend to receive the most attention), but it is also very largely the result of the fact that some minerals typically do not contain observable densities of planar defects, while other groups are quite prone to the types of structural disorder and intergrowth phenomena that can be studied with HRTEM methods. Because the publication of negative results is not very glamorous, TEM studies in which a mineral is found to be more-or-less perfect generally are not reported. In my opinion, however, it is an important observation that some structures are prone to perfection, others to imperfection.

Island Silicates (Nesosilicates)

OLIVINE Typical rock-forming olivines have been examined by a number of workers (White & Hyde 1983; the present author; L. Pierce, P. P. K. Smith, personal communications). Other than dislocations, no extended defects have been observed. As a result, there have been no published HRTEM studies on olivine. It should be noted, however, that laihunite, an oxidized Fe-rich olivine-type mineral, does possess superstructures due to vacancy ordering, an abundance of planar defects, and numerous fine-scale magnetite lamellae produced during oxidation (Kitamura et al 1984); these microstructures suggest that some more normal olivines might be susceptible to similar disorder under appropriate conditions. In addition, the Mn-olivine tephroite can contain Wadsley defects that can be interpreted as intergrown slabs of humite-group structure (see the section "Humite Group and Leucophoenicite" below).

OLIVINE SPINEL AND SPINELLOIDS Between approximately 50 and 125 kbar, depending on its chemical composition, olivine transforms to a spinel structure or, for magnesian compositions, to a structure referred to as "modified spinel" or beta phase. The silicate spinel structure is an island silicate and occurs naturally in shocked meteorites as ringwoodite. The beta phase is one of a family of spinel-related structures known as spinelloids and is a sorosilicate, since it contains pairs of corner-sharing tetrahedra. (It is considered in this section because of its close relationships to the island silicate olivine structures.) The beta structure recently has been described as the mineral wadsleyite. Price (1983) has presented HRTEM results on wadsleyite from a meteorite that apparently formed by transformation from ringwoodite. He found that the beta structure contained numerous faults corresponding to intergrown spinel structure, with spinel slab widths corresponding to up to four unit cells.

Davies & Akaogi (1983) used HRTEM to study a group of three synthetic spinelloids in the system $NiAl_2O_4$–Ni_2SiO_4. These structures contain T_3O_{10} groups, T_2O_7 groups, and mixed TO_4 and T_2O_7 groups (where T refers to a tetrahedrally coordinated cation; again, these are not all strictly island silicates but are considered here because of the close relationships to olivine and spinel structures). Davies & Akaogi reported a wealth of complex intergrowth structures among these phases.

GARNET Although a number of different garnet specimens of different chemistries have been examined (Turner 1978; the present author), no structural disorder or planar defects have been observed with HRTEM. As a result, no HRTEM garnet studies have been published.

HUMITE GROUP AND LEUCOPHOENICITE The humite group forms a polysomatic series extending from norbergite through chondrodite, humite, and clinohumite to the Mg-olivine forsterite; all of these structures can be represented as consisting of ordered mixtures of two types of slabs, one with olivine structure and the other with norbergite structure (Thompson 1978). Müller & Wenk (1978) first reported defects corresponding to mistakes in the sequence of these slabs in both clinohumite and chondrodite. More recently, White & Hyde (1982a,b) have performed an extensive survey of more than 30 specimens of both the humite group and its Mn-rich counterparts. They found that most of the Mg-rich specimens contain no detectable structural disorder, although there are some exceptions. On the other hand, many of the Mn-rich specimens exhibit extensive disorder of the two types of slabs, not only in the humite-group minerals, but also in the Mn-olivine tephroite. It would appear that structural disorder of this type, though not always present, is quite common in the humite group.

Leucophoenicite is another island silicate that is closely related to the humite group. Again, White & Hyde (1983) have shown that leucophoenicite can contain numerous errors in the sequence of structural slabs, similar to the defects found in humite-group structures. An example of a disordered leucophoenicite crystal is shown in Figure 3.

CHLORITOID This mineral is an island silicate in which the tetrahedra are linked by Al, Fe, and Mg octahedra to form relatively strong layers. From X-ray diffraction studies it is known that chloritoid can occur in a one-layer, triclinic structure, as well as in two distinct two-layer, monoclinic structures. Extremely intimate intergrowth of these structures, as well as a variety of planar defects, has been demonstrated with HRTEM (Jefferson & Thomas 1978).

ALUMINOSILICATES Wenk (1980) investigated intergrowths of kyanite and staurolite with some high-resolution imaging, but he primarily used conventional TEM techniques. In addition to the intergrowths, he observed stacking faults, microtwins, and narrow intergrown lamellae of chlorite. With the higher resolutions available on modern TEM instruments, these kyanite-staurolite intergrowths perhaps should be reexamined to determine the interface structures more exactly. Veblen examined kyanite from Gassets, Vermont, and found no structural disorder, other than the twinning that is commonly observed in petrographic thin section, as well as deformation twins produced during sample preparation.

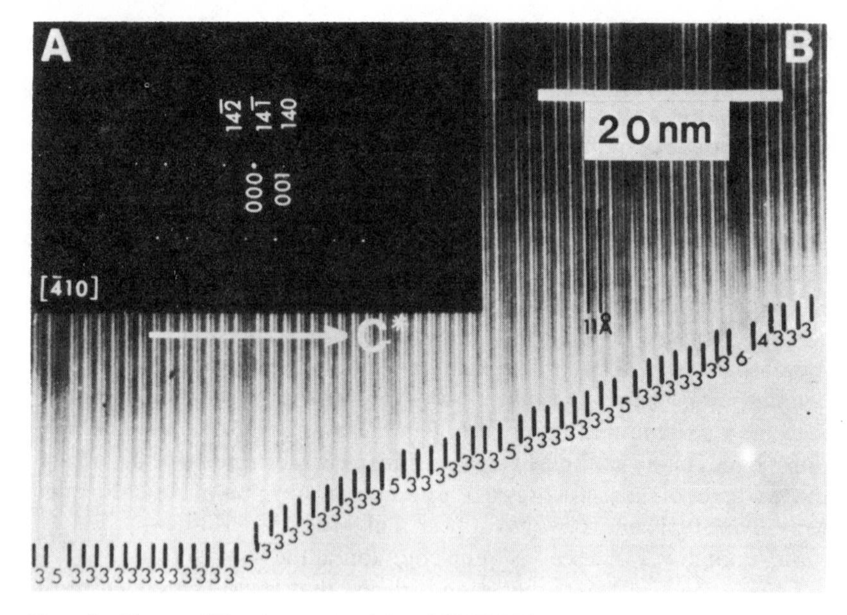

Figure 3 Electron diffraction pattern (*A*) and HRTEM image (*B*) of leucophoenicite. The normal structure contains 3 adjacent slabs having Ni_2In structure type. Faults where 4, 5, and 6 such slabs occur together are indicated (White & Hyde 1983; reproduced with permission from *The American Mineralogist*).

Nakajima et al (1975) studied a specimen of sillimanite with HRTEM and found no extended defects. Two specimens of sillimanite also have been examined by Veblen (reported in Rossman et al 1982). No planar defects were recognized, although exsolved needles of titaniferous hematite were observed in one of the specimens. There apparently have been no HRTEM studies of andalusite.

Mullite, a nonstoichiometric aluminosilicate containing varying numbers of oxygen vacancies, has been studied with HRTEM by Nakajima et al (1975) and Nakajima & Ribbe (1981a). Both the sillimanite-like substructure and a modulated, noncommensurate superstructure due to vacancy ordering have been imaged in these studies. In addition, in the latter study it was shown that extremely fine-scale twinning accounts for many of the diffraction effects previously observed for mullite. More recently, Ylä-Jääski & Nissen (1983) used HRTEM to show the compositional dependence of the orientations of periodic antiphase boundaries in mullite. They also presented simulated images that supported their interpretations.

Sorosilicates

VESUVIANITE Besides some of the spinelloids (see the section "Olivine Spinel and Spinelloids"), vesuvianite is the only sorosilicate (to my knowledge) for which a HRTEM study has been published, and vesuvianite is really a mixed silicate having both linked tetrahedral pairs and isolated tetrahedra. Buseck & Iijima (1974) presented structure images of vesuvianite, but no structural disorder was detected. O'Keefe et al (1978) later used computer-simulated images to show that the unit cell assigned by Buseck & Iijima (1974) was displaced by half of a unit cell translation from its correct position, an important demonstration of the utility of such image calculations.

Ring Silicates

TOURMALINE In one of the first published examples of structure imaging on a mineral, Iijima et al (1973) presented images of tourmaline similar to the one shown in Figure 4. Although no extended defects were observed, it was noted that there are variations in the image intensities of equivalent crystallographic sites. These variations were attributed tentatively to variations in the distribution of different atomic species on these sites.

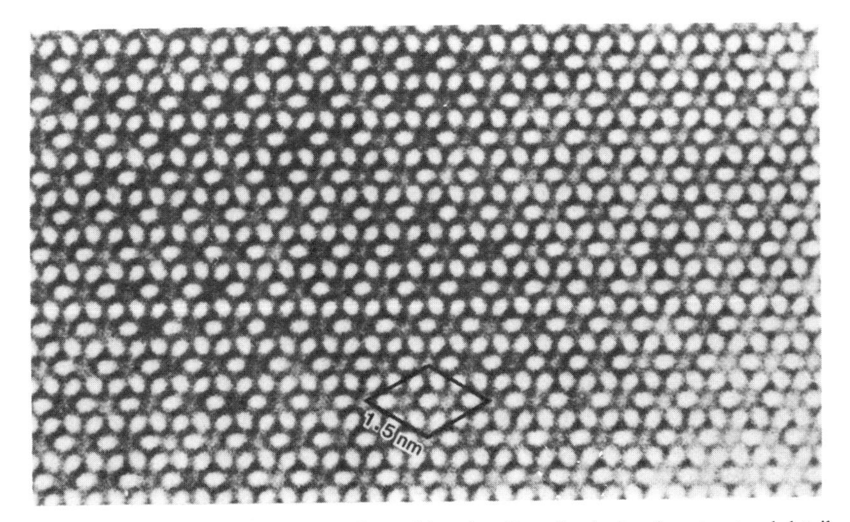

Figure 4 HRTEM image of tourmaline, with unit cell outlined, showing structural detail within one cell. Viewing at a low angle reveals nonperiodicity, with some corresponding parts of the image darker, possibly as a result of the nonuniform distribution of cations as discussed in text. (Reproduced courtesy of Sumio Iijima.)

BERYL Images of beryl viewed down the *c*-axis were presented by Buseck & Iijima (1974). Again, no defects were reported. S. Turner (personal communication) has observed regions of muscovite mica finely intergrown with beryl.

Chain Silicates

There have been more published HRTEM studies of the chain silicates (including the pyroxenes, amphiboles, wide-chain pyriboles, chain-width-disordered pyriboles, and pyroxenoids) than for any other mineral group. This is no doubt the case because the group includes many important rock-formers, but it is also because the group as a whole appears to be particularly prone to the types of structural disorder and intergrowths that are profitably studied with HRTEM.

PYROXENES High-resolution studies on pyroxenes have focused primarily on two distinct types of structural disorder: (*a*) nonperiodicity of the stacking of (100) layers in the pyroxene structure, which can be related to chemical variations or be wholly isochemical; and (*b*) the intercalation parallel to (010) of slabs (Wadsley defects) having silicate chains with widths other than 1, i.e. double chains, triple chains, etc. I consider these two types of disorder in turn. The reader is also directed to Buseck et al (1980) for a more extensive review of HRTEM studies on pyroxenes.

Stacking variations The pyriboles (pyroxenes, amphiboles, and other closely related structures) all contain layers of octahedrally coordinated cations parallel to the (100) planes. These layers can have either of two orientations with respect to the *c*-axis, usually designated + and −. [See Thompson (1970), Veblen et al (1977), or Buseck et al (1980) for a more detailed explanation.] The clinopyroxenes (and other clinopyriboles) have the same orientation in all octahedral layers of the structure, i.e. the "stacking sequence" . . . + + + + + + + + + . . . or . . . − − − − − − − − − . . ., which can be abbreviated (+) or (−), the sequence within one repeat of the structure. The orthopyroxenes (and other orthopyriboles) have the sequence . . . + + − − + + − − . . ., or simply (+ + − −). The protopyroxenes have the stacking sequence . . . + − + − + − + − . . ., or (+ −) for short. These are the only stacking sequences that have been demonstrated definitely to occur in ordered single-phase crystals.

Exsolution relations among augite (high-calcium clinopyroxene), pigeonite (low-calcium clinopyroxene), and orthopyroxene (low-calcium) are very commonly observed in a wide variety of igneous and metamorphic rock types. For these exsolution reactions to occur, not only must there be diffusion of Ca (and Mg and/or Fe) from or to a growing exsolution lamella, but in order to exsolve orthopyroxene from clinopyroxene or vice versa, there also must be a structural reorganization to change the stacking sequence. Several studies have employed HRTEM methods to infer the details of these structural changes. All such studies have relied on the spacings of (100) fringes, which under appropriate imaging conditions show 0.45-nm periodicity for augite, 9 nm for pigeonite, and 1.8 nm for orthopyroxene.

Vander Sande & Kohlstedt (1974), Champness & Lorimer (1974), and Kohlstedt & Vander Sande (1976) used (100) fringe spacings to elucidate the process of augite exsolution from orthopyroxene. It was shown that augite lamellae grow by the nucleation and growth of narrow ledges that are one augite unit cell high. More recently, Isaacs & Peacor (1982) combined HRTEM methods with X-ray analytical TEM to show that the exsolution of orthopyroxene from augite can take place by a two-step process involving (*a*) diffusion of Ca, Mg, and Fe to form zones of low-calcium clinopyroxene in the augite, followed by (*b*) structural reconstruction that transforms the low-calcium clinopyroxene to orthopyroxene. Nord (1982) used HRTEM and analytical TEM to elucidate the dislocation structures along the interfaces of augite exsolution lamellae in orthopyroxene. He noted that the dislocations can act as nucleation sites for narrow amphibole lamellae that form by hydration of the augite, with the dislocation cores acting as conduits for hydrating fluids. Nord (1980) also used HRTEM and electron diffraction to suggest a structural model for extremely narrow exsolution lamellae in orthopyroxenes, called Guinier-Preston zones.

The combination of analytical TEM and HRTEM has also been applied successfully to intergrowths of augite (high-Ca clinopyroxene) and pigeonite (low-Ca clinopyroxene) formed by exsolution. It is now widely recognized that these two pyroxenes can possess a great variety of intergrowth geometries; it has been shown that structural misfit between the two structures can be absorbed by stacking faults in the pigeonite, which may also play an important role in the eventual inversion of pigeonite to orthopyroxene (Nobugai & Morimoto 1979, Kitamura et al 1981a,b, Rietmeijer & Champness 1982, Livi 1983).

As noted above, nonperiodicity in pyroxene stacking sequences can occur not only in conjunction with compositional differences, but also in chemically homogeneous pyroxenes. In an extensive HRTEM study, Iijima & Buseck (1975) and Buseck & Iijima (1975) explored the stacking variations in both naturally occurring and heat-treated pyroxenes with compositions close to that of enstatite ($MgSiO_3$). They found that stacking faults are a common feature in orthoenstatite. Furthermore, they showed that the cooling of protoenstatite [(+ −)], the high-temperature polymorph, results in an intimately intergrown mixture of twinned clinoenstatite [(+) and (−)] and orthoenstatite [(+ + − −)]. They also established criteria, later modified by Buseck et al (1980), for using HRTEM to distinguish among enstatite crystals that have been formed in different ways.

Chain-width defects High-resolution microscopy has shown that pyroxenes, which ideally contain only single silicate chains, can contain intergrown slabs having anomalous chain width. These chain-width errors, which can be thought of as Wadsley or polysomatic defects, can consist of amphibole-type (double) chains, jimthompsonite-type (triple) chains, or even wider chains. Submicroscopic lamellae of amphibole were first observed with TEM by Smith (1977), and they now have been observed in pyroxenes with various chemical compositions from a variety of geological environments. The occurrence of chain-width errors in pyroxenes has been reviewed by Buseck et al (1980), Veblen (1981), and Veblen & Buseck (1981).

The most common chain-width defects in pyroxenes involve triple-chain and double-chain material. Figure 5 (*left*) shows an augite crystal with a number of terminating triple-chain lamellae that are one chain wide (1.35 nm). An amphibole lamella showing a ledge (arrowed) in orthopyroxene is shown in Figure 5 (*right*). Such ledges, which are generally two amphibole chains wide (1.8 nm), nucleate and grow in the process of widening these lamellae. It has been suggested that amphibole lamellae in pyroxenes may form by exsolution (Smith 1977, Yamaguchi et al 1978),

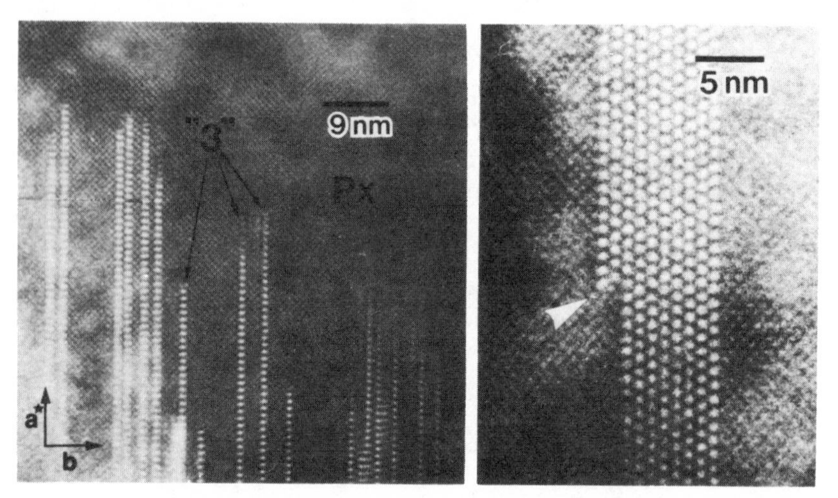

Figure 5 Chain-width disorder in pyroxenes. (*left*) Triple-chain lamellae ("3") in augite (Px). (*right*) An amphibole lamella in orthopyroxene, with a ledge in the interface (arrowed) that is two amphibole chains wide (Veblen & Buseck 1981; reproduced with permission of *The American Mineralogist*).

but Veblen & Buseck (1981) argued that all reported occurrences of chain-width faults in pyroxene are at least consistent with an alteration origin. Nakajima & Ribbe (1980, 1981b) also explained amphibole and triple-chain silicate in augite as resulting from alteration.

Two more recent reports of amphibole, triple-chain silicates, and sheet silicates replacing pyroxene are from rocks in which the replacement clearly is the result of alteration. Eggleton & Boland (1982) showed that the weathering of enstatite can proceed by a complex mechanism involving the growth of chain-width errors in the pyroxene, followed by replacement by sheet silicates. Akai (1982) studied the higher-temperature replacement of augite in an ore deposit. Again, the pyroxene was replaced first by hydrous chain silicates, which were in turn replaced by sheet silicates.

AMPHIBOLES AND WIDE-CHAIN SILICATES Just as in the pyroxenes, the amphiboles and wide-chain silicates (those with triple or wider chains) can occur with different stacking variations. For example, ortho-, clino-, and protoamphiboles have stacking sequences identical to the analogous pyroxene structures. However, to date there has not been much study of these stacking variations with HRTEM in amphiboles and wide-chain silicates (but see Hutchison et al 1975); this is probably a fertile area for future research.

Chain-width disorder in amphiboles and wide-chain pyriboles has, however, been studied extensively with HRTEM. This area has been reviewed by Veblen (1981), and consequently only a brief description of the results are presented here, along with some of the key references to this area.

The first report of chain-width defects in amphiboles was that of Chisholm (1973), who described the occurrence of some triple silicate chains in amphibole asbestos, based on a conventional TEM study. Subsequent studies have borne out the fact that chain-width defects are a very common form of disorder in amphibole asbestos (see, for example, Jefferson et al 1978, Veblen 1980, Cressey et al 1982, Chisholm 1983). The same appears to be true of nephrite, the jade variety of actinolite (Mallinson et al 1980). In general, defects with triple and wider chains appear to predominate, with single-chain defects being rare or absent.

While most, if not all, amphibole asbestos and jade appears to possess at least some chain-width disorder, this does not appear to be true of amphiboles that crystallize in more normal growth habits. However, chain-width defects can be abundant even in amphiboles with relatively normal morphology that have partially undergone solid-state hydration (alteration) reactions. Figure 6 shows an example of part of an amphibole crystal that has been altered; narrow lamellae of anomalous chain width have grown into the amphibole structure. The use of HRTEM to observe the structures, terminations, and distributions of such defects has been important for our understanding of how such solid-state replacement reactions take place (Veblen et al 1977, Veblen & Buseck 1980, 1981). For example, Figure 7 shows the termination of a sextuple-chain lamella in anthophyllite. It is thought that the replacement of anthophyllite by wide-chain silicates can take place by the growth of lamellae at such terminations.

Like some amphiboles, wide-chain silicates such as jimthompsonite (triple chains) and chesterite (alternating double and triple chains) commonly exhibit chain-width disorder. In addition, these minerals can occur too finely intergrown with each other and with anthophyllite to be observable with light microscopy. Finally, these minerals can contain regions in which small amounts of long-period ordered structures occur (Veblen & Buseck 1979a). Figure 8 shows a HRTEM image of a structure having the sequence of double and triple chains (2333). Such structures can

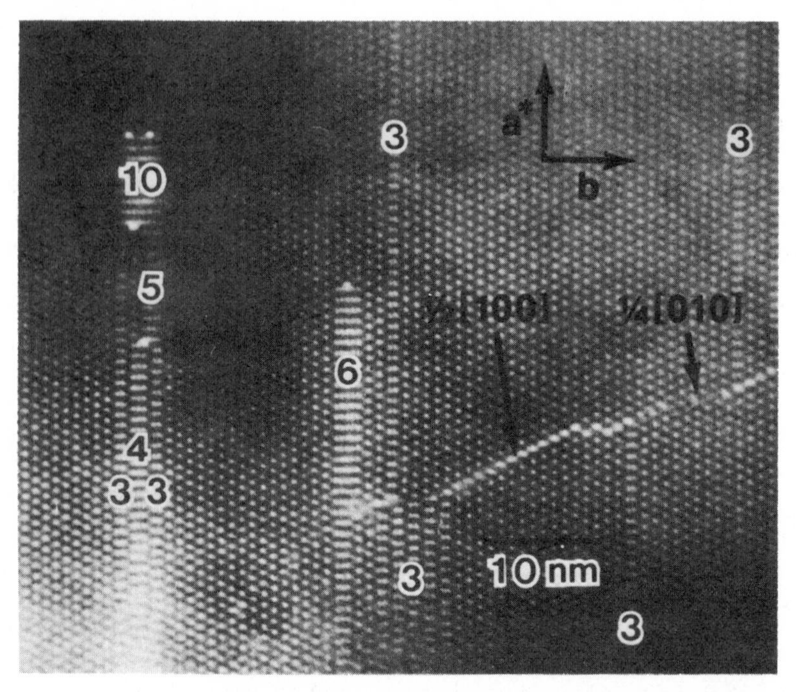

Figure 6 Amphibole that has begun to react to wide-chain silicate. Triple, quadruple, quintuple, and wider chains are now intergrown with the predominantly double-chain structure. Two displacive planar faults with projected displacements $\frac{1}{2}[100]$ and $\frac{1}{4}[010]$ are also shown (Veblen & Buseck 1981; reproduced with permission from *The American Mineralogist*).

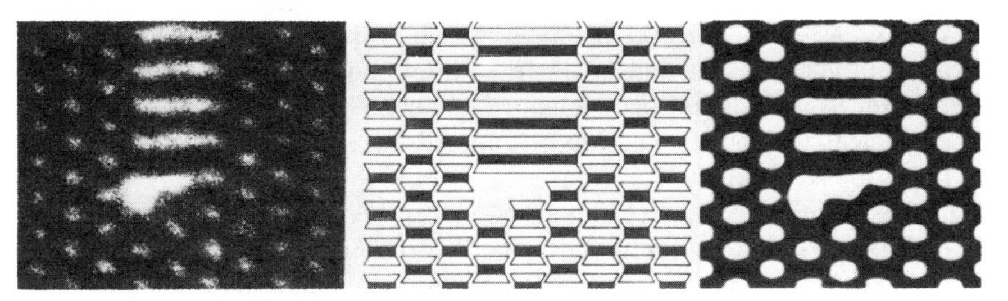

Figure 7 Termination of a sextuple-chain slab in anthophyllite. (*left*) Experimental HRTEM image. (*center*) Model of the defect shown in "I-beam" representation (see Veblen et al 1977). (*right*) Computer-simulated image based on the model. Each amphibole I-beam is 0.9 nm wide (Veblen & Buseck 1980; reproduced with permission of *The American Mineralogist*).

be discovered and determined only through the use of electron microscopy, since they occur in fine-grained intergrowth with other minerals and in crystals that are too small to isolate using more standard X-ray diffraction techniques. Finally, chain silicate "crystals" having essentially complete disorder in chain width also can occur, and again HRTEM is the only technique that can be used to determine the structures of these disordered materials.

PYROXENOIDS The pyroxenoids, like pyroxenes, are single-chain silicates, but with chains that repeat after every 3, 5, 7, or 9 tetrahedra. (Pyroxenes repeat every two tetrahedra.) HRTEM investigations have made important contributions to our understanding of pyroxenoid crystal structure, revealing both stacking disorder and disorder in the periodicities of the silicate chains. This disorder can occur in both synthetic and natural samples (Wenk et al 1976b, Palomino et al 1977, Czank & Liebau 1980,

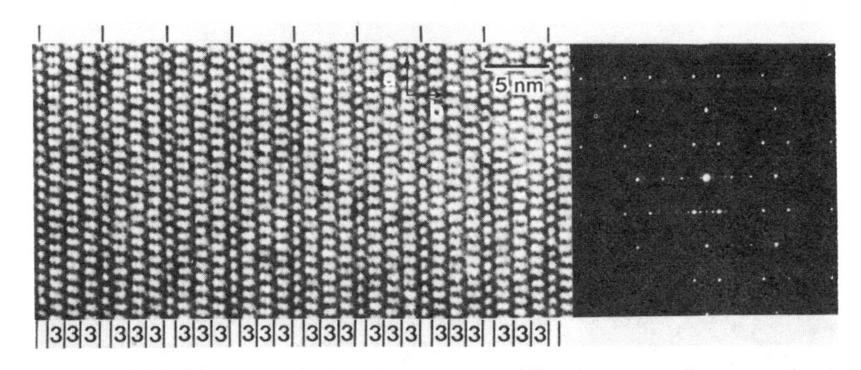

Figure 8 HRTEM image and selected-area electron diffraction pattern from an ordered chain silicate having the chain sequence (2333). Double-chain slabs are not labeled (Veblen & Buseck 1979a; reproduced with permission from *The American Mineralogist*).

Czank 1981, Ried & Korekawa 1980, Alario Franco et al 1980, Jefferson et al 1980, Jefferson & Pugh 1981). In addition, both synthetic and natural pyroxenes can contain occasional offsets in their silicate chains that are equivalent to the intercalation of unit-cell scale lamellae of pyroxenoid (Ried 1984, Veblen 1985b).

In synthetic pyroxenoids, the structural disorder is presumably the result of direct crystal growth from the melt or hydrothermal fluid. An example of synthetic ferrosilite III (a 9-repeat pyroxenoid) is presented in Figure 9, which shows both chain-periodicity disorder and stacking disorder (Czank & Simons 1983). In natural samples, however, disorder in the chain periodicities may result from crystal growth or from complex solid-state replacement reactions (Veblen 1985b).

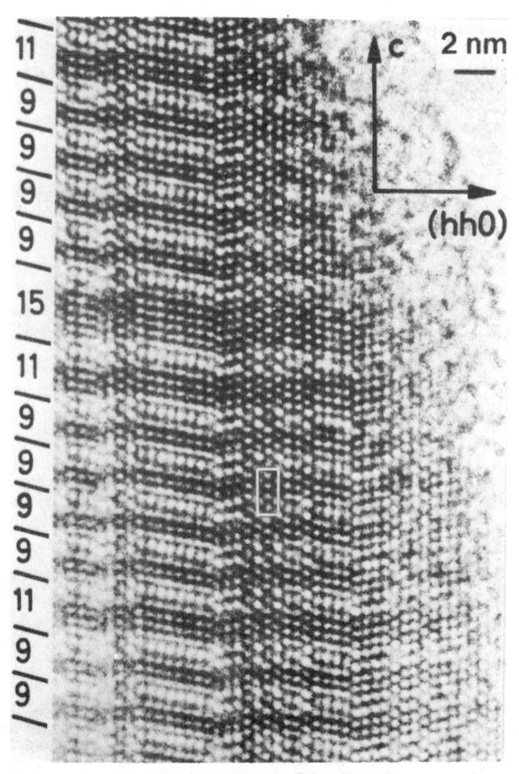

Figure 9 HRTEM image of the synthetic pyroxenoid ferrosilite III, showing disorder in the normally 9-repeat chains. (Chain repeats of 11 and 15 are indicated.) The image also shows stacking disorder parallel to *hhO* (Czank & Simons 1983; reproduced with permission from *Physics and Chemistry of Minerals*).

Sheet Silicates

Two important types of structural disorder are known to occur in the sheet silicates. The first is stacking disorder, which occurs when the rotations or shifts of subsequent layers in a sheet structure are nonperiodic. The second is mixed-layering disorder, which occurs when two or more different types of layers are intergrown in the same structure. In addition, the basic structures of some important sheet silicates are still not well understood. An example is greenalite, for which Guggenheim et al (1982) used HRTEM combined with X-ray diffraction data to propose a possible structure. It is expected that HRTEM may make major contributions to the crystal chemistry of such complex sheet structures in the future.

STACKING DISORDER Sheet silicates, such as micas and chlorites, consist of layers that can be rotated or shifted in various ways relative to adjacent layers. Periodic repetition of a sequence of rotations and/or shifts results in an ordered polytype, and polytypism is a common and important phenomenon in this entire mineral group (e.g. Smith & Yoder 1956, Ross et al 1966, Bailey & Brown 1962). Furthermore, it has long been recognized that nonperiodic stacking can occur in sheet silicates, resulting in disordered polytypes.

While X-ray experiments can detect the presence of stacking disorder in sheet silicates, they cannot be used to determine the exact stacking sequence of a disordered crystal. However, HRTEM has now been used to determine the exact stacking in a number of cases. Iijima & Buseck (1978) demonstrated the feasibility of obtaining stacking sequences in disordered micas. In an extensive study employing computer-simulated images of a variety of mica polytypes, Amouric et al (1981) showed that HRTEM images can indeed be used to yield the exact stacking, but great care must be exercised in experimental technique to avoid images that can be interpreted incorrectly. Olives Baños et al (1983) used HRTEM to determine the effects on stacking of deformation in biotite, and Amouric & Baronnet (1983) studied the effects of nucleation conditions on stacking in muscovite. Although sheet silicate groups other than the micas also exhibit polytypism and stacking disorder, only the chlorites have been studied with HRTEM. Spinnler et al (1984) discussed the theory of chlorite polytypism, which involves layer shifts as well as rotations, and they presented image calculations and experimental images of a disordered clinochlore that could be interpreted in terms of the exact stacking sequence.

MIXED LAYERING AND OTHER INTERGROWTH TYPES As noted above, mixed layering involves the intergrowth within an individual crystal of two or more different types of layers. This mixed layering can occur in a periodic fashion, but disordered mixed layering is also common in sheet silicates, having been recognized from X-ray studies of clay minerals for many years. There have been a number of reports of finely intergrown sheet silicates of different layer types and of occasional isolated layers of one type in a crystal that consists primarily of another type of layer (for example, brucite-like layers intergrown in talc; see references in Veblen 1983, Page & Wenk 1979). More recently, however, it has been recognized that extensive mixed layering can occur in apparently well-crystallized sheet silicates from metamorphic and igneous rocks (Veblen 1983, Veblen & Ferry 1983). Figure 10 shows two examples of intimate mixed layering between chlorite (C) and a sodium mica (M). [The chlorite and mica layers are defined nonconventionally in this figure—see Veblen (1983) for details.] Mixed layering of this sort is well suited to study by HRTEM and may be especially important because it alters the stoichiometry of the host crystal.

In another HRTEM study, Olives Baños et al (1983) showed that brucite-like layers intergrown in biotite grew on interlayer planes on which slip had occurred during deformation. In this example, the mixed layering was a secondary phenomenon due to a solid-state reaction, rather than primary crystal growth. Likewise, it has been shown that the replacement of biotite by chlorite can take place via a complex series of reactions involving mixed-layer intermediate structures (Veblen & Ferry 1983, Eggleton & Banfield 1984). It was shown in these studies that the specific mechanisms of reaction

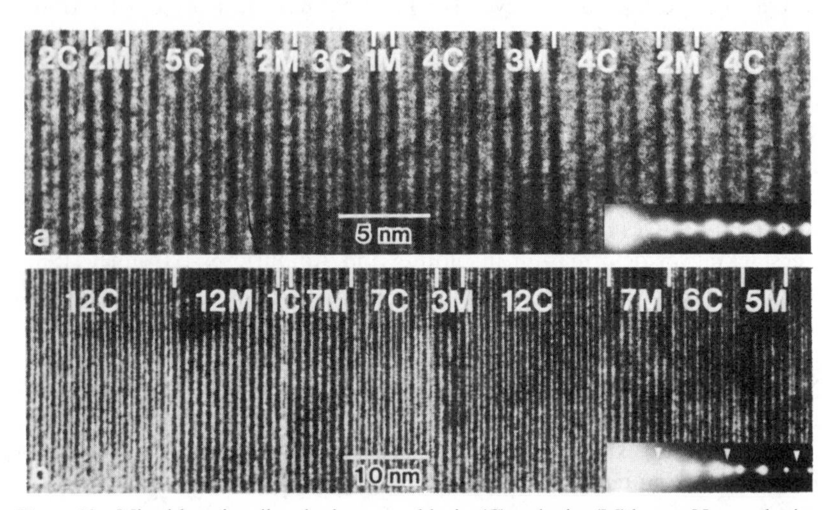

Figure 10 Mixed-layering disorder between chlorite (C) and mica (M) layers. Numerals give the numbers of adjacent layers of these two types, and c^* is horizontal. (*top*) Very intimate intermixture of C and M layers. (*bottom*) Mixing of wider slabs of C and M structure (Veblen 1983; reproduced with permission from *The American Mineralogist*).

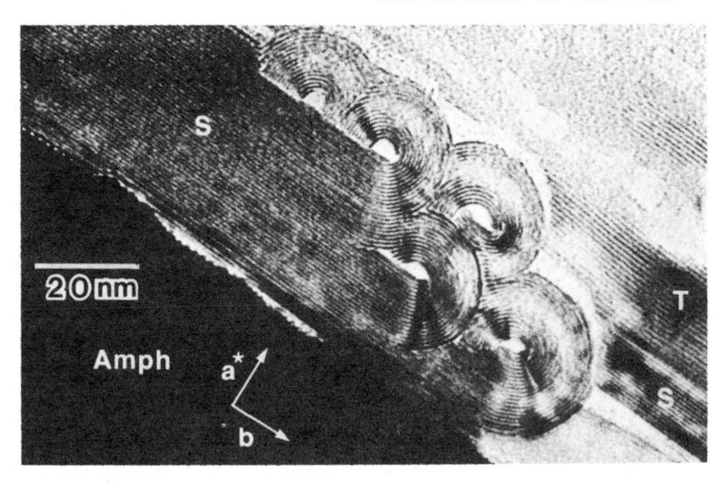

Figure 11 Intergrowth of the serpentine structures lizardite (planar layers) and chrysotile (curved layers). The layers are viewed on edge, and the serpentine (S) is intergrown with talc (T) and amphibole (Amph) (Veblen & Buseck 1979b; reproduced with permission from *Science*).

are intimately related to the macroscopically observable physical and chemical details of the reactions and that the structural mechanisms are probably determined by the chemical environment of the crystal during reaction.

In addition to mixed layering, other types of intergrowth are possible in sheet silicates. In his classic HRTEM work on chrysotile asbestos, Yada (1967, 1971) showed how the layers of this complex curled structure fit together concentrically or are rolled up like a carpet. Iijima & Zhu (1982) used HRTEM to elucidate the structural details in an unusual intergrowth of muscovite and biotite. In another example, Veblen & Buseck (1979b) presented HRTEM evidence of the intimate intergrowth of two polymorphs of serpentine [lizardite (planar layers) and chrysotile (curved layers)], as shown in Figure 11.

Framework Silicates

FELDSPARS In spite of their importance, the number of HRTEM studies on feldspars is limited, with the exception of studies that have imaged simple fringes associated with the modulated structure of intermediate plagioclases. Hashimoto et al (1976) have presented images of labradorite, but no defects other than the modulated structure were observed. More recent high-resolution images of intermediate plagioclase structures (Nakajima et al 1977, Kumao et al 1981), though visually pleasing, also have not resolved the long-standing questions about the details of these structures. Indeed, it is unlikely that HRTEM will prove to be an adequate tool for determining such structural subtleties, at least at its present state of development.

The HRTEM method has been applied successfully to alkali feldspars. Eggleton & Buseck (1980a,b) studied both the weathering processes and the inversion from orthoclase to microcline in potassium-rich feldspar. They found that this inversion from monoclinic to triclinic symmetry can result in extremely fine-scale twin domains, which can inhibit the complete inversion to maximum microcline. Likewise, Brown & Parsons (1984) used a combination of conventional TEM and HRTEM to deduce details of the development of exsolution, ordering, and twinning in a suite of slowly cooled cryptoperthites.

ZEOLITES In spite of problems associated with severe radiation damage in the electron microscope, there have been a few successful HRTEM studies of zeolites. For example, Bursill et al (1980) found a number of different types of structural defects in synthetic sodium zeolite A and were able to suggest possible vitrification mechanisms for this industrially important structure. Cartlidge et al (1983) were able to image the channels in a synthetic chabazite, and they showed that extensive fine-scale twinning can be an important aspect of this structure. By replacing Al in the framework of offretite with Si, Millward & Thomas (1984) improved the stability of this zeolite in the electron beam and were able to image faults in the structure. Similarly, Terasaki et al (1984) produced impressive images of defects in dealuminated zeolite L, in which large structural blocks were rotated by 32.2° with respect to the host structure. Spatial ordering of these rotated blocks locally produced a new structure with very long period.

SUMMARY

I have examined briefly the principles of HRTEM and the types of defects that can occur in minerals. I have also reviewed some of the contributions that HRTEM studies have made to our understanding of silicate minerals. Based on this work, mineralogical crystal chemists now have a reasonable idea of which structures are particularly prone to structural disorder. However, the mere fact that there are no HRTEM studies published on a particular mineral should not be taken to mean that it may not contain severe disorder under appropriate conditions.

Although there has been much progress in understanding the defect states of many of the common rock-forming silicates, future work can be expected to continue on as-of-yet unexplored minerals and on different occurrences from those already examined. In addition, it is likely that HRTEM will find new or expanded applications in areas such as the characterization of structural disorder in run products and starting materials from petrological experiments; the determination of interface structures, such as grain boundaries; and the determination of mechanisms in solid-state reactions and their relationship to reaction chemistry. Perhaps the most important trend will be the application of HRTEM techniques to more geologically oriented problems. Now that much of the initial survey work on natural silicate systems has been done, many Earth scientists are beginning to realize that HRTEM can provide them with an important tool for understanding geological processes on the submicroscopic scale.

ACKNOWLEDGMENTS

I thank P. R. Buseck, M. Czank, S. Iijima, and T. J. White for HRTEM images reproduced in this paper. Support from the National Science Foundation, Division of Earth Sciences, grant EAR83-06861, is gratefully acknowledged.

Literature Cited

Akai, J. 1982. Polymerization process of biopyribole in metasomatism at the Akatani ore deposit, Japan. *Contrib. Mineral. Petrol.* 80:117–31

Alario Franco, M., Jefferson, D. A., Pugh, N. J., Thomas, J. M. 1980. Lattice imaging of structural defects in a chain silicate: the pyroxenoid mineral rhodonite. *Mater. Res. Bull.* 15:73–79

Amouric, M., Baronnet, A. 1983. Effect of early nucleation conditions on synthetic muscovite polytypism as seen by high resolution transmission electron microscopy. *Phys. Chem. Miner.* 9:146–59

Amouric, M., Mercuriot, G., Baronnet, A. 1981. On computed and observed HRTEM images of perfect mica polytypes. *Bull. Minéral.* 104:298–313

Bailey, S. W., Brown, B. E. 1962. Chlorite polytypism: I. Regular and semi-random one-layer structures. *Am. Mineral.* 47: 819–50 (initially published incorrectly as Brown & Bailey)

Brown, W. L., Parsons, I. 1984. Exsolution and coarsening mechanisms and kinetics in an ordered cryptoperthite series. *Contrib. Mineral. Petrol.* 86:3–18

Bursill, L. A., Lodge, E. A., Thomas, J. M. 1980. Zeolitic structures as revealed by high-resolution electron microscopy. *Nature* 286:111–13

Buseck, P. R. 1984. Imaging of minerals with the TEM. *Bull. Electron Microsc. Soc. Am.* 14:47–53

Buseck, P. R., Cowley, J. M. 1983. Modulated and intergrowth structures in minerals and electron microscope methods for their study. *Am. Mineral.* 68:18–40

Buseck, P. R., Iijima, S. 1974. High resolution electron microscopy of silicates. *Am. Mineral.* 59:1–21

Buseck, P. R., Iijima, S. 1975. High resolution electron microscopy of enstatite II: geological application. *Am. Mineral.* 60:771–84

Buseck, P. R., Veblen, D. R. 1981. Defects in minerals as observed with high-resolution transmission electron microscopy. *Bull. Minéral.* 104:249–60

Buseck, P. R., Nord, G. L. Jr., Veblen, D. R. 1980. Subsolidus phenomena in pyroxenes. In *Pyroxenes, Mineral. Soc. Am. Rev. Mineral.*, ed. C. T. Prewitt, 7:117–211

Cartlidge, S., Wessicken, R., Nissen, H.-U. 1983. Electron microscopy study of zeolite ZK-14; a synthetic chabazite. *Phys. Chem. Miner.* 9:139–45

Champness, P. E., Lorimer, G. W. 1974. A direct lattice-resolution study of precipitation (exsolution) in orthopyroxene. *Philos. Mag.* 30:357–66

Chisholm, J. E. 1973. Planar defects in fibrous amphiboles. *J. Mater. Sci.* 8:475–83

Chisholm, J. E. 1983. Transmission electron microscopy of asbestos. In *Asbestos*, ed. S. S. Chissick, R. Derricott, 2:85–167. New York: Wiley

Cowley, J. M. 1975. *Diffraction Physics.*

Amsterdam/New York: North-Holland/ Am. Elsevier. 410 pp.
Cowley, J. M. 1979. Principles of image formation. See Hren et al 1979, pp. 1–42
Cowley, J. M., Iijima, S. 1976. The direct imaging of crystal structures. See Wenk et al 1976a, pp. 121–36
Cowley, J. M., Cohen, J. B., Salamon, M. B., Wuensch, B. J., eds. 1979. *Modulated Structures—1979, Am. Inst. Phys. Conf. Proc. No. 53.* New York: Am. Inst. Phys. 432 pp.
Cressey, B. A., Whittaker, E. J. W., Hutchison, J. L. 1982. Morphology and alteration of asbestiform grunerite and anthophyllite. *Mineral. Mag.* 46:77–87
Czank, M. 1981. Chain periodicity faults in babingtonite, $Ca_2Fe^{2+}H[Si_5O_{15}]$. *Acta Crystallogr. Sect. A* 37:617–20
Czank, M., Liebau, F. 1980. Periodicity faults in chain silicates: a new type of planar lattice fault observed with high resolution electron microscopy. *Phys. Chem. Miner.* 6:85–93
Czank, M., Simons, B. 1983. High resolution electron microscopic studies on ferrosilite III. *Phys. Chem. Miner.* 9:229–34
Davies, P. K., Akaogi, M. 1983. Phase intergrowths in spinelloids. *Nature* 305:788–90
Eggleton, R. A., Banfield, J. F. 1984. The alteration of granitic biotite to chlorite. *Am. Mineral.* In press
Eggleton, R. A., Boland, J. N. 1982. Weathering of enstatite to talc through a sequence of transitional phases. *Clays Clay Miner.* 30:11–20
Eggleton, R. A., Buseck, P. R. 1980a. High resolution electron microscopy of feldspar weathering. *Clays Clay Miner.* 28:173–78
Eggleton, R. A., Buseck, P. R. 1980b. The orthoclase-microcline inversion: a high-resolution transmission electron microscope study and strain analysis. *Contrib. Mineral. Petrol.* 74:123–33
Gjonnes, J., Moodie, A. F. 1965. Extinction conditions in the dynamic theory of electron diffraction. *Acta Crystallogr.* 19:65–67
Guggenheim, S., Bailey, S. W., Eggleton, R. A., Wilkes, P. 1982. Structural aspects of greenalite and related minerals. *Can. Mineral.* 20:1–18
Hashimoto, H., Nissen, H.-U., Ono, A., Kumao, A., Endoh, H., Woensdregt, C. F. 1976. High-resolution electron microscopy of labradorite feldspar. See Wenk et al 1976a, pp. 332–44
Hirth, J. P., Lothe, J. 1982. *Theory of Dislocations.* New York: Wiley. 857 pp. 2nd ed.
Hren, J. J., Goldstein, J. I., Joy, D. C., eds. 1979. *Analytical Electron Microscopy.* New York/London: Plenum. 601 pp.
Hutchison, J. L., Irusteta, M. C., Whittaker, E. J. W. 1975. High resolution electron microscopy and diffraction studies of fibrous amphiboles. *Acta Crystallogr. Sect. A* 31:794–801
Hutchison, J. L., Jefferson, D. A., Thomas, J. M. 1977. The ultrastructure of minerals as revealed by high resolution electron microscopy. In *Surface and Defect Properties of Solids,* ed. M. W. Roberts, J. M. Thomas, 6:320–58. London: Chem. Soc.
Iijima, S., Buseck, P. R. 1975. High resolution electron microscopy of enstatite I: twinning, polymorphism, and polytypism. *Am. Mineral.* 60:758–70
Iijima, S., Buseck, P. R. 1978. Experimental study of disordered mica structures by high-resolution electron microscopy. *Acta Crystallogr. Sect. A* 34:709–19
Iijima, S., Zhu, J. 1982. Muscovite-biotite interface studied by electron microscopy. *Am. Mineral.* 67:1195–1205
Iijima, S., Cowley, J. M., Donnay, G. 1973. High resolution electron microscopy of tourmaline crystals. *Tschermaks Mineral. Petrogr. Mitt.* 20:216–24
Isaacs, A. M., Peacor, D. R. 1982. Orthopyroxene exsolution in augite: a two-step, diffusion-transformation process. *Science* 218:152–53
Jefferson, D. A., Pugh, N. J. 1981. The ultrastructure of pyroxenoid chain silicates. III. Intersecting defects in a synthetic iron-manganese pyroxenoid. *Acta Crystallogr. Sect. A* 37:281–86
Jefferson, D. A., Thomas, J. M. 1978. High resolution electron microscopic and X-ray studies of non-random disorder in an unusual layered silicate (chloritoid). *Proc. R. Soc. London Ser. A* 361:399–411
Jefferson, D. A., Mallinson, L. G., Hutchison, J. L., Thomas, J. M. 1978. Multiple-chain and other unusual faults in amphiboles. *Contrib. Mineral. Petrol.* 66:1–4
Jefferson, D. A., Pugh, N. J., Alario-Franco, M., Mallinson, L. G., Millward, G. R., Thomas, J. M. 1980. The ultrastructure of pyroxenoid chain silicates. I. Variation of the chain configuration in rhodonite. *Acta Crystallogr. Sect. A* 36:1058–65
Kitamura, M., Yasuda, M., Morimoto, N. 1981a. A study of the fine textures of Bushveld augite by 200 kV analytical electron microscopy. *Bull. Minéral.* 104: 278–84
Kitamura, M., Yasuda, M., Morimoto, N. 1981b. Morphology change of exsolution lamellae of pigeonite in Bushveld augite—an electron microscopic observation. *Proc. Jpn. Acad. Ser. B* 57: 183–87
Kitamura, M., Shen, B., Banno, S., Morimoto, N. 1984. Fine textures of laihunite,

a nonstoichiometric distorted olivine-type mineral. *Am. Mineral.* 69:154–60

Kohlstedt, D. L., Vander Sande, J. B. 1976. On the detailed structure of ledges in an augite-enstatite interface. See Wenk et al 1976a, pp. 234–37

Kröger, F. A. 1985. Point defects in solids—physics, chemistry and thermodynamics. See Schock 1985, pp. 1–17

Kumao, A., Hashimoto, H., Nissen, H.-U., Endoh, H. 1981. Ca and Na positions in labradorite feldspar as derived from high-resolution electron microscopy and optical diffraction. *Acta Crystallogr. Sect. A* 37:229–38

Lasaga, A. C. 1981. The atomistic basis of kinetics: defects in minerals. In *Kinetics of Geochemical Processes, Mineral. Soc. Am. Rev. Mineral.*, ed. A. C. Lasaga, R. J. Kirkpatrick, 8:261–319

Livi, K. J. T. 1983. *Electron microscope investigations of exsolved augites from the ferromonzonite of the Laramie anorthosite complex, Wyoming.* MS thesis. State Univ. N.Y., Stony Brook. 172 pp.

Mallinson, L. G., Jefferson, D. A., Thomas, J. M., Hutchison, J. L. 1980. The internal structures of nephrite: experimental and computational evidence for the coexistence of multiple-chain silicates within an amphibole host. *Philos. Trans. R. Soc. London* 295:537–52

McConnell, J. D. C. 1983. A review of structural resonance and the nature of long-range interactions in modulated mineral structures. *Am. Mineral.* 68:1–10

Millward, G. R., Thomas, J. M. 1984. Real-space imaging of offretite and the identification of other coexistent zeolitic structures. *J. Chem. Soc. Chem. Commun.* 1984:77–79

Müller, W. F., Wenk, H.-R. 1978. Mixed-layer characteristics in real humite structures. *Acta Crystallogr. Sect. A* 34:607–9

Nakajima, Y., Ribbe, P. H. 1980. Alteration of pyroxenes from Hokkaido, Japan, to amphibole, clays, and other biopyriboles. *Neues Jahrb. Mineral. Monatsh.* 6:258–68

Nakajima, Y., Ribbe, P. H. 1981a. Twinning and superstructure of Al-rich mullite. *Am. Mineral.* 66:142–47

Nakajima, Y., Ribbe, P. H. 1981b. Texture and structural interpretation of the alteration of pyroxene to other biopyriboles. *Contrib. Mineral. Petrol.* 78:230–39

Nakajima, Y., Morimoto, N., Watanabe, E. 1975. Direct observation of oxygen vacancy in mullite, $1.89Al_2O_3 \cdot SiO_2$ by high resolution electron microscopy. *Proc. Jpn. Acad.* 51:173–78

Nakajima, Y., Morimoto, N., Kitamura, M. 1977. The superstructure of plagioclase feldspars: electron microscopic study of anorthite and labradorite. *Phys. Chem. Miner.* 1:213–25

Nobugai, K., Morimoto, N. 1979. Formation mechanism of pigeonite lamellae in Skaergaard augite. *Phys. Chem. Miner.* 4:361–71

Nord, G. L. Jr. 1980. The composition, structure, and stability of Guinier-Preston zones in lunar and terrestrial orthopyroxene. *Phys. Chem. Miner.* 6:109–28

Nord, G. L. Jr. 1982. Analytical electron microscopy in mineralogy: exsolved phases in pyroxenes. *Ultramicroscopy* 8:109–20

O'Keefe, M. A., Buseck, P. R., Iijima, S. 1978. Computed crystal structure images for high resolution electron microscopy. *Nature* 274:322–24

Olives Baños, J., Amouric, M., De Fouquet, C., Baronnet, A. 1983. Interlayering and interlayer slip in biotite as seen by HRTEM. *Am. Mineral.* 68:754–58

Page, R., Wenk, H.-R. 1979. Phyllosilicate alteration of plagioclase studied by transmission electron microscopy. *Geology* 7:393–97

Palomino, J. R., Jefferson, D. A., Hutchison, J. L., Thomas, J. M. 1977. Linear and planar defects in wollastonite. *J. Chem. Soc. Dalton Trans.* 1977:1834–36

Price, G. D. 1983. The nature and significance of stacking faults in wadsleyite, natural β-$(Mg, Fe)_2SiO_4$ from the Peace River meteorite. *Phys. Earth Planet. Inter.* 33:137–47

Ried, H. 1984. Intergrowth of pyroxene and pyroxenoid; chain periodicity faults in pyroxene. *Phys. Chem. Miner.* 10:230–35

Ried, H., Korekawa, M. 1980. Transmission electron microscopy of synthetic and natural fünferketten and siebenerketten pyroxenoids. *Phys. Chem. Miner.* 5:351–65

Rietmeijer, F. J. M., Champness, P. E. 1982. Exsolution structures in calcic pyroxenes from the Bjerkreim-Sokndal lopolith, SW Norway. *Mineral. Mag.* 45:11–24

Ross, M., Takeda, H., Wones, D. R. 1966. Mica polytypes: systematic description and identification. *Science* 151:191–93

Rossman, G. R., Grew, E. S., Dollase, W. A. 1982. The colors of sillimanite. *Am. Mineral.* 67:749–61

Scherzer, O. 1949. The theoretical resolution limit of the electron microscope. *J. Appl. Phys.* 20:20–29

Schock, R. N., ed. 1985. *Point Defects in Minerals.* Washington DC: Am. Geophys. Union. 232 pp.

Smith, J. V., Yoder, H. S. 1956. Experimental and theoretical studies of the mica polymorphs. *Mineral. Mag.* 31:209–35

Smith, P. P. K. 1977. An electron microscope

study of amphibole lamellae in augite. *Contrib. Mineral. Petrol.* 59:317–22

Spence, J. C. H. 1981. *Experimental High-Resolution Electron Microscopy.* Oxford: Clarendon. 370 pp.

Splinnler, G. E., Self, P. G., Iijima, S., Buseck, P. R. 1984. Stacking disorder in clinochlore chlorite. *Am. Mineral.* 69:252–63

Steeds, J. W. 1979. Convergent beam electron diffraction. See Hren et al 1979, pp. 387–422

Terasaki, O., Thomas, J. M., Ramdas, S. 1984. A new type of stacking fault in zeolites: presence of a coincidence boundary ($\sqrt{13}.\sqrt{13}$ R32.2° superstructure) perpendicular to the tunnel direction in zeolite L. *J. Chem. Soc. Chem. Commun.* 1984:216–17

Thomas, J. M. 1984. New ways of characterizing layered silicates and their intercalates. *Philos. Trans. R. Soc. London Ser. A* 311:271–85

Thomas, J. M., Jefferson, D. A., Mallinson, L. G., Smith, D. J., Crawford, E. S. 1979. The elucidation of the ultrastructure of silicate minerals by high resolution electron microscopy and X-ray emission microanalysis. *Chem. Scr.* 14:167–79

Thompson, J. B. Jr. 1970. Geometrical possibilities for amphibole structures: model biopyriboles. *Am. Mineral.* 55:292–93

Thompson, J. B. Jr. 1978. Biopyriboles and polysomatic series. *Am. Mineral.* 63:239–49

Turner, S. 1978. *High-resolution transmission electron microscopy of some manganese oxides and silicate minerals.* MS thesis. Ariz. State Univ., Tempe

Vander Sande, J. B., Kohlstedt, D. L. 1974. A high-resolution electron microscopy study of exsolution in enstatite. *Philos. Mag.* 29:1041–49

Veblen, D. R. 1980. Anthophyllite asbestos: microstructures, intergrown sheet silicates, and mechanisms of fiber formation. *Am. Mineral.* 65:1075–86

Veblen, D. R. 1981. Non-classical pyriboles and polysomatic reactions in biopyriboles. In *Amphiboles and Other Hydrous Pyriboles—Mineralogy, Mineral. Soc. Am. Rev. Mineral.*, ed. D. R. Veblen, 9A:189–236

Veblen, D. R. 1983. Microstructures and mixed layering in intergrown wonesite, chlorite, talc, biotite, and kaolinite. *Am. Mineral.* 68:566–80

Veblen, D. R. 1985a. Extended defects and vacancy non-stoichiometry in rock-forming minerals. See Schock 1985, pp. 122–31

Veblen, D. R. 1985b. TEM study of a pyroxene-to-pyroxenoid reaction. *Am. Mineral.* In press

Veblen, D. R., Buseck, P. R. 1979a. Chain-width order and disorder in biopyriboles. *Am. Mineral.* 64:687–700

Veblen, D. R., Buseck, P. R. 1979b. Serpentine minerals: intergrowths and new combination structures. *Science* 206: 1398–1400

Veblen, D. R., Buseck, P. R. 1980. Microstructures and reaction mechanisms in biopyriboles. *Am. Mineral.* 65:599–623

Veblen, D. R., Buseck, P. R. 1981. Hydrous pyriboles and sheet silicates in pyroxenes and uralites: intergrowth microstructures and reaction mechanisms. *Am. Mineral.* 66:1107–34

Veblen, D. R., Ferry, J. M. 1983. A TEM study of the biotite-chlorite reaction and comparison with petrologic observations. *Am. Mineral.* 68:1160–68

Veblen, D. R., Buseck, P. R., Burnham, C. W. 1977. Asbestiform chain silicates: new minerals and structural groups. *Science* 198:359–65

Wenk, H.-R. 1980. Defects along kyanite-staurolite interfaces. *Am. Mineral.* 65:766–69

Wenk, H.-R., Champness, P. E., Christie, J. M., Cowley, J. M., Heuer, A. H., et al, eds. 1976a. *Electron Microscopy in Mineralogy.* Berlin/Heidelberg/New York: Springer-Verlag. 564 pp.

Wenk, H.-R., Müller, W. F., Liddell, N. A., Phakey, P. P. 1976b. Polytypism in wollastonite. See Wenk et al 1976a, pp. 326–31

White, T. J., Hyde, B. G. 1982a. Electron microscope study of the humite minerals: I. Mg-rich specimens. *Phys. Chem. Miner.* 8:55–63

White, T. J., Hyde, B. G. 1982b. Electron microscope study of the humite minerals: II. Mn-rich specimens. *Phys. Chem. Miner.* 8:167–74

White, T. J., Hyde, B. G. 1983. An electron microscope study of leucophoenicite. *Am. Mineral.* 68:1009–21

Yada, K. 1967. Study of chrysotile asbestos by a high resolution electron microscope. *Acta Crystallogr.* 23:704–7

Yada, K. 1971. Study of microstructure of chrysotile asbestos by high resolution electron microscopy. *Acta Crystallogr. Sect. A* 27:659–64

Yamaguchi, Y., Akai, J., Tomita, K. 1978. Clinoamphibole lamellae in diopside of garnet lherzolite from Alpe Arami, Bellinzona, Switzerland. *Contrib. Mineral. Petrol.* 66:263–70

Ylä-Jääski, J., Nissen, H.-U. 1983. Investigation of superstructures in mullite by high resolution electron microscopy and electron diffraction. *Phys. Chem. Miner.* 10:47–54

Ann. Rev. Earth Planet. Sci. 1985. 13 : 147–73

COSMIC DUST: COLLECTION AND RESEARCH

D. E. Brownlee

Department of Astronomy, University of Washington, Seattle, Washington 98195

INTRODUCTION

The term "cosmic dust" as used here refers to particulate material that exists or has existed in the interplanetary medium as bodies smaller than 1 mm. The particles can be collected both in space and in the terrestrial environment, and they are a valuable resource of meteoritic material. The dust samples are complementary to the traditional meteorites that were much larger meteoroids in space. Because of their size, the analysis of collected dust particles is more difficult and limited than studies of meteorites. Nevertheless, collected dust samples are being vigorously investigated for two fundamentally important reasons. The first is that the most friable extraterrestrial materials can *only* be collected in the form of dust. Highly fragile materials cannot survive hypervelocity entry into the atmosphere in chunks as large as conventional meteorites. Most cometary meteoroids are known to be fragile, and dust collection is the only Earth-based technique for obtaining typical cometary solids.

The second reason why dust is important is that it is very abundant. Meteorites are exceedingly rare, but cosmic dust is so common that quite literally every footstep a person takes contacts a fragment of cosmic dust. Dust is so abundant that, with appropriate techniques, it can be recovered from historical deposits in deep-sea sediments and collected in real time in space and in the stratosphere. Stratigraphic layers in sediments record a continuous history of the terrestrial accretion of space debris that extends beyond 10^8 yr ago. This record can be searched for temporal changes in the meteoroid complex, such as might occur during fluctuations in the number of comets in the inner solar system or during passage of the solar system through an interstellar cloud. Effects of isolated events, such as the terrestrial impact of a large body, are also contained in sediments. The flux

0084–6597/85/0515–0147$02.00

of particles in space is high enough that it should be possible to design a collector that measures orbital parameters of the particles being collected. This would allow collection of material from identifiable sources, a project that was tried for conventional meteorites with very limited success because the collection rate was only one per decade (McCrosky et al 1971). Dust is abundant not only in terms of numbers but also in terms of mass. The bulk of the total mass of extraterrestrial material that is annually accreted by the Earth is in the 0.1–1 mm size range (Hughes 1978). Because of this and because of its rapid orbital evolution after release from parent bodies, the dust at 1 AU is likely to be a more representative sample of the entire meteoroid complex than are conventional meteorites.

This paper reviews the aspects of cosmic dust that are important to its collection and utilization as a resource of extraterrestrial material. Most of the collected dust particles appear to be primitive and sometimes unique solar system materials that contain important clues to early solar system processes and environments. Collected dust can also be studied to investigate selected properties of the interplanetary medium and the terrestrial environment. The treatment of the origin and evolution of dust in the solar system is focused here on aspects important to the interpretation of sample results. The bulk of the paper reviews the results of laboratory studies of dust samples collected in the stratosphere and deep-sea sediments.

ORIGIN AND EVOLUTION

Dust particles in the interplanetary medium are transient bodies with individual lifetimes that are much shorter than the age of the solar system. Radiation pressure drag (the Poynting-Robertson effect) and self-collisions remove particles from the inner solar system on a time scale of less than 10^6 yr (Dohnanyi 1978). It is not known how stable the dust inventory is, but hypervelocity microimpact craters found on meteoritic and lunar grains, which had ancient exposures to space, indicate that dust has existed in the interplanetary medium for billions of years (Brownlee & Rajan 1973). Presumably, the dust complex is maintained in some approximation to quasi-equilibrium, with new dust generated at a rate sufficient to replace the dust lost by the sink mechanisms.

All of the solid bodies in the solar system can inject particulates into the interplanetary medium, including planets like the Earth that release material during major impact events. For most of the lifetime of the solar system, however, the only two important dust sources within the solar system have been asteroids and comets. These are small bodies with low surface gravities and no permanent atmospheres. They are also by far the most abundant bodies in the planetary system. The solar system contains $>10^{11}$ comets and $\sim 10^6$ asteroids larger than a kilometer in diameter. Both of these parent materials appear to be relic planetesimals preserved since the time of planet formation. The asteroids are bodies ranging in size up to 1000 km. Presumably, they are fragments of planetesimals that formed in the Jupiter-Mars region of the solar nebula but that were never incorporated into a planet. Comets are enigmatic bodies that range in size up to at least 50 km. They are ice-rich bodies that sublime at temperatures above 150 K, and they are unstable in the terrestrial planet region of the solar system. The most popular theory for the origin of the comets is that they are ice/dust planetismals that formed in the Uranus-Pluto region of the nebula and that have been stored in much larger orbits for most of their lifetimes.

Comets produce visible dust tails and the annual meteor showers. They are the only proven source of dust, and they are widely believed to be the major source of both dust (Whipple 1967, Millman 1972) and the larger Earth-crossing meteoroids (Wetherill 1974). When the ice in a comet sublimes, it releases dust from the nucleus and propels it outward by gas drag. The dust then forms a tail, whose shape is determined by the interplay of gravity and light pressure (Sekanina 1980). Asteroids generate particles only as a by-product of collisions. A fraction of the dust at 1 AU certainly must be debris from main-belt asteroids, but no model of this source has ever suggested that it dominates comet production.

While comets are widely accepted as the most important dust source, the details of the supply mechanism are poorly understood. Long-period (10^6 yr) comets produce large quantities of dust, but the particles are not trapped in bound orbits and they are lost from the solar system. Comets with highly elliptical orbits have an almost perfect balance between kinetic and potential energy when they are close to the Sun and are generating particles. When sunlight shines on a newly liberated particle, its radial force due to momentum transfer counteracts part of the Sun's gravitational pull, the total energy of the particle becomes positive, and the particle escapes the solar system on a hyperbolic path (Harwit 1963). For parabolic comets passing close to the Sun, even particles of centimeter size whose ratio of light pressure force to gravity is only 10^{-5} can be "blown out" of the solar system by this mechanism.

Short-period comets have orbital periods of three to tens of years, and they can directly inject dust into bound solar orbits. Unfortunately, observed short-period comets fail by at least an order of magnitude in supplying the 10^7 g s^{-1} (Whipple 1967) of dust lost in the solar system by collisions and Poynting-Robertson drag (Kresak 1976, 1980, Roser 1976). It may be that most of the dust is released from rare and highly active comets. For example, both Whipple and Kresak suggest that the now nearly devolatilized Comet Encke must have been much more active in the past and may have itself produced the bulk of dust now in the inner solar system. Because the productive lifetimes of short-period comets are probably less than 10^4 yr, while the survival lifetimes of generated particles are more like 10^5 yr, this generation of dust in "pulses" could still result in a rather uniform concentration of particles in the solar system in time. Leinert et al (1983) suggest that most of the particles in the dust size range are debris from short-period comets, but that they were originally released as somewhat larger particles that later fragmented as a result of collisions.

In terms of interpreting laboratory studies of collected dust samples, there are many frustrating uncertainties about the origin of interplanetary dust. While comets are probably the major source, this has not been proven, and there is not a really good estimate of the ratio of asteroidal to cometary particles in the near-Earth environment. If comets are the major source, it is not known whether most particles originate from a single comet or thousands of comets. The nature and number of parents as well as the importance of collisions in dust evolution can fortunately be directly studied by laboratory analyses of collected samples. It is hoped that sample studies, combined with future data from spacecraft missions to comets, will shed new light on these problems.

COLLECTION IN SPACE

Space would seemingly be the ideal environment for collection of cosmic dust. Space collection does offer some unique opportunities, but as a general technique it is limited because of low collection rates and high collection velocities. The impact rate of 10 μm and larger particles on an Earth satellite is only 1 m^{-2} day^{-1}, and the typical impact velocity is probably ~ 15 km s^{-1}. The lowest impact velocity possible is ~ 3 km s^{-1}. This occurs when a particle approaches at the minimum velocity (escape velocity) and impacts the trailing end of the spacecraft on a trajectory that is parallel to the spacecraft orbit. At all of these velocities, however, the kinetic energy of the particle is higher than its binding energy, and the particle is largely destroyed upon "collection."

In the future it may be possible to nondestructively collect micrometeoroids in space by use of a gradual deceleration technique. At the present time the only practical method for collecting material is by approaches that essentially destroy most of the structural and chemical information in the particles but that retain valuable information on its elemental and isotopic composition. One technique is to have the particle impact onto an appropriate solid surface that favors retention of residue in the bottoms of hypervelocity impact craters formed in it. Simulation experiments and experience with microcraters on lunar samples indicate that silicates and probably most brittle materials are very poor substrates, but gold, aluminum, and probably other soft metals appear to be fairly effective because they couple the meteoroid's energy into the target without either vaporizing the sample or otherwise ejecting it from the crater. The first undisputed crater collection of this type was from a Hemenway et al (1967) experiment on *Gemini 10* where craters were collected in stainless steel. Residue found in several craters in more recent experiments has a composition that suggests an origin as high-velocity spacecraft or rocket debris (Clanton et al 1980). The only clear case of extraterrestrial material as residue in a crater was the residue with unfractionated chondritic elemental abundances in a 100 μm crater in aluminum exposed on *Skylab* (Brownlee et al 1974). The crater residue technique for collection in space works well with the correct collection substrate, and it will undoubtedly be used for future experiments.

The other space collection technique uses capture cells. This concept was invented by Herbert Zook at the NASA Johnson Space Center, and it is essentially an adaptation of the Whipple "meteor bumper" designed to shield spacecraft from destructive impacts. In a capture cell a meteoroid penetrates a thin diaphragm, which causes the meteoroid to vaporize. The vapor then condenses on the enclosed volume beneath the diaphragm. In theory, all condensable vapors are trapped in the cell and then can be later analyzed either by dissolving the sample out of the cell or by studying it in place with surface techniques like the ion microprobe. The first successful flight of a capture cell was made on a recent space shuttle flight (McDonnell et al 1984), and several capture cells, as well as crater collection experiments, are presently being exposed for a year on the *LDEF* satellite. Simulation studies by Zinner et al (1983b) indicate that high-quality isotopic measurements for selected elements can be made with the capture cell approach.

Space collections can at minimum determine the elemental compositions of individual particles in sizes up to ~ 100 μm and can provide some precise isotopic measurements. Space collections serve several unique purposes, one of which results from the fact that space collections are totally unbiased: all meteoroids are collected with equal efficiency. An additional advantage of space collections is that, in principle, they can be used to collect particles from known sources. A capture cell experiment to be flown on *Salyut* by Bibring et al (1983) will attempt to capture small particles from meteor showers by exposing the collector at the times of annual cometary meteor showers. The most convincing experiment along this line would precisely measure the velocity and impact direction of each particle, so that samples could be absolutely tied to a particular source. This active collection approach could also be used to identify the rare interstellar grains that constitute a small but finite fraction of the dust complex. The most direct approach for collecting cometary particles is to send an array of capture cells and cratering substrates directly through the dust coma that surrounds a cometary nucleus and return it to Earth. This type of mission, using a low-cost spacecraft launched on a free Earth return trajectory, was discussed by Tsou et al (1984).

ATMOSPHERIC ENTRY

All particles entering the atmosphere have kinetic energies sufficient to cause total vaporization if deceleration occurs on a short enough time scale. The miraculous aspect of dust interaction with the atmosphere is that the deceleration from hypervelocity is gentle and allows particles not only to be slowed without severe heating, but also to be slowed without severe mechanical stress.

Particles with bound solar orbits enter the atmosphere with velocities ranging from the escape velocity of 11.2 km s^{-1} to the 72.8 km s^{-1} velocity of a head-on impact with a body on a parabolic orbit. Observations of radio meteors show that 0.1–1.0 mm dust particles enter with typical velocities only a few kilometers per second above the escape velocity (Southworth & Sekanina 1979). These low entry velocities are due, in part, to orbit circularization by Poynting-Robertson drag (Dohnanyi 1978). Micron-sized and smaller particles may have much higher entry velocities as a result of light pressure effects at the time of particle generation (Zook & Berg 1975).

The entry of dust into the atmosphere differs significantly from the entry of the larger meteoroids that produce conventional meteorites. The airflow around centimeter-sized and larger bodies is complicated by the development of gas caps on their leading edges, while for smaller objects the interaction is free molecular flow. Air molecules simply impact the particle, with no aerodynamic effects before impact (Öpik 1958). If we assume that each impacting air molecule transfers all of its relative momentum to the particle, then the velocity of the decelerating meteoroid is given by

$$V = V_0 \, e^{-m'/m},$$

where V_0 is the initial velocity, m is the mass of the particle, and m' is the mass of the air column that has been encountered by the cross-sectional area of the particle. A particle can usually be considered to have been decelerated from its cosmic velocity after it encounters an air column equal to twice the particle mass. As can be seen in Figure 1, the high-velocity interaction with the atmosphere for dust occurs exclusively at altitudes above 70 km, while larger objects retain high velocities down to atmospheric levels below 40 km. For free molecular flow, the dynamic ram pressure on a meteoroid with unity accommodation of air molecules is $\rho_a V^2$, where ρ_a is the ambient air density. Dust loses its high velocity before it reaches dense air, and accordingly it is not subjected to high dynamic

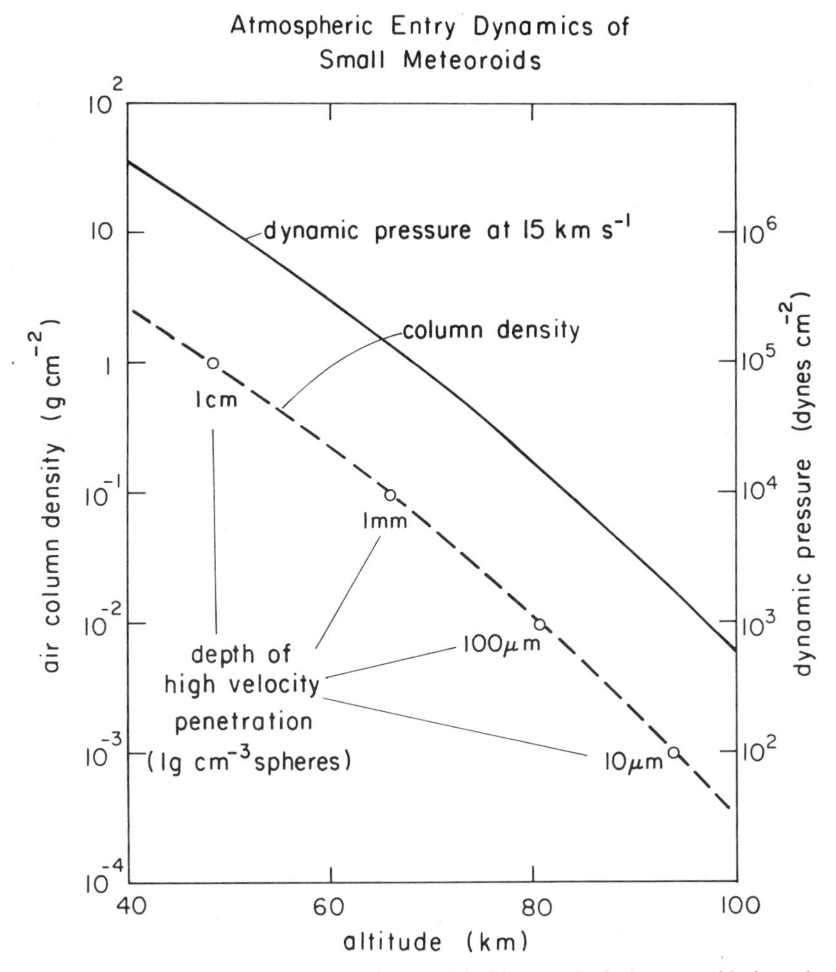

Figure 1 The depth of penetration of normal incidence spherical meteoroids into the atmosphere, under the assumption of no mass loss. The dynamic ram pressure is shown for 15 km s^{-1} travel at the various altitudes.

pressures. As can be seen in Figure 1, entering dust particles are not subjected to mechanical stress above 10^4 dyn cm^{-2} ($\sim$0.15 lb in^{-2}). This is the fundamental reason why dust is an important resource of primitive meteoritic material. Fragile materials with crushing strengths of only 10^2 dyn cm^{-2} can survive entry as dust, while only strong rocks can survive as pieces large enough to be found as conventional meteorites. This selection effect prevents typical cometary meteoroids from becoming meteorites. Meteor studies show that most bodies with cometary orbits fragment when the ram pressure exceeds 2×10^4 dyn cm^{-2} (Verniani 1969). Wetherill & ReVelle (1982) have shown that uncommon cometary meteoroids survive atmospheric stresses of $\sim 10^7$ dyn cm^{-2} and may be as strong as the weakest recovered meteorites.

Although rare and unusual cometary meteoroids may have survived to become meteorites, it is clear that the bulk of cometary materials can only survive in the form of millimeter-sized and smaller particles. As described by Verniani (1969), "Most meteors are of cometary origin and are porous crumbly objects composed of loosely conglomerate, spongelike material." Thus it appears that the only collectable samples of these objects are in the dust size regime.

Fragile materials can survive atmospheric entry in the form of dust, but they are all heated. Heating is actually more serious for dust than for conventional meteorites because dust particles are heated uniformly in their interiors and cannot have thermal gradients like in larger particles, where cool interiors coexist with molten exteriors. All dust particles are more strongly heated than the interiors of stony meteorites. Small particles decelerate at high enough altitudes that the power density generated by collision with air molecules can be matched by thermal radiation without the particle reaching its melting point ($\sim$1300°C for chondrites). Particles that do not melt are called "micrometeorites," a term coined by F. L. Whipple, who developed the theory of their survival (Whipple 1951). The temperature of an entering particle is given by

$$T = [\rho_a V^3/8\sigma\varepsilon]^{1/4},$$

where σ is the Stefan-Boltzmann constant and ε is the emissivity. The "maximum" diameter of a micrometeorite is usually considered to be $\sim$100 μm, but survival without melting is highly dependent on the particle density, the melting point, and the velocity and angle of entry into the atmosphere (Figure 2). Factors that favor survival are low density, high melting point, low entry velocity, and low entry angle. The "trick" for atmospheric entry without strong heating is to dissipate kinetic energy at high altitudes, where the power density input ($\frac{1}{2}\rho_a V^3$) is small. Particles as large as a millimeter can survive as micrometeorites if they enter at grazing incidence into the atmosphere; at the other extreme, metallic iron particles of 1 μm diameter will melt if they enter at a velocity of 70 km s^{-1} and vertical incidence. Because of uncertainty in entry angle and velocity, it is not possible to predict what maximum temperature was experienced for any single collected particle. Fraundorf (1980) and Fraundorf et al (1982c) have calculated that the typical 10 μm diameter micrometeorites that are collected in the stratosphere are heated to temperatures of 500–800°C. Typical heating time scales are 5–15 s. For the stratospheric micrometeorites (< 50 μm) that have been collected, less than 5% of the particles with chondritic elemental composition melted to form spheres. In contrast, more than 90% of the extraterrestrial particles larger than 200 μm that have been collected in ocean sediments are spheres. The ocean collection technique is, however, biased toward melted materials. As a general rule of thumb, it appears that typical particles smaller than 50 μm do not melt, while those larger than 100 μm do.

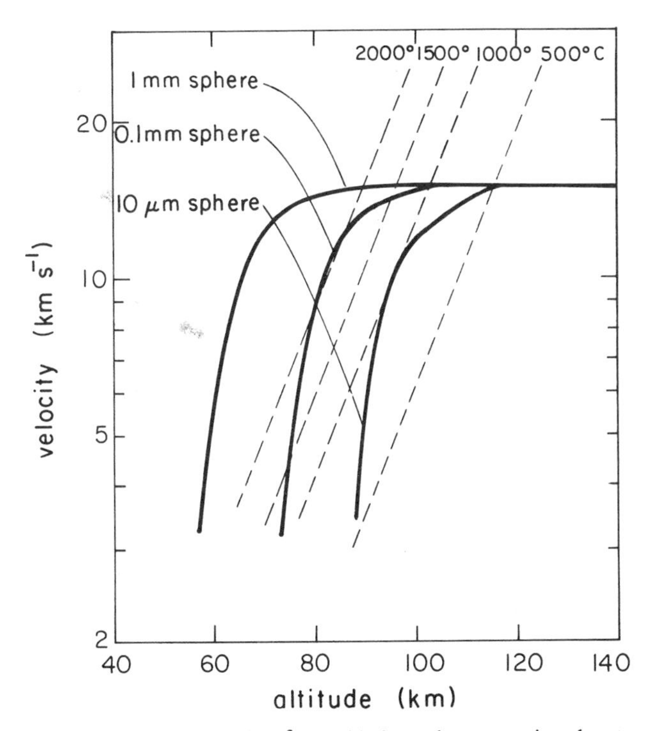

Figure 2 The deceleration of 3 g cm^{-3} nonablating spheres entering the atmosphere at normal incidence. The isotherms are calculated for perfect blackbody radiators.

An important consideration of atmospheric entry is whether survival is biased toward particles from particular sources. For a given size there is a strong bias against particles with high entry velocities. With only a few exceptions, visual meteors (>1 mm diameter) in the annual cometary showers have high entry velocities, and it would appear that atmospheric survival might favor asteroidal dust over comet dust. This, however, is probably not the case because dust particle orbits (not observable by meteor techniques) rapidly evolve as a result of light pressure effects and probably become relatively circular. This is consistent with the low entry velocities of the submillimeter particles observed as radio meteors. Orbital circularization reduces mean entry velocities for particles from parent bodies in the ecliptic plane, such as asteroids and short-period comets, but it cannot have a large effect on entry velocities for particles with high-inclination orbits. The entry process favors survival of particles from asteroids and comets with low-inclination orbits, and it is biased against particles from long-period comets and extrasolar grains.

COLLECTION IN THE TERRESTRIAL ENVIRONMENT

Roughly 10^4 tons of interplanetary dust impact the atmosphere each year. The flux of 10 μm particles is $\sim 1\ m^{-2}\ day^{-1}$, and that of 100 μm particles is $\sim 1\ m^{-2}\ yr^{-1}$ (McDonnell et al 1984, McDonnell 1978). Particles can be collected on land, from the ocean floor, and from the stratosphere. It may not be practical to collect large masses of dust, but it is feasible to collect millions of particles in the 10 μm to 1 mm size range.

Stratosphere

The most pristine dust particles are collected in the stratosphere. Collectable stratospheric particles are generally limited to the 2–50 μm diameter range because larger particles are too rare and smaller ones are swamped by the enormous background of submicron sulfate aerosol. Stratospheric particles of proven extraterrestrial origin were first collected in 1970 using a balloon-borne collector called the Vacuum Monster (Brownlee et al 1973, Brownlee & Hodge 1973). Successful collections using NASA U2 aircraft began in 1974 (Brownlee et al 1977), and at the present time routine collection and curation of particles for distribution to interested researchers is performed by the NASA Johnson Space Center in Houston, Texas (Clanton et al 1982).

In the stratosphere, 10-μm particles fall at a velocity of $\sim 1\ cm\ s^{-1}$ (Kasten 1968), a reduction of 10^6 from their velocity in space. Because flux is conserved, the concentration of 10-μm particles in the stratosphere is enhanced by a factor of 10^6 over the space value. The atmospheric density of 10 μm particles (Brownlee et al 1976) is 3×10^{-4} m^{-3}, and it is just barely possible to collect significant numbers of particles by high-volume air sampling onto very clean collection substrates. All collections have been made by inertial impaction of particles from a >200 m s^{-1} airstream onto clean plastic surfaces coated with several microns of a tarlike silicone oil. After collection, particles are individually picked off of collection surfaces and washed for analysis.

Stratospheric collections have been very successful, but future improvements are certainly possible. Collection with larger impactors would gather more of the rare large particles and would provide a greatly improved capability for sampling debris from events such as large meteors or showers. The mass increase in most showers is only minor (Erickson 1969), but it should be possible to collect material from the extraordinary showers that occur every few decades. To preserve the temporal signature of such events, it is necessary to collect large particles because their fall times to collection altitudes are short. Very high altitude collections (>40 km) may rise above the sulfate aerosol and allow collection of submicron extraterrestrial material. A particularly interesting possibility is the collection of small condensates from the fraction of the meteoroid complex that vaporizes in the atmosphere (Hunten et al 1980).

Deep-Sea Sediments

It is not really practical to collect particles larger than 100 μm in the atmosphere because of their low flux and high fall speed. It is quite simple, however, to collect such particles from deep-sea sediments, where low accumulation rates and long exposure times allow particles to collect in significantly greater concentrations than in any other terrestrial environment. For common Pacific red clay with accumulation rates of 2×10^{-6} m yr^{-1}, the total concentration of extraterrestrial material should be on the order of several parts per million by weight. At least 20 ppb of this material is recoverable in the form of spheres larger than 200 μm (Murrell et al 1980). Because of the size range sampled, of weathering problems, and of collection biases, most particles recovered from sediments are spheres in the 50 μm to 2 mm size range that formed by melting of particles too large to survive as micrometeorites. Totally unmelted particles up to a millimeter in diameter are rare in comparison with spheres, but they can be collected in large numbers (Brownlee et al 1980).

The majority of particles larger than 100 μm are ferromagnetic because of magnetite formation during atmospheric entry. Particles that melt usually crystallize to form spheres composed of olivine, magnetite, and glass, a material very similar to the fusion crusts of chondritic meteorites

(Blanchard et al 1980, Brownlee et al 1975). Large terrestrial magnetic spheres are rare in deep abyssal sediments far from continents and islands, and meteoric spheres can easily be collected with simple magnets. Spheres were first collected over a century ago by running a hand magnet through recovered abyssal clays (Murray & Renard 1891). Large-scale recovery of spheres from tons of recovered sediment was made by Millard & Finkelman (1970); spheres were also collected directly from the seafloor by use of towed sleds or rakes (Bruun et al 1955, Brownlee et al 1979). A remarkable collection effort by Finkelman (1972) showed that the denser iron-type spheres are concentrated in manganese nodules by factors of more than a hundred above that in the surrounding sediment.

Deep-sea collections have been reviewed by Petterson & Fredriksson (1957), Hodge (1981), Parkin & Tilles (1968), and Brownlee (1981). Approximately a million cosmic spheres have been collected from the ocean floor, and they are perhaps the least biased collection of interplanetary material, because once materials melt and form spheres, structural properties for atmospheric survival become equalized. All deep-sea particles were strongly heated during atmospheric entry, and they are exposed to chemical alteration in the sediments for typical time scales that are much longer than the few-thousand-year time scale for weathering of stony meteorites on land. Spheres are affected by heating, but they are changed in predictable ways; fortunately, weathering on the frigid ocean floor is simple and quite slow. The oldest recovered stony spheres are 50 Myr old, and all that remains is magnetite and the iron-rich rims of olivine grains. Typical younger stony spheres have an unaltered core surrounded by a rim where glass has been dissolved out, and sometimes the Mg-rich cores of olivine grains are also attacked. It is likely that the spheres survive in recognizable forms even in the oldest seafloor sediments.

Surface Collections

It is possible to collect extraterrestrial dust on the surface of the Earth, although it is much more difficult than in deep-sea sediments because the concentration is much lower and the abundance of terrestrial magnetic particles is high. Particles are collectable in polar ices (Wagstaff & King 1981) and also in much more complicated "contamination" environments, such as beach sands (Marvin & Einaudi 1967) and even desert soil (Fredriksson & Gowdy 1963). Surface collections are valuable in that they may provide material from specific sources, such as the Tunguska event as reported by Zbik (1984) and Ganapathy (1983). Correlation of particles with events might be done by taking particles from specific areas, from dated cores, or from material eroding out of specific units (such as was the source of the beach sand deposit studied by Marvin & Einaudi).

PARTICLE TYPES

Particles collected in the stratosphere are well preserved but small, whereas those collected from deep-sea sediments are large but significantly altered. It is by no means straightforward to compare particle types from both collections. If particle types from both collections can be correlated, then the deep-sea materials can provide milligram particles of materials that can be collected in the stratosphere only as nanogram samples. These large sample masses are critical for several important isotopic and trace-element analyses.

Deep-Sea Particles

The obvious extraterrestrial particles in sediments are magnetic spheres. These are usually considered to be melt products of atmospheric entry, but some may have originated in space by collisions (Parkin et al 1977). Most of the collected spheres are either the iron (I) type composed of FeNi metal, magnetite, and wustite (Figure 3) or the stony type (S) composed of olivine, magnetite, and glass with approximately chondritic elemental abundances (Figure 4). The two sphere types occur with approximately equal abundance at the 300 μm size. In fairly common Pacific red clay, nearly all magnetic spheres larger than 300 μm are extraterrestrial, although some sediments do contain large numbers of spheres of terrestrial origin. It certainly cannot be assumed that typical spheres in deep-sea sediments are extraterrestrial. Fortunately, the cosmic spheres have distinctive properties, and they can usually be absolutely identified when examined in polished section in a scanning electron microscope (SEM). The extraterrestrial origin for the spheres was first proposed in 1876 by Sir J. Murray because the I spheres had metallic iron cores with no obvious terrestrial origin. A large amount of supportive evidence implies that both the type I and S spheres are extraterrestrial. This evidence includes the occurrence of wustite in the I spheres (Marvin & Einaudi 1967), major- and trace-element composition for the S spheres (Ganapathy et al 1978), high $^{87}Sr/^{86}Sr$ ratios (Papanastassiou et al 1983), and oxygen isotopic composition (Mayeda et al 1983). Perhaps the most conclusive evidence has been the detection of the cosmogenic isotopes ^{53}Mn (Nishiizumi 1983) and ^{26}Al and ^{10}Be (Raisbeck et al 1983) that imply space exposure of on the order of 10^6 yr. Shimamura et al (1977) also reported anomalous potassium isotopic ratios that they attributed to cosmic-ray spallation.

The iron spheres are composed of magnetite and wustite that commonly enclose either a small metallic FeNi core or a smaller nugget of platinum group elements (Brownlee et al 1984). The spheres with FeNi cores are in an intermediate state of oxidation, while those with the Pt nuggets are nearly completely oxidized, with all platinum group elements concentrated into a bead only a few percent the size of the entire sphere. When the initially pure molten metal particle begins to oxidize, Ni and other siderophiles concentrate in the remaining metal phase. Spheres with FeNi metal cores often have Ni/Fe ratios in the metal that are much higher than the chondritic ratio. With nearly complete oxidation, Ni goes into the oxide, leaving only a nugget of platinum group elements whose diameter is only a few percent of the size of the entire sphere. The elemental composition of most of the nuggets matches chondritic abundances for the platinum group elements, with some depletion of the more volatile ones.

The iron spheres may have been derived from original metal in meteoroids, or they may have formed by reduction from fine-grained, carbon-rich meteoroids pulse heated at the top of the atmosphere. Certainly some, if not most, of the iron spheres form by reduction followed by inertial separation of stony and metallic phases (Brownlee et al 1983). This process explains why most stony spheres are depleted in Ni (and Cr),

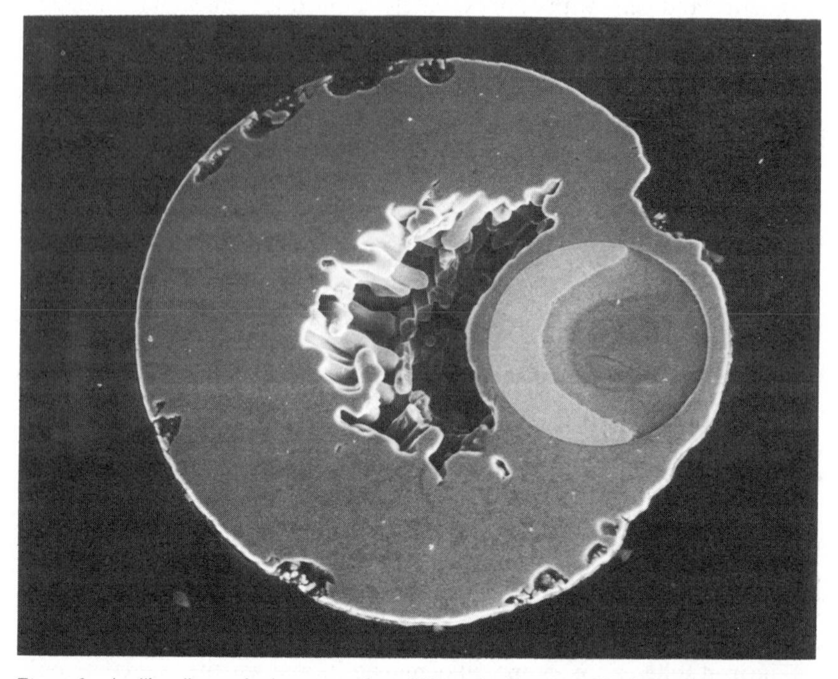

Figure 3 An "iron" cosmic deep-sea sphere 250 μm in diameter. The bulk of the particle is magnetite and wustite. The crystal-lined central vug is fairly common in the smaller spheres. The off-center core is metal and contains most of the nickel in the sphere. The interior of the core has been altered by weathering.

why I and S spheres are found in similar abundance, and why the ^{53}Mn (a lithopile) is depleted in the iron cosmic spheres relative to the stony ones (Nishiizumi 1983).

The stony spheres are composed almost exclusively of olivine, magnetite, and glass. They most commonly have barred textures and less frequently have porphyritic textures. Rare spheres contain a NiFe metal bead at one end or two beads at opposite ends of a prolate spheroid. Some spheres contain troilite spheroids or relict mineral grains [usually either forsterite, enstatite, or chromite (Blanchard et al 1980)]. Roughly 5% of the stony spheres larger than 200 μm contain relict grains, and a smaller number are essentially true unmelted micrometeorites (Brownlee et al 1980). The unmelted particles are usually composed of either forsterite, sulfide, and enstatite (with no fine-grained matrix) or forsterite, sulfide, and enstatite grains embedded in a submicron matrix with chondritic composition. The unmelted particles are best found in polished sections and can be identified by their composition.

The bulk composition of the stony spheres is chondritic, except for Na

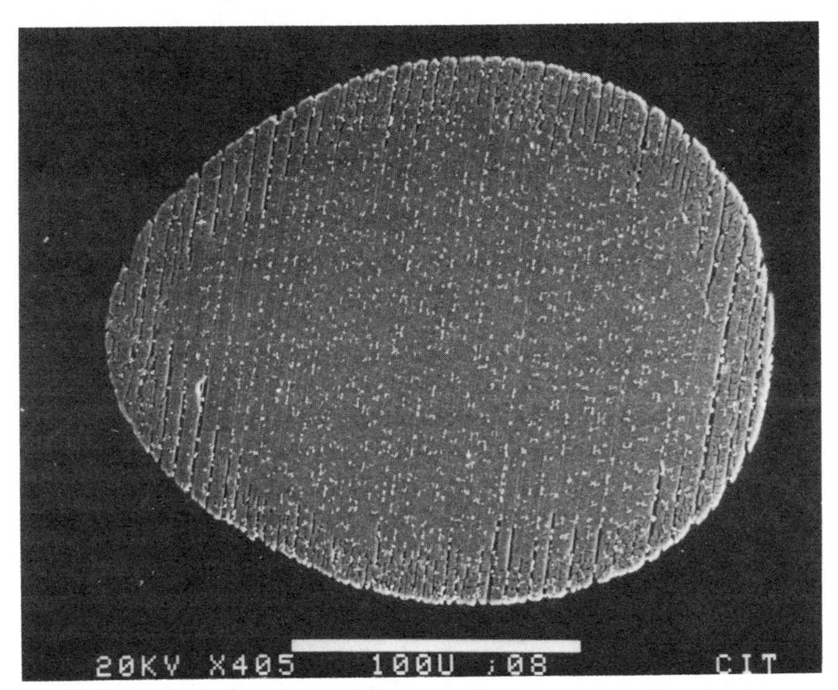

Figure 4 A barred stony deep-sea sphere composed of olivine, magnetite, and glass. The small bright grains are magnetite. Glass has weathered out of the outer 10 μm of the particle, leaving only olivine bars and magnetite (scale bar = 100 μm).

and S (which are presumably depleted by volatilization) and Ni and Cr (which are probably partially lost by formation and ejection of a metal or sulfide droplet). In the stony spheres with a metal bead at one end, Ni is completely concentrated in the metal phase.

Stratospheric Particles

Collectable stratospheric particles are usually smaller than 50 μm. On good collections when the stratosphere is not dominated by volcanic particles, extraterrestrial particles constitute roughly 30% of all irregular particles larger than 5 μm. Most extraterrestrial particles are easily identified because they have chondritic elemental abundances or contain fragments of such material. Only a relatively minor number of particles cannot be identified as having either an extraterrestrial or an obviously terrestrial origin on the basis of their elemental composition. It is very unlikely that any abundant type of micrometeorite would be overlooked in the stratosphere.

Because the stratospheric particles are small, it is difficult to classify them in ways that are both meaningful and operationally useful. Unfortunately, the most meaningful taxonomy schemes are likely to be so complex that they cannot be used on routine samples. Classification schemes have been presented by Brownlee (1978), Fraundorf et al (1982d), Brownlee et al (1982), MacKinnon et al (1982), and Kordesh et al (1983). Basically, the particles can be split into two categories: those that obviously melted (ablation spheres), and those that did not melt (micrometeorites). Simple subdivisions can be made on the basis of bulk composition, mineralogy, and structure. Relative abundances of particle types determined by bulk processing of a collection surface were reported by Zolensky et al (1984).

The ablation spheres in the stratosphere are rare and comprise less than 10% of 10-μm particles. In addition to the familiar iron and chondritic stony spheres found in deep-sea sediments, sulfide and nonchondritic silicate spheres are found in the stratosphere. The sulfide spheres are the most common type; oddly, they have never been found in deep-sea sediments, even though sulfides clearly survive weathering as relict grains in unmelted particles. It is likely that they do not occur in sizes as large as collectable deep-sea particles. The sulfide spheres are composed of magnetite and pyrrhotite and have been called FSN because they are composed of iron, sulfur, and nickel. Silicate spheres, depleted in Fe and enriched in Ca, Al, and Ti relative to chondritic abundances, have been found and are called CAT or CAS particles. Unlike the chondritic composition spheres, these are transparent. The only real evidence that the CAT or CAS spheres are extraterrestrial is the 11‰ isotopic fractionation in Mg in one such particle measured by Esat et al (1979). These spheres may be the residues of meteoroids that were so strongly heated that Fe, Si, and possibly Mg were depleted by volatilization. The final type of ablation particle is really intermediate between a micrometeorite and a true stony ablation sphere. These are the MMS (metal mound silicate) particles, composed of rounded or spheroidal silicates covered with small mounds of FeNi metal. Sometimes these occur in groups embedded in a lacy carbonaceous material. The MMS particles have been perfectly simulated by taking the most common type of chondritic micrometeorite (metal free) and flash melting them in vacuum with an electron beam. The MMS particles apparently form during atmospheric entry and dramatically illustrate the formation of metal by reduction from carbon-rich particles.

True micrometeorites can be put into two groups: those that have chondritic abundances (within a factor of 3) for the 12 most abundant elements, and those that do not. The nonchondritic particles that have been identified are most often dominated by a single mineral grain, but they also usually have some chondritic material attached to their surfaces. These grains were apparently previously embedded in fine-grained chondritic material. Many lines of evidence indicate that chondritic micrometeorites are extraterrestrial and that there is little likelihood that any particle with a truly chondritic elemental composition could be terrestrial. The strongest evidence that these dust particles are truly interplanetary is the detection of very high abundances of rare gases that almost certainly are implanted by the solar wind (Rajan et al 1977, Hudson et al 1981) and the detection of solar flare tracks (Bradley et al 1984b).

The chondritic particles are very fine grained, and there is considerable variation between samples at the submicron scale (Fraundorf 1981), which indicates the existence of several subgroups. All have very similar bulk elemental compositions. A complicating factor in interpreting these particles is that the effects of atmospheric heating are poorly understood. Infrared transmission spectroscopy (Fraundorf et al 1980, Sandford 1983) has shown that the chondritic particles fall into three groups, with infrared spectra similar to olivine, pyroxene, and hydrated silicates. Brownlee et al (1982) separated the chondritic particles into two main morphological groups discernible by SEM observation. This is a convenient (but crude) scheme that cannot be unambiguously used for all particles. One group is porous (CP, or chondritic porous) on the micron scale, and the other major group (CS) is smooth on the micron scale and yet has a chondritic composition on that same size scale. Both types are optically black. The CP particles have a classic cluster-of-grapes morphology that is distinct from all known meteoritic materials (Figure 5). The morphology is not unique, however, and is similar to many powders composed of submicron, fairly equidimensional grains. The majority of the CP particles are generally anhydrous and are composed of large numbers of mineral grains, ranging in size from ~50 Å to several microns. In the SEM, the particles appear as clusters of grains averaging ~0.3 μm in diameter. Examination in the transmission electron microscope (TEM) shows, however, that in some particles many of these submicron grains are single minerals, while in others most of the grains are composed of enormous numbers of much smaller crystals embedded in amorphous material. Most of the particles contain significant amounts of amorphous carbonaceous material and possibly amorphous silicates as well.

The major minerals larger than a micron are enstatite, olivine, and pyrrhotite. Less-abundant phases are pentlandite, NiFe carbide, diopside, albite, kamacite, magnetite, chromite, Ca carbonate, and Mg and Ca phosphate. The elemental composition of the CP particles is similar to CI meteorites, including carbon, but the mineralogy and texture of the particles are distinctly different from those of any classified meteorite. The

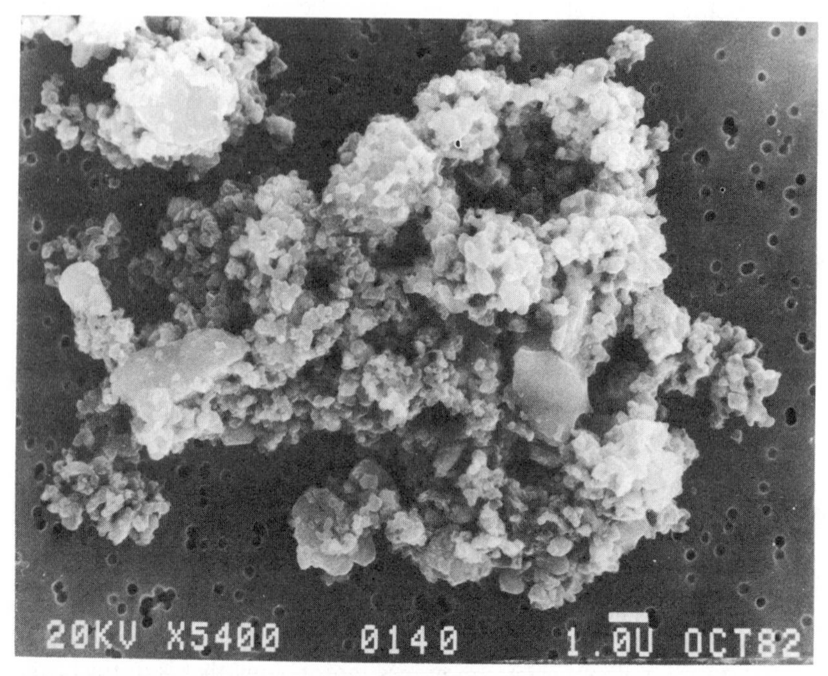

Figure 5 A porous chondritic composition (CP) particle that is an aggregate of a large number of submicron grains. This particle is unlike established meteorite groups and is a new type of chondritic material. Of all extraterrestrial materials, this particle type most commonly matches the physical properties of cometary meteoroids.

measured densities range from 0.7 to 2.2 g cm^{-3} and the CP particles are also far more porous than any conventional meteorite (Fraundorf et al 1982b). Rare particles crush on collection and cover areas as large as 300 μm with fine-grained debris. It is possible that these unusual chondritic particles had densities $\ll 1$ g cm^{-3} before collection. The CP particles are apparently a new type of carbonaceous chondrite that has no analog in conventional meteorites. They are the first known case of anhydrous, carbon-rich chondrites. MacKinnon & Rietmeijer (1983), however, indicate that at least one particle in this class does contain hydrated phases.

The CS particles (Figure 6) have smooth regions that have chondritic compositions, and when viewed in the SEM they do not look like porous aggregates of rather equidimensional grains. Smooth areas must be "matrix" and not simply large mineral grains. Some of the CS particles are depleted in Ca, and some have been shown to be composed of hydrated silicates (Tomeoka & Buseck 1984, Sandford 1984). Morphologically, the

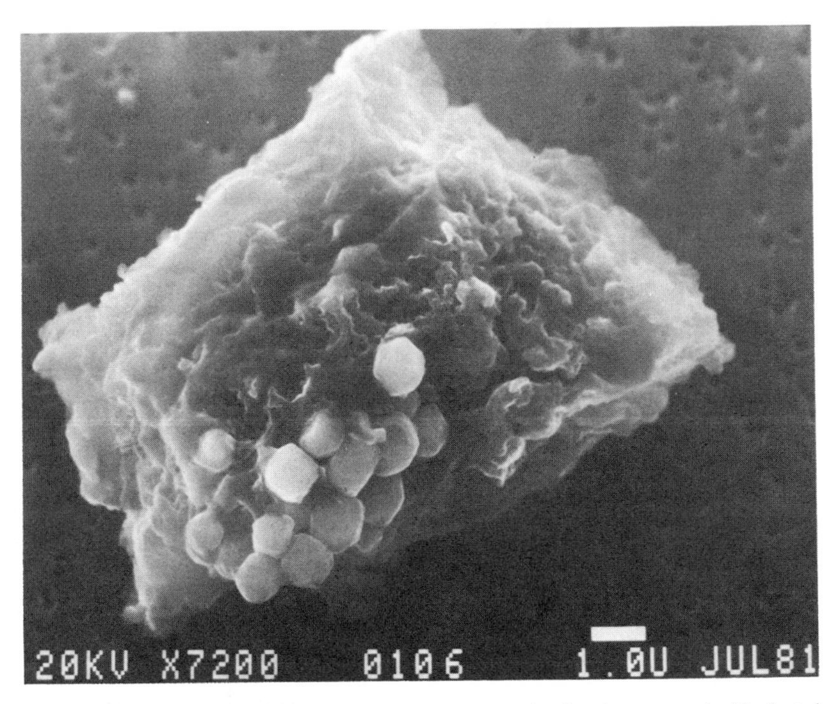

Figure 6 A smooth chondritic (CS) stratospheric meteorite that is composed of hydrated silicates. The 1-μm grains in the front of the particle are magnetite occurring in a cluster characteristic of CI chondrites. This particle is probably CI material.

CS particles are similar to fragments of the phyllosilicate matrix of CI and CM meteorites.

The most common nonchondritic micrometeorites are pyrrhotite, enstatite, forsterite, kamacite, phosphate, and carbonate, in approximate order of abundance. To be positively identified as extraterrestrial, these particles usually must be associated with chondritic material. This is probably not a serious selection effect, because a large proportion of particles suspected of being extraterrestrial on mineralogical grounds do have coatings of fine-grained chondritic material. The black chondritic surface material varies from isolated patches to heavy encrustations. Encrusted particles sometimes appear to be chondritic aggregate particles that significantly deviate from chondritic composition. Nearly all of the nonchondritic particles appear to be simply large mineral grains or clumps of large grains that were previously embedded in CP or CS material. If collisional evolution is important in the meteoritic complex, then these comparatively strong particles may be overabundant in the interplanetary medium in relation to their abundance inside the original parent bodies.

ANALYSIS IMPLICATIONS

Parent Materials

Almost all of the dust particles that have been analyzed appear to be samples of primitive solar system material. The bulk of the stratospheric micrometeorites are extremely fine-grained chondritic materials whose only possible meteoritic analogs are CIs and CMs. The fine-grained material in deep-sea particles is not preserved, but the bulk compositions and minerals of large relict grains provide powerful constraints on the parent meteoroids. Analysis of the major refractory elements (Mg, Ca, Al, and Ti) in hundreds of stony spheres show that $\sim 85\%$ of the spheres have abundances that match CI/CM values and are distinct from other types of meteorites. The abundances of these elements are not changed during entry or in the seafloor for unweathered material, and they are very diagnostic of the parent material. Analysis of minor-element abundances in forsterite relict grains match CM forsterite and are distinct from measurements of other terrestrial and extraterrestrial materials (Smith et al 1984). Abundant evidence indicates that the great majority of interplanetary dust is compositionally similar to CM and CI meteorites. This frequency is dramatically different from normal meteorites, where less than 3% of falls are CI or CM meteorites (Wasson 1974), the only true carbon-rich chondrites. Either the composition of the meteoroids is size dependent, or else conventional meteorites are a very poor sampling of the meteoroid complex.

The elemental abundances of most analyzed dust samples are similar to CI and CM chondrites, but this does not imply an origin from CI and CM parent bodies. Certainly, some of the dust particles must come from the same sources as CI and CM meteorites, but other parent bodies appear to be represented as well. Some of the hydrated stratospheric particles could be CI or CM material, as could some of the unmelted deep-sea particles composed of mafic grains and sulfides embedded in a fine-grained chondrite matrix. The oxygen isotopic composition of the stony spheres is compatible with the CM anhydrous component (Mayeda et al 1983). The previously mentioned minor-element abundances in forsterite are consistent with a CM origin, as are the high Cr contents in metal spheres found in some forsterite grains. It is difficult to reliably identify particles as genuine CI or CM material unless they contain unique features, such as the distinctive forsterite grains or magnetite framboids. Future work, both on the meteorites and the dust, should provide a good estimate of the fraction of the dust complex that is actually conventional carbonaceous chondrite material. Evidence that not all of the hydrated particles are actually related to known meteorites comes from the TEM work of Tomeoka & Buseck (1984). They identified the dominant hydrated phase in one particle as Mg smectite. The major hydrated phase in carbonaceous chondrites is serpentine, and smectite has not been previously identified in meteorites.

The anhydrous CP particles are aggregates of submicron grains, and they are clearly a type of material that has not been seen in meteorites. A simple and fundamental distinction is that the elemental abundances of the particles match the bulk CI value, and yet no fine-grained material in meteorites has chondritic composition on the size scale of the stratospheric particles. Chondrites, by definition, have bulk chondritic composition for major and minor elements, but the submicron matrix regions of all major groups of carbonaceous and unequilibrated chondrites have significant deviations from the bulk composition. In the case of the CI and CM meteorites, these deviations have been attributed to element redistribution by aqueous alteration (McSween 1979). Redistribution in all of the meteorites is likely the result of processes that occurred inside the parent bodies. This implies that the stratospheric particles may be the least altered solar system material.

Other unique properties of the CP particles also suggest that they are truly well-preserved materials. The occurrence of epsilon iron carbide and enstatite crystals with unusual growth habits suggests that some original solids that formed from the vapor phase have survived. Epsilon nickel-iron carbide has been identified by Christoffersen & Buseck (1983) and Bradley et al (1984a) using TEM techniques. It occurs in association with metal, magnetite, amorphous carbon, and (rarely) graphite. This phase had not been previously reported in natural systems, but it has been produced in laboratories by catalytic reactions involving carbon monoxide and grain surfaces. It is also considered a low-temperature phase.

Most of the particles also contain minor amounts of enstatite, which occurs either in the form of rods (whiskers), ribbons, or platelets. Some of the rods have axial screw dislocations, and the ribbons and platelets are flattened, or extended, along axes not consistent with cleavage, parting, or crystallization from a melt. The observed forms and growth habits are consistent with whisker and platelet growth from vapor (Bradley et al 1983). The survival of these fragile crystals suggests that many other vapor phase products may have also survived in the CP particles. One intriguing mystery is why the particles contain no metal. For common pressures discussed for the solar nebula, iron condenses at the same temperature as enstatite. The survival of thin enstatite platelets implies that the metal should have also survived without incorporation into oxidized phases, and yet the particles contain only traces of metal. If the enstatite condensed in equilibrium with solar nebula gas, then the metal condensation phase was apparently skipped. Some of the metal that did form was converted to carbide.

Some of the hydrated particles are similar to CI and CM chondrites, and like these meteorites, they may have been extensively altered by liquid water inside a parent body (Bunch & Chang 1980, McSween 1979). The porous aggregate particles are anhydrous and show no evidence of such processing. Although there is no proof, the porous anhydrous particles are probably cometary material. It is interesting to speculate that parent body alteration of this material might produce CI and CM meteorites as well as the hydrated micrometeorites.

Comets and Interstellar Grains

Fifteen years ago, Fred Whipple pointed out that there was no single collected particle that could be positively identified as interplanetary dust. There are now large numbers of collected interplanetary dust particles that have been positively identified, but the original question can now be applied to comets. Comet dust is abundant in space, and there is no obvious process that could prevent this dust from surviving as micrometeorites and ablation spheres. Comet dust should be a significant and possibly a major component of the dust collections. It is not presently known how to identify cometary materials, but analysis of individual particles on future comet missions may provide important clues. Sandford (1983) has shown that the shape of the infrared silicate feature measured for Comet Kohoutek can be matched by adding spectra of the three infrared dust groups.

The CP particles most closely match the common description of cometary dust as fragile, porous particles. Actually, it is possible that the most porous CP particles could not survive in an asteroidal body because of their collisional histories. If the CP particles are representative of cometary solids, then this would indicate that comet formation included a phase of gentle aggregation of submicron grains, along with ices and carbonaceous matter. The grains in CP particles have diverse properties and are anhydrous, which implies that they never equilibrated at low temperature with solar-nebula-type gas. The porous particles are black and would presumably be very dark, even when filled with ice. A comet nucleus made of CP material would have a low albedo and would have ice and dust mixed on a very fine scale.

Some presolar interstellar materials may have been well preserved in comets, and identification of such grains is a major rationale for the analysis of cometary solids. The most direct information that might indicate a presolar origin is isotopic composition. If the aggregate particles are mixtures of preserved interstellar grains, then this is a difficult measurement because most of their constituent grains are less than a micron in size. Isotopic analysis of CP particles for Mg by Esat et al (1979) showed a hint of a nonlinear effect at the 4‰ level, but the main result was that the Mg isotopic ratios were normal in the particles at the 1% level. Recent ion microprobe measurements of hydrogen in several micrometeorites have shown that whole particles are significantly enriched in deuterium and that some portions of particles are enriched by 2500‰ (Zinner et al 1983a, Zinner & McKeegan 1984). Only a unique carbonaceous chondrite (Renazzo) contains such fractionated hydrogen. The high D/H ratios may be indicators of interstellar carbonaceous material and may be related to even higher ratios thought to be produced by ion-molecule reactions in molecular clouds. Another possible method of linking collected particles with interstellar grains is through their infrared signatures (Sandford 1984, Sandford & Walker 1984).

Information About the Earth

Cosmic dust in sediments records major events in the Earth's history when large amounts of extraterrestrial material were accreted into the atmosphere. Dust particles also contain records of other terrestrial processes. Isotopic measurements by Mayeda et al (1983) have shown that the atmospheric oxygen that forms cosmic iron spheres is enriched in ^{18}O by 50‰. This oxygen is picked up at ~80 km, where the particles melt, and it is evidence for possible large fractionation of atmospheric oxygen at this altitude. It has recently been discovered that platinum nugget formation in these iron spheres is critically dependent on the partial pressure of oxygen in the atmosphere (Brownlee et al 1984), and the first appearance of these extraterrestrial nuggets in the geological record may provide a marker for when oxygen in the atmosphere first rose to half its present value.

CONCLUSION

Small interplanetary particles are an important resource of primitive solar system material, because meteoroid types that are too fragile to survive atmospheric entry to become conventional meteorites can be collected in the form of dust. The most common cometary particles are almost certainly represented in the dust collections, and they may be abundant. The particles in the collections that are most likely to be cometary are black particles composed of porous aggregates of submicron grains whose bulk composition matches relative solar abundances. These particles are similar to carbon-rich chondrites in bulk elemental composition, but they differ in many significant ways, including mineralogy, morphology, and D/H ratio.

Dust studies are complicated by the small masses of available material, but it is likely that much of the important information in primitive materials is in micron and smaller grains anyway. Even if kilogram samples of some of the particle types were available, analyses would still be done at the smallest possible size scale. This is particularly true with reference to searches for presolar interstellar grains. Techniques being developed for analysis of collected dust samples are also being used for studying other extraterrestrial materials; these techniques will be very valuable for the analysis of the first comet samples that are directly collected from the nucleus of an active comet.

Literature Cited

Bibring, J.-P., Borg, J., Langevin, Y., Rocard, F., Vassent, B. 1983. The C.O.M.E.T. experiment. *Lunar Planet. Sci. XIV*, pp. 37–38 (Abstr.)

Blanchard, M. B., Brownlee, D. E., Bunch, T. E., Hodge, P. W., Kyte, F. T. 1980. Meteoroid ablation spheres from deep sea sediments. *Earth Planet. Sci. Lett.* 46:178–90

Bradley, J. P., Brownlee, D. E., Veblen, D. R. 1983. Pyroxene whiskers and platelets in interplanetary dust: evidence of vapor phase growth. *Nature* 301:473–77

Bradley, J. P., Brownlee, D. E., Fraundorf, P. 1984a. Carbon compounds in interplanetary dust: evidence for formation by heterogeneous catalysis. *Science* 223: 56–58

Bradley, J. P., Brownlee, D. E., Fraundorf, P. 1984b. Discovery of nuclear tracks in interplanetary dust. *Science* 226:1432–34

Brownlee, D. E. 1978. Microparticle studies by sampling techniques. See McDonnell 1978, pp. 295–336

Brownlee, D. E. 1981. Extraterrestrial components in deep sea sediments. In *The Sea*, ed. C. Emiliani, 7:733–62. New York: Wiley

Brownlee, D. E., Hodge, P. W. 1973. Ablation debris and primary micrometeoroids in the stratosphere. *Space Res.* 13:1139–51

Brownlee, D. E., Rajan, R. S. 1973. Micrometeorite craters discovered on chondrule-like objects from Kapoeta meteorite. *Science* 182:1341–44

Brownlee, D. E., Hodge, P. W., Bucher,

W. 1973. The physical nature of interplanetary dust as inferred by particles collected at 35 km. In *Evolutionary and Physical Properties of Meteoroids, IAU Colloq. No. 13*, ed. C. L. Hemenway, P. M. Millman, A. F. Cook, pp. 291–95. *NASA SP-319*

Brownlee, D. E., Tomandl, D. A., Hodge, P. W., Horz, F. 1974. Elemental abundances in interplanetary dust. *Nature* 252:667–69

Brownlee, D. E., Blanchard, M. B., Cunningham, G. C., Beauchamp, R. H., Fruland, R. 1975. Criteria for identification of ablation debris from primitive meteoric bodies. *J. Geophys. Res.* 80:4917–24

Brownlee, D. E., Ferry, G. V., Tomandl, D. 1976. Stratospheric aluminum oxide. *Science* 191:1270–71

Brownlee, D. E., Tomandl, D. A., Olszewski, E. 1977. Interplanetary dust: a new source of extraterrestrial material for laboratory studies. *Proc. Lunar Sci. Conf., 8th*, pp. 149–60

Brownlee, D. E., Pilachowski, L. B., Hodge, P. W. 1979. Meteorite mining on the ocean floor. *Lunar Planet. Sci. X*, pp. 157–58 (Abstr.)

Brownlee, D. E., Bates, B. A., Pilachowski, L. B., Olszewski, E., Siegmund, W. A. 1980. Unmelted cosmic materials in deep sea sediments. *Lunar Planet. Sci. XI*, pp. 109–11 (Abstr.)

Brownlee, D. E., Olszewski, E., Wheelock, M. 1982. A working taxonomy for micrometeorites. *Lunar Planet. Sci. XII*, pp. 71–72 (Abstr.)

Brownlee, D. E., Bates, B. A., Beauchamp, R. H. 1983. Meteor ablation spheres as chondrule analogs. In *Chondrules and Their Origins*, ed. E. A. King, pp. 10–25. Houston: Lunar Planet. Inst.

Brownlee, D. E., Bates, B. A., Wheelock, M. M. 1984. Extraterrestrial platinum group nuggets in deep sea sediments. *Nature* 309:693–95

Bruun, A. F., Langer, E., Pauly, H. 1955. Magnetic particles found by raking the deep sea bottom. *Deep-Sea Res.* 2:230–46

Bunch, T. E., Chang, S. 1980. Carbonaceous chondrites II: carbonaceous chondrite phyllosilicates and light element geochemistry as indicators of parent body processes and surface conditions. *Geochim. Cosmochim. Acta* 44:1543–77

Christoffersen, R., Buseck, P. R. 1983. Epsilon Carbide: a low temperature component of interplanetary dust particles. *Science* 222:1327–29

Clanton, U. S., Zook, H. A., Schultz, R. A. 1980. Hypervelocity impacts on Skylab IV/Apollo windows. *Proc. Lunar Planet. Sci. Conf., 11th*, pp. 2261–73

Clanton, U. S., Nace, G. A., Gabel, E. M., Warren, J. L., Dardano, C. B. 1982. Possible comet samples: the NASA cosmic dust program. *Lunar Planet. Sci. XIII*, pp. 109–10 (Abstr.)

Dohnanyi, J. S. 1978. Particle dynamics. See McDonnell 1978, pp. 527–605

Erickson, J. E. 1969. Mass influx and penetration rate of meteor streams. *J. Geophys. Res.* 74:576–85

Esat, T. M., Brownlee, D. E., Papanastassiou, D. A., Wasserburg, G. J. 1979. Magnesium isotopic composition of interplanetary dust particles. *Science* 206:190–97

Finkelman, R. B. 1972. Relationship between manganese nodules and cosmic spherules. *Mar. Technol. Soc. J.* 6:34–39

Fraundorf, P. 1980. The distribution of temperature maxima for micrometeorites decelerated in the Earth's atmosphere without melting. *Geophys. Res. Lett.* 7:765–68

Fraundorf, P. 1981. Interplanetary dust in the transmission electron microscope: diverse materials from the early solar system. *Geochim. Cosmochim. Acta* 45:915–43

Fraundorf, P., Patel, R. I., Shirck, J., Walker, R. M., Freeman, J. J. 1980. Optical spectroscopy of interplanetary dust collected in the Earth's stratosphere. *Nature* 286:866–68

Fraundorf, P., Brownlee, D. E., Walker, R. M. 1982a. Laboratory studies of interplanetary dust. In *Comets*, ed. L. L. Wilkening, pp. 383–409. Tucson: Univ. Ariz. Press

Fraundorf, P., Hintz, C., Lowry, O., McKeegan, K. D., Sandford, S. A. 1982b. Determination of the mass, surface density, and volume of individual interplanetary dust particles. *Lunar Planet. Sci. XIII*, pp. 225–26 (Abstr.)

Fraundorf, P., Lyons, T., Schubert, P. 1982c. The survival of solar flare tracks in interplanetary dust silicates on deceleration in the Earth's atmosphere. *J. Geophys. Res.* 87:A409–12 (*Proc. Lunar Planet. Sci. Conf., 13th*)

Fraundorf, P., McKeegan, K. D., Sandford, S. A., Swan, P., Walker, R. M. 1982d. An inventory of particles from stratospheric collectors, extraterrestrial and otherwise. *J. Geophys. Res.* 87:A403–8 (*Proc. Lunar Planet. Sci. Conf., 13th*)

Fredriksson, K., Gowdy, R. 1963. Meteoritic debris from the southern California desert. *Geochim. Cosmochim. Acta* 27:241–43

Ganapathy, R. 1983. The Tunguska explosion of 1908: discovery of meteoritic debris near the explosion site and at the South Pole. *Science* 220:1158–61

Ganapathy, R., Brownlee, D. E., Hodge, P. W. 1978. Silicate spherules from deep sea

sediments: confirmation of extraterrestrial origin. *Science* 201:1119–21
Harwit, M. 1963. Origins of the zodiacal dust cloud. *J. Geophys. Res.* 68:2171–80
Hemenway, C. L., Hallgren, D. S., Kerridge, J. F. 1967. Results from the Gemini S-10 and S-12 experiments. *SpaceRes.* 8:521–35
Hodge, P. W. 1981. *Interplanetary Dust.* New York: Gordon & Breach. 280 pp.
Hudson, B., Flynn, G. J., Fraundorf, P., Hohenberg, C. M., Shirck, J. 1981. Noble gases in stratospheric dust particles: confirmation of extraterrestrial origin. *Science* 211:383–86
Hughes, D. W. 1978. Meteors. See McDonnell 1978, pp. 123–85
Hunten, D. M., Turco, R. P., Toon, O. B. 1980. Smoke and dust particles of meteoric origin in the mesosphere and stratosphere. *J. Atmos. Sci.* 37:1342–57
Kasten, F. 1968. Falling speed of aerosol particles. *J. Appl. Meteorol.* 7:944–47
Kordesh, K. M., MacKinnon, I. D. R., McKay, D. S. 1983. A new classification and database for stratospheric dust particles. *Lunar Planet. Sci. XIV*, pp. 389–90 (Abstr.)
Kresak, L. 1976. Orbital evolution of dust streams released from comets. *Bull. Astron. Inst. Czech.* 27:35–46
Kresak, L. 1980. Sources of interplanetary dust. In *Solid Particles in the Solar System, IAU Symp. No. 90*, ed. I. Halliday, B. A. McIntosh, pp. 211–22. Boston: Reidel
Leinert, C., Roser, S., Buitrago, J. 1983. How to maintain the spatial distribution of interplanetary dust. *Astron. Astrophys.* 118:345–57
MacKinnon, I. D. R., McKay, D. S., Nace, G., Isaacs, A. 1982. Classification of the Johnson Space Center stratospheric dust collection. *J. Geophys. Res.* 87:A413–21 (*Proc. Lunar Planet. Sci. Conf., 13th*)
MacKinnon, I. D. R., Rietmeijer, F. J. M. 1983. Layer silicates and a bismuth phase in chondritic aggregate W7029*A. *Meteoritics* 18:343–44
Marvin, U. B., Einaudi, M. T. 1967. Black magnetic spherules from Pleistocene and recent beach sands. *Geochim. Cosmochim. Acta* 31:1871–84
Mayeda, T. K., Clayton, R. N., Brownlee, D. E. 1983. Oxygen isotopes in micrometeorites. *Meteoritics* 18:349–50
McCrosky, R. E., Posen, A., Schwartz, G., Shao, C.-Y. 1971. Lost City meteorite—its recovery and a comparison with other fireballs. *J. Geophys. Res.* 76:4090–108
McDonnell, J. A. M., ed. 1978. *Cosmic Dust.* New York: Wiley. 693 pp.
McDonnell, J. A. M., Carey, W. C., Dixon, D. G. 1984. Cosmic dust collection by capture cells. *Nature* 309:237–40
McSween, H. Y. 1979. Are carbonaceous chondrites primitive or processed? *Rev. Geophys. Space Phys.* 17:1059–78
Millard, H. T., Finkelman, R. B. 1970. Chemical and mineralogical compositions of cosmic and terrestrial spherules from a marine sediment. *J. Geophys. Res.* 75: 2125–33
Millman, P. M. 1972. Cometary meteoroids. In *From Plasma to Planet, Nobel Symp. No. 21*, ed. A. Elvius, pp. 157–68. New York: Wiley
Murray, J., Renard, A. F. 1891. *Rep. Sci. Results Voyage H.M.S. Challenger 3.* Edinburgh: Neill & Co.
Murrell, M. T., Davis, P. A., Nishiizumi, K., Millard, H. T. 1980. Deep sea spherules from Pacific clay: mass distribution and influx rate. *Geochim. Cosmochim. Acta* 44:2067–74
Nishiizumi, K. 1983. Measurement of ^{53}Mn in deep sea iron and stony spherules. *Earth Planet. Sci. Lett.* 63:223–28
Öpik, E. 1958. *Physics of Meteor Flight in the Atmosphere, Intersci. Tracts Phys. Astron. No. 6.* New York: Interscience. 174 pp.
Papanastassiou, D. A., Wasserburg, G. J., Brownlee, D. E. 1983. Chemical and isotopic study of extraterrestrial particles from the ocean floor. *Earth Planet. Sci. Lett.* 64:341–55
Parkin, D. W., Tilles, D. 1968. Influx measurements of extraterrestrial material. *Science* 159:936–46
Parkin, D. W., Sullivan, R. A. L., Andrews, J. N. 1977. Cosmic spherules as rounded bodies in space. *Nature* 266:515–17
Petterson, H., Fredriksson, K. 1957. Magnetic spherules in deep sea deposits. *Pacific Sci.* 11:71–81
Raisbeck, G. M., Yiou, F., Klein, J., Middleton, R., Yamakoshi, Y., et al. 1983. ^{26}Al and ^{10}Be in deep sea stony spherules: evidence for small parent bodies. *Lunar Planet. Sci. XIV*, pp. 622–23 (Abstr.)
Rajan, R. S., Brownlee, D. E., Tomandl, D., Hodge, P. W., Farrar, H., Britten, R. A. 1977. Detection of ^{4}He in stratospheric particles gives evidence of extraterrestrial origin. *Nature* 267:133–34
Roser, S. 1976. Can short-period comets maintain the zodiacal cloud? *Lect. Notes Phys.* 48:319–22
Sandford, S. A. 1983. Spectral matching of astronomical data from Comet Kohoutek with infrared data on collected interplanetary dust. *Meteoritics* 18:391
Sandford, S. A. 1984. Laboratory infrared spectra of meteorites and interplanetary dust from 2.5 to 25 microns. *Lunar Planet. Sci. XV*, pp. 715–16 (Abstr.)
Sandford, S. A., Walker, R. M. 1984. Links between astronomical observations of

protostellar clouds and laboratory measurements of interplanetary dust: the 6.8 micron carbonate band. *Meteoritics*. In press

Sekanina, Z. 1980. Physical characteristics of cometary dust from dynamical studies. In *Solid Particles in the Solar System, IAU Symp. No. 90*, ed. I. Halliday, B. A. McIntosh, pp. 237–50. Boston: Reidel

Shimamura, T., Arai, O., Kobayashi, K. 1977. Isotopic ratios of potassium in magnetic spherules from deep sea sediments. *Earth Planet. Sci. Lett.* 36:317–21

Smith, J. V., Steele, I. M., Brownlee, D. E. 1984. Minor elements in relict olivine grains of deep sea spheres: match with Mg-rich olivines from C2 meteorites. *Nature*. In press

Southworth, R. B., Sekanina, Z. 1979. Physical and dynamical studies of meteors. *NASA CR-2316*. Washington DC: GPO

Tomeoka, K., Buseck, P. R. 1984. Transmission electron microscopy of Low-Ca; a hydrated interplanetary dust particle. *Earth Planet. Sci. Lett.* 69:243–54

Tsou, P., Brownlee, D. E., Albee, A. 1984. Comet flyby sample return experiment. In *Cometary Exploration II*, ed. T. I. Gombosi, pp. 215–23. Budapest: Hung. Acad. Sci.

Verniani, F. 1969. Structure and fragmentation of meteoroids. *Space Sci. Rev.* 10: 230–61

Wagstaff, J., King, E. A. 1981. Micrometeorites and possible cometary dust from Antarctic ice cores. *Lunar Planet. Sci. XII*, pp. 1124–26 (Abstr.)

Wasson, J. T. 1974. *Meteorites*. New York: Springer-Verlag. 316 pp.

Wetherill, G. W. 1974. Solar system sources of meteorites and large meteoroids. *Ann. Rev. Earth Planet. Sci.* 2:303–31

Wetherill, G. W., ReVelle, D. O. 1982. Relationships between comets, large meteors and meteorites. In *Comets*, ed. L. L. Wilkening, pp. 297–319. Tucson: Univ. Ariz. Press

Whipple, F. L. 1951. The theory of micrometeorites. Part II. In heterothermal atmospheres. *Proc. Natl. Acad. Sci. USA* 37: 19–30

Whipple, F. L. 1967. On maintaining the meteoritical complex. In *The Zodiacal Light and the Interplanetary Medium*, ed. J. L. Weinberg, pp. 409–26. *NASA SP-150*. Washington DC: GPO

Zbik, M. 1984. Morphology of the outermost shells of the Tunguska black magnetic spherules. *J. Geophys. Res.* 89:B605–11 (*Proc. Lunar Planet. Sci. Conf., 14th*)

Zinner, E., McKeegan, K. D. 1984. Ion probe measurements of hydrogen and carbon isotopes in interplanetary dust. *Lunar Planet. Sci. XV*, pp. 961–62 (Abstr.)

Zinner, E., McKeegan, K. D., Walker, R. M. 1983a. Laboratory measurements of D/H ratios in interplanetary dust. *Nature* 305:119–21

Zinner, E., Pailer, N., Kuczera, H. 1983b. LDEF: chemical and isotopic measurement of micrometeoroids by SIMS. *Adv. Space Res.* 2:251–53

Zolensky, M. E., MacKinnon, I. D. R., McKay, D. S. 1984. Towards a complete inventory of stratospheric dust particles, with implications for their classification. *Lunar Planet. Sci. XV*, pp. 963–64 (Abstr.)

Zook, H., Berg, O. E. 1975. A source for hypervelocity cosmic dust particles. *Planet. Space Sci.* 23:1391–97

Ann. Rev. Earth Planet. Sci. 1985. 13: 175–99

THE APPALACHIAN-OUACHITA CONNECTION: Paleozoic Orogenic Belt at the Southern Margin of North America

William A. Thomas

Department of Geology, University of Alabama, University, Alabama 35486

INTRODUCTION

Late Paleozoic orogenic belt structures exposed in the Appalachian Mountains of Alabama and in the Ouachita Mountains of Arkansas plunge from opposite directions beneath postorogenic Mesozoic-Cenozoic strata in the Mississippi Embayment of the Gulf Coastal Plain (Figure 1). The Paleozoic stratigraphic sequence and details of structural style in the Ouachita outcrops contrast strongly with those in the Appalachian outcrops. Furthermore, straight-line projection of structural strike from the outcrops does not lead to a simple connection of structures beneath the Coastal Plain. Concealed by the Coastal Plain cover are answers to questions such as the following: Are the orogenic belts continuous, discontinuous, offset, or intersecting? How are along-strike changes in stratigraphy and structural style expressed? What is the sequence of tectonic evolution?

Early attempts to interpret the subsurface relationships between the Appalachian and Ouachita structures were of necessity based on the first few scattered wells that were drilled through Coastal Plain strata into Paleozoic rocks. Subsequent studies have enjoyed the availability of progressively more numerous deep wells and a growing accumulation of geophysical data. Various stages in the evolution of thought were outlined by King (1950, 1961, 1975) and by Thomas (1973, 1976). This review summarizes available subsurface data in a pre-Mesozoic paleogeologic

0084–6597/85/0515–0175$02.00

(subcrop) map, compares subsurface and outcrop data, and synthesizes the tectonic history of the Appalachian-Ouachita orogen.

In the outcrops in Alabama, the Appalachian orogenic belt includes a foreland fold-thrust belt on the northwest and a complex internal metamorphic belt on the southeast (Figures 2, 3). The stratigraphic sequence of Cambrian to Pennsylvanian age in the fold-thrust belt records depositional environments ranging from shallow-marine carbonate shelf to delta plain. The fold-thrust belt is characterized by large-scale, internally coherent thrust sheets in which structural style is controlled by a thick competent layer of Cambrian-Ordovician carbonate rocks. In the metamorphic belt, grade increases southeastward across postmetamorphic thrust faults.

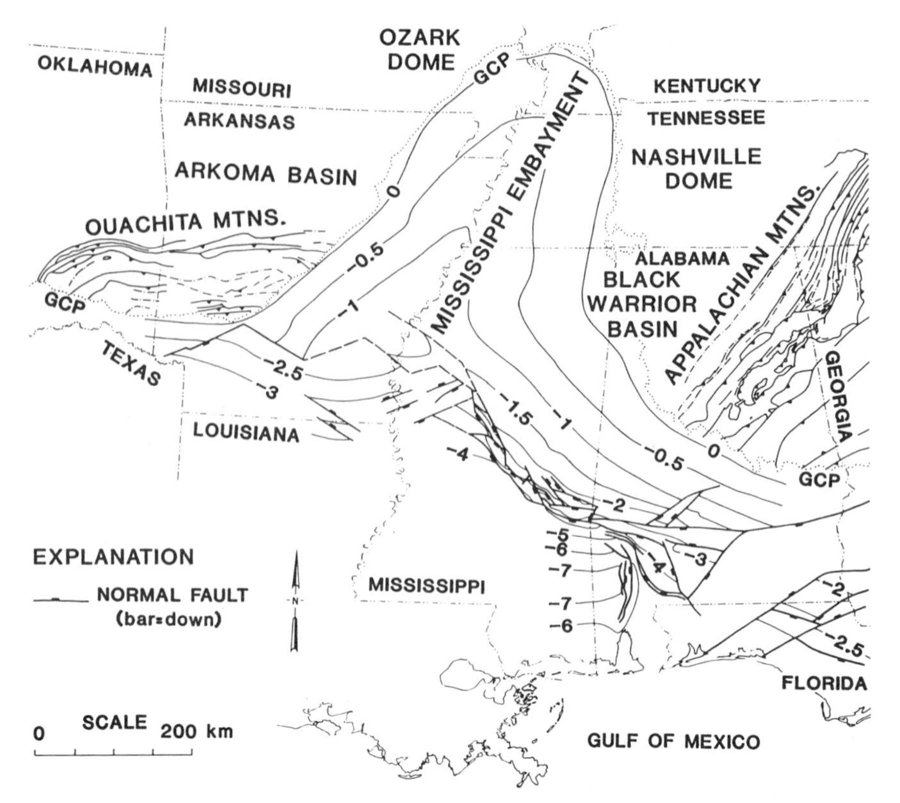

Figure 1 Regional map of exposed Paleozoic orogen in Appalachian and Ouachita Mountains, structures of southern part of North American craton, and Mesozoic-Cenozoic Gulf Coastal Plain; and structure contour map (contours in kilometers) of the base of post-Paleozoic strata of the Gulf Coastal Plain. Dotted line (labeled GCP) marks the limit of the Gulf Coastal Plain.

In the Arkansas outcrops (Figures 2, 3), strata in the Ouachita fold-thrust belt range in age from Ordovician to Pennsylvanian and are interpreted as off-shelf deep-water sediments. Internally complex thrust sheets and disharmonic structures contrast with exposed Appalachian structures and are consistent with the lack of a competent layer as thick and extensive as that in the Appalachians. Parts of the Ouachita fold-thrust belt are characterized by slaty cleavage, and locally the rocks are cut by numerous quartz veins.

On the southern part of the North American craton, the Arkoma and Black Warrior foreland basins border the Ouachitas and Appalachians, respectively (Figures 1, 2). In the foreland basins, the Cambrian to Pennsylvanian sequence includes shallow-marine carbonate shelf facies and shallow-marine to deltaic clastic sediments.

Subsurface data from 195 wells drilled into pre-Mesozoic rocks are synthesized in a pre-Mesozoic paleogeologic (subcrop) map (Figure 2). Data from 82 wells were obtained through my detailed study of drill cuttings, cores, and petrophysical well logs; for the other wells, rock descriptions were taken from publications and industry reports. The detailed rock descriptions provide the basis for identification of the base of the Coastal Plain sequence and for lithostratigraphic correlation of the rocks beneath the Coastal Plain both to exposed sections in the Appalachians and Ouachitas and to the thoroughly documented subsurface sections in the Black Warrior and Arkoma basins. Outcrop map patterns from exposed parts of the Appalachian and Ouachita orogenic belts have been used as a guide in extrapolating subcrop map patterns from the well data. In a few places, closely spaced wells define the geometry of individual structures, and in other places, reflection seismic profiles illustrate structural geometry.

The paleogeologic map shows that a continuous belt of deformed rocks extends beneath the Coastal Plain from the exposed Appalachian structures to the exposed Ouachita structures (Figure 2). The belt curves beneath the Coastal Plain in a pattern analogous to curves of salients and recesses in the exposed Appalachians and other mountain belts. The trace of the orogenic belt defines a salient (convex cratonward curve) in the Ouachita Mountains and a recess (concave cratonward curve) in the Alabama Appalachians.

Mesozoic-Cenozoic downwarping of the Gulf Coastal Plain, including the broad southward-plunging syncline of the Mississippi Embayment, has been imposed upon the Paleozoic structures of the Appalachian-Ouachita orogen and the adjacent foreland basins (Figure 1). Faults associated with the Mesozoic opening of the Gulf of Mexico crosscut some Paleozoic structures of the Appalachian-Ouachita orogenic belt. The Mesozoic faults

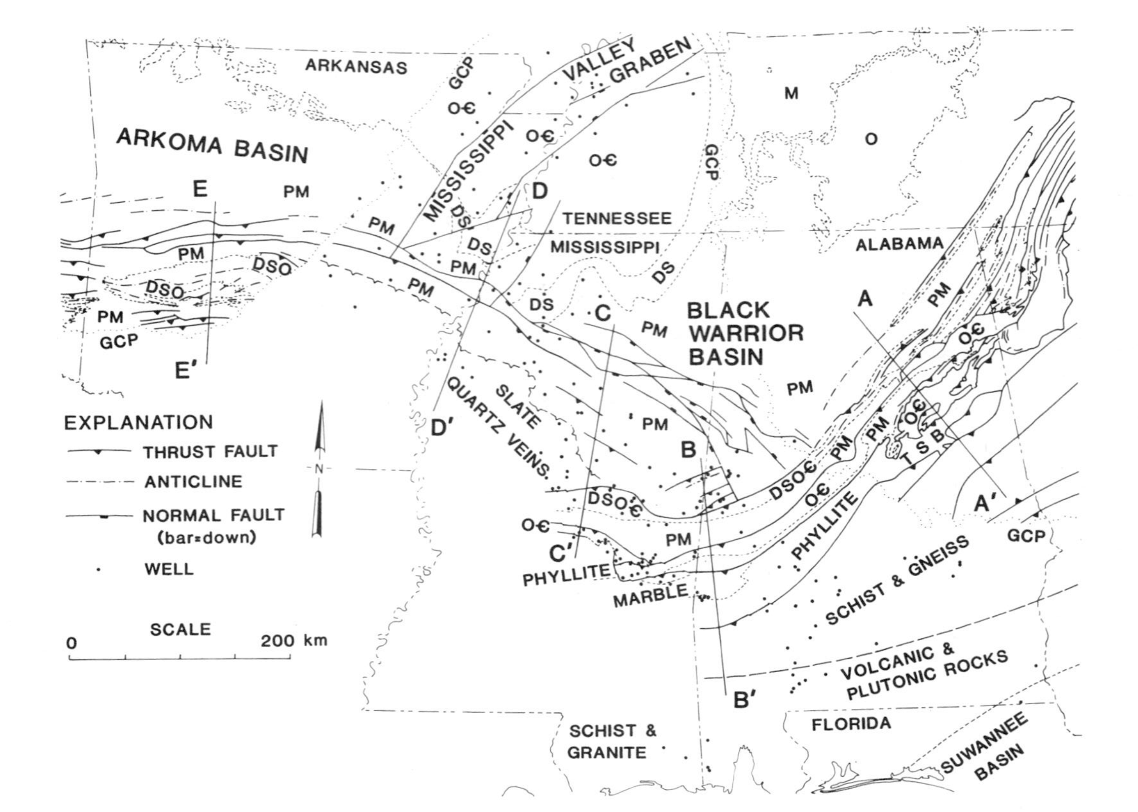

Figure 2 Pre-Mesozoic paleogeologic map of Appalachian-Ouachita orogen and adjacent foreland basins beneath Gulf Coastal Plain. Standard geologic symbols designate age or ages of rocks within outcrop and subcrop areas. Abbreviations: TSB = Talladega slate belt, GCP = Gulf Coastal Plain. Lines of structural cross sections labeled as in Figure 3. Numerous wells in Black Warrior and Arkoma basins are not shown.

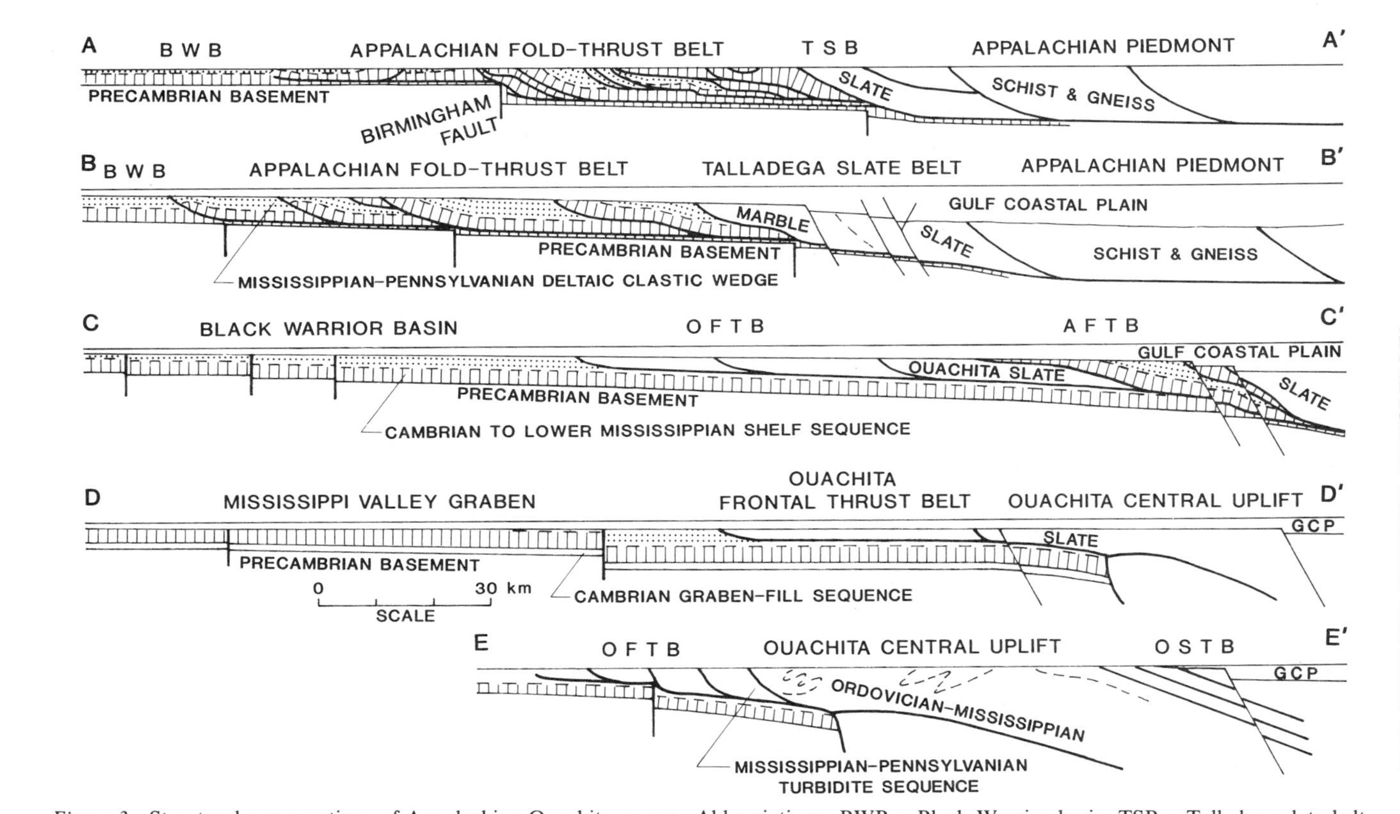

Figure 3 Structural cross sections of Appalachian-Ouachita orogen. Abbreviations: BWB = Black Warrior basin, TSB = Talladega slate belt, OFTB = Ouachita frontal thrust belt, AFTB = Appalachian fold-thrust belt, GCP = Gulf Coastal Plain, OSTB = Ouachita southern thrust belt. Numerous documented smaller-scale thrust faults and folds are not shown on cross section E–E′. Lines of cross section are shown in Figure 2.

affect pre-Mesozoic paleogeologic map patterns, because erosion of Mesozoic upthrown blocks has exposed deeper levels of Paleozoic structures that now appear on the map adjacent to higher levels of Paleozoic structures preserved on Mesozoic downthrown blocks. In parts of the Mississippi Embayment, Mesozoic plutons intrude Paleozoic rocks of the Ouachita orogenic belt and the foreland.

REGIONAL STRUCTURAL GEOLOGY

Foreland Basins

The Black Warrior and Arkoma basins are defined by homoclines dipping away from the craton and extending beneath the cratonward-directed frontal structures of the Appalachian-Ouachita fold-thrust belt (Figures 2, 3). On the cratonward side (north), the Black Warrior basin is bordered by the Nashville dome, and the Arkoma basin by the Ozark dome (Figure 1). A broad nose of the Ozark dome plunges southeastward and extends beneath Mesozoic-Cenozoic strata in the Mississippi Embayment. Beneath the Mississippi Embayment, a more narrow Paleozoic arch plunges southward between the Black Warrior and Arkoma basins (Figure 2). Because of southward dip from the Nashville and Ozark domes into the Black Warrior and Arkoma basins, progressively younger Paleozoic rocks are preserved toward the south (Figure 2).

A complex system of steep basement faults in the subsurface of Kentucky outlines the Rome trough and the Rough Creek graben and extends westward beneath the northern end of the Mississippi Embayment (Harris 1975, Soderberg & Keller 1981). Geophysical data (Kane et al 1981) and sparse well data indicate that a similar basement graben, the Mississippi Valley graben, extends southwestward along the southward-plunging Paleozoic arch beneath the Mississippi Embayment, probably to the subsurface trace of the Ouachita orogen (Figure 2).

The Black Warrior basin homocline has an average southwestward dip of less than 2°. A northwest-trending system of normal faults displaces the homocline down-to-southwest by more than 2 km (Figures 2, 3). On the southeast, the fault system intersects the front of the Appalachian fold-thrust belt at an angle of approximately 90°. The fault system extends northwestward entirely across the Black Warrior basin, and farther west along the Arkoma basin, the system of large-scale, down-to-south normal faults extends beneath and approximately parallel with the frontal thrust faults of the Ouachitas (Figures 2, 3). Forward ramps in Ouachita thrust faults are associated with underlying steep faults (Buchanan & Johnson 1968). The youngest rocks preserved in the Black Warrior basin (Early to

Middle Pennsylvanian) are displaced by the faults. Synsedimentary normal fault movement during Atokan deposition has been documented for the Arkoma basin (Buchanan & Johnson 1968, Haley 1982).

Appalachian Outcrops

The exposed Appalachian fold-thrust belt is characterized by large-scale, internally coherent thrust sheets detached near the base of the Paleozoic sedimentary sequence (Figures 2, 3; Neathery & Thomas 1983). A thick Cambrian-Ordovician carbonate competent unit controls structural style. Most forward ramps rise from the basal décollement to the surface, but part of the belt includes multiple-level décollements. The frontal part of the belt includes broad flat-bottomed synclines and narrow ramp anticlines above a relatively shallow basement. The basement surface beneath the allochthonous cover is displaced more than 3 km down-to-southeast by the steep Birmingham fault (Figure 3), and on the southeast, thrust ramps have greater relief than those to the northwest (Thomas 1982). The southeastern part of the fold-thrust belt includes multiple-level stacked low-angle thrust sheets.

On the southeast, sedimentary rocks and structures of the fold-thrust belt are overridden by the Talladega slate belt, a thrust sheet of greenschist facies metasedimentary and metavolcanic rocks (Tull 1982). The slate belt consists mainly of a single southeast-dipping stratigraphic sequence that includes correlatives of part of the Paleozoic section farther northwest. Southeast of the Talladega slate belt, the Appalachian Piedmont consists of higher grade metamorphic rocks including schist, gneiss, quartzite, amphibolite, and plutonic rocks (Neathery & Thomas 1983).

Ouachita Outcrops

The frontal part of the exposed Ouachitas exhibits large-scale, cratonward-directed thrust faults in Carboniferous rocks (Figures 2, 3). A central uplift of older rocks is characterized by penetrative, polyphase fold systems (Viele 1979). The boundary between the central uplift and the frontal thrust faults is marked by a linear zone of deformed Pennsylvanian shales containing blocks of sandstone. This chaotic zone has been interpreted as a part of a subduction complex (Viele 1979); alternative interpretations emphasize tectonic deformation and submarine sedimentary slumping (Stone & McFarland 1982). Slaty cleavage is common in pelitic rocks, and quartz veins are abundant in parts of the Ouachitas (Miser 1959). South of the central uplift, southward-dipping Carboniferous strata are in part imbricated by southward-dipping thrust faults. South of the Ouachita outcrops beneath the Gulf Coastal Plain, late-orogenic or postorogenic

Desmoinesian and younger Pennsylvanian and Permian fluvial to shallow-marine strata unconformably overlie deformed Ouachita rocks (Woods & Addington 1973).

REGIONAL STRATIGRAPHY

Foreland Basins and Adjacent Craton

Paleozoic strata of the Black Warrior and Arkoma foreland basins and the adjacent craton overlie Precambrian basement rocks of North American continental crust. The Paleozoic section includes four major subdivisions: a basal clastic sequence of Cambrian age; a thick carbonate-shelf facies of Cambrian-Ordovician age; a thinner shelf sequence of Ordovician to Mississippian carbonate, chert, and minor clastic units; and Mississippian-Pennsylvanian clastic wedge sediments and equivalent carbonate facies.

The age of the basal clastic unit ranges from Early to Late Cambrian and illustrates northwestward transgression onto the craton (base of the Sauk sequence of Sloss 1963). Regionally, a thin sandstone (generally less than 50 m) overlies Precambrian basement and is overlain by a transgressive carbonate facies. Southeastward toward the Appalachian orogen, the clastic unit is thicker and includes mudstone (Figure 4*A*). Both the sandstone and mudstone had sources on the craton (Palmer 1971). In contrast to the regionally extensive, relatively thin, transgressive basal sandstone, Early to Middle Cambrian clastic sequences more than 1 km thick are localized in downthrown fault blocks of the Rome–Rough Creek–Mississippi Valley graben system (Figure 4*A*; Woodward 1961). These sequences of sandstone, siltstone, and mudstone are overlain by transgressive Late Cambrian carbonate rocks that extend across the graben boundary faults.

The thick (generally more than 1.3 km), regionally extensive Upper Cambrian–Lower Ordovician shallow-marine shelf facies of limestones and dolostones is unconformably overlain by equally extensive, thinner (generally less than 400 m) Middle Ordovician carbonate rocks. Biostratigraphic data suggest that the cratonwide pre–Middle Ordovician unconformity (between the Sauk and Tippecanoe sequences of Sloss 1963) represents a progressively smaller time span southward away from the craton. South of the Ozark dome, the carbonate sequence includes rounded quartz sand in both sandstone and sandy carbonate beds. To the east, quartz sand is less common. The texturally and compositionally mature quartz sands reflect cratonic sources. In the area of the Mississippi Valley graben, the Cambrian-Ordovician carbonate sequence includes dark-colored rocks of probable deeper or outer shelf facies.

In contrast to the thick, widespread Upper Cambrian to Middle Ordovician carbonate facies, the Upper Ordovician through Lower Mississippian section is relatively thin (generally less than 250 m), and units are laterally less continuous (Figure 4*A*). The succession includes carbonate rocks, chert, mudstone, and sandstone, and it is interrupted by cratonwide and subregional unconformities. In the eastern part of the Black Warrior basin, the Upper Ordovician carbonate facies grades eastward into a westward-prograding clastic wedge that is thicker farther east in the Appalachian fold-thrust belt (Chowns & McKinney 1980). Chert and carbonate rocks characterize the Silurian and Devonian and extend throughout the region in the Lower Mississippian.

The upper part of the Paleozoic section includes a relatively thick clastic facies that contrasts with the underlying carbonate-dominated section (Figure 4). In the Black Warrior basin, the Upper Mississippian–Pennsylvanian consists of a northeastward-prograding clastic wedge of deltaic, barrier, and shallow-marine sediments (Figure 4*B*; Thomas 1974, 1979). The clastic wedge progrades northeastward over and intertongues with a shallow-marine carbonate facies. The oldest clastic wedge sediments overlying the carbonate shelf sequence are of late Meramecian age. The entire Mississippian clastic facies grades northeastward into a carbonate facies, but the Pennsylvanian clastic sequence is more extensive, has no equivalent carbonate facies, and merges with a separate southwestward-prograding clastic wedge (Thomas 1974). Sandstone petrography indicates a provenance that included a sedimentary and metamorphic fold-thrust belt, subduction complex, and volcanic arc (Mack et al 1983). The sediment dispersal direction interpreted from regional facies distribution and paleogeographic reconstruction of depositional systems indicates that the sediment source was located southwest of the Black Warrior basin (Thomas 1972, 1979, Thomas & Mack 1982).

The Mississippian-Pennsylvanian section in the Arkoma basin, like that in the Black Warrior basin, includes Mississippian carbonate and clastic facies, as well as a coarser and more extensive Pennsylvanian clastic facies (Glick 1975, 1979, Haley 1982). The section thickens southward toward very thick equivalent units in the Ouachita Mountains (Figure 4*B*). Mississippian shallow-marine mudstones, sandstones, and limestones grade southward to dark-colored mudstones (Glick 1979, Haley 1982). Distribution of clastic sediments in the Pennsylvanian is interpreted in the context of a southward-prograding delta system along the northern limb of the Arkoma basin (Glick 1975, Haley 1982). The dispersal system of these clastic sediments differs from that of the Pennsylvanian in the Black Warrior basin. Pennsylvanian deltaic sediments grade southward across

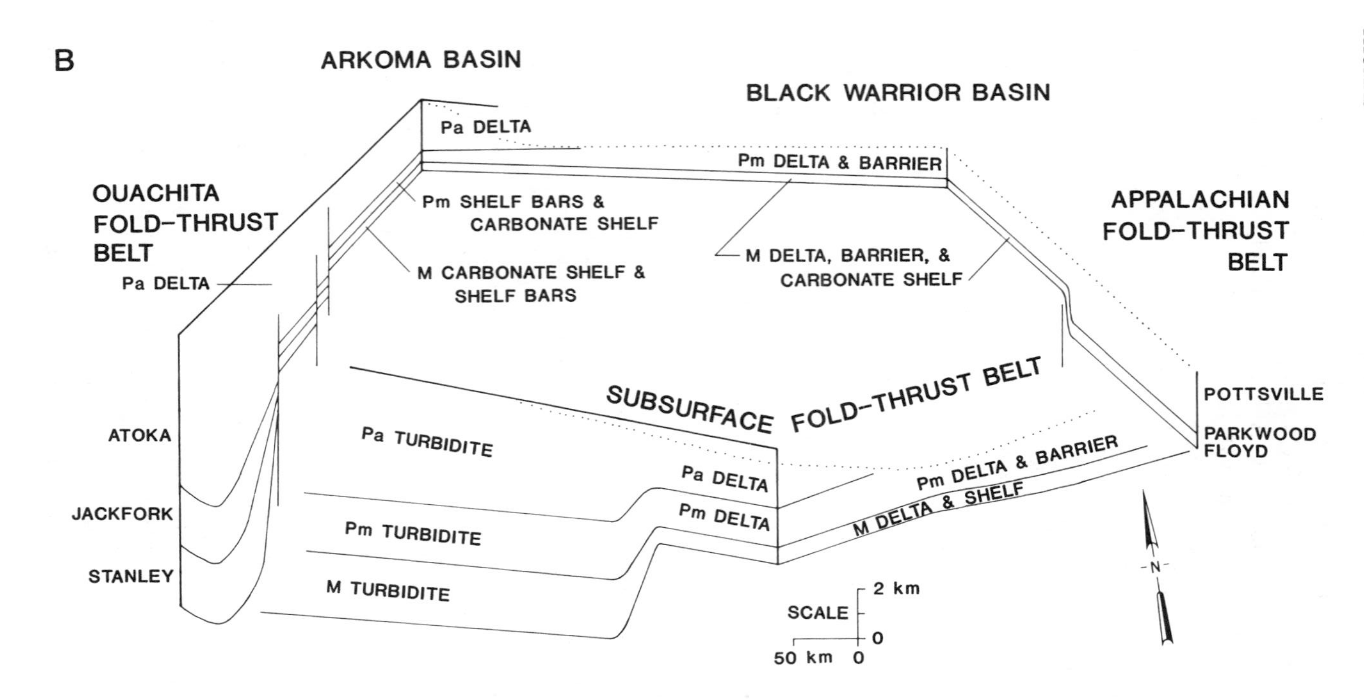
B
ARKOMA BASIN
BLACK WARRIOR BASIN
Pa DELTA
Pm DELTA & BARRIER
OUACHITA FOLD-THRUST BELT
Pa DELTA
Pm SHELF BARS & CARBONATE SHELF
M CARBONATE SHELF & SHELF BARS
M DELTA, BARRIER, & CARBONATE SHELF
APPALACHIAN FOLD-THRUST BELT
SUBSURFACE FOLD-THRUST BELT
POTTSVILLE
PARKWOOD
FLOYD
ATOKA
JACKFORK
STANLEY
Pa TURBIDITE
Pm TURBIDITE
M TURBIDITE
Pa DELTA
Pm DELTA
Pm DELTA & BARRIER
M DELTA & SHELF
SCALE
2 km
0
50 km
0
N

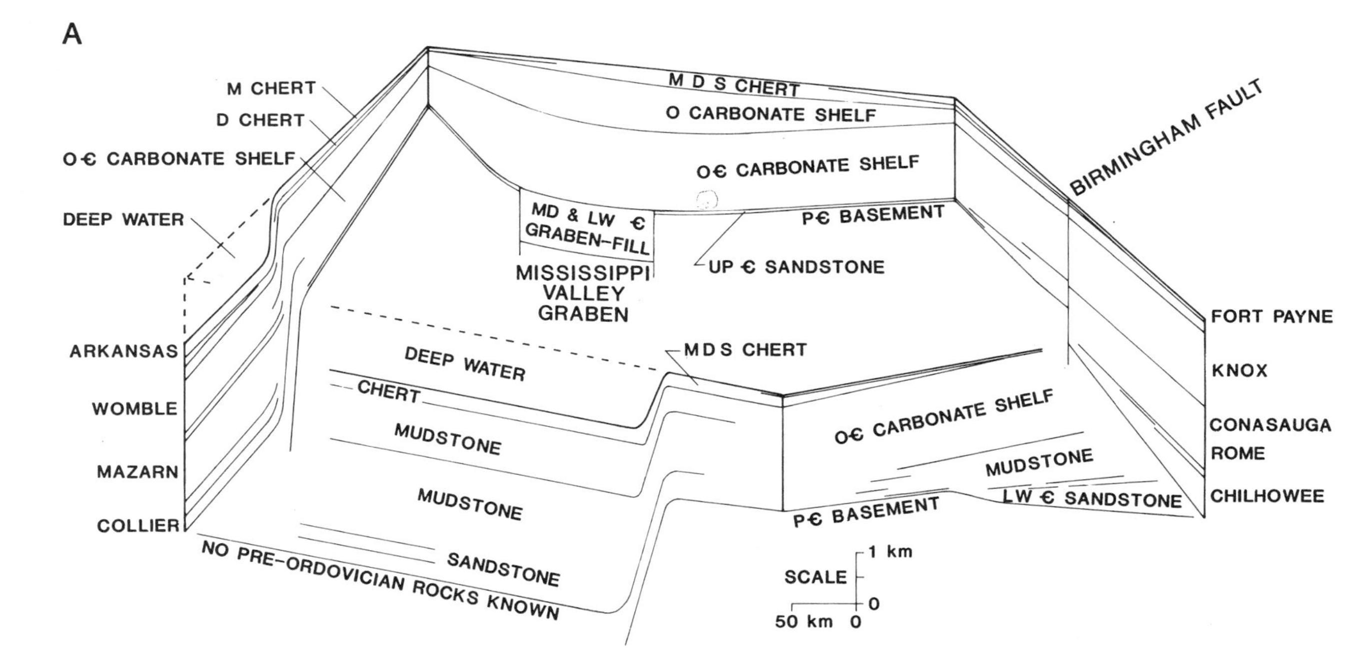

Figure 4 Schematic panel diagrams of Paleozoic stratigraphy and depositional framework of Black Warrior and Arkoma foreland basins and Appalachian-Ouachita orogen restored to prethrusting configuration. The ages of units are indicated by standard geologic symbols. Abbreviations: LW = Lower, MD = Middle, UP = Upper, Pm = Pennsylvanian Morrowan, Pa = Pennsylvanian Atokan. *A*. Cambrian–Lower Mississippian passive-margin facies. *B*. Upper Mississippian–Pennsylvanian convergent-margin facies. (Dotted line at top indicates approximate post-Paleozoic erosion level.) Selected stratigraphic units are identified.

the Arkoma basin into deep-water deposits, but deltaic facies extend across the basin in the upper Atokan (Figure 4*B*).

Appalachian Outcrops

Paleozoic stratigraphy of the Appalachian outcrops in Alabama is generally comparable to that in the Black Warrior foreland basin. In the southeastern Appalachians, the basal clastic unit consists of sandstone and mudstone and includes some carbonate units (Figure 4*A*). The sequence ranges through Early and Middle Cambrian age and is much thicker than the basal clastic unit to the northwest (Kidd & Neathery 1976). The clastic sequence southeast of the Birmingham basement fault is significantly thicker and contains much more sandstone than that northwest of the fault (Figure 4*A*; Thomas 1982).

The Cambrian-Ordovician carbonate unit persists throughout the Appalachian fold-thrust belt. No significant facies variations have been recognized, and therefore the position of the Cambrian-Ordovician shelf edge is southeast of the palinspastically restored location of the present trailing edge of the fold-thrust belt. Possible shelf-edge facies are contained in the Talladega slate belt on the southeast (Tull 1982).

The Middle Ordovician through Lower Mississippian section is relatively thin, incomplete because of unconformities, and laterally variable (Figure 4*A*; Thomas 1982). On the southeast, Middle Ordovician graptolite-bearing black shale overlies shelf-carbonate rocks and suggests local subsidence, possibly near the shelf edge. The Ordovician includes a clastic wedge of redbeds and sandstone that progrades westward over Middle Ordovician limestones into the Upper Ordovician (Chowns & McKinney 1980) and constitutes the most southerly record of the Taconic orogeny in the Appalachian-Ouachita foreland. Similarly distributed Silurian clastic sediments grade westward and northwestward to carbonate. The Devonian includes chert and locally distributed shallow-marine sandstone. Local unconformities indicate episodic low-amplitude synsedimentary structural movement, but Lower Mississippian shallow-marine chert and carbonate indicate widespread shelf deposition.

Mississippian-Pennsylvanian clastic wedge sediments are similar in lithology, facies distribution, depositional environments, sediment dispersal, and provenance to those of the Black Warrior basin (Thomas 1974). The facies distribution demonstrates northeastward progradation of the clastic wedge parallel with subsequent Appalachian structural strike. Clastic wedge sediments southeast of the Birmingham fault are significantly thicker than in the Black Warrior basin, suggesting synsedimentary fault movement (Figure 4*B*; Thomas 1974). The pattern of northeastward sediment transport and progradation is modified in the middle Lower

Pennsylvanian by the addition of coarse clastic sediments derived from the southeast (Horsey 1981), the earliest indication of an orogenic sediment source along that sector of the continental margin.

The Talladega slate belt includes a metasedimentary sequence of clastic rocks above and below a marble unit (Tull 1982). The marble is equivalent to at least part of the Cambrian-Ordovician carbonate unit of the fold-thrust belt, and the upper clastic unit is overlain by Lower Devonian chert similar to that in the Black Warrior basin. Above the chert is a greenstone that is unique in the region.

Ouachita Outcrops

The Paleozoic stratigraphic sequence in the Ouachita Mountains differs significantly from that in the Arkoma foreland basin and that in the Appalachian outcrops. No strata older than Ordovician and no Precambrian basement have been recognized. The Ordovician through Lower Mississippian section consists mainly of mudstone and chert. The section is approximately 3.5 km thick, signifying slow sedimentation rates. In contrast, the Mississippian-Pennsylvanian sequence is a very thick (approximately 10 km) succession of turbidites (Figure 4). The change in sedimentation rate in the Mississippian is contemporaneous with the change from dominantly carbonate to dominantly clastic sedimentation in the foreland basins and the Appalachian fold-thrust belt.

Off-shelf, deep-water, graptolite-bearing, dark-colored mudstones characterize the Ordovician in the Ouachita outcrops; however, the sequence includes quartzose sandstones, carbonate boulders, and carbonate turbidites that indicate supply from the shelf (Morris 1974a, Stone & McFarland 1982). Granite and meta-arkose clasts in Ordovician sandstone suggest a source from continental basement rocks, perhaps from scarps along the continental margin (Stone & Haley 1977). Chert units in the upper Middle Ordovician and the Devonian-Lower Mississippian signify a diminished supply of clastic sediment. The Silurian is thin to very thin; however, it includes a southward-thickening sandstone that may have had a southern source (Morris 1974a). The Devonian–Lower Mississippian Arkansas Novaculite of the Ouachita Mountains includes equivalents of Devonian chert units that pinch out northward in the Arkoma basin and northeastward in the Black Warrior basin, and equivalents of shallow-marine Lower Mississippian chert and carbonate units that extend throughout the foreland basins (Thomas 1972, 1976). The Arkansas Novaculite is underlain and overlain by deep-water deposits and is commonly interpreted as deep-water chert; however, some sedimentary features have been interpreted as of shallow-water origin.

The thick deep-water Carboniferous turbidite facies in the Ouachita

Mountains is dominated by mudstone in the lower part and consists of sandstone and mudstone in the upper part (Morris 1974a). Stratigraphically upward and northward from the Ouachitas, the deep-water facies grades to more shallow-water facies (Figure 4*B*; Haley 1982).

Mississippian strata are mainly dark-colored mudstones and include sandstone, tuff, and siliceous mudstone. Vertical and lateral distributions of distal and proximal turbidites indicate supply from the south or southeast (Niem 1976). Sandstone units are generally thicker and more numerous in the southern Ouachitas, and sandstone petrography indicates a provenance of metamorphic, sedimentary, and volcanic rocks (Morris 1974a). Paleocurrent data indicate northward and northwestward current flow through the southern and eastern Ouachitas and more westward flow in the western Ouachitas. The distribution of volcanic tuff indicates a southern source (Niem 1977).

The Pennsylvanian (Morrowan-Atokan) section consists of interbedded dark-colored mudstone and sandstone. The distribution of sandstone compositional types has been interpreted to indicate supply from two sources: more quartzose sand from the craton, and more feldspathic and lithic sand from an orogenic provenance on the south or southeast (Morris 1974b). Paleocurrent data show generally westward current flow along the length of the Ouachitas, compatible with longitudinal westward transport along the trough of sediment introduced from both north and south. The section includes boulder conglomerates containing clasts of carbonate rocks derived from the pre-Pennsylvanian shelf facies to the north (Shideler 1970, Gordon & Stone 1977).

SUBSURFACE APPALACHIAN-OUACHITA OROGENIC BELT

Introduction

Most wells drilled through Gulf Coastal Plain sediments have penetrated a limited thickness (generally no more than a few tens of meters) of the pre-Mesozoic section, and a vertical sequence of lithologic units is defined in only a few wells. Identification of Paleozoic stratigraphic units is based on lithologic correlation to units of exposed sections or in the succession defined from numerous wells in the foreland basins. Identification of stratigraphic units provides the basic data for preparation of the pre-Mesozoic paleogeologic map, as well as for description of regional stratigraphy in the subsurface orogenic belt.

Rock types and paleogeologic map patterns indicate that stratigraphy and structural style generally like those of the exposed Appalachians extend as far west as central Mississippi (Figure 2). A subsurface belt of slaty pelitic rocks like those exposed in the central uplift of the Ouachitas extends southeastward as far as central Mississippi.

"Appalachian-Style" Stratigraphy and Structure

The oldest rocks drilled in the subsurface Appalachian fold-thrust belt of eastern Mississippi and western Alabama are part of the Cambrian-Ordovician carbonate-shelf sequence. The thickness of the carbonate sequence is similar to that in the adjacent Black Warrior foreland basin. No rocks identified with the basal clastic sequence have been drilled; however, for the stratigraphically lowest carbonate rock penetrated, the age is undetermined and may be the same as that of part of the basal clastic sequence farther east in the Appalachian outcrops (Figure 4*A*). In the southwestern part of the subsurface fold-thrust belt, the carbonate sequence includes at least one quartzose sandstone similar to those of the Arkoma basin sequence.

Few wells have penetrated the Upper Ordovician–Lower Mississippian sequence; however, a Devonian chert unit similar to that in the Black Warrior basin extends along the northern part of the fold-thrust belt. The westernmost wells in the northern part of the fold-thrust belt drilled dark-colored calcareous mudstones and limestones containing Silurian brachiopods (King, 1961, Thomas 1972). The rocks are unlike any other known Silurian in the region and suggest a possible outer shelf or slope facies. The few penetrations of this interval recall the thin, locally absent units of the Appalachian outcrops.

Carboniferous rocks in the fold-thrust belt are carbonaceous mudstones and sandstones similar to those in the adjacent Black Warrior basin; however, not enough of the sequence has been drilled to confirm interpretations of sediment dispersal or details of depositional environments. The rock types are consistent with deltaic sedimentation.

No rocks certainly younger than Ordovician have been drilled along the southern (more interior) part of the subsurface fold-thrust belt, a circumstance that is analogous to the Appalachian outcrops where younger rocks commonly are not preserved in the more interior structures. Furthermore, Silurian-Devonian rocks are unconformably absent in some of the more southeasterly exposed structures and may be so also in the subsurface fold-thrust belt.

In western Alabama and eastern Mississippi, the subsurface fold-thrust belt consists of two extensive thrust sheets and an areally restricted frontal thrust sheet (Figures 2, 3). The paleogeologic map defines two elongate south-dipping panels of Paleozoic strata that strike southwestward in Alabama and curve westward in Mississippi (Figure 2). The map pattern, confirmed by seismic data, indicates south-dipping thrust sheets exposed at forward ramps on thrust faults that dip southward to the regional basal décollement. The two large-scale thrust sheets are along strike from exposed Appalachian structures, and the forward ramps of the subsurface faults are comparable in magnitude and stratigraphic position to those of the exposed structures. Whether the geometry is laterally continuous or undergoes along-strike changes at the scale of those in the Appalachian outcrops is undetermined on the basis of available data. One well evidently drilled through the northernmost of the two large-scale thrust faults. Below the Mesozoic, the well drilled through an upright section from Mississippian to Cambrian; below the Cambrian carbonate rocks, it drilled into dark-colored mudstones and thin sandstones of probable Pennsylvanian age. Subcrop patterns suggest a similar stratigraphic separation along the southern thrust fault (Figures 2, 3).

The northwesternmost anticline in the Appalachian outcrops plunges southwestward and passes into an area of horizontal strata near the limit of the Coastal Plain cover, and thus the front of the fold-thrust belt shifts southeastward across strike (Figure 2). Farther southwest beneath the Coastal Plain, another frontal structure appears along the boundary of the fold-thrust belt and foreland basin. At the paleogeologic map surface, the anticline is entirely in Carboniferous rocks, but elevations of the base of the Mississippian clastic sequence drilled in several wells demonstrate approximately 2 km of structural relief. A seismic reflection profile confirms that the anticline is associated with a forward ramp on a décollement that rises from near the base of the Paleozoic sedimentary sequence (Figure 3). The anticline/thrust sheet ends northeastward abruptly at a northwest-trending tear fault or lateral ramp that is aligned with the trace of a northwest-trending steep basement fault in the Black Warrior basin (Figure 2). The southwest end of the frontal anticline is uncertain, but the structure plunges southwestward adjacent to the deepest part of the Black Warrior basin.

On the southeast and south, the trailing edge of the fold-thrust belt is bordered by low-grade metasedimentary rocks, including phyllite, slate, chlorite schist, quartzite, and marble (Figures 2, 3; Thomas 1973, Neathery & Thomas 1975). All of the rock types have analogues in the Talladega slate belt of the Appalachian outcrops along strike to the northeast.

The boundary between the subsurface Talladega slate belt and the fold-thrust belt trends southwestward and curves westward in westernmost Alabama and Mississippi. On the west in Mississippi, the trace of the boundary has been complicated where Mesozoic normal faults cut the Paleozoic structures (Figures 2, 3). On upthrown blocks of the Mesozoic faults, erosion has uncovered lower structural levels of the Paleozoic orogen, whereas higher structural levels of Paleozoic rocks are preserved in

Mesozoic downthrown blocks. This relationship is reflected in the paleogeologic map pattern, where lower Paleozoic carbonate rocks are now at higher elevations and are adjacent to phyllites of the Talladega slate belt.

Southeast of the subsurface Talladega slate belt, several wells have penetrated metamorphic rocks similar to those of the Alabama Piedmont outcrops. Rock types include schist, granitic gneiss, and mylonite. Subsurface analogues of the exposed Piedmont have been drilled in Alabama, but in parts of westernmost Alabama and adjacent Mississippi, the great thickness of post-Paleozoic cover exceeds the depth of drilling (Figure 1).

In southern Alabama and Florida, south of a prominent west-trending magnetic anomaly, wells have penetrated volcanic, plutonic, and metamorphic rocks, as well as Ordovician-Devonian sedimentary rocks of the Suwannee basin (Figure 2). In southwestern Alabama and southeastern Mississippi, metamorphic and plutonic rocks have been drilled beneath Mesozoic strata (Figure 2). These rocks are tentatively interpreted as exotic terranes accreted to the Appalachian North American continental margin.

"Ouachita-Style" Stratigraphy and Structure

The structure of the Ouachita central uplift is indicated on the scale of drill cuttings by slaty cleavage and vein quartz. These are recognizable in wells in a southeast-trending area between the Ouachita outcrops and central Mississippi (Figures 2, 3). In central Mississippi, dark-colored, slaty, pelitic rocks apparently extend north (on the foreland side) of the northward-directed large-scale thrust sheets of Appalachian style and stratigraphy. Sparse wells leave questions about details of the subcrop pattern, but evidently "Ouachita-style" slaty rocks are near and north or northwest of thrust sheets of "Appalachian-style." Rock types are compatible with an abrupt shelf-edge facies change from shelf carbonate and clastic rocks to deep-basin, dark-colored, pelitic rocks (Figure 4; Thomas 1972). The subcrop pattern suggests an abrupt facies boundary that has been displaced by imbricate thrust faults that strike at an oblique angle to the trend of the facies boundary (Figures 2, 3).

The subsurface Ouachita central uplift consists predominantly of dark-colored pelitic rocks and includes dark-colored siliceous mudstone, dark-colored chert, and sandstone. The ubiquity of dark-colored mudstones in the Ouachita stratigraphic section precludes definitive assignment of the subsurface rocks to stratigraphic units. The dark-colored siliceous pelitic rocks are similar to parts of the Mississippian turbidite sequence. The chert is compatible with assignment to Ordovician, Devonian, or Mississippian, but no thick chert unit comparable to the Devonian-Mississippian

Arkansas Novaculite has been drilled. Because of the sparsity of wells and the lack of biostratigraphic data, correlation to specific units is uncertain. However, the sequence is clearly like parts of that in the exposed Ouachitas and contrasts with that in the exposed Appalachians and in the foreland basins.

Northeast of the subsurface central uplift and southeastward along strike from the exposed frontal Ouachita thrust faults, several wells have cored beds showing dips of more than 15°. These attitudes suggest a frontal belt of thrust faults and folds, although no stratigraphic duplication can be demonstrated in the available well data. Seismic profiles are similar to seismic profiles of the exposed frontal Ouachitas. Analogy with the exposed structures and sparse well and seismic data suggest that a frontal thrust belt extends along the northeast side of the subsurface central uplift from the outcrops in central Arkansas to central Mississippi (Figures 2, 3). In central Mississippi, the frontal "Ouachita" thrust faults extend into the area of the frontal "Appalachian" thrust faults. Rocks along the subsurface frontal Ouachita structures are partly carbonaceous mudstones and sandstones similar to Mississippian-Pennsylvanian rocks of the Arkoma and Black Warrior foreland basins as well as the exposed frontal Ouachitas.

Southward along the west side of the Mississippi Embayment, the edge of the Coastal Plain cover curves westward, and the Coastal Plain strata lap northward onto south-dipping Ouachita structures south of the central uplift (Figures 1–3). However, eastward from the Ouachita outcrops, the most southerly wells drilled through Coastal Plain cover have encountered slaty rocks of the central uplift, and the eastward subsurface extent of southern Ouachita structures is unknown.

SUMMARY OF TECTONIC EVOLUTION

In foreland basins at the southern edge of the North American craton, Precambrian basement rocks are overlain by a Cambrian to Lower Mississippian passive-margin carbonate-shelf facies and an Upper Mississippian and Pennsylvanian synorogenic clastic wedge. The cratonic basement and cover extend southward beneath allochthonous rocks of the Appalachian-Ouachita orogenic belt, which curves from the Alabama recess to the Ouachita salient. The evolution of the passive and later convergent margin is commonly interpreted in the context of the opening and later closing of the Iapetus Ocean. The Paleozoic rocks and structures are truncated and overstepped by Mesozoic and Cenozoic strata of the Gulf Coastal Plain.

A late Precambrian opening of the Iapetus Ocean is indicated by rift-related sedimentary and volcanic rocks along the Appalachians as far south as Georgia (Rankin 1975); however, no such indicators have been recognized along the Ouachitas. Nevertheless, the distribution of early Paleozoic carbonate-shelf and deep-water facies suggests the trace of a shelf edge around the Ouachita region, and the depositional framework implies that a rifted margin of continental crust controlled the location of the shelf edge. In the Ouachita Mountains, the autochthonous carbonate-shelf facies extends southward beneath allochthonous deep-water mudstones in the Ouachitas. The continental-margin shelf edge is interpreted to be beneath the present location of the Ouachita central uplift, but alternative interpretations range from the north to the south border of the central uplift (Lillie et al 1983). The base of the transgressive shelf facies of the Sauk sequence is of Early Cambrian age in the Alabama Appalachians, but no shelf-facies strata older than Late Cambrian have been identified adjacent to the Ouachitas. Rift-related sedimentary and volcanic rocks along faults in southern Oklahoma are of Early to Middle Cambrian (535 ± 30 to 525 ± 25 Myr) age (Ham et al 1964), and the structure has been interpreted as an aulacogen (Hoffman et al 1974). Early (?) and Middle Cambrian extensional faulting along the Rome–Rough Creek–Mississippi Valley graben system is indicated by a thick graben-fill sedimentary sequence, and the graben system extends southwestward toward the continental margin (Figure 5*A*). Similar timing of movement is suggested for the Birmingham basement fault.

The late Precambrian–early Paleozoic continental margin describes large-scale orthogonal bends from the southern Appalachians to the Ouachitas and southward into Texas (Figure 5*A*). The shape is interpreted to be a result of transform offset of a northeast-trending rift system (Thomas 1976, 1977). The rift- and transform-bounded continental margin outlines the Ouachita embayment and the Alabama promontory of North America (Figure 5*A*). The interpreted trace of the continental margin is consistent with an abrupt change in magnetic signature at the transform fault (Zietz 1982). An alternative interpretation attributes the shape of the margin to triple junctions at which failed arms are represented by the Southern Oklahoma aulacogen (Burke & Dewey 1973, Hoffman et al 1974) and the Mississippi Valley (Reelfoot) graben (Ervin & McGinnis 1975). Instead, in the context of a transform-offset rifted margin, the Southern Oklahoma fault system may reflect extension of transform faults into the continent (Figure 5*A*) [as suggested for modern continental-margin basins by Francheteau & LePichon (1972), especially for the Benue Trough (Benkhelil & Robineau 1983, Popoff et al 1983)]. Furthermore, the Rome–Rough Creek–Mississippi Valley graben and the Birmingham half-graben are consistent with northwest-southeast extension across the Alabama promontory in association with a northwest-trending transform fault. Ages of rift-related rocks suggest later movement along the Ouachita margin than along the Appalachians; however, the orthogonally zigzag passive margin had evolved around the Ouachita embayment no later than the end of Middle Cambrian time. In contrast, a passive margin along the Appalachians from Alabama northeastward had been established by the beginning of the Cambrian.

From the Cambrian through the Early Mississippian, a passive continental margin persisted around the Alabama promontory and the Ouachita embayment. The passive-margin carbonate-shelf facies extends

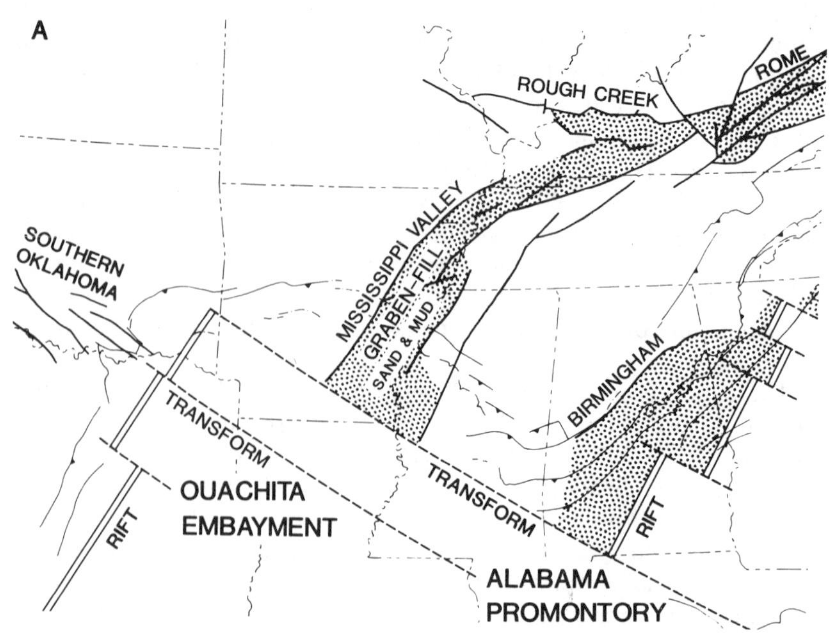

Figure 5 Paleogeographic reconstructions of phases in the tectonic evolution of the Appalachian-Ouachita orogen (state boundaries and present traces of late Paleozoic structures on all maps for location reference). *A*. Late Precambrian and Early to Middle Cambrian: rifted continental margin; graben-filling sediments on continental crust. *B*. Middle to Late Ordovician: passive margin around Alabama promontory and Ouachita embayment; carbonate shelf (including deeper or outer shelf in area of Mississippi Valley graben) on continental crust and off-shelf deep basin; clastic wedge prograding westward from Taconic orogen. *C*. Late Mississippian: arc-continent collision at Alabama promontory; clastic wedge prograding northeastward onto shallow shelf in Black Warrior basin and westward into deep basin in Ouachita embayment; shallow-marine shelf and passive margin around Ouachita embayment; separate clastic wedge prograding westward from Alleghanian orogenic source northeast of Alabama. *D*. Late Atokan: thrusting along Appalachian-Ouachita margin; clastic wedge prograding northward and cratonic delta prograding southward to fill Arkoma foreland basin along old shelf edge.

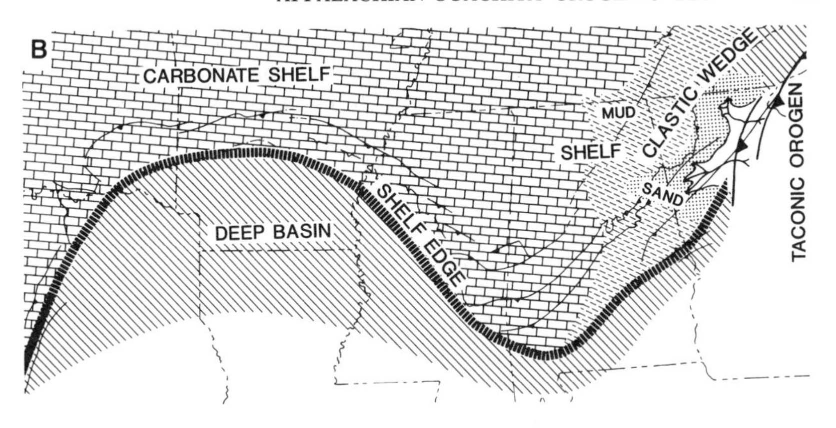
B
CARBONATE SHELF
MUD
SHELF
CLASTIC WEDGE
SAND
DEEP BASIN
SHELF EDGE
TACONIC OROGEN

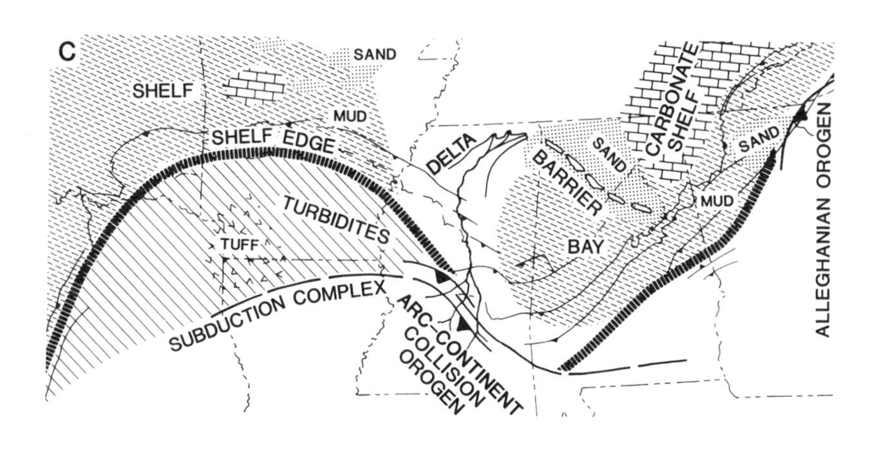
C
SAND
SHELF
MUD
SHELF EDGE
DELTA
SAND
BARRIER
CARBONATE SHELF
SAND
MUD
TURBIDITES
TUFF
BAY
SUBDUCTION COMPLEX
ARC-CONTINENT COLLISION OROGEN
ALLEGHANIAN OROGEN

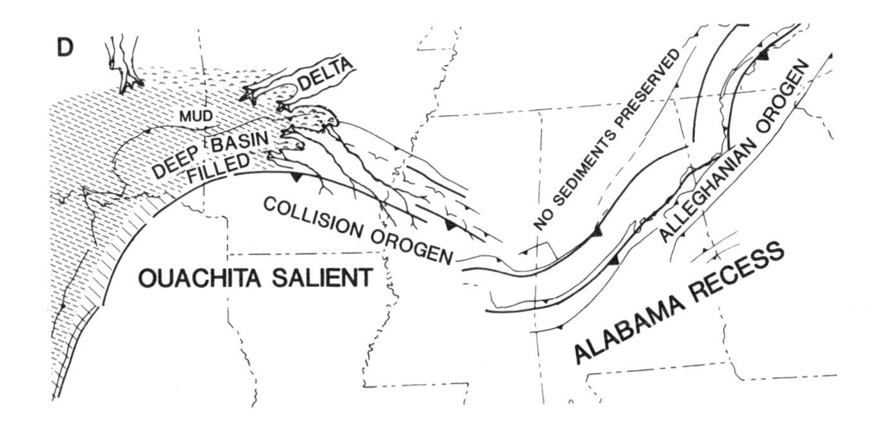
D
DELTA
MUD
DEEP BASIN FILLED
NO SEDIMENTS PRESERVED
ALLEGHANIAN OROGEN
COLLISION OROGEN
OUACHITA SALIENT
ALABAMA RECESS

throughout the foreland basins and the Appalachian fold-thrust belt, reflecting continental crust of the Alabama promontory (Figure 5*B*). Deeper or outer shelf facies are contained in the area of the Mississippi Valley graben. Within the Ouachita embayment, the off-shelf deep-basin facies represents deposition beyond the margin of continental crust or on attenuated continental crust. Stratigraphy of the passive-margin facies includes variations that suggest the effects of subsidence along the continental margin as well as eustatic sea-level changes. In contrast to the passive margin west of the Alabama promontory, a convergent-margin orogen north of the promontory supplied sediment to a Middle Ordovician clastic wedge, only the distal part of which prograded southwestward onto the carbonate facies on the Alabama promontory (Figure 5*B*). Ordovician facies thus demonstrate differences in timing of orogenic events along different parts of the continental margin.

A change in the regional tectonic framework from passive to convergent active margin is indicated by the initial progradation of clastic-wedge sediments onto the shelf in late Meramecian time (Figure 5*C*). Sediment dispersal patterns in the clastic wedge indicate sediment supply from a source southwest of the Alabama promontory and southeast to south of the Ouachita embayment (Figure 5*C*). The composition of clastic-wedge sediments indicates an orogenic provenance that resulted from an arc-continent collision (Mack et al 1983). The Black Warrior basin on the Alabama promontory is a peripheral basin (in the terminology of Dickinson 1974) associated with southward subduction of North American continental crust. Persistent deltaic to shallow-marine environments during the Late Mississippian and Early Pennsylvanian indicate that the sediment accumulation rate equaled the subsidence rate. The down-to-basin normal faults in the Black Warrior basin may have been initiated along with subsidence of the continental margin in response to tectonic loading.

In contrast to the deltaic and shallow-marine sediments deposited on the Alabama promontory, the equivalent clastic sequence in the Ouachita embayment consists of deep-water turbidites (Figure 5*C*). A deep marginal basin between a north-facing subduction complex and the continental margin around the Ouachita embayment remained open through the Late Mississippian and Early Pennsylvanian simultaneously with arc-continent collision along the Alabama promontory. Orogenesis migrated along the continental margin through time, and closing and filling of the marginal basin are indicated by the upward transition from deep-water turbidites to shallow-marine and deltaic sediments in the late Atokan of the Ouachitas (Figure 5*D*).

Orogenic activity culminated in large-scale cratonward thrusting. In the

Appalachian fold-thrust belt, both in outcrop and subsurface, allochthonous rocks include the lower Paleozoic carbonate-shelf sequence and the shallow-marine to deltaic upper Paleozoic clastic wedge of the Alabama promontory. In contrast, in the Ouachita fold-thrust belt, deep-basin off-shelf facies representing both the passive- and convergent-margin settings have been thrust over autochthonous passive-margin shelf facies (Figures 2, 3). Thus, between the Ouachita and Appalachian structures in central Mississippi, the frontal structures of the fold-thrust belt cross from the deep-basin facies eastward into carbonate-shelf facies; late Paleozoic structural strike intersects the early Paleozoic passive margin at an oblique angle (Figure 5). The curve of the late Paleozoic orogenic belt from the Alabama recess to the Ouachita salient mimics, but does not precisely duplicate, the shape of the early Paleozoic passive margin from the Alabama promontory to the Ouachita embayment (Figures 2, 5). The sinuous curve of the cratonward limit of allochthonous strata is not necessarily parallel with structural strike, because the frontal structures of the Ouachita salient may flatten eastward along strike into autochthonous strata of the Black Warrior basin (Thomas 1973).

Following Appalachian-Ouachita orogenesis, the opening of the Gulf of Mexico in the early Mesozoic is indicated by a system of northwest-trending faults that cut Paleozoic rocks; downthrown fault blocks are filled by Triassic-Jurassic clastic sediments and evaporites (Figures 1, 3). Postrift subsidence of the Gulf Coastal Plain toward the Gulf of Mexico is reflected in the present structural configuration of the Coastal Plain sedimentary sequence above the Paleozoic structures of the Appalachian-Ouachita orogen and the adjacent foreland basins.

Acknowledgments

Critical reviews of the manuscript by J. K. Arbenz, W. M. Caplan, J. L. Coleman, Jr., G. W. Colton, R. E. Denison, R. H. Groshong, Jr., and C. G. Stone are gratefully acknowledged. Rachel Thomas assisted in the preparation of the illustrations. Part of my research has been supported by grants from the National Science Foundation (EAR-8109470, EAR-8218604).

Literature Cited

Benkhelil, J., Robineau, B. 1983. Le Fossé de la Bénoué est-il un Rift? *Bull. Cent. Rech. Explor.-Prod. Elf-Aquitaine* 7:315–21

Buchanan, R. S., Johnson, F. K. 1968. Bonanza gas field—a model for Arkoma basin growth faulting. In *Geology of the Western Arkoma Basin and Ouachita Mountains, Oklahoma, Okla. City Geol. Soc. Guideb.*, ed. L. M. Cline, pp. 75–85

Burke, K., Dewey, J. F. 1973. Plume-generated triple junctions: key indicators in applying plate tectonics to old rocks. *J. Geol.* 81:406–33

Chowns, T. M., McKinney, F. K. 1980. Depositional facies in Middle-Upper Ordovician and Silurian rocks of Alabama and Georgia. In *Excursions in Southeastern Geology*, ed. R. W. Frey, 2:323–48. Falls Church, Va: Am. Geol. Inst.

Dickinson, W. R. 1974. Plate tectonics and sedimentation. In *Tectonics and Sedimentation, Soc. Econ. Paleontol. Mineral. Spec. Publ.*, ed. W. R. Dickinson, 22:1–27

Ervin, C. P., McGinnis, L. D. 1975. Reelfoot rift: reactivated precursor to the Mississippi embayment. *Geol. Soc. Am. Bull.* 86:1287–95

Francheteau, J., LePichon, X. 1972. Marginal fracture zones as structural framework of continental margins in South Atlantic Ocean. *Am. Assoc. Pet. Geol. Bull.* 56:991–1007

Glick, E. E. 1975. Arkansas and northern Louisiana. In *Paleotectonic Investigations of the Pennsylvanian System in the United States, US Geol. Surv. Prof. Pap. 853 (Part 1)*, coord. E. D. McKee, E. J. Crosby, pp. 157–75

Glick, E. E. 1979. Arkansas. In *Paleotectonic Investigations of the Mississippian System in the United States, US Geol. Surv. Prof. Pap. 1010 (Part 1)*, coord. L. C. Craig, C. W. Connor, pp. 125–45

Gordon, M. Jr., Stone, C. G. 1977. Correlation of the Carboniferous rocks of the Ouachita trough with those of the adjacent foreland. *Symp. Geol. Ouachita Mountains*, ed. C. G. Stone, 1:70–91. Little Rock: Ark. Geol. Comm.

Haley, B. R. 1982. Geology and energy resources of the Arkoma basin, Oklahoma and Arkansas. *Univ. Mo.-Rolla J.* 3:43–53

Ham, W. E., Denison, R. E., Merritt, C. A. 1964. Basement rocks and structural evolution of southern Oklahoma. *Okla. Geol. Surv. Bull. 95*. 302 pp.

Harris, L. D. 1975. Oil and gas data from the Lower Ordovician and Cambrian rocks of the Appalachian basin. *US Geol. Surv. Invest. Ser. Map I-917 D*

Hoffman, P., Dewey, J. F., Burke, K. 1974. Aulacogens and their genetic relation to geosynclines, with a Proterozoic example from Great Slave Lake, Canada. In *Modern and Ancient Geosynclinal Sedimentation, Soc. Econ. Paleontol. Mineral. Spec. Publ.*, ed. R. H. Dott, Jr., R. H. Shaver, 19:38–55

Horsey, C. A. 1981. Depositional environments of the Pennsylvanian Pottsville Formation in the Black Warrior basin of Alabama. *J. Sediment. Petrol.* 51:799–806

Kane, M. F., Hildenbrand, T. G., Hendricks, J. D. 1981. Model for the tectonic evolution of the Mississippi embayment and its contemporary seismicity. *Geology* 9:563–68

Kidd, J. T., Neathery, T. L. 1976. Correlation between Cambrian rocks of the southern Appalachian geosyncline and the interior low plateaus. *Geology* 4:767–69

King, P. B. 1950. Tectonic framework of southeastern United States. *Am. Assoc. Pet. Geol. Bull.* 34:635–71

King, P. B. 1961. The subsurface Ouachita structural belt east of the Ouachita Mountains. In *The Ouachita System, Univ. Tex. Publ. 6120*, P. T. Flawn, A. Goldstein, Jr., P. B. King, C. E. Weaver, pp. 83–98

King, P. B. 1975. The Ouachita and Appalachian orogenic belts. In *The Ocean Basins and Margins*, ed. A. E. M. Nairn, F. G. Stehli, 3:201–41. New York: Plenum

Lillie, R. J., Nelson, K. D., de Voogd, B., Brewer, J. A., Oliver, J. E., et al. 1983. Crustal structure of Ouachita Mountains, Arkansas: a model based on integration of COCORP reflection profiles and regional geophysical data. *Am. Assoc. Pet. Geol. Bull.* 67:907–31

Mack, G. H., Thomas, W. A., Horsey, C. A. 1983. Composition of Carboniferous sandstones and tectonic framework of southern Appalachian-Ouachita orogen. *J. Sediment. Petrol.* 53:931–46

Miser, H. D. 1959. Structure and vein quartz of the Ouachita Mountains of Oklahoma and Arkansas. *Symp. Geol. Ouachita Mountains*, ed. L. M. Cline, W. J. Hilseweck, D. E. Feray, pp. 30–43. Dallas: Dallas Geol. Soc., Ardmore Geol. Soc.

Morris, R. C. 1974a. Sedimentary and tectonic history of Ouachita Mountains. In *Tectonics and Sedimentation, Soc. Econ. Paleontol. Mineral. Spec. Publ.*, ed. W. R. Dickinson, 22:120–42

Morris, R. C. 1974b. Carboniferous rocks of the Ouachita Mountains, Arkansas: a study of facies patterns along the unstable slope and axis of a flysch trough. In *Carboniferous of the Southeastern United States, Geol. Soc. Am. Spec. Pap.*, ed. G. Briggs, 148:241–79

Neathery, T. L., Thomas, W. A. 1975. Pre-Mesozoic basement rocks of the Alabama coastal plain. *Gulf Coast Assoc. Geol. Soc. Trans.* 25:86–99

Neathery, T. L., Thomas, W. A. 1983. Geodynamics transect of the Appalachian orogen in Alabama. In *Profiles of Orogenic Belts, Am. Geophys. Union Geodyn. Ser.*, ed. N. Rast, F. M. Delany, 10:301–7

Niem, A. R. 1976. Patterns of flysch deposition and deep-sea fans in the lower Stanley Group (Mississippian), Ouachita Mountains, Oklahoma and Arkansas. *J. Sediment. Petrol.* 46:633–46

Niem, A. R. 1977. Mississippian pyroclastic flow and ash-fall deposits in the deep-marine Ouachita flysch basin, Oklahoma and Arkansas. *Geol. Soc. Am. Bull.* 88:49–61

Palmer, A. R. 1971. The Cambrian of the Appalachian and eastern New England regions, eastern United States. In *Cambrian of the New World*, ed. C. H. Holland, pp. 169–217. New York: Interscience

Popoff, M., Benkhelil, J., Simon, B., Motte, J.-J. 1983. Approche géodynamique du Fossé de la Bénoué (NE Nigéria) à partir des données de terrain et de télédétection. *Bull. Cent. Rech. Explor.-Prod. Elf-Aquitaine* 7:323–37

Rankin, D. W. 1975. The continental margin of eastern North America in the southern Appalachians: the opening and closing of the proto-Atlantic Ocean. *Am. J. Sci.* 275-A:298–336

Shideler, G. L. 1970. Provenance of Johns Valley boulders in late Paleozoic Ouachita facies, southeastern Oklahoma and southwestern Arkansas. *Am. Assoc. Pet. Geol. Bull.* 54:789–806

Sloss, L. L. 1963. Sequences in the cratonic interior of North America. *Geol. Soc. Am. Bull.* 74:93–114

Soderberg, R. K., Keller, G. R. 1981. Geophysical evidence for deep basin in western Kentucky. *Am. Assoc. Pet. Geol. Bull.* 65:226–34

Stone, C. G., Haley, B. R. 1977. The occurrence and origin of the granite—metaarkose erratics in the Ordovician Blakely Sandstone, Arkansas. *Symp. Geol. Ouachita Mountains*, ed. C. G. Stone, 1:107–11. Little Rock: Ark. Geol. Comm.

Stone, C. G., McFarland, J. D. III. 1982. Field guide to the Paleozoic rocks of the Ouachita Mountain and Arkansas Valley Provinces, Arkansas. *Ark. Geol. Comm. Guideb. 81-1.* 140 pp.

Thomas, W. A. 1972. Regional Paleozoic stratigraphy in Mississippi between Ouachita and Appalachian Mountains. *Am. Assoc. Pet. Geol. Bull.* 56:81–106

Thomas, W. A. 1973. Southwestern Appalachian structural system beneath the Gulf coastal plain. *Am. J. Sci.* 273-A:372–90

Thomas, W. A. 1974. Converging clastic wedges in the Mississippian of Alabama. In *Carboniferous of the Southeastern United States, Geol. Soc. Am. Spec. Pap.*, ed. G. Briggs, 148:187–207

Thomas, W. A. 1976. Evolution of Ouachita-Appalachian continental margin. *J. Geol.* 84:323–42

Thomas, W. A. 1977. Evolution of Appalachian-Ouachita salients and recesses from reentrants and promontories in the continental margin. *Am. J. Sci.* 277:1233–78

Thomas, W. A. 1979. Mississippian stratigraphy of Alabama. In *The Mississippian and Pennsylvanian (Carboniferous) Systems in the United States—Alabama and Mississippi, US Geol. Surv. Prof. Pap. 1110-I*, pp. I1–22

Thomas, W. A. 1982. Stratigraphy and structure of the Appalachian fold and thrust belt in Alabama. In *Appalachian Thrust Belt in Alabama: Tectonics and Sedimentation, Field Trip Guideb., Ann. Meet., Geol. Soc. of Am.*, ed. W. A. Thomas, T. L. Neathery, pp. 55–66. Tuscaloosa: Ala. Geol. Soc.

Thomas, W. A., Mack, G. H. 1982. Paleogeographic relationship of a Mississippian barrier-island and shelf-bar system (Hartselle Sandstone) in Alabama to the Appalachian-Ouachita orogenic belt. *Geol. Soc. Am. Bull.* 93:6–19

Tull, J. F. 1982. Stratigraphic framework of the Talladega slate belt, Alabama Appalachians. In *Tectonic Studies in the Talladega and Carolina Slate Belts, Southern Appalachian Orogen, Geol. Soc. Am. Spec. Pap.*, ed. D. N. Bearce, W. W. Black, S. A. Kish, J. F. Tull, 191:3–18

Viele, G. W. 1979. Geologic map and cross section, eastern Ouachita Mountains, Arkansas. *Geol. Soc. Am. Map Chart Ser. MC-28F.* 8 pp.

Woods, R. D., Addington, S. W. 1973. Pre-Jurassic geologic framework, northern Gulf Basin. *Gulf Coast Assoc. Geol. Soc. Trans.* 23:92–108

Woodward, H. P. 1961. Preliminary subsurface study of southeastern Appalachian interior plateau. *Am. Assoc. Pet. Geol. Bull.* 45:1634–55

Zietz, I., compiler. 1982. Composite magnetic anomaly map of the United States. Part A: conterminous United States. *US Geol. Surv. Geophys. Invest. Map GP-954-A*

Ann. Rev. Earth Planet. Sci. 1985. 13: 201–40

THE MAGMA OCEAN CONCEPT AND LUNAR EVOLUTION

Paul H. Warren

Institute of Geophysics and Planetary Physics, Department of Earth and Space Sciences, University of California, Los Angeles, California 90024

INTRODUCTION

Despite its small size overall, the Moon has an igneously differentiated crust roughly 55–75 km thick (Toksöz 1979). The oldest lunar crust is anorthositic, with an average (depending on petrogenetic model) of 75–95% plagioclase (Ca,Na-aluminosilicate). One way that anorthosite may form is if plagioclase, which has a density of 2.7 g cm^{-3}, floats as it crystallizes from denser basaltic magma. There is little doubt that plagioclase is buoyant in anhydrous lunar magmas (e.g. Walker & Hays 1977, Warren & Wasson 1979b). The global distribution and purity of the anorthosites, produced during a time span of <300 Myr, suggest flotation over a magma "ocean"—a global, near-surface shell of magma, tens or hundreds of kilometers thick.

"Magma ocean" is probably a hyperbole. Among other things, "ocean" implies the system is virtually 100% liquid, with a mainly gas/liquid upper surface and a waterlike viscosity of the order of 10^{-2} poise. Warren (1984b) suggests that in most contexts the more general term "magmasphere" would be preferable. Nevertheless, massive primordial magmatism influenced almost all phases of the Moon's evolution. The Moon's bulk composition and origin are constrained largely by data for its crust; much of this crust formed by primordial magmatism. Studying the primordial lunar differentiation is an indirect means to constrain the origin of the solar system and of the Earth in particular (Ringwood 1979), because primordial heating probably affected the Earth in a similar fashion. It is important from a purely terrestrial point of view to study the nature and extent of the lunar primordial differentiation.

0084–6597/85/0515–0201$02.00

Beginnings of Magma Oceanography (1864–1976)

The science of magma oceanography is immature but not altogether new. In the nineteenth century, the planets were presumed to have formed from incandescent matter; this assumption implied that the entire Earth was once molten. Lord Kelvin (Thomson 1864), noting that silicate adiabats "most probably" have small dT/dP (T = temperature, P = pressure) relative to melting curves, inferred that crystallization of a magma ocean would "commence at the bottom, or at the center, if there is no solid nucleus to begin with, and would proceed outwards."

Most US Apollo (manned) and Russian Luna (unmanned) missions landed within maria—"seas" of comparatively young basalt flows. Representative portions of the ancient highlands were only sampled by *Luna 20* (February 1972) and *Apollo 16* (April 1972). Nevertheless, a primordial magma ocean was postulated on the basis of a few anorthositic particles of *Apollo 11* soil (Wood et al 1970, Smith et al 1970). After the initial round of Moon rock studies (the last US lunar mission, *Apollo 17*, flew in December 1972), detailed theoretical analyses of magma ocean crystallization began to appear (Binder 1974, Wood 1975, Walker et al 1975, Drake 1976). More refined treatments still tend to favor some sort of magma ocean. But skepticism seems to be increasing (Walker 1983, Longhi & Ashwal 1984), mainly because it is not clear how sufficient heat could have accumulated to form a magma ocean, and because a single magma seems inadequate to account for the diversity of the (growing) data base for ancient pristine rocks.

CLASSIFICATION OF LUNAR ROCKS

The Pristine Rock Concept

The sampled portion of the lunar crust contains few relics of rocks that formed before the waning of intense meteoritic bombardment due to late planetary accretion, about 4.0 Gyr ago. Consequently, most rocks from the ancient highlands are polymict impact breccias, produced by mixing of older rocks. Any rock fragment (or large breccia clast) that escaped mixing, and retains an endogenetic igneous or metamorphic composition, is classified as "pristine." Except for mare basalts, the rate of discovery of pristine samples (mostly as clasts within polymict breccias) has been slow, roughly 10 per year. For an understanding of the ancient, endogenetic crust, pristine samples are indispensable.

The complex problem of how to distinguish pristine from nonpristine rocks was reviewed by Warren & Wasson (1977) and Norman & Ryder

(1979) [see also Ryder et al (1980), Shervais et al (1983), and Warren et al (1983b)]. Texture is often crucial, but even among the limited set of ancient samples with pristine compositions, pristine *textures* are rare. The least ambiguous single means for identifying pristine rocks is by analysis of siderophile elements. Most meteorites have high siderophile contents, and polymict impact breccias almost always contain small but telltale meteoritic components. Of the limited set of lunar rocks that on the basis of other indicators (such as texture) are unambiguously pristine, virtually all have extremely low siderophile contents.

Mare Basalts

Mare basalts are distinctive dark, ferroan lava flows that are concentrated within great impact basins. They cover about 17% of the lunar surface (Hörz 1978). Recently, a 4.2-Gyr-old mare basalt was discovered (L. A. Taylor et al 1983), but most are between 2.9 and 3.9 Gyr old. Pristine mare basalts are common, and except for regolith breccias, polymict impact breccias with large mare components are rare.

KREEP Lithologies

The key characteristic of KREEP rocks is that they contain high concentrations of incompatible elements (such as *K*, *R*are *E*arth *E*lements, and *P*), always in the same pattern of element : element ratios (e.g. the ratio of the lightest REE, La, to the heaviest REE, Lu, is always about 2.2 times the chondritic value). The few known pristine KREEP rocks are about 3.9 Gyr old (Table 1) and are generally basaltic in major element composition and texture (see reviews by Meyer 1977, Warren & Wasson 1979b). A KREEP component dominates trace element patterns of most highlands polymict breccias. The ages of the pristine KREEP samples may be unrepresentative, however. Pristine KREEP emplaced into the upper crust before 3.9 Gyr ago was vulnerable to destruction by mixture into polymict breccias; and most KREEP rocks are polymict breccias.

Ancient Lithologies

Mare basalt probably amounts to less than 1% of the lunar crust (Taylor 1982a), and KREEP is comparably minor. Anorthositic materials cover most of the ancient, heavily cratered highlands and probably dominate the upper half of the crust. Most nonmare (or highlands) rocks are dominated by plagioclase, orthopyroxene, and olivine. Early papers (Keil et al 1972) classified them as a single group, the ANT (*A*northosite-*N*orite-*T*roctolite) suite. It later became apparent that the anorthosites are fundamentally unrelated to the norites and troctolites (the A probably formed separately

Table 1 Sm–Nd and Rb–Sr isotopic age data for pristine rocks[a]

Sample	Lithology	Sm–Nd age	Rb–Sr age[b]	Initial $^{87}Sr/^{86}Sr$
Range	mare basalts (100)	—	2.9–3.8[c]	0.6991–0.6997
Range	KREEP basalts (8)	—	3.83–3.93	0.6994–0.7006
14321c	granite	4.11 ± 0.20	4.00 ± 0.11	0.703 ± 8
15455c	Mg-rich norite	4.48 (?)	4.48 ± 0.12	0.69896 ± 3
67667	Mg-rich lherzolite	4.18 ± 0.07	4.18 (?)	0.69905 (?)
72255c	Mg-rich norite	no data	4.08 ± 0.05	0.69913 ± 7
72417	Mg-rich dunite	no data	4.45 ± 0.10	0.69900 ± 7
73255c	Mg-rich norite	4.23 ± 0.05	no data	no data
76535	Mg-rich troctolite	4.26 ± 0.06	4.51 ± 0.07	0.69900 ± 3
77215	Mg-rich norite	4.37 ± 0.07	4.33 ± 0.04	0.69901 ± 7
78236	Mg-rich norite	4.34, 4.43	4.29 ± 0.02	0.69901 ± 2
Mean[d]	ferroan anorthosites (5)	no data	no data	0.698949 ± 11

[a] Data source: review by Nyquist (1982), except for 14321c (Nyquist et al 1983), basalts, and ferroan anorthosites (Nyquist 1977). Note: Nyquist (1977) lists one of the KREEP basalts (72275,171) as a pigeonite basalt.

[b] Throughout this article, the time constant for decay of ^{87}Rb is assumed to be 0.0142 Gyr^{-1}. Ages reported using 0.0139 Gyr^{-1} are adjusted here by multiplying times 0.979.

[c] One basalt of mare affinity from *Apollo 14* has a Rb–Sr age of 4.14 ± 0.05 Gyr (L. A. Taylor et al 1983), but this age is highly exceptional, at least among sampled mare basalts.

[d] Least-squares-weighted mean and uncertainty of mean, based on 16 data for 15415, 60015, 60025, 61016c, and 64423,13,1. The range of the data is from 0.69887 ± 7 to 0.69910 ± 12. The listed mean does not include 4 data from Nunes et al (1974), which would drive it down to 0.698925 ± 10.

from the N and the T), and unusual rocks like dunite and gabbro had to be classified. In this paper, all nonmare lithologies except KREEP are termed "ancient."

FERROAN ANORTHOSITES Roughly 50% of all ancient rocks are ferroan anorthosites. Dowty et al (1974) first noted the distinctiveness of ferroan anorthosites, so-called because their sparse (generally $<5\%$) mafic silicates have consistently low Mg/(Mg + Fe), or *mg*, by ancient lunar standards. Equally important, their plagioclase is consistently Ca-rich (Na-poor). Their textures are usually cataclastic (brecciated, yet monomict). The few that are unbrecciated have cumulate or coarse granulitic (plutonic-metamorphic) textures, and it is widely assumed that the entire group originated as cumulates.

MG-RICH ROCKS Warner et al (1976) first noted the distinctiveness of the Mg-rich group, which accounts for nearly all of the remaining 50% of ancient rocks. Unbrecciated Mg-rich rocks have cumulate or coarse granulitic textures. Presumably, the entire group originated as cumulates. The Mg-rich group comprises mainly norites and troctolites, but also includes several gabbros, a dunite, and a lherzolite. The distinction between Mg-rich rocks and ferroan anorthosites is best illustrated by plotting *mg* vs Ca/(Ca + Na) (Figure 1). The distinct clustering into two groups, with little or no overlap, militates against both groups originating from a single magma.

Mg-rich rocks are distinctly less anorthositic than ferroan anorthosites (Figure 2). Warren & Wasson (1980a) suggest that it is unlikely that a series of rocks with modest plagioclase contents such as the Mg-rich rocks could have floated over any reasonable density magma. Thus, the ferroan anorthosite group is more plausibly a series of flotation cumulates than is the Mg-rich group; and if the magma ocean concept has any validity, the

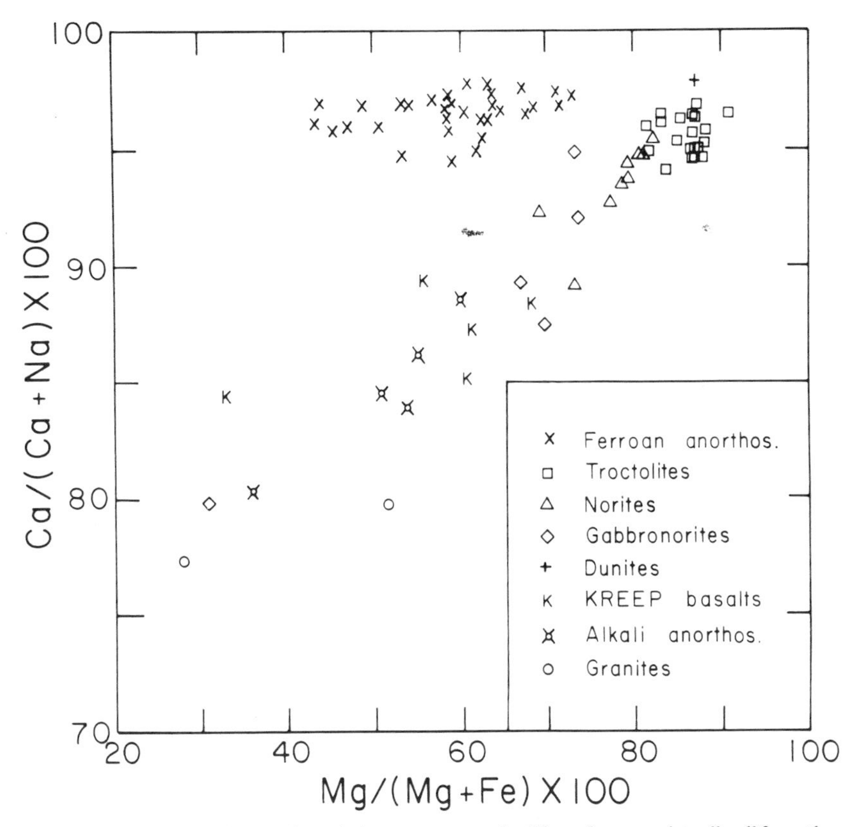

Figure 1 Ca/(Ca + Na) vs *mg* for pristine nonmare rocks. These data are virtually all from the compilations of Ryder & Norman (1978, 1979) and from the present author's "foray" papers (Warren et al 1983b, and references therein). Note the clear separation of ferroan anorthosites from all other rocks.

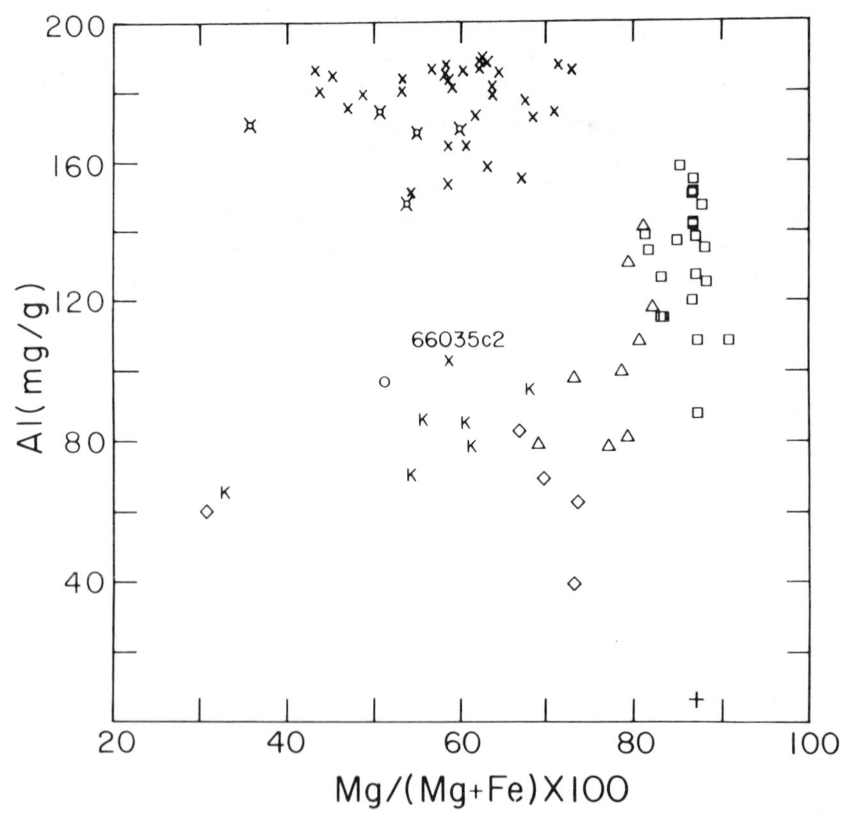

Figure 2 Al vs *mg* for pristine nonmare rocks. The symbols and data sources are the same as for Figure 1. "Ferroan" rocks are systematically more anorthositic (Al-rich) than Mg-rich rocks. One exception, 66035c2, is a small, coarse-grained clast that is probably unrepresentative for Al.

Mg-rich rocks probably come from intrusions that were emplaced into an older, ferroan anorthosite portion of the crust. The possibility that a single magma can account for Mg-rich rocks and ferroan anorthosites is discussed further in a later section.

MISFITS Rare granites (Warren et al 1983a) and alkali anorthosites (Warren et al 1983b, Shervais et al 1983) may be extreme differentiates related to Mg-rich rocks and/or to KREEP. James & Flohr (1983) propose a genetic distinction between a subset of Mg-rich rocks (gabbronorites), which have important amounts of high-Ca pyroxene, and the majority of the Mg-rich rocks, which have little or no high-Ca pyroxene.

EVIDENCE FOR A PRIMORDIAL, DEEP, GLOBAL DIFFERENTIATION

1. Global, Highly Anorthositic Crust

The best argument for the lunar magma ocean is often overlooked in favor of subtler, more ingenious ones. Except for thin, scattered veneers of mare basalt, the Moon's surface material averages about 75% plagioclase. Remote-sensing data for Fe, Mg, Al/Si, and Mg/Si (Adler & Trombka 1977) indicate that the highlands in general are at least as anorthositic as the *Apollo 16* site, where soils and polymict breccias average 75% plagioclase. The only large areas where the plagioclase content is $\ll 75\%$ are maria, basaltic veneers that appear to be generally <2 km thick (Hörz 1978). Interpretation of the remote-sensing data in terms of plagioclase abundance is borne out by regolith data for three other highlands sites: *Surveyor 7* (Turkevich 1973), *Luna 20*, and the source of the ALHA81005 meteorite (e.g. Kallemeyn & Warren 1983). A generous estimate of the amount of Al and Ca in the bulk Moon (Taylor 1982a) implies a total of about 16% plagioclase. Thus, the upper crust is at least 5 times richer than the bulk Moon in plagioclase.

The average lower crust might be far less anorthositic than the average surface composition. Seismic data are equally consistent with a gabbroic composition (Liebermann & Ringwood 1976). But remotely sensed compositions of fallback deposits in mare-free basins such as Mendeleev, Hertzsprung, and Korolev (Adler & Trombka 1977, Davis 1980) and of ejecta from Orientale basin (Spudis et al 1984) imply that plagioclase contents like those of the *Apollo 16* site prevail to the depths "sampled" by the impacts. In big impacts, the maximum source depth of ejecta is not to be confused with the depth of the transient crater, which is probably much greater. Head et al (1975) concluded that ejecta from the Imbrium basin probably originated no deeper than 27 km. But other estimates range from 60 (Grieve 1980) to 330 (O'Keefe & Ahrens 1978) km. Much of the fallback deposit of Hertzsprung and the ejecta from Orientale probably came from $\gg 10$ km deep.

Gravity data appear to confirm that the plagioclase content of the upper few tens of kilometers is comparable to the surface abundance. The nearside highlands stand an average of 2.6 km above the circular maria, which display large positive gravity anomalies (mascons). Most of this offset appears to be isostatically compensated (Haines & Metzger 1980). The mascons are probably not significantly denser than average lunar material ($3.34\ g\ cm^{-3}$; Taylor 1982a). Thus, isostasy implies that a low-density phase (plagioclase) is enriched in the nonmascon crust. Kaula et al (1974) calculated a lower limit of 34 km for the global depth equivalence of a layer of pure plagioclase; for the nearside only, Wood (1983) derived a lower limit of 29 km. Most gravity-based estimates (e.g. Haines & Metzger 1980) are higher than these limits.

Several early post-Apollo models invoked heterogeneous accretion to account for the enrichment of Al and Ca in the crust. Brett (1973) and Taylor (1975) cite ample reasons for dismissing this notion. Furthermore, oxygen isotopic ratios imply similar ultimate sources for typical anorthosites and mare basalts generated by partial melting of the deep interior (Clayton & Mayeda 1975). Widespread production of anorthositic magma is ruled out by phase equilibria. Thus, a global surficial layer of cumulate anorthosite suggests flotation of plagioclase over a global shell of basaltic magma. Models in which anorthosite cumulates form piecemeal are discussed in a later section.

2. *Extreme Antiquity of Many Pristine Nonmare Rocks*

The ages of ancient lunar rocks are reviewed by Carlson & Lugmair (1981a) and Nyquist (1982). Most of these "ages" are better interpreted as lower limits for crystallization ages, because most of the oldest lunar rocks are severely shocked. Also, the measured ages might indicate times when plutonic rocks were excavated by impacts, ending periods of postcrystallization subsolidus isotopic equilibrium (Carlson & Lugmair 1981b). As Nyquist's (1982) Figure 1 shows, Ar–Ar ages seem more susceptible to resetting by shock than Rb–Sr or Sm–Nd ages. A whole-rock Sm–Nd isochron for four pristine Mg-rich rocks suggests a common age of 4.33 ± 0.08 Gyr (Carlson & Lugmair 1981b). Eight pristine Mg-rich rocks have been dated by Rb–Sr and/or Sm–Nd internal isochrons (Table 1). Three are more than 4.4 Gyr old. The average (4.30–4.33 Gyr, depending on which age is accepted for 76535), is biased toward low ages. Rocks formed at 4.5 Gyr were less likely to survive the final stages of accretion than rocks formed at 4.1 Gyr (Hartmann 1980). Model ages for nonpristine rocks using the complex U–Pb system "strongly suggest" a major differentiation at about 4.47–4.51 Gyr (Oberli et al 1979).

The magma ocean model implies that ferroan anorthosites are among the oldest lunar rocks. Unfortunately, pristine ferroan anorthosites are extremely poor in Rb and REE, so no Rb–Sr or Sm–Nd internal isochron age has been measured. The highest Ar–Ar age for what is definitely a ferroan anorthosite is 4.19 Gyr (for 60025; Schaeffer & Husain 1974). Among *Apollo 16* rocks (Maurer et al 1978, Stöffler et al 1981), those with over 90% plagioclase tend to have higher Ar–Ar ages (mostly 4.12–4.26 Gyr) than the rest (mostly 3.85–4.04 Gyr), a trend that suggests that ferroan anorthosites are generally older than Mg-rich rocks.

These data leave little doubt that the Moon began producing diverse cumulates within about 100 Myr of the origin of the solar system. Conceivably, some of these cumulates formed in giant pools of impact melt (Delano & Ringwood 1978), but the detailed characteristics of the pristine rocks weigh against this interpretation (Warren & Wasson 1979c, Norman & Ryder 1979, Wolf et al 1979). Moreover, if prior magmatism had not already produced a differentiated crust, the impact melts would have been exclusively ultrabasic, and most of their cumulates would have been dunites or peridotites; and the bigger the melt pool, the more preexisting crust is required to avoid having the pool be ultrabasic. (Of course, if accretional energy supplied most of the primordial heat, the magma ocean was a type of impact melt pool.) Had the Moon heated exclusively by radioactivity from U, Th, and K, about 2 Gyr would have passed before the onset of internal melting (Wood 1972). Clearly, heating accompanied formation, and the Moon became comparatively quiescent within <500 Myr. Regardless of the exact nature of the "magma ocean," the Moon definitely experienced primordial differentiation.

3. *Complementary Patterns of Europium Anomalies*

None of the rare earth elements are volatile or siderophile, so the bulk Moon presumably has a REE pattern roughly parallel to those of chondritic meteorites. Most mare basalts have pronounced negative Eu anomalies (Figure 3), which are probably inherited from mantle source regions. The estimated depths of the sources vary from 100–200 km (Binder & Lange 1980) to 300–500 km (Delano & Livi 1981). The anomalies imply fractionation of plagioclase, but experimental petrology studies (Kesson & Lindsley 1976) indicate that plagioclase was seldom a residual phase,

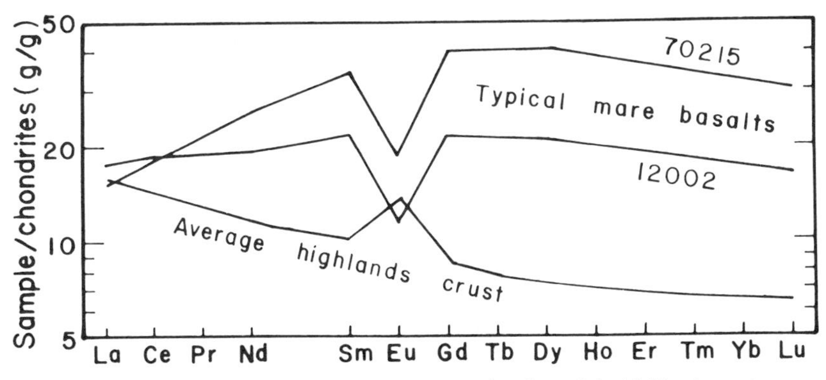

Figure 3 Rare earth element patterns of typical mare basalts and the highlands crust appear complementary. The data are from Taylor (1982a).

implying that it was removed before the sources formed. The overall composition of the highlands is anorthositic with a positive Eu anomaly (Figure 3). The straightforward inference (e.g. Taylor & Jakes 1974, Wakita et al 1975, Shih & Schonfeld 1976, Nyquist et al 1981) is that the mare basalt source regions were complementary mafic cumulates of the same magma(s) that produced the ancient anorthosite crust. As reviewed by Taylor (1975, 1982a), many alternatives have been proposed (see also Walker 1983), but the cumulate remelting model is widely accepted.

4. *Uniformity of the KREEP Component*

As reviewed by Warren & Wasson (1979a), the ratios of incompatible elements among lunar rocks rich in such elements virtually all conform to a single, uniform pattern ("the" KREEP pattern) throughout the sampled portion of the Moon; this uniformity suggests that these rocks derived from the residuum ("urKREEP") of a single, global magma. Particularly impressive are the precisely determined Sm/Nd elemental ratios and $^{143}Nd/^{144}Nd$ isotopic ratios for KREEP (Lugmair & Carlson 1978, Carlson & Lugmair 1979). Warren & Wasson (1979a), and more emphatically Walker et al (1979), cautioned that most KREEP rocks are too mafic-basaltic in major composition to be simple, *unaltered* samples of magma ocean residuum. But detailed crystallization models appear consistent with the residuum having a KREEP pattern of *incompatible* elements (Longhi 1980, Gromet et al 1981, Fujimaki & Tatsumoto 1984). Of course, errors accumulate in such models as differentiation proceeds, and urKREEP forms last of all.

5. *Petrochemistry/Longitude Patterns*

Patterns of petrochemistry correlated with longitude suggest a global differentiation process. The most obvious such pattern is the nearside-farside asymmetry in the distribution of mare basalts. Remote-sensing data indicate that KREEP follows a similar distribution (Adler & Trombka 1977). Subtler petrochemistry-longitude correlations are manifested by minor- and trace-element data for Apollo and Luna samples (Warren & Wasson 1980b, Warren et al 1983b).

6. *High Content of Incompatible Elements in the Crust*

Taylor (1982a) suggests that a magma ocean is required to account for the crustal enrichment of incompatible elements. This argument implies (*a*) that the overall incompatible element content of the crust is similar to the mean surface content, and (*b*) that incompatible elements could not be similarly enriched through a series of partial-melting processes. Regarding (*a*), note that KREEP rocks tend to have basaltic composition and (in pristine form) texture, which implies that they were emplaced mainly as surface flows.

CONSTRAINTS ON THE DEPTH OF THE MOLTEN ZONE

Kopal (1977) contends that the departure of the Moon's interior from hydrostatic equilibrium rules out a magma ocean >20 km deep. There is no consensus about the depth of the magma ocean!

1. *Mass Balance to Account for Crustal Plagioclase*

This argument assumes that nonmare surface compositions are at least modestly representative of the bulk of the crust. To form a 60-km layer that is 75% plagioclase, the mass of Al required is 1.9 wt% of the Moon. Petrogenetic constraints suggest that only the ferroan anorthosites, which constitute roughly 50% of the pristine upper crust, formed from the magma ocean. Even so, the implied Al requirement is 1.0 wt% of the Moon. Because Al, Th, and U are refractory lithophile elements, they are probably in roughly chondritic proportion to one another in the bulk Moon (e.g. Morgan et al 1978, Taylor 1982a). Concentrations of Th and U (and by implication Al) in the bulk Moon are constrained by heat flow data. As interpreted by Langseth et al (1976), the heat flow data imply that the bulk Moon has an Al concentration of about 3.8 wt%. Conel & Morton (1975), however, contend that the two sites where heat flow was measured are probably unrepresentative. If so, the true Al concentration in the bulk Moon is probably $\ll 3.8$ wt%. An "average" of recent estimates is 2.8 wt% (Warren 1983a). For our purposes, a high value is a conservative one. In order for 1.0/3.8 of the Al to wind up as magma ocean cumulates, at least 26 wt% of the Moon has to be melted, assuming total Al extraction. If we take 2.8 wt% as a more realistic bulk Moon Al content, the implied minimum fraction melted becomes 36 wt%, a fraction corresponding to a shell from the surface down to 250 km (Figure 4). Of course, if $<50\%$ of the bulk crust is magma-ocean-derived anorthosite, the implied fraction of the Moon melted would be commensurably lower.

2. *Compressional Structures in the Crust*

As the Moon cooled, it contracted, and compressive stresses developed in the crust. The stresses, in proportion to the degree of primordial melting, produced fault scarps. Solomon & Chaiken (1976; supported by Cassen et al 1979) estimated from the scarcity of crustal fault scarps that the Moon's radius has changed by <1 km over the last 3.8 Gyr, and they concluded that

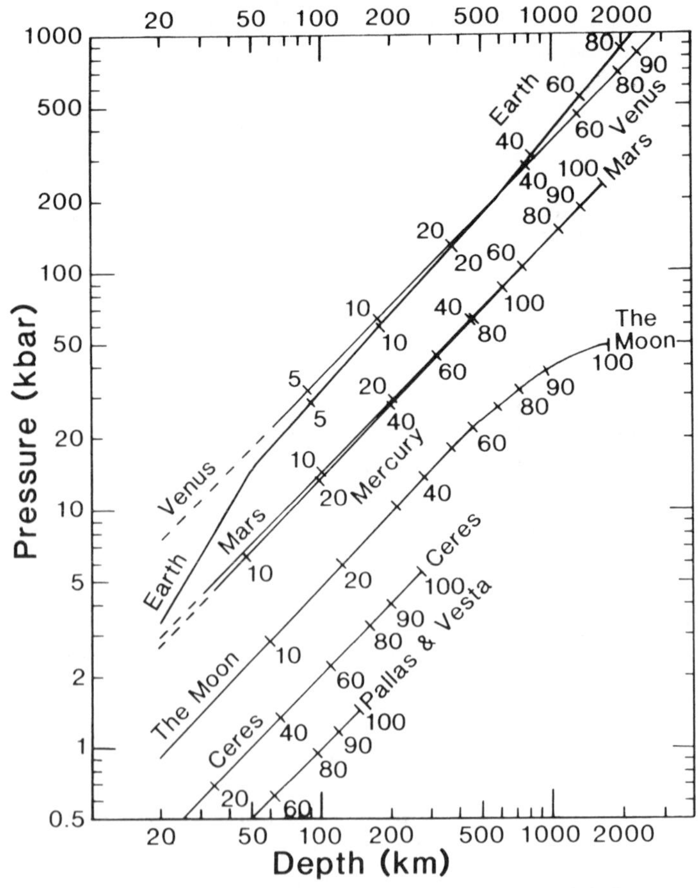

Figure 4 Pressure vs depth inside the terrestrial planets, the Moon, and the largest asteroids. The numbers along the curves refer to the volume percentage of noncore portion above depth. The values for Earth are after Dziewonski et al (1975), and the trends for the other bodies were calculated using Equation (2-65) of Turcotte & Schubert (1982). The sizes of cores are from Solomon (1980), except assumed to be 15 vol% for the asteroids and 0 vol% for the Moon. (In order to simplify calculations, the Moon's mantle was treated as a "core," and its crust as a "mantle.") Except for the Earth and Moon, density gradients within noncore portions are ignored; for larger planets, the actual pressure-depth relationships probably parallel that for the Earth more closely than indicated here.

the magma ocean (assuming a sharp interface between a zone of complete melting and a lower zone without any melting) was never more than 200 $\pm$100 km deep. Solomon (1980) later revised this estimate to 300$\pm$100 km. Binder & Lange (1980) and Binder (1982b), however, argue that the decrease in radius subsequent to the total melting of the Moon would be a factor of three lower than the value of Solomon & Chaiken (1976), and that the actual change of radius is ≫1 km. According to Binder (1982b), the abundance of fault scarps in the highlands indicates that the Moon was initially totally molten.

3. *Depths of Source Regions of Mare Basalts*

As mentioned above, most models for mare basalt petrogenesis call for remelting of magma ocean cumulates, and 100 km is a very conservative lower limit on the depth of the mare basalt source regions. Even Binder & Lange's (1980) estimated depth range includes 200 km. Before the magma ocean produced the mare basalt sources, it probably deposited a thick series of more magnesian, less readily remelted cumulates (e.g. Taylor & Jakes 1974, Binder 1976). Therefore, accepting the cumulate-remelting model for mare basalt petrogenesis, we obtain a conservative lower limit for the depth of the magma ocean of 250 km.

4. *Rapid Cooling of the Magma Ocean*

Brett (1977) argues that the extremely ancient model ages of many lunar basalts indicate that primordial melting involved only the outer few hundred kilometers of the Moon. This hypothesis assumes that the cumulate pile below the magma ocean takes several hundred million years to cool enough to halt subsolidus convection, which causes continuing isotopic fractionation.

5. *Absence of Mafic Cumulates Complementary to Ferroan Anorthosites*

If the ferroan anorthosites were produced by the magma ocean, the complementary cumulates, which are presumably ultramafic but with "ferroan" mineralogy, are evidently buried below the upper part of the crust from which our samples were derived (Warren & Wasson 1980a, Longhi 1980). The earliest complementary cumulates are probably at least twice as deep as the topmost ones. Thus, a limit on the depth of the ocean should be at least twice as great as a lower limit on the depth plumbed by impact craters. If we accept the most conservative estimate for the depth "sampled" by the Imbrium impact (Head et al 1975), maximum magma ocean depths <50 km can be ruled out. Seismic data also indicate that ultramafic cumulates are rare throughout the crust (Toksöz et al 1973, Liebermann & Ringwood 1976).

6. *Near-Solidus Temperatures in the Present Deep Interior*

Nearly all the K, Th, and U in the magma ocean probably concentrated into the residual liquid just below the crust, and most of the "ocean" zone probably lost its heat sources and did not maintain near-melting temperatures (Toksöz & Solomon 1973). Electrical conductivity data (e.g. Hood & Sonett 1982) seem to require near-solidus temperatures below roughly 500 km, while seismic data suggest (Toksöz 1979, Nakamura et al 1982) incipient melting below about 1000 km. Goins et al (1979) note that the seismic data might also be explained by invoking a lower *mg* for the mantle below 1000 km, but this too would suggest that primordial melting did not penetrate below about 1000 km. Thus, the maximum depth of the differentiated zone was probably <1000 km.

7. *High Content of Incompatible Elements in the Crust*

Taylor's (1978) model suggests that a magma ocean about 500 km deep is implied by crustal incompatible element enrichments.

Synthesis: What Was the Maximum Depth?

Obviously these supposed constraints do not all accurately gauge "the" maximum depth, for some of them contradict one another. The depth of the fully molten zone is far more difficult to gauge than the depth of the entire zone that was differentiated including whatever partially molten portion it contained (see below). The depth of the entire differentiated zone was probably >250 km (based mainly on the depth of mare basalt sources, and backed up by mass balance for Al) and <1000 km (based mainly on present deep interior temperatures). Defining the magma ocean loosely as a convective, largely molten shell (a magmasphere), it apparently encompassed roughly 60 vol% of the Moon.

EFFECTS OF PRESSURE ON MINERAL STABILITY RELATIONSHIPS

Olivine-Pyroxene Relationships

The initial magma ocean is usually modeled as a total melt of bulk Moon composition. Realistically, primordial melting was partial, not total. Even if the ocean was a total melt, crystallization probably occurred mainly along the bottom, where pressures were greatest, because for silicate compositions, dT/dP is far greater for melting curves than for adiabats (Thomson 1864). Whatever physical scenario is preferred for the "ocean," the pressures where most of the primordial lunar melting and crystallization occurred were of the order of 10 kbar (Figure 4).

In anhydrous basaltic magmas, the ratio of olivine to pyroxene and the corresponding composition of the melt phase are highly sensitive to pressure (Figure 5). At pressures of the order 10 kbar, melts tend to be less

SiO_2-rich (and solids tend to be more SiO_2-rich) than at surface pressure. Thus, if the bulk Moon is similar in major element composition to anhydrous ordinary chondritic silicates, pyroxene is more refractory than olivine at all pressures >10 kbar. By the same token, a melt parcel saturated with plagioclase and pyroxene at depth is likely to be saturated with plagioclase and olivine upon ascending (via convection or otherwise) to the base of the crust. In most models these effects are ignored, and the only cumulus phase in the deepest portion of the magma ocean is assumed to be olivine (e.g. Taylor & Jakes 1974, Drake 1976, Longhi 1977, Minear & Fletcher 1978, Taylor 1978, Fujimaki & Tatsumoto 1984). Such models underestimate the (Mg + Fe)/Si ratio of the melts that eventually crystallize plagioclase to form the crust, and in some cases needlessly imply a hyperchondritic (Al + Ca)/Si ratio for the bulk composition of the Moon (Warren & Wasson 1979b).

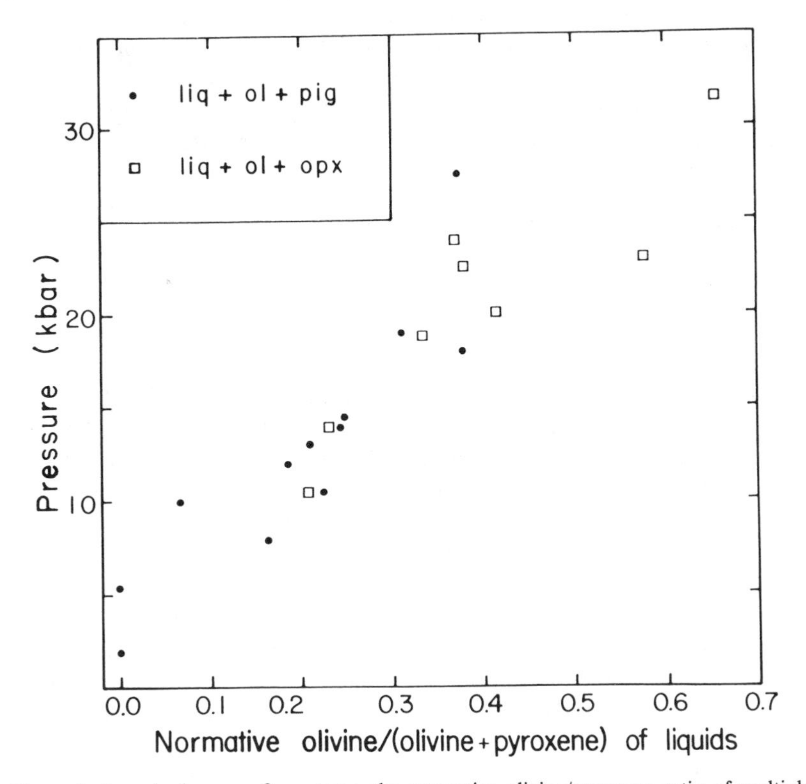

Figure 5 In anhydrous mafic systems, the normative olivine/pyroxene ratio of multiply saturated melts is diminished by pressure, i.e. the pyroxene/olivine ratio of solids is enhanced by pressure. Data are from experiments using natural lunar compositions. Abbreviations: liq = liquid, ol = olivine, pig = pigeonite, opx = orthopyroxene. After Warren (1984a).

High-Density Phases

Plagioclase reacts to form garnet at pressures in excess of roughly 12 kbar (Ringwood & Essene 1970). Buck & Toksöz (1980) calculate that the Moon's lower mantle (assumed to be of peridotitic, bulk-Moon composition) probably contains about 10% garnet, 5% Tschermak pyroxenes, and 1% jadeite. Mg-spinel might also be stabilized by pressure. Garnet has a strong affinity for heavy REE, and Shih (1977) suggests that garnet crystallization in the magma ocean helped to establish the light-REE-enriched REE pattern of KREEP. Too much garnet would conflict with basic data on the Moon's density.

Plagioclase–Mafic Silicate Relationships

As noted by Morse (1982), the stability of plagioclase relative to mafic silicates diminishes with increasing pressure. The influence of pressure on this ratio is not nearly as pronounced as on the pyroxene/olivine ratio, but it probably reinforced the effect of plagioclase buoyancy [and possibly other factors (Herbert et al 1977b, Morse 1982)] to cause the upper surface of the magma ocean to form pure plagioclase adcumulates.

WAS THE OCEAN MAINLY LIQUID, OR MAINLY "MAGMIFER"?

Potential Heat Sources

The origin(s) of primordial heat is among the greatest unresolved problems of planetology. There are several recent reviews of this topic (Sonett & Reynolds 1979, Hostetler & Drake 1980, Basaltic Volcanism Study Project 1981, pp. 1144–51) but none of these are geared toward the Moon. Leading contenders among the possible heat sources are (*a*) accretion, i.e. conversion of kinetic energy into heat during large impacts, (*b*) electromagnetic induction due to an enhanced solar wind (Herbert et al 1977a), (*c*) short-lived radioactive species, such as ^{26}Al ($t_{1/2} = 0.7$ Myr) or "superheavy" elements, (*d*) enhanced solar luminosity, (*e*) tidal dissipation, and (*f*) heat inherited from the Earth if the Moon formed by fission (Binder 1978). Core formation did not raise the Moon's mean temperature more than about 10 K (Solomon 1980). Mechanisms (*a*) and (*b*) would add heat chiefly within roughly 400 km of the surface; (*c*), (*e*), and (*f*) would heat the interior more or less evenly throughout; and (*d*) would heat at the surface. None, except perhaps (*c*) and (*f*), would likely have melted the whole Moon.

In the lunar literature, accretion is the most popular mechanism, but only because accretion is less implausible than any one of the rest. Accretional heat is initially buried at a depth of about two projectile radii. Unless a projectile's radius is larger than roughly 40 km, the heat tends to be lost when subsequent impacts eject material across the lunar surface (Wetherill 1976). Accretion could only have produced extensive melting if the time interval for accretion was shorter and/or the bodies accreted were larger than models (Safronov 1978, Wetherill 1980, 1981, Ransford 1982) indicate. In Wood's (1983) model, the bodies accreted were larger than previously supposed. To assume accretion as the sole primordial heating mechanism for the Moon is to imply that some other mechanism worked for meteorites, because accretion is probably not an adequate heat source for widespread melting on small meteorite parent bodies (Sonett & Reynolds 1979). However, neither mechanism (*c*) (short-lived radioactive species) nor (*d*) (enhanced solar luminosity) is effective unless the time interval for accretion is short, in which case accretional heating suffices anyway.

Tendency of Convection to Maintain a Mostly Solid System

Many physical models of the primordial differentiation (e.g. Solomon & Longhi 1977, Minear & Fletcher 1978) assume that the magma ocean was a total melt of the outer Moon. But if accretion was the main heat source, relatively little of the Moon could have been fully melted (e.g. Wetherill 1975, 1981, Alvén & Arrhenius 1976, Hess et al 1977). As the Moon heated, its interior became partially molten on the way to becoming fully molten. Even a few percent of partial melt lowers the viscosity of peridotite by roughly a factor of 10 (Bussod & Christie 1983). If the viscosity is sufficiently low, convection will transport heat toward the surface. As partial melting continues, the viscosity continues to fall, causing more rapid heat loss. Thus, the system "self-regulates" (Stevenson 1980), such that the temperature remains close to the solidus. Longhi (1981) suggested that fully molten material coalesced into a "mini-magma ocean" >60 km deep atop a convective, partially molten mantle. He noted that petrogenetically such a system would not be much different from a deeper magma ocean formed by total melting. Frenkel (1983) suggested that the vehicle for primordial differentiation was a "magma generating system," and Shirley (1983) developed the first detailed model along these lines, coining the term "magmifer" for the convective, partially molten portion of the mantle.

In Shirley's (1983) model, the initial crust is derived from shallow intrusive magma chambers over upwelling zones in the convective magmifer, a process akin to ophiolite genesis. Buoyant cumulate anorthosite layers tend not to be subducted; instead, they become crustal nuclei that eventually spread to cover the globe. As the Moon heats further, the magmifer expands downward, attaining a maximum depth of roughly 300–400 km. Meanwhile, the degree of partial melting in the magmifer increases, resulting in bigger influxes of magma into the near-surface magma chambers. The magma chambers spread laterally, eventually coalescing into a global molten shell. For roughly 20 Myr the global magma layer continues to be replenished by basaltic partial melts from the magmifer at nearly the same rate as it loses mass by crystallization. Depending upon numerous assumptions, the thickness of the melt layer never exceeds roughly 7 km.

Convection works to make a system adiabatic (of uniform heat content throughout). Because convection tended to cool the magmifer, the temperature at its bottom was probably never more than roughly 50 K above the bulk Moon solidus, and temperatures in its upper part were probably not far above adiabatic relative to the bottom. As a result of the latent heat of melting/crystallization, the value of adiabatic dT/dP in a partially molten system is about 4 K $kbar^{-1}$ (Cawthorn 1975, D. Shirley, personal communication, 1984), which is about 6 K $kbar^{-1}$ less than dT/dP for silicate melting. Thus, the increment of temperature above the solidus at the top of the magmifer is given by $T_{inc} = 50 + 6dP_m$, where dP_m is the pressure range in the magmifer. The fraction of melt f increases toward the top of the magmifer (Figure 6), which occurs where f (a function of T_{inc}) exceeds a certain critical value f_s such that melt buoyantly separates from crystals and rises into the fully molten layer. In a small system, f_s is of the order of 30% (Arndt 1977), but it is probably far lower within large source regions (Stolper et al 1981). Convective velocity in the magmifer must offset loss of partial melt to the molten basaltic layer (Stevenson 1980), or else the system degenerates into a fully molten layer atop a layer that is almost fully solid and comparatively stagnant.

In some models (e.g. Walker et al 1975, Herbert et al 1977b) a small magmifer occurs below a much larger molten layer. The geotherm is assumed to have a thermal maximum at roughly 200 km due to accretional heating, and the melt layer is essentially a total melt of the outer Moon. However, such a thermal maximum would probably be transitory and localized beneath epicenters of large impacts. As Walker et al (1975) noted, the system would convect above the thermal maximum. Convection would redistribute heat and continue to do so until the level of the maximum fell to the level of the bottom of the low-viscosity (partially molten) portion of the Moon. Also, crystallization would release latent heat at the cooler base of the system, and the fully molten layer would be stirred by impacts hitting the surface (Wood 1975, Hartmann 1980). These last three factors would tend to establish and maintain a monotonic increase of temperature with depth.

The range of pressure in the magmifer was most likely of the order 20 kbar, because (*a*) the primordial differentiation evidently involved roughly half of the Moon, and (*b*) f probably equals f_s when T_{inc} is roughly

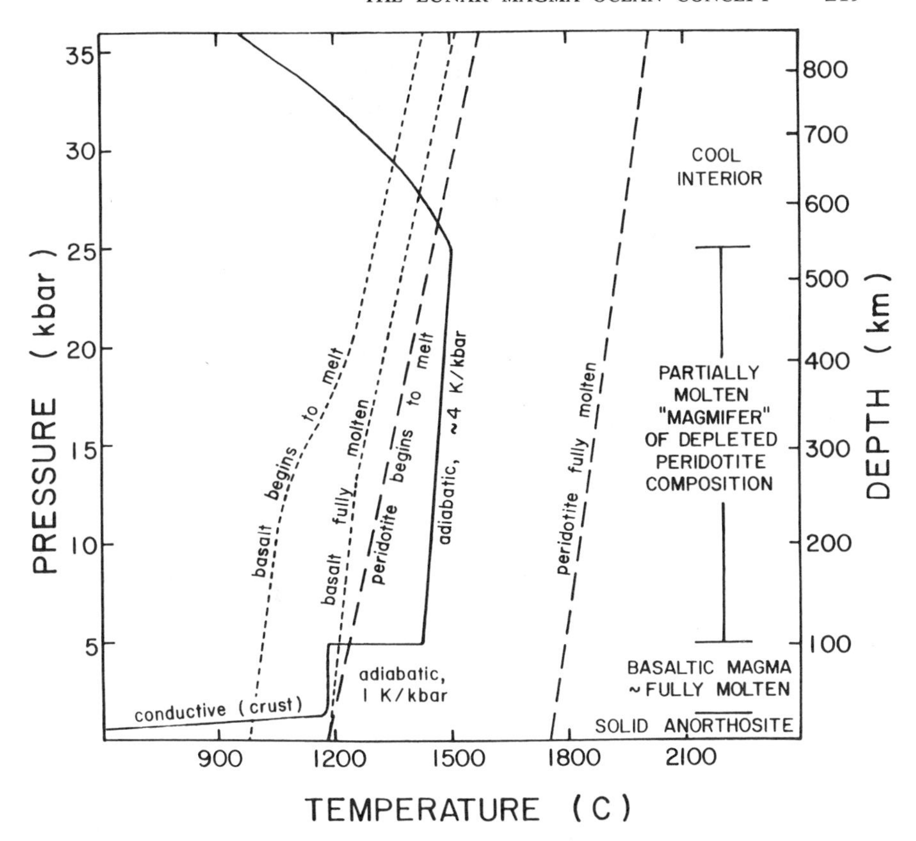

Figure 6 In a "magmifer" type of magma ocean, the geotherm is conductive (high dT/dP) in the anorthositic crust, adiabatic ($dT/dP = 1$ K kbar^{-1}) in the basaltic melt layer, discontinuous at the interface between the molten (basaltic) layer and the mostly solid (depleted peridotitic) magmifer, and adiabatic ($dT/dP = 4$ K kbar^{-1}) in the magmifer. If the magma ocean's bottom was at the pressure shown (25 kbar), then the magma ocean comprised about 67% of the Moon.

150–200 K. An increment of 190 K above the solidus results in about 40% partial melting of most peridotites (Jaques & Green 1980, Mysen & Kushiro 1977), or near-total melting of a peridotite rich in Ca and Al (Mysen & Kushiro 1977). Shirley's (1983) calculated maximum thickness for the fully molten layer (7 km) depends on many assumptions, but a "partially molten magma ocean" seems more plausible than simpler models in which nearly the entire system is molten.

Petrogenetic Implications of the Magmifer Model

In the partially molten magma ocean model, the molten layer is continually crystallizing, with anorthosites plating out along the top and mafic cumulates forming along the bottom (which is also the top of the magmifer). But the fully molten layer is continually replenished with fresh magma from below, and the situation approximates a steady state. In support of the partially molten magma ocean model, Shirley (1983) cites geochemical characteristics of the ferroan anorthosites: their nearly horizontal [constant Ca/(Ca + Na)] trend on Figure 1; and the near constancy of their Eu and Sr concentrations, in contrast to their wide range of concentrations for incompatible elements such as Sm. Actually, these data are consistent with closed-system fractional crystallization models (Longhi 1982, Warren 1983b).

There is probably greater heterogeneity in a largely solid magma ocean than in a purely liquid one. Longhi & Boudreau (1979) and Longhi (1980) suggested that the compositional evolution of the magma ocean was affected by the cyclical mixing of primitive magmas from central, mainly liquid parts of the ocean with evolved magmas in upper, largely crystalline parts. One of the many implications of cyclical magma mixing (O'Hara 1977) is that large incompatible element enrichments are accompanied by limited compositional evolution for major elements. Although the physical model is different from that of Longhi & Boudreau (1979), continual magma mixing is the essence of the partially molten magma ocean model. Such magma mixing would help explain the paradox that pristine KREEP basalts, named for their high incompatible element contents, are generally too mafic-basaltic in composition (Walker et al 1979) to be simple derivatives of the residuum of a magma ocean.

In terms of petrologic consequences involving major elements, there may be few dramatic differences between a partially molten magma ocean and a fully molten one. Warren & Wasson (1979b) noted that partial (as opposed to total) melting "would not greatly alter the course of magma ocean crystallization, except to (in effect) leave off the early stages . . . early solids would be orthopyroxene + olivine, either way." One advantage of the partially molten magma ocean model is that soon after the waning of the ocean, the production of Mg-rich magmas is expected as a consequence of continued (albeit less rapid) melt removal from the former magmifer; this aspect of the model is discussed in more detail below. The petrogenetic implications of a magmasphere that is only partially molten require further study.

PETROLOGIC EVIDENCE FOR MULTIPLE ANCIENT DIFFERENTIATION EPISODES

The Mg/(Mg+Fe) vs Ca/(Ca+Na) Diagram

Until the late 1970s, it was widely assumed that the magma ocean produced the entire spectrum of ancient cumulates. G. J. Taylor et al (1973) noted a

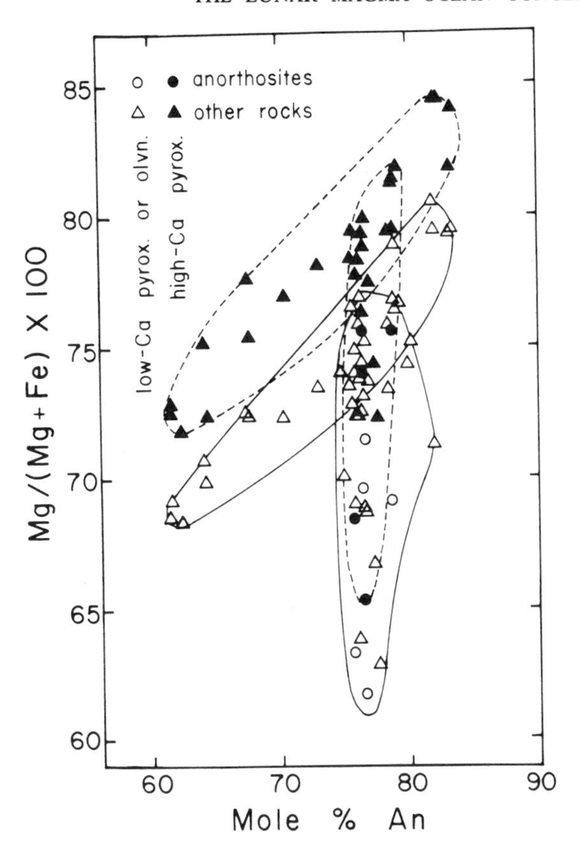

Figure 7 Cumulates from the Stillwater complex display Ca/(Ca + Na) vs *mg* trends similar to those of the ancient lunar cumulates (Figure 1). After Raedeke & McCallum (1980).

tendency for mineral compositions of *Luna 20* soil particles to cluster into two groups; and Roedder & Weiblen (1974) first inferred two separate trends on diagrams similar to Figure 1. But it was not until Warner et al (1976) plotted numerous "plutonic" (pristine) samples on a diagram similar to Figure 1 that the dichotomy between ferroan anorthosites and Mg-rich rocks was manifest. One terrestrial layered intrusion, the Stillwater complex, exhibits similar trends (Figure 7) but without any gap in between. McCallum (1983) argues that the Stillwater data show a gap as well, if a few samples with $<85\%$ plagioclase are excluded. But no such arbitrary exclusion is used to evince the gap in Figure 1. Calculations show that even without the gap, the spread of the data on Figure 1 would preclude derivation of all the ancient cumulates in any straightforward manner from a single magma (Longhi 1982).

Isotopic Data

Pristine ferroan anorthosites are extremely poor in Rb and REE, and no internal isochron Rb–Sr or Sm–Nd age is available. However, their measured $^{87}Sr/^{86}Sr$ ratios (Table 1) are often so low that the initial ratios can be constrained to be less than the initial $^{87}Sr/^{86}Sr$ ratios of most dated Mg-rich rocks (Warren & Wasson 1980a). These data preclude derivation of all ancient cumulates from a single magma, unless Mg-rich rocks crystallized later than ferroan anorthosites. In any normal course of crystallization, magnesian rocks form before ferroan ones.

A rare point of agreement among models of the magma ocean is that it probably completed 99% of its crystallization within 200 Myr (Krass & Fedeev 1975, Solomon & Longhi 1977, Herbert et al 1978, Minear 1980, Shirley 1983). Three of the eight pristine Mg-rich rocks dated via Sm–Nd or Rb–Sr internal isochrons have ages <4.23 Gyr (Table 1). The range among the eight is >400 Myr, and even if the oldest and the youngest rocks are excluded, the range is still >300 Myr. As discussed by Carlson & Lugmair (1981b), the modest ages are probably not due to shock-induced metamorphism, but they conceivably represent the termination of diffusional equilibrium in hot plutonic rocks upon excavation via impact to cool near-surface surroundings. However, one of the "young" Mg-rich rocks (67667) has strongly zoned plagioclase (Hansen et al 1980).

Model ages of younger basalts confirm that the magma ocean was short-lived. Model ages for mare basalts from Rb–Sr, Sm–Nd, and U–Th–Pb studies (reviewed by Shih & Schonfeld 1976) cluster at about 4.3–4.4 Gyr. For KREEP-rich samples (mostly nonpristine), Rb–Sr model ages cluster around 4.25–4.40 Gyr (Palme 1977, Warren & Wasson 1979a). Model Sm–Nd ages for four KREEP-rich rocks from four different sites average 4.35 ± 0.03 Gyr (Lugmair & Carlson 1978). Curiously, pristine KREEP basalts tend to have lower Rb–Sr model ages. Model Rb–Sr ages of seven pristine *Apollo 15* KREEP basalts average 4.21 Gyr (Nyquist 1977). For one of the same rocks, the Sm–Nd model age is 4.40 Gyr (Lugmair & Carlson 1978). The Rb–Sr model age of an *Apollo 17* pristine KREEP basalt is only 4.03 Gyr (Gray et al 1974). Nevertheless, if we assume that Rb, Sr, Sm, and Nd were fractionated as the magma ocean degenerated into "urKREEP" residuum, the model ages of most KREEP-rich samples imply that the residuum formed by 4.35 Gyr at the latest. The younger Mg-rich rocks are therefore too young to be products of the magma ocean.

Other Distinctive Types of Pristine Rocks

By 1980, it was apparent (Warren & Wasson 1979b, 1980a, James 1980) that the ancient cumulates, especially the Mg-rich rocks, are too diverse to be all from a single magma system. Since then, lithologies of bewildering variety have been discovered. James & Flohr (1983) review six distinctive Mg-rich rocks with major contents of high-Ca pyroxene (the gabbronorites), of which five were identified since 1976, three since 1979. Warren et al (1983a) discuss six distinctive granitic lithologies (five identified since 1976, three since 1979). Warren et al (1983b) and Shervais et al (1983) review numerous anorthosites that have far higher contents of Na and other alkalis than ferroan anorthosites. This "alkali anorthosite" group was virtually unstudied before 1980.

"New" rock types are being deliberately hunted by petrologists (for example, by concentrating on samples from *Apollo 14*, the most KREEP-rich highlands site). These exotic "new" lithologies have not yet raised doubts about the classification of most ancient lithologies as either Mg-rich or ferroan anorthosite rocks; some of the new rock types might be end-member types of Mg-rich rocks. Nevertheless, it is scarcely credible that such compositional diversity can be explained under the assumption that the entire nonmare, nonKREEP crust formed through differentiation of one magma, even if it was a magma "ocean."

Diversity of Trace Element Patterns

Consider anorthosites, which are all nearly pure plagioclase and have similar *mg*. Most ferroan anorthosites contain <0.1 $\mu g\ g^{-1}$ Sm (five of them contain <0.05 $\mu g\ g^{-1}$), whereas most alkali anorthosites contain >4 $\mu g\ g^{-1}$, and two whitlockite-rich anorthosites contain 46–86 $\mu g\ g^{-1}$ (Warren et al 1983b). Magnesian troctolites exhibit comparable diversity of REE contents. Furthermore, there is a marked tendency for the most REE-rich anorthosites and troctolites to come from one landing site—*Apollo 14* (Warren et al 1983b). This trend cannot be attributed to sampling errors except in the sense that anorthosites and troctolites produced by the magma ocean in the *Apollo 14* region might be radically different from the types it produced 20–30° of longitude to the east, which would imply incredible heterogeneity of magma composition.

The ratios Ti/Sm (McKay et al 1978), Sc/Sm (Norman & Ryder 1980), and Mn/Sm (Warren et al 1983b) are markedly nearer to chondritic in ferroan anorthosites than in all but a few Mg-rich rocks (Figure 8). These trends are consistent with formation of the ferroan anorthosites directly from the magma ocean, but suggest a more complex origin for the Mg-rich rocks (Norman & Ryder 1980). Minor element mineral composition data (Steele et al 1980) give similar indications.

The range for Eu/Al ratios is not so great, but even harder to explain via a single-magma model (Warren et al 1984). Both Eu and Al have crystal/liquid distribution coefficients slightly greater than 1 for plagioclase

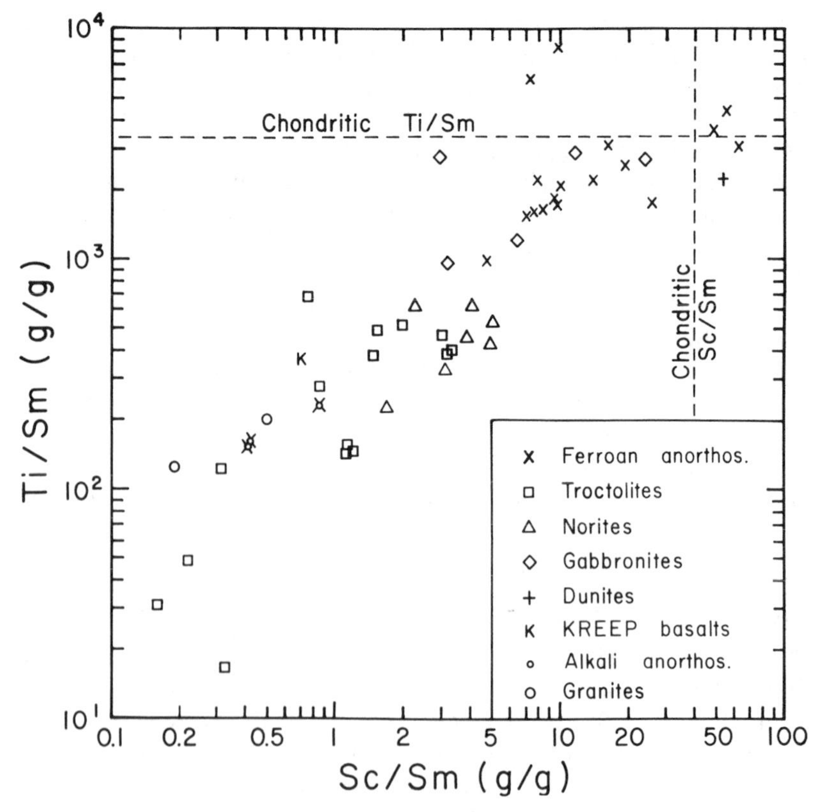

Figure 8 Sc/Sm vs Ti/Sm for pristine nonmare rocks. Comagmatic rocks of similar major element composition should have roughly similar Sc/Sm and Ti/Sm ratios. The data sources are as for Figure 1.

(specifically, about 1.1 for Eu and 1.9–2.5 for Al) and much less than 1 for all other major lunar minerals. These distribution patterns imply that sampling errors are minimal for Eu/Al, so long as samples are plagioclase-rich, and that Eu/Al should increase during fractional crystallization. There should be an inverse correlation between *mg* and Eu/Al, among cumulates from a single magma. The data (Figure 9) show that Mg-rich rocks follow such a trend, but the lowest Eu/Al ratios occur almost exclusively among rocks with low *mg* (the ferroan anorthosites). Absolute values of the ferroan anorthosite Eu/Al ratios are consistent with a magmasphere origin, assuming that the bulk Moon's Eu/Al ratio is chondritic (Warren et al 1984).

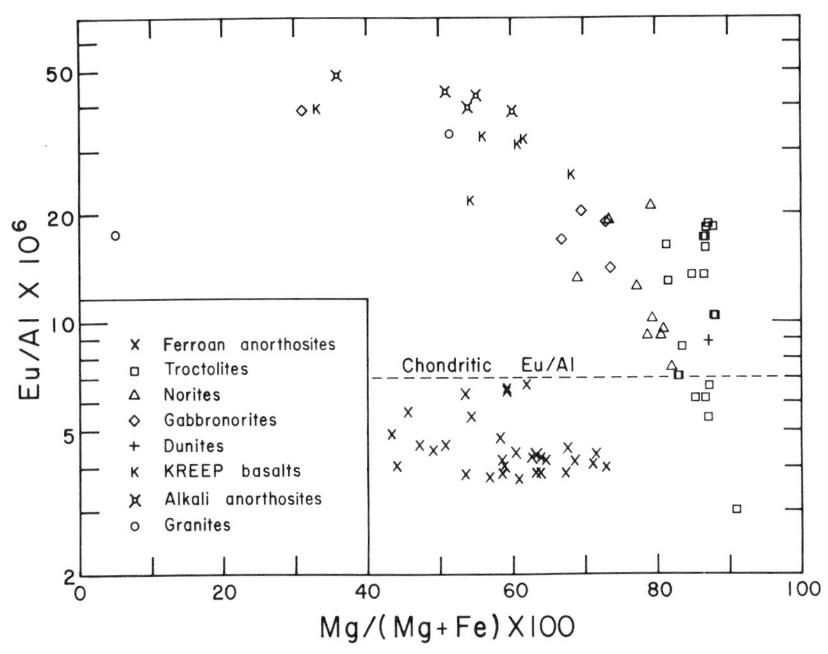

Figure 9 Eu/Al vs *mg* for pristine nonmare rocks. Comagmatic rocks of similar *mg* should have similar Eu/Al ratios (see text). The data sources are as for Figure 1.

Single Differentiation Episode Models

The accumulating evidence for multiple ancient magmas notwithstanding, a magma ocean seems to be required to explain *at least* the ferroan anorthosites, and some petrologists still favor a single magma for the entire spectrum of ancient cumulates. The most elaborate model for single-magma ancient cumulates genesis (Longhi & Boudreau 1979) invokes magma mixing (parts of the magma ocean mixing with other parts) to account for the geochemical paradox that Mg-rich rocks are generally far richer in incompatible elements than are ferroan anorthosites. The ferroan nature of the anorthosites is attributed to trapped liquid. But the larger mafic silicates in the anorthosites are probably mostly cumulus (Warren & Wasson 1980a, Ryder 1982). Longhi (1982) concedes that mineral composition trends (Figure 1) alone demonstrate that the two groups are fundamentally unrelated. However, if Longhi (1981) and Shirley (1983) are correct about the largely solid nature of the magma "ocean," the magma mixing effects outlined by Longhi & Boudreau (1979) must have been important in the evolution of the urKREEP residuum.

McCallum (1983) proposed that the Mg-rich rocks formed from the magma ocean, but that they had their trace element compositions altered through metasomatism by the urKREEP residuum; ferroan anorthosites were not metasomatized because they formed higher in the crust. In its present form, this model does not explain why there were only ferroan anorthosites in the upper crust and only Mg-rich rocks in the lower crust (in terms of *mg*, the reverse of the stratigraphy expected if the crust formed above a magma ocean).

Models That Invoke No Magma Ocean

Wetherill (1975, 1976, 1981) points out that accretional heating might not melt a large fraction of the Moon. He suggests that instead of a magma ocean, primordial differentiation involved regional magma chambers. The Moon is zone-refined by intermittent impact melting during accretion. As the Moon grows, large impacts produce localized, near-surface plutons, which differentiate so that incompatible elements and low-density minerals (plagioclase) concentrate upward. Melting is discontinuous in space and time, but by reiteration of vertical differentiation each time a new impact occurs in the same region, the Moon is eventually covered with a global layer of anorthosite. Similar models have been proposed by Alfvén & Arrhenius (1976) and Hess et al (1977). In the most recent and elaborate model of this general type (Longhi & Ashwal 1984), ferroan anorthosites form by diapiric uprise of portions of solidified mafic layered intrusions. Partial remelting due to pressure release would be expected as the diapirs rise, consistent with the suggestion of Haskin et al (1981) that ferroan anorthosites are products of "episodic partial remelting." A few ferroan samples have unannealed cumulate textures, however [e.g. 66035c2 (Warren & Wasson 1980b)].

The bulk upper lunar crust averages about 75% plagioclase, but ferroan anorthosites average more than 95% plagioclase. Complementary mafic cumulates, which must be comparably ferroan (e.g. Longhi 1982), are absent from the sampled crust. The purity of the ferroan anorthosites implies highly efficient differentiation. According to Alfvén & Arrhenius (1976, p. 207), it is "doubtful," for a body as small as the Moon, that any given volume of material will be melted more than once before being buried too deeply to be melted again. As noted above, until prior magmatism has produced a differentiated crust, impact melts will be ultrabasic, and their cumulates will be mainly dunites or peridotites. Unless the intrusions are assumed to all be coeval (tantamount to a magma ocean), the anorthosites have to separate—cleanly—from associated mafic cumulates and rise through rigid, older rocks that trapped the parent melts at depth. Thus, the yield and purity of anorthosite would be small. The uniformity of the

KREEP component is another constraint poorly satisfied by no-ocean models. From diverse plutons, there should have emerged diverse REE-rich liquids.

Walker (1983) reviewed the evidence for multiple ancient differentiation episodes, and suggested that the lunar crust formed by serial magmatism as a series of flows and plutons. The question is, Did *all* of the lunar crust form by serial magmatism, or was the ferroan anorthosite portion (at least) formed by a magma ocean? Walker (1983) states that a magma ocean is implied "if the bulk crust . . . (is) significantly more feldspathic than internally generated partial melt." The normative plagioclase contents of internally generated partial melts are constrained by phase equilibria (Walker et al 1973) to be less than about 55 wt%. As discussed earlier, the plagioclase content of the bulk *upper* crust does appear to be significantly greater than 55 wt%.

Walker (1983) contends that pro-magma ocean arguments based on complementary Eu anomalies suffer from model dependency. He suggests that the Eu anomalies of the mare basalts formed by plagioclase fractionation in holding chambers. Before the magmas erupted, they mixed with fresh pulses of primitive magma. The resultant mixed magmas were not saturated with plagioclase, but they nonetheless bore trace element signatures of plagioclase fractionation. Experimental petrology data indicate that most, if not all, mare basalts were not saturated with plagioclase in their source regions. Thus, Walker's (1983) model requires a delicate timing pattern: Eruption from the holding chambers must consistently occur soon after mixing between primitive and fractionated magma (before renewal of plagioclase saturation).

Smith (1982) discusses the origin of the lunar crust by extensive primordial melting, but he does not specify whether a magma ocean is required or not. This model is similar in many respects to the Alfvén & Arrhenius (1976) model and to Shirley's (1983) magmifer model. In this model, ferroan anorthosites form earliest, and Mg-rich rocks are slightly younger.

SYNTHESIS: THE PRIMORDIAL DIFFERENTIATION AND ITS AFTEREFFECTS

The model that follows is only what seems most likely, based on the limited data and interpretations summarized above. Caveat emptor!

Formation of the First Stable (Anorthosite) Crust

A thick, global layer of ferroan anorthosite formed by plagioclase flotation over a primordial magmasphere that encompassed roughly half of the

Moon. Heat budget limitations suggest that the entire system was never simultaneously molten, however: Most of it was probably only a partial melt. It has been suggested (Wasson & Warren 1980) that uneven crystallization in the magmasphere contributed to the genesis of the Moon's gross hemispheric asymmetries. The crust suffered intense meteoritic bombardment even as it formed (Hartmann 1980). The surface was probably covered with a megaregolith of breccias then, as now. Impacts often ruptured the crust, strewing ocean magma across the surface, but all ancient extrusives were quickly brecciated to smithereens. The magmasphere may have lost much of its sodium by volatilization via ruptures (Steele & Smith 1973, Warren & Wasson 1980a), which would explain the high Ca/Na ratios of the ferroan anorthosites (Figure 1). Beneath the anorthosite layer, pockets of KREEPlike residual liquid formed.

Formation of the Mg-Rich Rocks

Most Mg-rich rocks are probably products of postmagmasphere magmas. But some of them formed < 100 Myr after the formation of the Moon itself. The transition from the magma "ocean" to serial magmatism was probably not abrupt: Early postmagmasphere magmas ascending from the mantle tended to mix with pockets of magma ocean residuum (urKREEP). The assimilation of urKREEP may explain the KREEPlike REE patterns inferred for Mg-rich magmas by numerous authors (Warren & Wasson 1980a). Differing degrees of assimilation may have enhanced diversity within the Mg-rich group (James & Flohr 1983), particularly because far more urKREEP seems to have collected near the longitude of the *Apollo 14* site than elsewhere (Warren et al 1983b). Surviving pristine Mg-rich rocks are strictly intrusive cumulates, but extrusive relatives may have been obliterated by impacts.

The sources of the Mg-rich magmas may have been (*a*) primitive regions of the deep interior, little affected by the magmasphere, or (*b*) magmasphere cumulates. Most Mg-rich rocks have high *mg* ratios (Figure 1) compared with other lunar rock types, which implies that the sources were magnesian and the degree of melting was often high. Yet nearly all Mg-rich rocks contain cumulus plagioclase. The parent magmas may have assimilated plagioclase (Al, Ca, and Si, but not much else) from ferroan anorthosites. Binder (1980) suggests that the sources were magma ocean cumulates of the deep crust. Melting near the interface between the primitive interior and the bottom of the differentiated zone appears more likely. In the fully molten magma ocean model, the deepest cumulates are both the most magnesian and the least dense. This stratigraphy is likely to lead to convective overturn, in which case the earliest cumulates would tend to melt because of adiabatic decompression as well as gravitational energy release (Herbert 1980). In Shirley's (1983) model, the bottom of the magmifer migrates upward as the Moon cools and melt removal causes viscosity to increase beyond a threshold for convection. At the maximum depth reached by the magmifer, the temperature has been raised to near the solidus but no melt has been removed, so a full lunar complement of K, Th, and U generates heat. This region was likely to partially melt soon after the magmasphere epoch.

Subsequent Fate of urKREEP

The magmasphere residuum (urKREEP) was probably parental to KREEP. Based on KREEP model ages, urKREEP formed by 4.35 Gyr ago. The KREEP component seems to be absent from polymict breccias older than about 4.2 Gyr (Warner et al 1977). The > 150 Myr lapse before KREEP began extruding may be due to the moderate density and high viscosity of urKREEP, which resulted in only slow upward migration through the crust (Shirley & Wasson 1981). The residuum had tremendous concentrations of heat sources (K, Th, and U) and generated heat at a rate of roughly 1.1 J g^{-1} Myr^{-1} (Shirley & Wasson 1981). The last few tenths of a percent of magmasphere liquid might never have crystallized had KREEP basalt eruptions not transported some of that heat-generating capacity to the surface. As KREEP slowly percolated upward through the crust, it altered some of the deeper cumulates, either by remelting (Binder 1980) or metasomatism (McCallum 1983). KREEP was probably often assimilated by (or mixed with) magmas rising from the mantle.

Formation of Mare Basalts

Most models for mare basalt sources invoke negative Eu anomalies that are produced by plagioclase removal from the magma ocean. Binder (1982a) suggests that mare basalt magmas assimilated small but important urKREEP components. The details of mare basalt genesis are beyond the scope of this review, but in all models, the timing and the location of melting are predetermined by earlier events, and the primordial upward concentration of incompatible elements is pivotal. Had the Moon's K, Th, and U been concentrated even closer to the surface, there would have been less insulation above, and consequently less heat retention below. The interior might never have melted again. On the other hand, had K, Th, and U not been concentrated so close to the surface, the era of basaltic volcanism would have lasted longer, and the melt compositions would have been different.

MAGMA OCEANOGRAPHY GENERALIZED

Whatever mechanism supplied primordial heat to the Moon probably had a similar effect on all the terrestrial planets (Hostetler & Drake 1980). A priori, a magmasphere is more likely to form on a large planet than on the Moon for the following reasons: (*a*) Cooling is related to the surface area/mass ratio (Table 2), so a larger planet is more likely to build up sufficient heat to form a magmasphere. (*b*) Core formation can release considerable heat (Solomon 1980), but the Moon has a small or nonexistent core. (*c*) A dense primordial atmosphere may help insulate a magmasphere (T. Matsui & Y. Abe, in preparation), and atmosphere mass is a function of escape velocity. (*d*) It has been suggested (e.g. Nisbet & Walker 1982, Turcotte & Pflugrath 1984) that because silicate melts may be denser than ultramafic rocks at pressures greater than roughly 100 kbar (i.e. pressures found at depths below roughly 300 km in the Earth, and nowhere in the Moon), heat buildup in the mantle might be facilitated.

Beyond some thermal threshold, probably comparable to the threshold for creation of a magmasphere, volatiles such as H_2O and CO_2 are degassed from the main mass of a planet into its atmosphere, or, if it has no appreciable atmosphere, into space. Lunar materials are notoriously volatile poor. Isotopic data indicate that the Earth's upper mantle was once degassed, less than about 50 Myr after it formed (Allègre et al 1983), Volatiles may be restored to degassed mantles by either (*a*) slowly mixing upward from a deeper, undegassed layer, or (*b*) being added to the surface (e.g. as a late accretionary veneer of cometlike material) and subsequently mixed downward. Mechanism (*b*), suggested for the Earth by numerous

Table 2 Basic data for noncore portions of terrestrial planets and the Moon

	The Moon	Mercury	Mars	Venus	Earth
Volume (% of planet)[a]	99	57	86	85	85.6
Volume (km^3 × 10^{10})	2.2	3.5	14	80	90
Mass (kg × 10^{22})	7	10	50	350	400
100-km pressure (kbar)	5	13	14	28	31
Median pressure (kbar)[b]	17	35	80	360	410
Bottom pressure (kbar)	47	85	230	1100	1350
Median depth (km)[b]	360	260	580	1010	1052
Bottom depth (km)	1500	600	1700	2800	2890
Surface area/mass (μm^2 kg^{-1})[c]	517	207	226	95	85
Distance from Sun (AU)	1.0	0.39	1.5	0.72	1.0

[a] See Solomon (1980).
[b] Depth and pressure at which 50 vol% of noncore portion is below and 50 vol% is above.
[c] Mass of whole planet, including metallic portion.

authors (e.g. Wasson 1971, Anders & Owen 1977), probably requires plate tectonics to subduct lithosphere.

Terrestrial Planets

ASTEROIDS Anorthosites are only a minor component of the howardite parent body, even though mafic cumulates are common (Bunch 1975). An asteroid's small size would not favor production of a magma ocean, because small bodies tend to cool rapidly. Also, the pressure range from the surface to the bottom of the mantle of the largest asteroid is only about 5 kbar (Figure 4). If the bottom of the mantle were 100 K above the solidus (sufficiently melted to enable convection even in the weak gravity of an asteroid), the adiabatic temperature at the surface would be only 130 K above the solidus (the comparable value for the Moon is about 330 K). Thus, barring a tremendous pulse of heat, the degree of partial melting in an asteroid is probably never high, and no "ocean" capable of elutriating plagioclase ever forms.

MERCURY Pressures in Mercury's mantle are similar to pressures in the Moon (Figure 4, Table 2). Ransford (1982) suggests that Mercury received relatively little accretional heat, but Solomon (1980) infers from the abundance of fault scarps on Mercury that its radius decreased 1–2 km after a primordial differentiation more volumetrically extensive than that of the Moon. Unless the bulk noncore compositions of the two bodies are entirely different, Mercury's primordial differentiation probably formed a thick, anorthositic crust similar to that of the Moon. The visible reflectance spectrum of Mercury bears a strong enough resemblance to spectra from lunar highlands soils to suggest (Balsaltic Volcanism Study Project 1981) that its crust is similarly anorthositic.

MARS Solomon (1980) infers from the dominance of extensional tectonics on Mars that primordial melting there was considerably less extensive than on the Moon. If Mars had a magmasphere, the ancient martian highlands may consist in part of primordial anorthositic cumulates. But the yield of anorthosite from a martian magmasphere would probably not be commensurate with the yield from the Moon's magmasphere, because the higher pressures in the martian mantle imply that much of the Al would be retained in garnet.

VENUS Venus is similar to the Earth in size, density, and heliocentric distance; presumably these two planets had similar primordial differentiations. The different tectonic style of present-day Venus may be due in part to a lack of H_2O, or a preponderance of CO_2 over H_2O, in its mantle (Warner & Morrison 1978), as well as the insulation of its atmosphere.

The Earth

Few authors have had the audacity to apply magma oceanography to a planet as large and complex as the Earth. The Earth apparently never formed a thick, feldspathic crust commensurate to the Moon's. Smith (1981) doubts that the Earth ever had a global magma ocean. Warner (1979) and Taylor (1982a) suggest that because terrestrial magmas are H_2O-rich, they are not dense enough to float feldspar. But it seems unlikely that H_2O would remain in any shallow layer hot enough to constitute a magmasphere. A better explanation is that pressures in the Earth's magmasphere led to garnet crystallization at depth, leaving less Al to form feldspar near the surface (Warren & Wasson 1979b, Anderson 1981, Taylor 1982b).

Warren (1984b) adapted Shirley's (1983) lunar model to the Earth, as follows: If equal percentages of the noncore portions of the Earth and the Moon were incorporated into their respective magmaspheres, the pressure range in the terrestrial magmasphere was roughly 20 times that in its lunar counterpart (Figure 4). If we assume that the lower, magmifer portion of the magmasphere has an adiabatic (4 K $kbar^{-1}$) P-T profile, the magmifer gives way to a fully molten layer at a fixed pressure (dP_m, roughly 20 kbar) above the bottom of the system; if anything, dP_m is smaller in a large system (the outer Earth) than in a small one (the outer Moon). Thus, a terrestrial magmasphere would probably have a relatively thick fully molten layer, and a relatively thin magmifer, compared with its lunar counterpart. Convective heat transfer is in general far more efficient in a fully molten layer than in a relatively viscous "magmifer." Thus, the heat input required to increase the total system (magmifer layer plus upper, fully molten layer) thickness beyond the maximum magmifer thickness (roughly 20 kbar) is far greater than that required to thicken it up to the maximum magmifer thickness. The system tends to self-regulate, such that its total thickness is unlikely (barring overwhelming heat input) to ever become much greater than the maximum magmifer thickness. As a result, the fraction of the noncore portion of the Earth encompassed by its magmasphere was probably considerably smaller than the corresponding fraction for the Moon, other factors in favor of greater primordial heat input and retention for the Earth notwithstanding.

Anderson (1981) suggests that a terrestrial magma ocean formed a layer of eclogite in the mantle between roughly 300 and 670 km depth, and that this layer is the source region for mid-ocean ridge basalts (MORBs). He suggests that complementary incompatible element patterns of MORBs versus continental basalts (plus kimberlites) are analogous to the complementary Eu anomalies often cited as evidence for a lunar magma ocean. This analogy should not be carried too far. The Eu anomalies of the mare basalts suggest massive fractionation of plagioclase from the deep interior of the Moon, yet their major element compositions indicate that the source regions retained no plagioclase. Without this paradox, we might conclude that plagioclase fractionation occurred as the sources melted. As Anderson (1981) notes, vertical stratification of the mantle is not compatible with most models of mantle convection. Convection appears to extend down at least 700 km (Condie 1982).

An ingenious, if not convincing, model by Nisbet & Walker (1982) suggests that the Earth's magma ocean was still molten during the period (as recently as 2.7 Gyr ago) when ultramafic komatiites formed. Nisbet & Walker (1982) suggest that ultramafic melts are denser than magnesian olivine at pressures in excess of roughly 40 kbar. Neutral density at roughly 130 km depth forms a barrier to convective heat transport, so a magma ocean persists for several Gyr below that depth. Komatiites form when ocean magma occasionally "breaches" the buoyant, mostly solid layer above. The estimate of 130 km for the neutral density depth appears low by at least a factor of two (Hess 1983, Ohtani 1983, Herzberg 1983, Rigden et al 1984). Subsolidus convection as the Earth heated may have forestalled extensive melting at depths $\gg$200 km. Today the mantle is virtually all solid. What caused the mantle below the neutral density level to ever cool? What caused melt from below the neutral level to ever rise? Or, once erupting, to ever stop erupting? Warren (1984a) suggests a less direct link between primordial heating and the origin of komatiites.

Hofmeister (1983) pioneered a detailed model for crystallization of a terrestrial magma ocean. Her main conclusion was that a global magma ocean could not have produced the Archean sialic crust. Age data alone show that none of the Archean crust formed directly from a primordial magmasphere. However, the compositional discrepancy between the primordial crust predicted in Hofmeister's (1983) model and the actual Archean crust would have been smaller if she had modeled a magmasphere thicker than 120 km because of pressure effects on garnet/pyroxene/olivine stability [important for Ca and Si (cf. the lunar model of Warren & Wasson 1979b)] and simple mass balance considerations (for potassium). It seems likely that the Earth experienced extensive primordial melting, and that a substantial buoyant, felsic crust resulted (Goles & Seymour 1978), perhaps by one of the mechanisms proposed (Herbert et al 1977b, Shirley 1983) to produce initial nuclei for the lunar crust. Much of the first buoyant crust was probably recycled into the mantle, but some of it may have survived in metamorphosed form.

Conclusions

At least the outer one third of the Moon was extensively melted at the time of planet formation, but most of the differentiated zone was probably never fully molten. Not all of the Moon's ancient cumulates formed from its magmasphere. In fact, only the ferroan anorthosites appear to be of direct magmaspheric origin. Pressures in the interiors of larger planets cause Al to be retained in garnet and other high-density phases instead of plagioclase. A magmasphere influences subsequent planetary evolution, particularly for small, rapidly cooled bodies like the Moon. The Earth's first buoyant crust presumably comprised late differentiates of its magmasphere.

Future Research

Recent models that invoke a partially molten "ocean" seem to plausibly account for heat limitations. Improvements should be possible using geophysical fluid dynamics, preferably with an appreciation of petrologic constraints. By continuing to study pristine lunar rocks, we can better constrain the diversity of ancient cumulates. How many ancient magma types were there? What were the temporal relationships among them? For magmaspheres on bigger planets like the Earth, high-pressure experimental petrology data (Ohtani 1983, Rigden et al 1984) allow some constraints, but more such data are needed.

ACKNOWLEDGMENTS

I am grateful to J. Longhi, G. Ryder, D. N. Shirley, D. Walker, and J. T. Wasson for constructive reviews. My research is supported by NASA grant NAG 9-87.

Literature Cited

Adler, I., Trombka, J. I. 1977. Orbital chemistry—lunar surface analysis from the X-ray and gamma ray remote sensing experiments. *Phys. Chem. Earth* 10:17–43

Alfvén, H., Arrhenius, G. 1976. *Evolution of the Solar System.* Washington DC: NASA. 599 pp.

Allègre, C. J., Staudacher, T., Sarda, P., Kury, M. 1983. Constraints on the evolution of Earth's mantle from rare gas systematics. *Nature* 303:762–66

Anders, E., Owen, T. 1977. Mars and Earth: Origin and abundance of volatiles. *Science* 198:453–65

Anderson, D. L. 1981. Hotspots, basalts, and the evolution of the mantle. *Science* 213:82–89

Arndt, N. T. 1977. Ultrabasic magmas and high-degree melting of the mantle. *Contrib. Mineral. Petrol.* 64:205–21

Basaltic Volcanism Study Project. 1981. *Basaltic Volcanism on the Terrestrial Planets.* Houston: Lunar Planet. Inst. 1286 pp.

Binder, A. B. 1974. On the origin of the Moon by rotational fission. *Moon* 11:53–76

Binder, A. B. 1976. On the implications of an olivine dominated upper mantle on the development of a Moon of fission origin. *Moon* 16:159–73

Binder, A. B. 1978. On fission and the devolatilization of a Moon of fission origin. *Earth Planet. Sci. Lett.* 41:381–85

Binder, A. B. 1980. On the origins of lunar pristine crustal rocks. *Proc. Conf. Lunar Highl. Crust*, pp. 71–79

Binder, A. B. 1982a. The mare basalt magma source region and mare basalt magma genesis. *J. Geophys. Res.* 87:A37–53

Binder, A. B. 1982b. Post-Imbrian global lunar tectonism: evidence for an initially totally molten Moon. *Moon Planets* 26:117–33

Binder, A. B., Lange, M. A. 1980. On the thermal history, thermal state, and related tectonism of a Moon of fission origin. *J. Geophys. Res.* 85:3194–3208

Brett, R. 1973. The lunar crust: a product of

heterogeneous accretion or differentiation of a homogeneous Moon? *Geochim. Cosmochim. Acta* 37:2697–2703

Brett, R. 1977. The case against early melting of the bulk of the Moon. *Geochim. Cosmochim. Acta* 41:443–45

Buck, W. R., Toksöz, M. N. 1980. The bulk composition of the Moon based on geophysical constraints. *Proc. Lunar Planet. Sci. Conf., 11th*, pp. 2043–58

Bunch, T. E. 1975. Petrography and petrology of basaltic achondrite polymict breccias (howardites). *Proc. Lunar Sci. Conf., 6th*, pp. 469–92

Bussod, G. Y., Christie, J. M. 1983. The effect of partial melting on the mechanical properties of spinel lherzolite. *Eos, Trans. Am. Geophys. Union* 64:849 (Abstr.)

Carlson, R. W., Lugmair, G. W. 1979. Sm–Nd constraints on early lunar differentiation and the evolution of KREEP. *Earth Planet. Sci. Lett.* 45:123–32

Carlson, R. W., Lugmair, G. W. 1981a. Time and duration of lunar highlands crust formation. *Earth Planet. Sci. Lett.* 52:227–38

Carlson, R. W., Lugmair, G. W. 1981b. Sm–Nd age of lherzolite 67667: implications for the processes involved in lunar crustal formation. *Earth Planet. Sci. Lett.* 56:1–8

Cassen, P., Reynolds, R. T., Graziani, F., Summers, A., McNellis, J., Blalock, L. 1979. Convection and lunar thermal history. *Phys. Earth Planet. Inter.* 19:183–96

Cawthorn, R. G. 1975. Degrees of melting in mantle diapirs and the origin of ultrabasic liquids. *Earth Planet. Sci. Lett.* 27:231–38

Clayton, R. N., Mayeda, T. K. 1975. Genetic relations between the Moon and meteorites. *Proc. Lunar Sci. Conf., 6th*, pp. 1761–69

Condie, K. C. 1982. *Plate Tectonics and Crustal Evolution.* New York: Pergamon. 310 pp.

Conel, J. E., Morton, J. B. 1975. Interpretation of lunar heat flow data. *Moon* 14:263–89

Davis, P. A. Jr. 1980. Iron and titanium distribution on the Moon from orbital gamma ray spectrometry with implications for crustal evolutionary models. *J. Geophys. Res.* 85:3209–24

Delano, J. W., Livi, K. 1981. Lunar volcanic glasses and their constraints on mare petrogenesis. *Geochim. Cosmochim. Acta* 45:2137–49

Delano, J. W., Ringwood, A. E. 1978. Siderophile elements in the lunar highlands: nature of the indigenous component and implications for the origin of the Moon. *Proc. Lunar Planet. Sci. Conf., 9th*, pp. 111–59

Dowty, E., Prinz, M., Keil, K. 1974. Ferroan anorthosite: a widespread and distinctive lunar rock type. *Earth Planet. Sci. Lett.* 24:15–25

Drake, M. J. 1976. Evolution of major mineral compositions and trace element abundances during fractional crystallization of a model lunar composition. *Geochim. Cosmochim. Acta* 40:401–11

Dziewonski, A. M., Hales, A. L., Lapwood, E. R. 1975. Parametrically simple Earth models consistent with geophysical data. *Phys. Earth Planet. Inter.* 10:12–48

Frenkel, M. Ya. 1983. The phase convection in relation to mechanism of the highland lunar rocks formation. *Lunar Planet. Sci. XIV*, pp. 213–14 (Abstr.)

Fujimaki, H., Tatsumoto, M. 1984. Lu–Hf constraints on the evolution of lunar basalts. *J. Geophys. Res.* 89:B445–58

Goins, N. R., Toksöz, M. N., Dainty, A. M. 1979. The lunar interior: a summary report. *Proc. Lunar Planet. Sci. Conf., 10th*, pp. 2421–39

Goles, G. G., Seymour, R. S. 1978. Terrestrial magmatic ocean: an inexact lunar analog. *Eos, Trans. Am. Geophys. Union* 59:1214 (Abstr.)

Gray, C. M., Compston, W., Foster, J. J., Rudowski, R. 1974. Rb–Sr ages of clasts from within Boulder 1, Station 2, Apollo 17. In *Interdisciplinary Studies of Samples from Boulder 1, Station 2, Apollo 17*, ed. J. Wood, 2:VII.1–10. Cambridge, Mass: Smithsonian Astrophys. Obs. 210 pp.

Grieve, R. A. F. 1980. Cratering in the lunar highlands: some problems with the process, record and effects. *Proc. Conf. Lunar Highl. Crust*, pp. 173–96

Gromet, L. P., Hess, P. C., Rutherford, M. J. 1981. An origin for the REE characteristics of KREEP. *Proc. Lunar Planet. Sci. Conf., 12th*, pp. 903–13

Haines, E. L., Metzger, A. E. 1980. Lunar highland crustal models based on iron concentrations: isostasy and center-of-mass displacement. *Proc. Lunar Planet. Sci. Conf., 11th*, pp. 689–718

Hansen, E. C., Smith, J. V., Steele, I. M. 1980. Petrology and mineral chemistry of 67667, a unique feldspathic lherzolite. *Proc. Lunar Planet. Sci. Conf., 11th*, pp. 523–33

Hartmann, W. K. 1980. Dropping stones in magma oceans: effects of early lunar cratering. *Proc. Conf. Lunar Highl. Crust*, pp. 155–71

Haskin, L. A., Lindstrom, M. M., Salpas, P. A., Lindstrom, D. J. 1981. On compositional variations among lunar anorthosites. *Proc. Lunar Planet. Sci. Conf., 12th*, pp. 41–66

Head, J. W., Settle, M., Stein, R. S. 1975.

Volume of material ejected from major lunar basins and implications for the depth of excavation of lunar samples. *Proc. Lunar Sci. Conf., 6th,* pp. 2805–29

Herbert, F. 1980. Time-dependent lunar density models. *Proc. Lunar Planet. Sci. Conf., 11th,* pp. 2015–30

Herbert, F., Sonett, C. P., Wiskerchen, M. J. 1977a. Model "zero-age" lunar thermal profiles resulting from electrical induction. *J. Geophys. Res.* 82:2054–60

Herbert, F., Drake, M. J., Sonett, C. P., Wiskerchen, M. J. 1977b. Some constraints on the thermal history of the lunar magma ocean. *Proc. Lunar Sci. Conf., 8th,* pp. 573–82

Herbert, F., Drake, M. J., Sonett, C. P. 1978. Geophysical and geochemical evolution of the lunar magma ocean. *Proc. Lunar Planet. Sci. Conf., 9th,* pp. 249–62

Herzberg, C. T. 1983. Solidus and liquidus temperatures and mineralogies for anhydrous garnet-lherzolite to 15 GPa. *Phys. Earth Planet. Inter.* 32:193–202

Hess, P. C. 1983. Plagioclase suspensions and the origin of the mare basalt source regions. *Lunar Planet. Sci. XIV,* pp. 307–8 (Abstr.)

Hess, P. C., Rutherford, M. J., Campbell, H. W. 1977. Origin and evolution of LKFM. *Proc. Lunar Sci. Conf., 8th,* pp. 2357–73

Hofmeister, A. M. 1983. Effect of a hadean terrestrial magma ocean on crust and mantle evolution. *J. Geophys. Res.* 88: 4963–83

Hood, L. L., Sonett, C. P. 1982. Limits on the lunar temperature profile. *Geophys. Res. Lett.* 9:37–40

Hörz, F. 1978. How thick are lunar mare basalts? *Proc. Lunar Planet. Sci. Conf., 9th,* pp. 3311–31

Hostetler, C. J., Drake, M. J. 1980. On the early global melting of the terrestrial planets. *Proc. Lunar Planet. Sci. Conf., 11th,* pp. 1915–29

James, O. B. 1980. Rocks of the early lunar crust. *Proc. Lunar Planet. Sci. Conf., 11th,* pp. 365–93

James, O. B., Flohr, M. K. 1983. Subdivision of the Mg-suite noritic rocks into Mg-gabbronorites and Mg-norites. *J. Geophys. Res.* 88:A603–14

Jaques, A. L., Green, D. H. 1980. Anhydrous melting of peridotite at 0–15 Kb pressure and the genesis of tholeiitic basalts. *Contrib. Mineral. Petrol.* 73:287–310

Kallemeyn, G. W., Warren, P. H. 1983. Compositional implications regarding the lunar origin of the ALHA81005 meteorite. *Geophys. Res. Lett.* 10:833–36

Kaula, W. M., Schubert, G., Lingenfelter, R. E. 1974. Apollo laser altimetry and inferences as to lunar structure. *Proc. Lunar Sci. Conf., 5th,* pp. 3049–58

Keil, K., Kurat, G., Prinz, M., Green, J. 1972. Lithic fragments, glasses and chondrules from Luna 16 fines. *Earth Planet. Sci. Lett.* 13:243–56

Kesson, S. E., Lindsley, D. H. 1976. Mare basalt petrogenesis—a review of experimental studies. *Rev. Geophys. Space Phys.* 14:361–73

Kopal, Z. 1977. Dynamical arguments which concern melting of the Moon. *Philos. Trans. R. Soc. London Ser. A* 285:561–68

Krass, M. S., Fedeev, V. E. 1975. Thermal convection and phase separation of substances in planetary interiors. *Astron. Vestn.* 9:152–61

Langseth, M. G., Keihm, S. J., Peters, K. 1976. Revised lunar heat-flow values. *Proc. Lunar Sci. Conf., 7th,* pp. 3143–71

Liebermann, R. C., Ringwood, A. E. 1976. Elastic properties of anorthite and the nature of the lunar crust. *Earth Planet. Sci. Lett.* 31:69–74

Longhi, J. 1977. Magma oceanography 2: chemical evolution and crustal formation. *Proc. Lunar Sci. Conf., 8th,* pp. 601–21

Longhi, J. 1980. A model of early lunar differentiation. *Proc. Lunar Planet. Sci. Conf., 11th,* pp. 289–315

Longhi, J. 1981. Preliminary modeling of high-pressure partial melting: implications for early lunar differentiation. *Proc. Lunar Planet. Sci. Conf., 12th,* pp. 1001–18

Longhi, J. 1982. Effects of fractional crystallization and cumulus processes on mineral composition trends of some lunar and terrestrial rock series. *J. Geophys. Res.* 87: A54–64

Longhi, J., Ashwal, L. D. 1984. A two-stage model for lunar anorthosites: an alternative to the magma ocean hypothesis. *Lunar Planet. Sci. XV,* pp. 491–92 (Abstr.)

Longhi, J., Boudreau, A. E. 1979. Complex igneous processes and the formation of the primitive lunar crustal rocks. *Proc. Lunar Planet. Sci. Conf., 10th,* pp. 2085–2105

Lugmair, G. W., Carlson, R. W. 1978. The Sm–Nd history of KREEP. *Proc. Lunar Planet. Sci. Conf., 9th,* pp. 689–704

Maurer, P., Eberhardt, P., Geiss, J., Grogler, N., Stettler, A., et al. 1978. Pre-Imbrian craters and basins: ages, compositions and excavation depths of Apollo 16 breccias. *Geochim. Cosmochim. Acta* 42:1687–1720

McCallum, I. S. 1983. Formation of Mg-rich pristine rocks by crustal metasomatism. *Lunar Planet. Sci. XIV,* pp. 473–74 (Abstr.)

McKay, G. A., Wiesmann, H., Nyquist, L. E., Wooden, J. L., Bansal, B. M. 1978. Petrology, chemistry, and chronology of 14078: chemical constraints on the origin

of KREEP. *Proc. Lunar Planet. Sci. Conf., 9th*, pp. 661–87
Meyer, C. 1977. Petrology, mineralogy and chemistry of KREEP basalt. *Phys. Chem. Earth* 10:239–60
Minear, J. W. 1980. The lunar magma ocean: a transient lunar phenomenon? *Proc. Lunar Planet. Sci. Conf., 11th*, pp. 1941–55
Minear, J. W., Fletcher, C. R. 1978. Crystallization of a lunar magma ocean. *Proc. Lunar Planet. Sci. Conf., 9th*, pp. 263–83
Morgan, J. W., Hertogen, J., Anders, E. 1978. The Moon: composition determined by nebular processes. *Moon Planets* 18:465–78
Morse, S. A. 1982. Adcumulus growth of anorthosite at the base of the lunar crust. *J. Geophys. Res.* 87:A10–18
Mysen, B. O., Kushiro, I. 1977. Compositional variations among coexisting phases with degree of melting of peridotite in the upper mantle. *Am. Mineral.* 62:843–65
Nakamura, Y., Latham, G. V., Dorman, H. J. 1982. Apollo lunar seismic experiment—final summary. *J. Geophys. Res.* 87:A117–23
Nisbet, E. G., Walker, D. 1982. Komatiites and the structure of the Archaean mantle. *Earth Planet. Sci. Lett.* 60:105–13
Norman, M. D., Ryder, G. 1979. A summary of the petrology and geochemistry of pristine highlands rocks. *Proc. Lunar Planet. Sci. Conf., 10th*, pp. 531–59
Norman, M. D., Ryder, G. 1980. Geochemical constraints on the igneous evolution of the lunar crust. *Proc. Lunar Planet. Sci. Conf., 11th*, pp. 317–31
Nunes, P. D., Knight, R. J., Unruh, D. M., Tatsumoto, M. 1974. The primitive nature of the lunar crust and the problem of initial Pb isotopic compositions of a lunar rock: a Rb–Sr and U–Th–Pb study of Apollo 16 samples. *Lunar Sci. V*, pp. 559–61 (Abstr.)
Nyquist, L. E. 1977. Lunar Rb–Sr chronology. *Phys. Chem. Earth* 10:103–42
Nyquist, L. E. 1982. Radiometric ages and isotopic systematics of pristine plutonic lunar rocks. In *Workshop on Magmatic Processes of Early Planetary Crusts*, ed. D. Walker, I. S. McCallum, pp. 114–20. Houston: Lunar Planet. Inst. 234 pp. (*Tech. Rep. 82-01*)
Nyquist, L. E., Wooden, J. L., Shih, C.-Y., Wiesmann, H., Bansal, B. M. 1981. Isotopic and REE studies of lunar basalt 12038: implications for petrogenesis of aluminous mare basalts. *Earth Planet. Sci. Lett.* 55:335–55
Nyquist, L. E., Shih, C.-Y., Bansal, B., Wiesmann, H., Wooden, J. 1983. Formation of a lunar granite 4.1 AE ago. *Lunar Planet. Sci. XIV*, pp. 576–77 (Abstr.)
Oberli, F., Huneke, J. C., Wasserburg, G. J. 1979. U–Pb and K–Ar systematics of cataclysm and precataclysm lunar impactites. *Lunar Planet. Sci. X*, pp. 940–42 (Abstr.)
O'Hara, M. J. 1977. Geochemical evolution during fractional crystallization of a periodically refilled magma chamber. *Nature* 266:503–7
Ohtani, E. 1983. Melting temperature distribution and fractionation in the lower mantle. *Phys. Earth Planet. Inter.* 33:12–25
O'Keefe, J. D., Ahrens, T. J. 1978. Impact flows and crater scaling on the Moon. *Phys. Earth Planet. Inter.* 16:341–51
Palme, H. 1977. On the age of KREEP. *Geochim. Cosmochim. Acta* 41:1791–1801
Raedeke, L. D., McCallum, I. S. 1980. A comparison of fractionation trends in the lunar crust and the Stillwater complex. *Proc. Conf. Lunar Highl. Crust*, pp. 133–53
Ransford, G. A. 1982. The accretional heating of the terrestrial planets: a review. *Phys. Earth Planet. Inter.* 29:209–17
Rigden, S. M., Ahrens, T. J., Stolper, E. M. 1984. Liquid silicate densities at high pressures: first shock wave measurements. *Lunar Planet. Sci. XV*, pp. 689–90 (Abstr.)
Ringwood, A. E. 1979. *Origin of the Earth and Moon*. New York: Springer. 295 pp.
Ringwood, A. E., Essene, E. 1970. Petrogenesis of Apollo-11 basalts, internal constitution and origin of the Moon. *Proc. Apollo 11 Lunar Sci. Conf.*, pp. 769–99
Roedder, E., Weiblen, P. W. 1974. Petrology of clasts in lunar breccia 67915. *Proc. Lunar Sci. Conf., 5th*, pp. 303–18
Ryder, G. 1982. Lunar anorthosite 60025, the petrogenesis of lunar anorthosites, and the composition of the Moon. *Geochim. Cosmochim. Acta* 46:1591–1601
Ryder, G., Norman, M. 1978. *Catalog of Pristine Non-Mare Materials, Part 2, Anorthosites*. Houston: Curator, NASA Johnson Space Cent. 86 pp.
Ryder, G., Norman, M. 1979. *Catalog of Pristine Non-Mare Materials, Part 1, Non-Anorthosites, Revised*. Houston: Curator, NASA Johnson Space Cent. 147 pp.
Ryder, G., Norman, M., Score, R. A. 1980. The distinction of pristine from meteorite-contaminated highlands rocks using metal compositions. *Proc. Lunar Planet. Sci. Conf., 11th*, pp. 471–79
Safronov, V. S. 1978. The heating of the Earth during its formation. *Icarus* 33:3–12
Schaeffer, O. A., Husain, L. 1974. Chronology of lunar basin formation. *Proc. Lunar Sci. Conf., 5th*, pp. 1541–55
Shervais, J. W., Taylor, L. A., Laul, J. C. 1983. Ancient crustal components in the Fra

Mauro breccias. *J. Geophys. Res.* 88: B177–92
Shih, C.-Y. 1977. Origins of KREEP basalts. *Proc. Lunar Sci. Conf., 8th*, pp. 2375–2401
Shih, C.-Y., Schonfeld, E. 1976. Mare basalt genesis: a cumulate-remelting model. *Proc. Lunar Sci. Conf., 7th*, pp. 1757–92
Shirley, D. N. 1983. A partially molten magma ocean model. *J. Geophys. Res.* 88: A519–27
Shirley, D. N., Wasson, J. T. 1981. Mechanism for the extrusion of KREEP. *Proc. Lunar Planet. Sci. Conf., 12th*, pp. 965–87
Smith, J. V. 1981. The first 800 million years of Earth's history. *Philos. Trans. R. Soc. London Ser. A* 301: 401–22
Smith, J. V. 1982. Heterogeneous growth of meteorites and planets, especially the Earth and Moon. *J. Geol.* 90: 1–48
Smith, J. V., Anderson, A. T., Newton, R. C., Olsen, E. J., Wyllie, P. J., et al. 1970. Petrologic history of the Moon inferred from petrography, mineralogy, and petrogenesis of Apollo 11 rocks. *Proc. Apollo 11 Lunar Sci. Conf.*, pp. 897–925
Solomon, S. C. 1980. Differentiation of crusts and cores of the terrestrial planets: lessons for the early Earth? *Precambrian Res.* 10: 177–94
Solomon, S. C., Chaiken, J. 1976. Thermal expansion and thermal stress in the Moon and terrestrial planets: clues to early thermal history. *Proc. Lunar Sci. Conf., 7th*, pp. 3229–43
Solomon, S. C., Longhi, J. 1977. Magma oceanography: 1. Thermal evolution. *Proc. Lunar Sci. Conf., 8th*, pp. 583–99
Sonett, C. P., Reynolds, R. T. 1979. Primordial heating of asteroidal parent bodies. In *Asteroids*, ed. T. Gehrels, pp. 822–48. Tucson: Univ. Ariz. Press. 1181 pp.
Spudis, P. D., Hawke, B. R., Jackowski, T. 1984. Geochemical mixing-model studies of ejecta from lunar farside basins: implications for crustal models. *Lunar Planet. Sci. XV*, pp. 812–13 (Abstr.)
Steele, I. M., Smith, J. V. 1973. Mineralogy and petrology of some Apollo 16 rocks and fines: general petrologic model of Moon. *Proc. Lunar Sci. Conf., 4th*, pp. 519–36
Steele, I. M., Hutcheon, I. D., Smith, J. V. 1980. Ion microprobe analysis and petrogenetic interpretations of Li, Mg, Ti, K, Sr, Ba in lunar plagioclase. *Proc. Lunar Planet. Sci. Conf., 11th*, pp. 571–90
Stevenson, D. J. 1980. Self-regulation and melt migration (can magma oceans exist?). *Eos, Trans. Am. Geophys. Union* 61: 1021 (Abstr.)
Stöffler, D., Ostertag, R., Reimold, W. U., Borchardt, R., Malley, J., Rehfeldt, A. 1981. Distribution and provenance of lunar highland rock types at North Ray Crater, Apollo 16. *Proc. Lunar Planet. Sci. Conf., 12th*, pp. 185–207
Stolper, E. M., Walker, D., Hager, B. H., Hays, J. 1981. Melt segregation from partially molten regions: the importance of melt density and source region size. *J. Geophys. Res.* 86: 6261–71
Taylor, G. J., Drake, M. J., Wood, J. A., Marvin, U. B. 1973. The Luna 20 lithic fragments, and the composition and origin of the lunar highlands. *Geochim. Cosmochim. Acta* 37: 1087–1106
Taylor, L. A., Shervais, J. W., Hunter, R. H., Shih, C.-Y., Bansal, B. M., et al. 1983. Pre-4.2 AE mare basalt volcanism in the lunar highlands. *Earth Planet. Sci. Lett.* 66: 33–47
Taylor, S. R. 1975. *Lunar Science: A Post-Apollo View*. New York: Pergamon. 372 pp.
Taylor, S. R. 1978. Geochemical constraints on melting and differentiation of the Moon. *Proc. Lunar Planet. Sci. Conf., 9th*, pp. 15–23
Taylor, S. R. 1982a. *Planetary Science: A Lunar Perspective*. Houston: Lunar Planet. Inst. 481 pp.
Taylor, S. R. 1982b. Lunar and terrestrial crusts: a contrast in origin and evolution. *Phys. Earth Planet. Inter.* 29: 233–41
Taylor, S. R., Jakes, P. 1974. The geochemical evolution of the Moon. *Proc. Lunar Sci. Conf., 5th*, pp. 1287–1305
Thomson, W. 1864. On the secular cooling of the Earth. *Trans. R. Soc. Edinburgh* 23: 157–69
Toksöz, M. N. 1979. Planetary seismology and interiors. *Rev. Geophys. Space Phys.* 17: 1641–55
Toksöz, M. N., Solomon, S. C. 1973. Thermal history and evolution of the Moon. *Moon* 7: 251–78
Toksöz, M. N., Dainty, A. M., Solomon, S. C., Anderson, K. R. 1973. Velocity structure and evolution of the Moon. *Proc. Lunar Sci. Conf., 4th*, pp. 2529–47
Turcotte, D. L., Pflugrath, J. C. 1984. Was the early Earth completely molten? *Lunar Planet. Sci. XV*, pp. 870–71 (Abstr.)
Turcotte, D. L., Schubert, G. 1982. *Geodynamics: Applications of Continuum Physics to Geological Problems*. New York: Wiley. 450 pp.
Turkevich, A. L. 1973. Average chemical composition of the lunar surface. *Moon* 8: 365–67
Wakita, H., Laul, J. C., Schmitt, R. A. 1975. Some thoughts on the origin of lunar

ANT-KREEP and mare basalts. *Geochem. J.* 9:25–41
Walker, D. 1983. Lunar and terrestrial crust formation. *J. Geophys. Res.* 88:B17–25
Walker, D., Hays, J. F. 1977. Plagioclase floatation and lunar crust formation. *Geology* 5:425–28
Walker, D., Longhi, J., Grove, T. L., Stolper, E., Hays, J. F. 1973. Experimental petrology and origin of rocks from the Descartes Highlands. *Proc. Lunar Sci. Conf., 4th*, pp. 1013–32
Walker, D., Longhi, J., Hays, J. F. 1975. Differentiation of a very thick magma body and implications for the source regions of mare basalts. *Proc. Lunar Sci. Conf., 6th*, pp. 1103–20
Walker, D., Stolper, E. M., Hays, J. F. 1979. Basaltic volcanism: the importance of planet size. *Proc. Lunar Planet. Sci. Conf., 10th*, pp. 1995–2015
Warner, J. L. 1979. The role of water in planetary tectonics. In *Workshop on Ancient Crusts of the Terrestrial Planets*, pp. 79–82. Houston: Lunar Planet. Inst. 98 pp.
Warner, J. L., Morrison, D. A. 1978. Planetary tectonics I: The role of water. *Lunar Planet. Sci. IX*, pp. 1217–19 (Abstr.)
Warner, J. L., Simonds, C. H., Phinney, W. C. 1976. Genetic distinction between anorthosites and Mg-rich plutonic rocks: new data from 76255. *Lunar Sci. VII*, pp. 915–17 (Abstr.)
Warner, J. L., Phinney, W. C., Bickel, C. E., Simonds, C. H. 1977. Feldspathic granulitic impactites and pre-final bombardment lunar evolution. *Proc. Lunar Sci. Conf., 8th*, pp. 2051–66
Warren, P. H. 1983a. Models of bulk Moon composition: a review. In *Workshop on Pristine Lunar Highlands Rocks and the Early History of the Moon*, ed. J. Longhi, G. Ryder, pp. 75–79. Houston: Lunar Planet. Inst. 92 pp. (*Tech. Rep. 83-02*)
Warren, P. H. 1983b. Al–Sm–Eu–Sr systematics of eucrites and Moon rocks: implications for planetary bulk compositions. *Geochim. Cosmochim. Acta* 47: 1559–71
Warren, P. H. 1984a. Primordial degassing, lithosphere thickness, and the origin of komatiites. *Geology* 12:335–38
Warren, P. H. 1984b. Earth's primordial differentiation, and its aftereffects. In *Workshop on the Early Earth: The Interval from Accretion to the Older Archean*, pp. 74–76. Houston: Lunar Planet. Inst. 81 pp.
Warren, P. H., Wasson, J. T. 1977. Pristine nonmare rocks and the nature of the lunar crust. *Proc. Lunar Sci. Conf., 8th*, pp. 2215–35
Warren, P. H., Wasson, J. T. 1979a. The origin of KREEP. *Rev. Geophys. Space Phys.* 17:73–88
Warren, P. H., Wasson, J. T. 1979b. Effects of pressure on the crystallization of a "chrondritic" magma ocean and implications for the bulk composition of the Moon. *Proc. Lunar Planet. Sci. Conf., 10th*, pp. 2051–83
Warren, P. H., Wasson, J. T. 1979c. The compositional-petrographic search for pristine nonmare rocks: third foray. *Proc. Lunar Planet. Sci. Conf., 10th*, pp. 583–610
Warren, P. H., Wasson, J. T. 1980a. Early lunar petrogenesis, oceanic and extraoceanic. *Proc. Conf. Lunar Highl. Crust*, pp. 81–99
Warren, P. H., Wasson, J. T. 1980b. Further foraging for pristine nonmare rocks: correlations between geochemistry and longitude. *Proc. Lunar Planet. Sci. Conf., 11th*, pp. 431–70
Warren, P. H., Taylor, G. J., Keil, K., Shirley, D. N., Wasson, J. T. 1983a. Petrology and chemistry of two "large" granite clasts from the Moon. *Earth Planet. Sci. Lett.* 64:175–85
Warren, P. H., Taylor, G. J., Keil, K., Kallemeyn, G. W., Shirley, D. N., Wasson, J. T. 1983b. Seventh foray: whitlockite-rich lithologies, a diopside-bearing troctolitic anorthosite, ferroan anorthosites, and KREEP. *J. Geophys. Res.* 88:B151–64
Warren, P. H., Kallemeyn, G. W., Wasson, J. T. 1984. Pristine rocks (8th foray): genetic distinctions using Eu/Al and Sr/Al ratios. *Lunar Planet. Sci. XV*, pp. 894–95 (Abstr.)
Wasson, J. T. 1971. Volatile elements on the Earth and the Moon. *Earth Planet. Sci. Lett.* 11:219–25
Wasson, J. T., Warren, P. H. 1980. Contribution of the mantle to the lunar asymmetry. *Icarus* 44:752–71
Wetherill, G. W. 1975. Possible slow accretion of the Moon and its thermal and petrological consequences. In *Conference on the Origins of Mare Basalts and Their Implications for Lunar Evolution*, pp. 184–88. Houston: Lunar Planet. Inst. 204 pp.
Wetherill, G. W. 1976. The role of large bodies in the formation of the Earth and Moon. *Proc. Lunar Sci. Conf., 7th*, pp. 3245–57
Wetherill, G. W. 1980. Formation of the terrestrial planets. *Ann. Rev. Astron. Astrophys.* 18:77–113
Wetherill, G. W. 1981. Nature and origin of basin-forming projectiles. In *Multi-Ring Basins*, ed. P. H. Schultz, R. B. Merrill, pp. 1–18. New York: Pergamon. 295 pp.
Wolf, R., Woodrow, A., Anders, E. 1979. Lunar basalts and pristine highland rocks: comparison of siderophile and volatile

elements. *Proc. Lunar Planet. Sci. Conf., 10th*, pp. 2107–30

Wood, J. A. 1972. Thermal history and early magmatism in the Moon. *Icarus* 16:229–40

Wood, J. A. 1975. Lunar petrogenesis in a well-stirred magma ocean. *Proc. Lunar Sci. Conf., 6th*, pp. 1087–1102

Wood, J. A. 1983. The lunar magma ocean, thirteen years after Apollo 11. In *Workshop on Pristine Lunar Highlands Rocks and the Early History of the Moon*, ed. J. Longhi, G. Ryder, pp. 87–89. Houston: Lunar Planet. Inst. 92 pp. (*Tech. Rep. 83-02*)

Wood, J. A., Dickey, J. S., Marvin, U. B., Powell, B. N. 1970. Lunar anorthosites and a geophysical model of the Moon. *Proc. Apollo 11 Lunar Sci. Conf.*, pp. 965–88

Ann. Rev. Earth Planet. Sci. 1985. 13:241–68

AMINO ACID RACEMIZATION DATING OF FOSSIL BONES

Jeffrey L. Bada

Amino Acid Dating Laboratory, Scripps Institution of Oceanography, University of California, San Diego, La Jolla, California 92093

INTRODUCTION

Only L-amino acids usually occur in the proteins of living organisms, although some D-amino acids are present in certain bacteria. A system containing exclusively L-amino acids is thermodynamically unstable, since under conditions of chemical equilibrium D- and L-amino acids are present in equal amounts. The amino acid racemization reaction shown in Figure 1 will, after a period of time that is characteristic of each amino acid, eventually convert pure L-amino acids into a racemic mixture (i.e. one in which the D- and L-enantiomers are present in equal amounts).

It has been known for nearly a century that amino acids are rapidly racemized in acidic and basic solutions heated at elevated temperatures (Neuberger 1948). Within the last 15 years it has become established that racemization also takes place in the neutral pH region at rates that are comparable with those in dilute acid and base (Bada 1984b). Moreover,

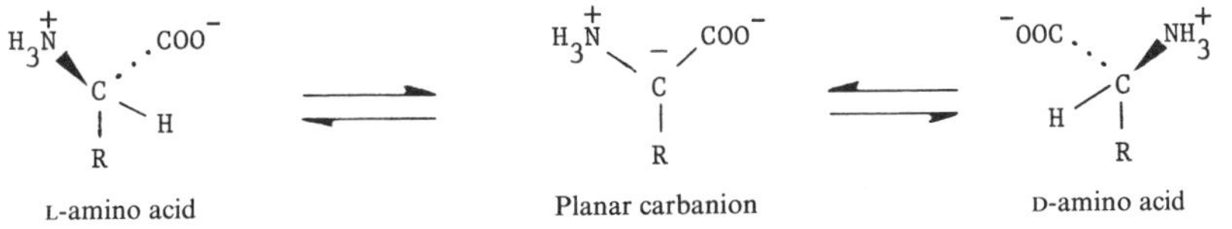

L-amino acid — Planar carbanion — D-amino acid

Figure 1 The mechanism of amino acid racemization, showing the formation of the carbanion intermediate. Base abstracts the α-proton; this is the rate-limiting step in the reaction. Readdition of the proton occurs by the reaction of water with the carbanion. This mechanism is applicable to both free and peptide-bound amino acids. Taken from Neuberger (1948) and Bada (1984a).

0084–6597/85/0515–0241$02.00

racemization has been detected in fossils and in the metabolically stable proteins of living mammals, and it has been suggested that racemization might form the basis of a dating method that can be used to determine the age of amino acids in these systems.

Racemization is a reversible first-order reaction that can be written as

$$\text{L-amino acid} \underset{k_i'}{\overset{k_i}{\rightleftarrows}} \text{D-amino acid}, \tag{1}$$

where the k_i values are the first-order rate constants for interconversion of the amino acid enantiomers. The kinetic equation for this reaction is (Bada 1984b)

$$\ln\left[\frac{1+(\text{D/L})}{1-K(\text{D/L})}\right] - \ln\left[\frac{1+(\text{D/L})}{1-K(\text{D/L})}\right]_{t=0} = (1+K)k_i t, \tag{2}$$

where D/L is the amino acid enantiomeric ratio at a particular time t, and $K = k_i'/k_i$. For amino acids with one center of asymmetry, the value of K is equal to unity (since $k_i = k_i'$), while for diastereomeric amino acids[1] such as isoleucine and 4-hydroxyproline, K is different from unity. The $t = 0$ term in Equation (2) is needed to account for the fact that the initial D/L ratio in the system under investigation may not be zero, and because some racemization may occur during sample processing (for example, during acid hydrolysis). When the extent of racemization is less than about 0.15, racemization can be considered an irreversible first-order reaction, and Equation (2) can be simplified to

$$\ln\,(1+\text{D/L}) - \ln\,(1+\text{D/L})_{t=0} = k_i t. \tag{3}$$

Investigations of free amino acids in buffered aqueous solutions heated at elevated temperatures indicate that the racemization reaction obeys the expected reversible first-order rate law (Bada 1984b, and references therein). However, in some peptides and proteins and in certain types of fossils (notably carbonates), the amino acid racemization kinetics are more complex. The rate of racemization ($k_{rac} = 2k_i$) depends on the particular amino acid. Amino acids (e.g. aspartic acid) with R-groups that enhance the stability of the carbanion intermediate have faster racemization rates than those amino acids with R-groups that destabilize the carbanion (e.g. the leucines). Racemization rates in aqueous solution are also influenced by pH, ionic strength, and metal ion chelation (Bada 1984b).

Amino acid racemization rates have now been determined in a variety of systems on the Earth, ranging from aqueous solutions heated at elevated

[1] For diastereomeric amino acids, the reaction is more properly referred to as epimerization. However, for simplicity the reaction is termed racemization here for all amino acids.

temperatures to known age fossils collected from various environments. Some representative half-lives are given in Table 1. At 100°C the various amino acids would be totally racemized on a time scale ranging from a few days to years, depending on the amino acid. At lower temperatures the rates are slower, which is expected, since chemical reactions are temperature dependent. In fossils the racemization half-lives are on the order of 10^4–10^5 yr, which indicates that the racemization reaction can be used to estimate the age of Holocene and Upper Pleistocene fossils. Fossils from these geologic periods are often difficult to date by other geochronological methods. Amino acid racemization dating can be used to estimate the age of fossils that are too old (> 30,000–40,000 yr) for radiocarbon dating. Even within the period datable by radiocarbon, racemization can be used to verify radiocarbon ages. The racemization-based method is particularly useful for dating fossil bones. Radiometric methods other than ^{14}C are generally not applicable to bones (Fleming 1976), since they either are based upon the complex incorporation of a radionuclide during the burial history of the bone (uranium-series dating) or require closed system behavior with regard to the daughter isotope (K–Ar dating).

The number of publications in the field of amino acid racemization dating is large. This review focuses on the racemization dating of fossil bones and teeth, since this application has generated the most intense debate about the validity of racemization-derived ages. The kinetics and mechanism of racemization in bones are discussed, as are some of the dating results obtained using the racemization-based method.

AMINO ACIDS IN FOSSIL BONES

The bones of living animals contain about 25–30% organic material, which consists primarily of proteins (mainly collagen; Hare 1980, Skelton 1983). Upon the death of an animal, the proteinaceous material in bones begins to decompose according to the pathways shown in Figure 2. The proteins first hydrolyze at labile peptide bonds such as those containing serine and aspartic acid residues. The peptides that are produced are then further hydrolyzed via peptide bond cleavage and internal aminolysis. Eventually this hydrolytic process produces free amino acids.

The protein hydrolysis steps in Figure 2 can be observed by analyzing various known-age fossils. In Holocene and Pleistocene calcareous fossils the various protein decomposition products shown in Figure 2 have been observed, and in Miocene fossil shells only small peptides and free amino acids are present (Hare & Mitterer 1966, Bada & Man 1980, Steinberg 1982). The dense carbonate matrix evidently prevents loss of even free amino acids. In fossil bones, however, the low-molecular-weight hydrolysis

Table 1 Amino acid racemization half-lives in years, except as indicated[a]

	Half-life			References
	Aspartic acid (asp)	Alanine (ala)	Isoleucine (iso)	
Aqueous solution 100°C, pH 7–8				
free amino acids	30 days	120 days	300 days	Bada 1984a,b
proteins	1–3 days	—	—	Masters 1982a
In vivo				
mammalian teeth	350	—	—	Bada 1984a
Holocene and Pleistocene fossils				
bones and teeth				
Egypt and Sudan	~3500	~1 × 10⁴	—	Bada et al 1979a, Bada & Shou 1980
East Africa	0.5–20 × 10⁴	—	1–3 × 10⁵	Hare et al 1978, Bada 1981
Southern California Coastal	~3 × 10⁴	~10⁵	—	Ike et al 1979, Bada et al 1979a
La Brea Tar Pits	~10⁵	—	—	McMenamin et al 1982
Saskatchewan, Canada	~10⁵	—	—	Bada & Helfman 1975
shells				
Southern Florida[b]	—	—	6 × 10⁴	Mitterer 1975
Bermuda[c]	—	—	~10⁵	Harmon et al 1983
Southern California[b]	—	—	~10⁵	Karrow & Bada 1980
Lake Bonneville, Utah[c]	—	—	4 × 10⁴–10⁵	Scott et al 1983
Canadian Arctic[b]	—	—	3 × 10⁵	Miller & Hare 1980
plant material				
Seeds, Southern California	2 × 10⁴	—	—	Engel et al 1978
Bristlecone pine, White Mountains, California	3 × 10⁴	—	—	Zumberge et al 1980
Deep-ocean sediments				
Siliceous	—	—	2 × 10⁶	King 1974
Carbonates	10⁵	2 × 10⁵	2 × 10⁵	Kvenvolden et al 1973, Bada & Man 1980
HCl-insoluble fraction	10⁶	2 × 10⁶	>5 × 10⁶	Bada & Man 1980

[a] One half-life = ln $2/k_{rac}$, where $k_{rac} = 2k_i$. One half-life is the time required to reach a D/L ratio of 0.33. [b] Bivalve mollusk. [c] Gastropod.

products (i.e. small peptides and free amino acids) are present only in trace quantities, indicating that these components are lost to the surrounding environment (Matsu'ura & Ueta 1980). Because of their porous nature and tendency to undergo fossilization (Behrensmeyer 1978, White & Hannus 1983), bones are not a closed system with respect to amino acids. Not only are free amino acids and small peptides lost, but also secondary amino acids can be added to fossil bones during their depositional history.

The rate of loss of the low-molecular-weight hydrolysis products from a fossil bone is apparently rapid (on the geologic time scale). "Leaching" of bones by groundwaters is probably the principal mechanism by which the low-molecular-weight products are lost (Hare 1980), although diffusion may also be an important factor.

The rate of protein hydrolysis is temperature dependent, so collagen is best preserved in fossil bones found in cool environments (Tuross et al 1980). Although collagenlike structures have been observed microscopically (Wyckoff 1972) and collagen has been detected immunologically (Lowenstein 1980) in bones more than 1 Myr old, it is apparent, based on

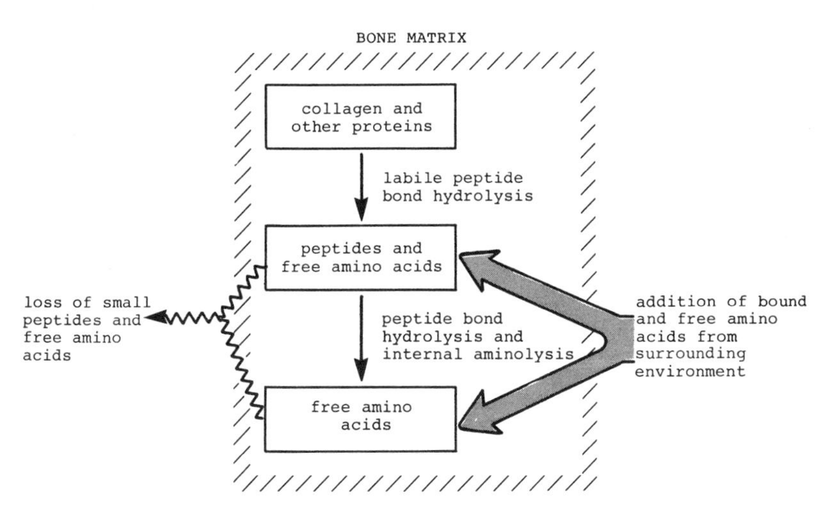

Figure 2 A generalized diagenetic scheme for proteins and amino acids in bones. The first step in collagen decomposition may be denaturation to gelation (Weiner et al 1980, Hare 1980), but this step is likely rapid on the geological time scale. The degree of preservation (or loss) of collagen and its hydrolysis products depends on the fossil bone's age, and it also may be a function of the extent of "leaching" of the sample by groundwaters (Hare 1980). This diagenetic scheme is equally applicable to both bones and teeth. Based on results given in van Kleef et al (1975), Hedges & Wallace (1980), Tuross et al (1980), Hare (1980), Matsu'ura & Ueta (1980), Steinberg (1982), Steinberg & Bada (1983), and Armstrong et al (1983).

the amino acid composition of fossil bones (Matter & Miller 1972, Dungworth et al 1974, Hare 1980, Armstrong et al 1983), that collagen and its hydrolysis products are not preserved for periods in excess of a million years, even in cool environments. In temperate environments, the high-molecular-weight protein components in bones are likely extensively hydrolyzed in less than 10^3–10^4 yr.

The amino acid 4-hydroxyproline is unique to collagen, so its presence can be used as an indication of the degree of preservation of collagen and its hydrolysis products. Many fossil bones have been found to contain 4-hydroxyproline (Wyckoff 1972, Kessels & Dungworth 1980, Hare 1980, Armstrong et al 1983), but in some, notably those with low total amino acid contents, this amino acid is either absent or present only in trace quantities. It is generally observed that with increasing geological age, bones eventually lose their collagenlike amino acid signature and have a composition that is dominated by amino acids such as serine and aspartic and glutamic acids, which are not major amino acids in collagen (Wyckoff 1972, Hare 1980). It has been suggested that the noncollagen amino acid composition of fossil bones with low amino acid contents arises because some of the minor constituent proteins present in modern bone are more stable with respect to hydrolysis than is collagen (Hare 1980). An alternative explanation is that since the bone mineral matrix (i.e. hydroxyapatite) has an affinity for acidic amino acids and proteins (Bernardi et al 1972, Gorbunoff 1984), the noncollagen composition results from the absorption of secondary amino acids from the surrounding environment.

KINETICS AND MECHANISM OF RACEMIZATION IN BONES

The scheme shown in Figure 2 has important implications concerning amino acid racemization dating, since the amino acids in the various hydrolysis products have been found to have different racemization rates. In fossil carbonates, the most extensively racemized components are free amino acids (Bada & Man 1980), which is likely a consequence of the internal aminolysis-hydrolysis pathway (Steinberg & Bada 1983). In bones, this highly racemized component is lost and not preserved in the fossil. Actually, this situation is fortunate, because in carbonates the presence of the highly racemized free amino acids derived from internal aminolysis is likely one of the causes for deviations from reversible first-order kinetics in fossil shells (Bada & Man 1980).

The kinetics of amino acid racemization in bones can be studied using both laboratory simulation experiments and well-dated fossils collected from localities having similar present-day temperatures. The kinetics of aspartic acid racemization in heated modern bones and in some representative fossils are shown in Figure 3. As can be seen, the racemization of aspartic acid follows the expected reversible first-order kinetic expression at least to 1–2 half-lives. Reversible first-order kinetics (through at least 3–4 half-lives) have been observed for several other amino acids in heated modern bone (Bada 1972, Bada et al 1973, Hare 1974). In comparison with the racemization kinetics in bone, more complex kinetics are often observed in fossil shells, and in the latter system, racemization generally only obeys reversible first-order kinetics to about 1 half-life (Bada & Man 1980).

The addition of secondary amino acids into a fossil bone (see Figure 2) from the surrounding environment could produce anomalous racemization

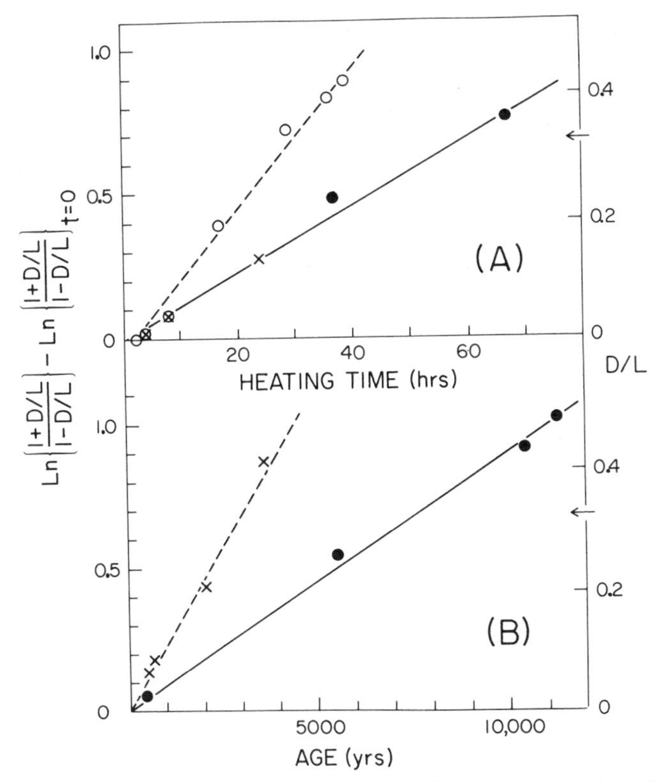

Figure 3 Kinetics of racemization of aspartic acid in bones. *A* Modern bone fragments heated at 105°C under continuous "leaching conditions" (○), at 96°C between pH 3 and 8 (●), and under boiling conditions (×). *B* Fossil bones from environments with present-day temperatures of about 26–27°C (Sudan and Philippines, ×) and about 18°C (Arizona, ●). The arrows indicate one half-life. The slopes of the lines are equal to 2 k_{asp}. Based on data given in Bada (1982), Bada & Shou (1980), Hare (1980), Shou (1979), and Skelton (1983).

kinetics. Secondary amino acid components should consist predominantly of the L-enantiomer, and thus the accumulation of externally derived amino acids would cause derivations from the kinetics described by Equation (2). In the fossil bones of dinosaurs, alloisoleucine has not been detected, although isoleucine is present, which indicates that the majority of the amino acids in these fossils are not indigenous (Wyckoff 1972, Matter & Miller 1972).

The rate of racemization is highly temperature dependent. As can be seen from Figure 3, the half-lives range from a few days at 100°C to thousands of years at 20–25°C. In order to establish the temperature dependence of amino acid racemization in bone, the rate constants derived from Figure 3 can be plotted in the form of the Arrhenius equation (Figure 4). These

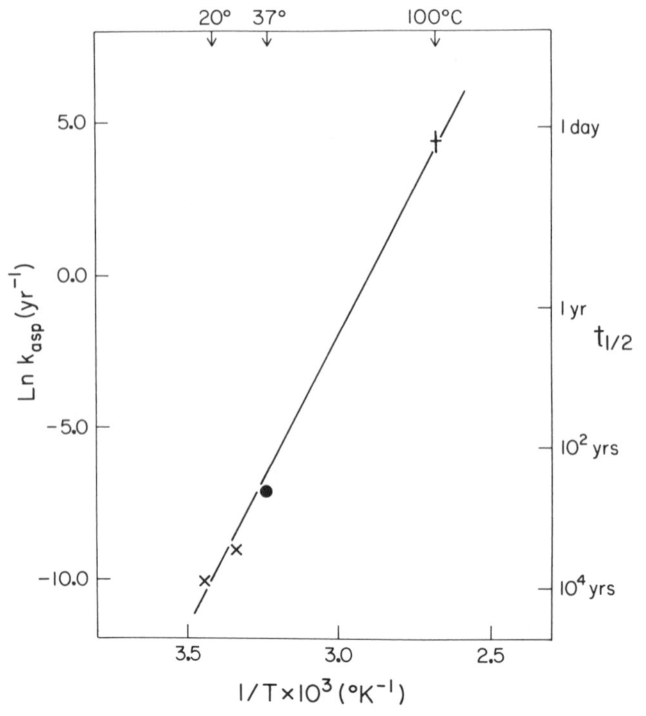

Figure 4 The k_{asp} values determined for the heated modern bone (about 100°C, +) and the fossil bones (×) shown in Figure 3, plotted in the form of the Arrhenius equation. Also included is the in vivo k_{asp} value (●) determined for human teeth (Bada 1984a). The slope of the line is equal to E_a/R, where E_a is the Arrhenius activation energy and R (=1.987) is the universal gas constant. A least-squares fit of these data yields the equation $\ln k_{asp}\,(\mathrm{yr}^{-1}) = 56.6 - 19.7 \times 10^3\, T^{-1}$.

results yield an Arrhenius activation energy (E_a) for aspartic acid racemization in bones of about 38 ± 3 kcal mol^{-1}. This E_a value is similar to the values determined for several other amino acids in heated modern bone samples (Bada 1972, Bada et al 1973). An activation energy of 35–40 kcal mol^{-1} implies that variations of $\pm 1°$ in the temperature of the depositional environment of a fossil bone would effect the rate of racemization in the fossil by about $\pm 25\%$.

In Holocene and Upper Pleistocene fossil bones, the predominant amino-acid-containing components are peptides, which probably have molecular weights of 10^4–10^5 (Armstrong et al 1983). The carbanion shown in Figure 1 predicts that for peptide-bound amino acids, the R-substituent should be the dominant factor determining the relative rates of racemization of the various amino acids. Aspartic acid would thus be predicted to racemize faster than the leucines. Results from both heated modern bones and fossil bones indicate that this is the case (Bada et al 1973, Bada & Shou 1980). Thus, it is apparent that the carbanion mechanism shown in Figure 1 is the pathway by which amino acids undergo racemization in bones.

A variety of factors can effect the carbanion intermediate stability, and these might potentially influence the rate of amino acid racemization in fossils. Changes in environmental pH, for example, could influence the rate of racemization, since the rate-limiting step in the formation of the carbanion (see Figure 1) is the extraction of the α-proton by either the hydroxide ion or some other base. In experiments with modern bone fragments heated at elevated temperatures (Shou 1979, Bada & Shou 1980), it has been found that in the range pH 3–8, racemization rates are not affected by pH (see Figure 3). Apparently the bone minerals effectively buffer the system involved in racemization. Thus, fluctuations in environmental pH do not significantly affect the amino acid racemization rates in bone. At mildly acidic and basic pH values, where the bone minerals begin to dissolve, pH could affect racemization rates in bone. However, bones are not preserved in these extreme pH environments, and therefore this effect would not be expected to directly influence the racemization in bones.

The carbanion stability can also be influenced by the position of an amino acid in a peptide chain (Bada 1984a,b). For example, amino acids that are at the N-terminal position are more rapidly racemized than those at the C-terminal end. It is conceivable that in the peptide hydrolysis products derived from bone proteins, some amino acids could occupy predominantly either terminal position. This could affect the racemization rate of that particular amino acid. Whether this is a factor in determining amino acid racemization rates in fossil bones has not been investigated.

Under conditions of extreme desiccation, racemization rates in bones are dramatically reduced at elevated temperature (Hare 1974). However, bones generally contain enough indigenous water that this is not an important factor in determining the racemization rates in fossil bones. In some environments, such as those associated with asphaltic deposits and mummified remains, the availability of water apparently does cause a reduction in racemization rates in comparison with those observed under normal environmental conditions (Hare 1980, McMenamin et al 1982). In order to date such samples, racemization rates must be evaluated for the characteristic environment (McMenamin et al 1982).

Hare (1974, 1980) has argued that differential "leaching" histories of fossil bones could affect racemization kinetics. However, investigations of this "leaching" effect at various sites throughout the world indicate that this factor does not significantly influence racemization rates in fossil bones and teeth (King & Bada 1979).

DATING FOSSIL BONES

Fossil bones from many sites throughout the world have been dated using the racemization method. Some of the dated samples include those from the following locations[2]: Klaises River Mouth Caves, South Africa (Bada & Deems 1975); Olduvai Gorge, Tanzania (Hare et al 1978, Bada 1981); Old Kingdom sites in Egypt (Bada et al 1979b); caves on Mt. Carmel, Israel (Masters 1982b); the Zhoukoudian or Peking Man deposits and other sites in China and Japan (Li & Lin 1979, Matsu'ura & Ueta 1980); various anthropological sites in Europe (Bada et al 1979b, Belluomini 1981); and paleoindian burial sites in North America (Bada et al 1974, Bada & Masters 1982, Skelton 1983). In general, the racemization ages were found to be consistent with dates based on either radiocarbon, historical records, or geological context.

Because it has one of the fastest amino acid racemization rates, aspartic acid has generally been used for dating Holocene and Upper Pleistocene fossil bones. Isoleucine, on the other hand, racemizes about 10 times slower than aspartic acid and is used to date older fossil bones.

Rate Constant (k_i) Evaluation

As is shown in Figure 3, the kinetics of amino acid racemization in fossil bones are described by Equation (2). Thus, the rate of racemization in a fossil bone is determined by the value of k_i, which in turn is primarily a function of the integrated exposure temperature of the fossil bone.

In order to date a fossil bone using amino acid racemization, we must first

[2] This list is only a representative sampling and is not meant to be a complete listing of the sites at which racemization has been used to date fossil bones.

determine a value of k_i for the sample. The value of k_i for a particular locality can be estimated using a "calibration" procedure (Bada & Protsch 1973), wherein the D/L ratio in a fossil bone of known age from the area of study is substituted into Equation (2). The resulting in situ k_i value is an average integrated value that incorporates variations in the temperature (and other environmental parameters) of the locality over the time period represented by the age of the "calibration" sample. Following "calibration," the k_i value can be used, with certain limitations, to date other samples from the general area.

Another technique for estimating the k_i value for a particular site is to establish the average integrated exposure temperature history of the fossil (Bischoff & Childers 1979). However, since accurate temperature histories (e.g. $\pm 2°C$) are difficult to establish accurately, this procedure yields only an approximate k_i estimate. The racemization ages determined using this procedure for k_i evaluation may have large (± 50–100%) uncertainties.

Experimental and Analytical Methods

A fossil bone sample (usually ~ 5 g) is prepared for analysis by ultrasonically cleaning the fragment in dilute HCl and water. Only dense, compact bone pieces (e.g. tibia, femur) are generally processed, because porous bones (such as ribs and vertebrae) generally contain large quantities of extraneous material and thus are difficult to clean. After cleaning, the sample is hydrolyzed in excess double-distilled 6 M HCl for either 6 or 24 hr. The hydrolysis time depends on the extent of racemization in the sample. In samples where the extent of racemization is small (< 0.1), the shorter hydrolysis time should be used in order to reduce the acid-catalyzed racemization. Following hydrolysis, the HCl solution is evaporated to dryness and the residue redissolved in double-distilled water. The sample then is desalted on Dowex 50W-X8 (100–200 mesh) resin that has been cleaned with NaOH, protonated with double-distilled HCl, and washed with doubly distilled water. Amino acids are eluted from the column with 1–2 M NH_4OH, and the effluent is evaporated to dryness. Aspartic acid is separated from the other amino acids by anion exchange chromatography (BioRad AG1-X8, 100–200 mesh, anion exchange resin) using 1 M acetic acid for elution.

A variety of techniques can be used to separate and quantify amino acid enantiomers. A review of these various methods is beyond the scope of this paper, and thus only the methods used in the Amino Acid Dating Laboratory at the Scripps Institution of Oceanography are described here. Aspartic acid D/L ratios are determined using the diastereomeric dipeptide method of Manning & Moore (1968). The aspartic acid isolated from a fossil bone is reacted with L-leucine-*N*-carboxy-anhydride (L-leu NCA) in order to synthesize the dipeptides L-leucyl-L-aspartic acid and L-leucyl-D-aspartic acid. These diastereomeric dipeptides are then separated on an automatic amino acid analyzer (Beckman Model 119C), which is interfaced with a DEC PDP-11 computer used for data acquisition and analysis. During a series of analyses, standards (i.e. derivatized solutions with known D/L aspartic acid ratios) are run repeatedly along with the samples under investigation. The reproducibility and precision of a D/L aspartic acid ratio is generally on the order of ± 5–10%. Some representative chromatograms obtained using this analytical scheme have been published elsewhere (Bada et al 1979b, Bada 1984a).

The D/L enantiomeric ratios of other amino acids such as alanine, glutamic acid, and leucine are determined by gas chromatography using the TPC[3] method (Hoopes et al 1978). The *N*-trifluoroacetyl-L-prolyl-DL-amino acid methyl esters are synthesized, and they are then separated using a fused silica capillary column (30 m × 0.25 mm) coated with OV101 on a Hewlett-Packard model 5710A gas chromatograph equipped with a flame ionization detector.

The alloisoleucine/isoleucine (allo/iso) ratio is determined directly on the automatic amino acid analyzer, as are the amino acid compositions of the samples.

Reliability of a Racemization Date

The D/L enantiomeric ratio of a particular amino acid in a fossil bone can, in general, be determined with high precision. Also, different analytical methods give comparable D/L ratios (Bada et al 1979a). However, the conversion of the measured D/L ratio into a reliable age estimate can only be accomplished if a suitable "calibration" (k_i) value has been determined for the site or locality where the bone was found. The accuracy of a racemization-based age is a direct function of the reliability of the age of the "calibration" sample used to calculate k_i. The k_i value used to date a particular sample should be compatible with its estimated exposure temperature. Using the procedures described elsewhere (Bada & Helfman 1975, Bada et al 1979b, Bada 1981, Belluomini 1981), it is possible to calculate an exposure temperature from a particular k_i value. Comparison of the racemization-derived temperature with that derived from other evidence provides an important check on the reliability of the racemization dates based on a certain k_i value. For Holocene age samples, the racemization-based temperature estimate should be comparable to the present-day temperature of the locality, whereas temperatures derived from Pleistocene samples will always be somewhat lower, since these samples

[3] TPC = *N*-trifluoroacetyl-L-prolyl chloride.

have been exposed to temperatures associated with both glacial and interglacial periods (Bada et al 1979b).

The presence of secondary amino acids in a fossil bone should also be evaluated, since this factor would affect the reliability of an amino acid racemization age. Amino acids introduced from the surrounding environment should consist predominantly of the L-enantiomers, so contamination would give rise to lower D/L ratios and thus racemization ages that are too young. The extent of amino acid contamination can be evaluated by measuring the extent of racemization of several amino acids. Elevated temperature experiments (Bada et al 1973) have shown that the extent of racemization in a bone follows the following pattern: D/L aspartic acid > D/L alanine ≅ D/L glutamic acid > D/L leucine ≅ alloisoleucine/isoleucine. The presence of this pattern in a fossil bone provides evidence that the sample has not been seriously contaminated (Bada et al 1973). It should also be possible to calculate the racemization age of a fossil using the extent of racemization of several amino acids. The racemization ages derived from different amino acids should be in good agreement if the sample is free of contamination (Bada et al 1979b).

RESULTS FROM OLDUVAI GORGE, TANZANIA (EAST AFRICA)

The Olduvai Gorge region in the northcentral Tanzanian Rift Valley offers an excellent opportunity to study the racemization reactions of amino acids in fossil bones and teeth over a time period extending from the Holocene back to several million years. The geology of the region has been extensively studied as a result of the discovery of numerous hominid fossils and artifacts (Hay 1976). Because the relative and absolute ages of the various stratigraphic units of the Olduvai region are well established, the rate of amino acid racemization can be evaluated over several geologic time periods. The conditions at Olduvai represent some of the most divergent chemical environments on the Earth. The present climate periodically alternates between very wet and arid conditions because of biannual monsoons. The numerous alkaline lakes in the region represent an extreme as far as the maximum pH values in natural environments on the Earth. In addition, present-day surface temperatures are sometimes so high that fossils lying on the surface are warm to the touch (Bada 1981).

Holocene and Upper Pleistocene Samples

Holocene and Upper Pleistocene samples from both the Olduvai Gorge and the Nasera rock shelter, located approximately 40 km northwest of Olduvai, have been dated by radiocarbon. Using these ages, the measured

D/L aspartic acid ratios in the samples, and Equation (2), we can calculate values for k_{asp} at these two sites. These results are summarized in Table 2.

The Holocene rate constants at the Nasera rock shelter and Olduvai are nearly identical, as are the Upper Pleistocene rate constants. Although their present-day temperatures are roughly similar, these two sites have very different physical characteristics. The Olduvai samples were collected from near-surface fossiliferous deposits, which are exposed to present-day environmental conditions that are often highly variable. On the other hand, the Nasera rock shelter is located in a protected overhang of a large rock outcropping. This locality has also had an extensive human occupational history. The concordance of the rate constants (k_{asp} values) determined for the two sites suggests that the different physical characteristics of the environment have little effect on aspartic acid racemization rates.

The Holocene Olduvai Gorge samples have an amino acid composition similar to collagen, whereas the Holocene Nasera rock shelter samples have a noncollagenous amino acid composition (King & Bada 1979). Studies of the extent of leaching in these two sample sets indicate that the Nasera rock shelter samples are much more extensively leached than the Olduvai samples (King & Bada 1979). Nevertheless, the rate constants for these two sample sets are the same, which suggests that at least in this case the rate of racemization of aspartic acid in bone is not necessarily affected by changes in the amino acid composition produced by the hydrolysis and subsequent leaching of the collagenous material originally present in the bone.

The k_{asp} Upper Pleistocene values given in Table 2 have been used to date other Upper Pleistocene samples from the Olduvai region (Table 3). The

Table 2 The k_{asp} values for the Olduvai Gorge region[a]

Site location	Radiocarbon age (yr)[b]	D/L Asp	k_{asp} (yr^{-1})
Nasera rock shelter			
level 3A	2180 ± 200 (ISGS 438)	0.212	6.7×10^{-5}
level 5A	21,600 ± 400 (ISGS 455)	0.428	1.8×10^{-5}
Olduvai Gorge			
Namorod ash	1360 ± 40 (LJ 2979)[c] 1350 ± 100 (LJ 3330)[d]	0.165	7.4×10^{-5}
Naisuisui Beds	17,500 ± 1000 (UCLA 1695)	0.32	1.5×10^{-5}

[a] Taken from Bada (1981).
[b] The numbers in parentheses are the radiocarbon laboratory identification numbers. Unless indicated, the dates were determined using the organic fraction ("collagen"?).
[c] Associated land snail.
[d] Amino acids isolated from the same bone used for racemization analysis.

Table 3 Ages of Upper Pleistocene bones from the Olduvai Gorge area[a]

Sample location	D/L Asp	Radiocarbon age (yr)[b]	Aspartic acid age (yr)[c]
Nasera rock shelter			
Level 4	0.406	—	22,000
Level 6	0.482	22,900 ± 400 (ISGS-425)	26,000
Olduvai Gorge			
Naisuisui Beds, base of type section at second fault	0.57	—	39,000
Upper Ndutu Beds			
rim of Gorge at FLK	0.72	> 29,000[d]	56,000
SE of second fault	0.51	3340 ± 800 (UCLA 1903)	33,000

[a] Taken from Bada (1981).
[b] The numbers in parentheses are the radiocarbon laboratory identification numbers. The dates for these samples were determined using the organic fraction ("collagen"?).
[c] Calculated from Equation (2) using the Upper Pleistocene k_{asp} values given in Table 2.
[d] Based on calcareous fossils and calcrete deposits.

aspartic acid ages obtained for the other levels in the Nasera rock shelter are in good agreement with the radiocarbon ages and the stratigraphy of the site. The aspartic acid ages for the Upper Ndutu Beds at Olduvai are also consistent with the estimated age of > 29,000 yr (Hay 1976). However, one of the samples from the Upper Ndutu Beds with an aspartic acid age of 33,000 yr was also dated by the UCLA Radiocarbon Laboratory, and the resulting ^{14}C age was 3340 ± 800 years (UCLA 1903). The radiocarbon age of this bone is thus considerably younger than the racemization age of the sample, and it is also inconsistent with other radiocarbon dates and geologically inferred ages for the Upper Ndutu Beds (Hay 1976). These results suggest that this Upper Ndutu bone sample was badly contaminated with secondary carbon components having a relatively recent radiocarbon age. This sample contained less than 0.05% organic nitrogen, which indicates that the original bone protein had been extensively decomposed. Even though this bone yielded an erroneous ^{14}C age, the aspartic racemization age of the bone is still compatible with the Upper Pleistocene age assigned to the Upper Ndutu Beds. It is apparent from these results that the racemization method provides an important check on radiocarbon-based ages of bones that have very low organic carbon contents.

Middle and Lower Pleistocene Samples

The k_{asp} values in Table 2 indicate that after 80,000 to 100,000 yr, the aspartic acid in fossil bones from the Olduvai Gorge region should be totally racemized, i.e. D/L aspartic acid = 1.0. However, analyses of bone samples from older stratigraphic units at Olduvai indicate that this is not the case. For example, a fossil bone from the lower unit of the Masek Beds, which range in age from ~450,000 to 600,000 yr, was found to have a D/L aspartic acid ratio of ~0.75 (Bada 1981). Samples from other older stratigraphic units showed even lower D/L aspartic acid ratios (Bada et al 1973). Also, in the stratigraphic section at the Nasera rock shelter, the D/L aspartic acid ratios below a depth of ~2 m reached a value of ~0.5 (Bada 1981) and did not change throughout the remaining lower sections of the deposit. (The total depth of this section is ~4–5 m.)

These results suggest that after about 100,000 yr, fossil bones in the Olduvai region contain only trace quantities of indigenous aspartic acid, and that secondary aspartic acid introduced from surrounding soils and percolating groundwaters is present in significant quantities. Because this contamination introduces primarily L-amino acids into a fossil, the D/L aspartic acid ratios are lower than expected. An age calculated from the extent of aspartic acid racemization in the contaminated samples is always too young.

Other amino acids may be less affected by contamination than aspartic acid and thus may be more suitable for dating the older stratigraphic units at Olduvai. Results obtained by Hare et al (1978) and Bada (1981) suggest that in tooth enamel, isoleucine still may be relatively free of contamination, and that the epimerization reaction of this amino acid can be used to date the older stratigraphic sections at Olduvai. In Table 4 are given some representative results. As can be seen, the ratio of alloisoleucine to isoleucine (allo/iso) steadily increases in the progressively older stratigraphic sections at Olduvai. Moreover, in the enamel fraction isolated from a fossil rhinocerous tooth from the oldest stratigraphic unit in the Olduvai region (i.e. the ~3.5 Myr old Laetolil Beds), the measured allo/iso ratio is close to the expected equilibrium value of about 1.3 (Bada 1972, Hare 1974). These results suggest that in the Olduvai area, isoleucine epimerization in teeth is effectively a closed system with respect to the introduction of secondary isoleucine contamination for periods in excess of 3–4 Myr.

Isoleucine may be less susceptible to contamination than aspartic acid because the peptide bonds containing isoleucine and other hydrophobic amino acids are more stable to hydrolysis than are the peptide bonds containing aspartic acid (Bada & Man 1980). Thus, aspartyl bonds in the original bone and teeth proteins likely are hydrolyzed more rapidly than isoleucine. Moreover, in the carbonate-rich soils, such as those in the Olduvai region, acidic amino acids likely are more abundant than hydrophobic amino acids (Mitterer 1972), so amino acids introduced into fossils from the surrounding environment may be richer in aspartic acid than in the hydrophobic amino acids. In addition, the bone mineral hydroxyapatite has an affinity for acidic amino acids and proteins (Bernardi et al 1972, Gorbunoff 1984), so these acidic amino acids are probably absorbed by fossil bones from the environment more readily than other amino acids.

Using the sample from the base of Bed I for "calibration" of the isoleucine epimerization rate (k_{iso}), we obtain ages for the other stratigraphic units that are in reasonable agreement with the ages estimated from stratigraphic and radiometric age determinations (Table 4). Even some bones yield reasonable ages. For example, a bone from the lower unit of the Masek Beds was found to have an epimerization age of about 500,000 yr, which is consistent with the age estimated for this stratigraphic unit from geological and paleomagnetic evidence. However, analyses of a bone sample from Bed IV, a stratigraphic section older than the Masek Beds, yielded an allo/iso ratio less than that of the lower unit Masek Bed sample. Also, a bone from the Laetolil Beds at Laetoli (J. L. Bada, unpublished) was found to have an

Table 4 Isoleucine ages of fossil teeth (enamel fraction) from the Olduvai Gorge region

Locality	Known or estimated age (yr)	allo/iso	allo/iso estimated age (yr)[c]
Olduvai Gorge			
Masek Beds, lower unit	$4.5\text{–}6 \times 10^5$	0.6[a]	5×10^5
Upper Bed IV	$6\text{–}7 \times 10^5$	0.82[a]	7×10^5
Bed III/IV	8×10^5	0.92[b]	8×10^5
Upper Bed II	$\sim 1.3 \times 10^6$	1.00[b]	1×10^6
Lower Bed II	1.65×10^6	1.17[b]	1.5×10^6
Top of Bed I	1.72×10^6	1.09[b]	1.2×10^6
Base of Bed I	1.8×10^6	1.21[b]	$k_{iso} = 1.3 \times 10^{-6}\ yr^{-1}$
Laetoli (~40 km SW of Olduvai)			
Laetolil Beds	$\sim 3.5 \times 10^6$	1.28[a]	—

[a] From J. L. Bada (unpublished data). The sample from the Masek Beds is a bone sample.
[b] Hare et al (1978).
[c] Calculated from $\ln[(1+\text{allo/iso})/(1-0.8\text{allo/iso})] = 1.8k_{iso}t$, where k_{iso} is determined using the indicated sample (Hare et al 1978).

allo/iso ratio much lower than that in the enamel fraction. This result implies that isoleucine in bones is also susceptible to contamination, and that isoleucine ages of bones from the older stratigraphic units should likely be considered minimum age estimates.

Radiocarbon-dated bones (Tables 2 and 3) from Upper Pleistocene levels at the Nasera rock shelter were found to have an allo/iso ratio of ~0.1 (J. L. Bada, unpublished). Using this ratio and the value $k_{iso} = 1.3 \times 10^{-6}$ yr^{-1} (see Table 4) yields an age of about 60,000 yr, which is older than the radiocarbon ages of the samples. This result indicates that the Lower Pleistocene k_{iso} value should not be used to date the Upper Pleistocene deposits, nor should Upper Pleistocene–based "calibration" constants be used for the isoleucine epimerization dating of older stratigraphic units.

Olduvai Paleotemperature Estimates

The Holocene and Upper Pleistocene k_{asp} values, as well as the Lower Pleistocene k_{iso} value, can be used to estimate the average integrated ground temperatures at Olduvai over various periods of time. Aspartic acid racemization–derived temperatures can be calculated from the equation (taken from Figure 4)

$$\ln k_{asp}\ (\mathrm{yr}^{-1}) = 56.6 - 19.7 \times 10^3\ T^{-1}, \tag{4}$$

where T is the exposure temperature in K, and k_{asp} is the value derived from the "calibration" sample. Substituting the Holocene Olduvai k_{asp} value into Equation (4) yields a temperature of 23°C, which is in good agreement with the present-day mean annual air temperature of about 23°C measured at Olduvai (Hay 1976). The racemization-calculated temperature is slightly lower than the soil temperatures [26–28°C (Bada 1981, and unpublished)] measured at Olduvai, however. The Upper Pleistocene k_{asp} value yields a temperature of 17°C, which is considerably cooler than present-day temperatures.

The exposure temperatures derived from isoleucine epimerization can be calculated from the equation

$$\ln k_{iso}\ (\mathrm{yr}^{-1}) = 44.7 - 16.9 \times 10^3\ T^{-1}, \tag{5}$$

which was obtained from elevated-temperature experiments using modern bone samples (Bada 1972). Substitution of the Lower Pleistocene k_{iso} value given in Table 4 into Equation (5) yields a temperature of 16°C, which is similar to that calculated from the Upper Pleistocene k_{asp} value.

In addition to these estimated exposure temperature calculations, the Holocene and Upper Pleistocene k_{asp} values can also be used as indicators of the difference in average temperatures ($\Delta\bar{T}$) to which the samples from these two geological periods have been exposed. Using the equation given elsewhere (Schroeder & Bada 1973) and the Holocene and Upper Pleistocene k_{asp} values shown in Table 2, we calculate a value of $\Delta\bar{T} \cong 7$–8°C, which suggests that there has been a substantial difference in the ground temperatures at Olduvai throughout the Holocene and Upper Pleistocene

Both the aspartic-acid- and isoleucine-derived temperatures, as well as the calculated $\Delta\bar{T}$ of the Holocene and Upper Pleistocene samples, imply that the ground temperatures in the Olduvai Gorge region were significantly cooler in the Pleistocene than during more recent times. This does not necessarily imply that the air temperatures in this region were dramatically lower in the past. At present, soil temperatures are 2–5°C higher than air temperatures (Hay 1976, J. L. Bada, unpublished). Changes in the albedo of the ground surface, which could arise from changes in the extent and type of vegetation cover, would dramatically affect the difference in soil and air temperatures. A decrease in the mean air temperature of $\sim$7°C during the Pleistocene, coupled with a decrease in the difference between the ground and air temperatures, would yield a ground temperature compatible with that derived from aspartic acid racemization. A temperature decrease of about 7°C in East Africa during the Upper Pleistocene has been suggested from other evidence (Butzer 1971).

RESULTS FOR CALIFORNIA PALEOINDIAN SKELETONS

The extent of amino acid racemization in fossil bones and teeth from the Olduvai Gorge region provides reasonable ages over a geologic period ranging from the Holocene back through the Pleistocene. Investigations similar to the Olduvai studies have been carried out in other regions, and in general the racemization-derived ages are consistent with independently derived ages. Nevertheless, a major controversy about the validity of racemization-derived ages has arisen because of some dates for Indian skeletons found in various localities in California. In 1974 and 1975, a series of aspartic acid racemization dates for several California paleoindians were determined, and the ages of skeletons from sites at Del Mar and Sunnyvale were estimated to be 40,000–60,000 yr (Bada et al 1974, Bada & Helfman 1975). These data seemed to provide some of the first direct evidence that human beings had migrated into the Americas considerably earlier than the generally accepted time of about 13,000 yr ago. These aspartic acid racemization dates were determined using a ^{14}C-dated skeleton (Laguna) for k_{asp} calibration, which was dated at 17,150 $\pm$ 1450 yr using the HCl-insoluble "collagen" fraction isolated from skull material (Berger et al 1971).[4]

Several Holocene radiocarbon-dated skeletons from California were subsequently dated by the amino acid racemization method, and the resulting ages generally agreed closely with the ^{14}C dates (Ike et al 1979, Bada & Masters 1982, Skelton 1983). Also, racemization analyses of Upper Pleistocene age horse bones collected from near the Del Mar (Bada & Masters 1982) and Sunnyvale (Lajoie et al 1980) sites also suggested that the human skeletons were Upper Pleistocene in age. The racemization dates for the Del Mar and Sunnyvale skeletons caused considerable controversy concerning the validity of aspartic acid racemization–based dates. Anthropologists were reluctant to accept these "early" dates for the colonization of the Americas, and as a consequence the validity of the racemization method was questioned (for example, see Gerow 1981).

In the decade that has elapsed since the racemization age determinations were published, several additional studies have been carried out. Bischoff & Rosenbauer (1981) have dated the Del Mar and Sunnyvale skeletons using the uranium decay series (U-series) method. The resulting ages were 8000 to 12,000 yr, four to five times younger than the estimated aspartic acid racemization ages of these skeletons. However, U-series ages always tend to be younger (by as much as two to three times) than the actual burial ages of Upper Pleistocene fossil bones (Szabo 1980, Bada & Finkel 1983). Thus, the U-series dates did not necessarily rule out an Upper Pleistocene age for the Del Mar and Sunnyvale skeletons. In addition, a new radiocarbon dating technique that utilizes accelerator mass spectrometry (AMS) has been used to determine the ^{14}C ages of many of the disputed California paleoindian skeletons. With AMS-based radiocarbon dating, only a few milligrams of carbon are required; thus, samples that were too precious or small to be dated by the conventional ^{14}C method (β-counting) are potentially datable by the AMS-based radiocarbon method (Hedges 1981). The results of the AMS-based radiocarbon age determinations of several of the California paleoindian skeletons, along with the measured D/L aspartic acid ratios in the skeletons and the k_{asp} values calculated from the ^{14}C ages, are given in Table 5.

The AMS-based ^{14}C dates indicate that all of the skeletons previously assigned an Upper Pleistocene age are in fact more likely to be Holocene in age. The original racemization dates for the Del Mar and Sunnyvale

[4] Actually, two dates were reported for this skeleton. Long bone fragments were also processed, but insufficient material was isolated for absolute dating. An age of >14,800 yr (UCLA-1233B) was obtained.

skeletons were based on the 17,150 ± 1450 yr age for the Laguna skeleton. However, as can be seen in Table 5, the AMS ^{14}C age for this skeleton indicates that it is only ~5000 yr old. Also, the AMS-based ^{14}C age for Los Angeles Man is considerably younger than the date of >23,600 yr determined by conventional radiocarbon dating about 15 years ago (Berger et al 1971). The AMS ^{14}C ages of these skeletons indicate that the k_{asp} value suitable for dating the other paleoindian skeletons in California is considerably greater than that used previously. Using the average k_{asp} value (i.e. $k_{asp} = 6.0 \pm 3 \times 10^{-5}\ yr^{-1}$) derived from the AMS ^{14}C ages for the Laguna and Los Angeles Man skeletons yields racemization ages for the Del Mar and Sunnyvale skeletons that are clearly Holocene, a result consistent with their AMS-based ^{14}C ages.

The results in Table 5 show that there is an apparent relationship between the k_{asp} values in the various California paleoindian skeletons and the level of preservation of amino acids in the bones. Figure 5 shows a plot of the log k_{asp} values for the various paleoindian skeletons vs the amount of amino acids preserved in the bones. As can be seen, in samples where the level of amino acid preservation is <5%, the k_{asp} value is clearly greater than when the degree of amino acid preservation is >5%. This suggests that when the collagen has undergone extensive hydrolysis, the peptide residues that remain have an aspartic acid racemization rate that is greater than that in bones in which the collagen is better preserved. During the process of collagen hydrolysis, aspartic acid residues may end up preferentially at the N-terminal position of the peptide hydrolysis products. This in turn could give rise to a more rapid racemization rate, since amino acids at this position in peptides have been found to have racemization rates greatly exceeding those of amino acid residues at other positions (Bada 1984b).

The California paleoindian skeletons that have low amino acid contents were found to be depleted in 4-hydroxyproline and enriched in acidic amino acids, although the general amino acid compositions of these samples still somewhat resemble that of collagen (Bada et al 1984). This indicates that the collagenous matter in these bones has undergone severe diagenesis. It is not known what factors have contributed to the difference in the extent of collagen preservation in the California paleoindian skeletons of approximately the same age, but they probably include leaching by groundwaters and/or heating effects resulting from either shallow burial depths or cultural burial practices. Also, many of these California paleoindian skeletons come from coastal shell middens in which the presence of carbonates in the burial soils may have accelerated the rate of collagen hydrolysis (Kriausakul & Mitterer 1980).

The results in Figure 5 also indicate that a particular calibration constant should only be used to date other fossil bones having amino acid contents

Table 5 Radiocarbon ages (in yr), organic nitrogen or amino acid contents, and aspartic acid racemization rates for some California paleoindian skeletons

Locality and sample name	Radiocarbon age[a]	% organic nitrogen or amino acids preserved	D/L Asp[g]	k_{asp} $(yr^{-1}) \times 10^5$	References
Southern California					
La Jolla					
W-12/76 #II	8330 ± 160[b]	17[e]	0.19	1.5	King & Bada 1979
SDM 16709	8360 ± 75[b] 8470 ± 140[d] (AMS)	5–11[e]	0.19	1.5	Ike et al 1979, Kessels & Dungworth 1980, Bada et al 1984
Los Angeles	3560 ± 220[b] (AMS) > 23,600[d]	6[f]	0.35	8.3 to ~1.3	Bada et al 1974, Bada & Helfman 1975, Berger et al 1971, Taylor et al 1985
Laguna	5100 ± 250[d] (AMS) > 14,800[b] 17,150 ± 1470[b]	5[f]	0.25	3.6 to 1.1	Berger et al 1971, Bada et al 1974, Bada et al 1984
Yuha	1650 ± 250[c] (AMS) 2820 ± 200[c] (AMS) 3850 ± 250[b] (AMS)	3[f]	0.5	29 to 12	Bada & Finkel 1983, Bischoff & Childers 1979, Payen et al 1978, Stafford et al 1984
San Jacinto	3020 ± 140[c]	0.2[e]	0.42	13	Taylor 1983, Kessels & Dungworth 1980, Bada et al 1979a
Del Mar	5400 ± 120[d] (AMS)	2[f]	0.48	8.4	Bada et al 1974, Bada & Helfman 1975, Bada et al 1984

Northern California					
Stanford I	4830 ± 150[d] (AMS) 4950 ± 130[d] (AMS) 5130 ± 70[b]	40[f]	0.14	1.5	Bada & Helfman 1975, Bada & Masters 1982, Bada et al 1984
Sunnyvale	3600 ± 600[c] (AMS) 4390 ± 600[c] 4650 ± 400[b] (AMS) 4850 ± 400[c] (AMS) 6300 ± 400[d] (AMS)	1[f]	0.5	13 to 7.6	Bada & Helfman 1975, Taylor et al 1983

[a] Accelerator mass spectrometer–based ages are indicated by (AMS).
[b] Radiocarbon age of the HCl-insoluble organic fraction isolated from the bones.
[c] Radiocarbon age of the HCl- or NaOH-soluble organic fraction isolated from the bones.
[d] Radiocarbon age of the total amino acid fraction isolated from the bones.
[e] Based on amino acids; modern bone = 1.7 to 2.5 mmol of amino acids g^{-1} (Kessels & Dungworth 1980, Hare 1980).
[f] Based on organic nitrogen; modern bone = 5.2% organic nitrogen (Miller & Martin 1968).
[g] The samples were all hydrolyzed for 24 hr; thus the $(\text{D/L Asp})_{t=0}$ value is 0.07 (Bada & Protsch 1973).

similar to that of the calibration samples. Previous studies have shown that the calibration and dated samples should have similar amino acid composition (King & Bada 1979, Kessels & Dungworth 1980). The use of a calibration sample that has an amino acid content and composition vastly different than that of the dated samples could yield anomalous racemization ages.

The racemization results of the California paleoindian skeletons demonstrate that one of the essential factors in determining the validity of the racemization date for any particular fossil bone is the reliability of the age estimate of the calibration sample. An erroneously dated calibration sample will obviously yield erroneous racemization dates. If the Laguna and Los Angeles Man skeletons had been originally assigned the Holocene ages recently determined using the AMS-based radiocarbon method, skeletons such as those from Del Mar and Sunnyvale never would have yielded racemization ages of great antiquity.

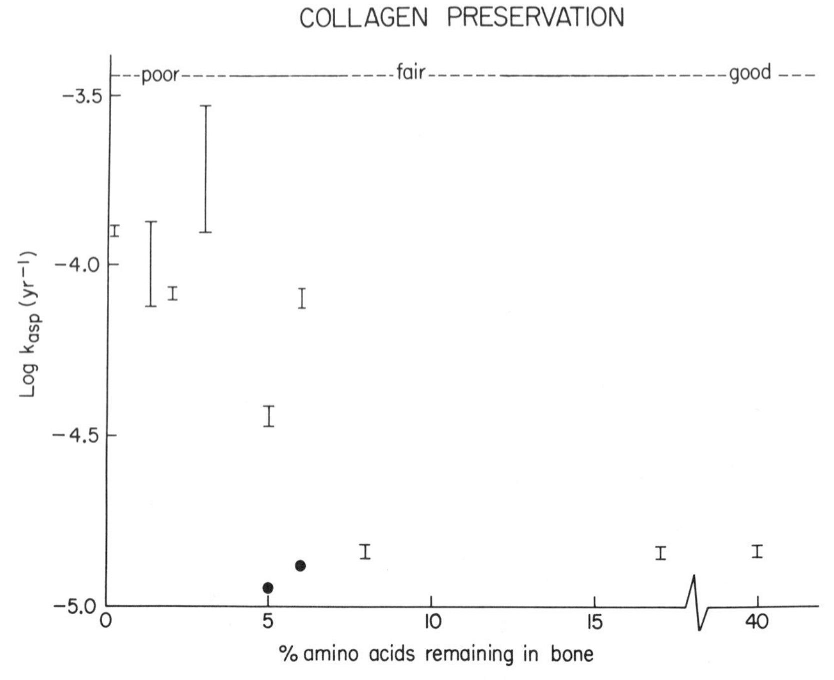

Figure 5 The log k_{asp} values for the various California paleoindian skeletons listed in Table 5 plotted vs the percentage of amino acids remaining in the bones; ● indicates the log k_{asp} values for the Laguna and Los Angeles Man skeletons derived using radiocarbon dates obtained ~15 years ago by conventional β-counting methods.

CONCLUSIONS

The ages derived from the extent of amino acid racemization in a fossil bone seem generally to be reasonably accurate, and this method provides an important geochronometer that can be used to estimate the age of fossil bones and teeth over a time period ranging from a few thousand to several million years. Since the racemization method utilizes an original constituent of the bones (i.e. amino acids), it provides a direct age indication of a fossil bone or tooth, whereas radiometric methods other than ^{14}C can generally only be used to date associated deposits and not the fossil itself.

There are some instances where racemization ages are older than those derived from radiocarbon measurements. This discrepancy apparently results either from problems with the application of radiocarbon dating to fossil bones having low organic carbon contents or from the use of unreliably dated bones for "calibration" of the racemization rate. These problems should eventually be resolved as the AMS-based radiocarbon dating method is more widely used.

Although the racemization reaction of several different amino acids has been studied in fossil bones, only aspartic acid and isoleucine have generally been used for age calculations. The racemization reactions of other amino acids should be utilized in the future, since the reliability of a racemization date would be enhanced if the extent of racemization of several amino acids were used to calculate an average racemization age. The amino acid 4-hydroxyproline, which is unique to collagen, has received only limited use in dating, even though this amino acid is one of the least prone to contamination problems. Some preliminary rate studies indicate 4-hydroxyproline has a racemization rate in bones that exceeds that of isoleucine but is less than that of aspartic acid (Bada & Schroeder 1975, Schroeder 1974). Thus, the racemization dates derived using this amino acid would provide important checks on those derived from other amino acids.

Literature Cited

Armstrong, W. G., Halstead, L. B., Reed, F. B., Wood, L. 1983. Fossil proteins in vertebrate calcified tissues. *Philos. Trans. R. Soc. London Ser. B* 301:301–43

Bada, J. L. 1972. The dating of fossil bones using the racemization of isoleucine. *Earth Planet. Sci. Lett.* 15:223–31

Bada, J. L. 1981. Racemization of amino acids in fossil bones and teeth from the Olduvai Gorge Region, Tanzania, East Africa. *Earth Planet. Sci. Lett.* 55:292–98

Bada, J. L. 1982. Racemization of amino acids in nature. *Interdisc. Sci. Rev.* 7:30–46

Bada, J. L. 1984a. In vivo racemization in mammalian proteins. *Methods Enzymol.* 106:98–115

Bada, J. L. 1984b. Racemization of amino acids. In *Chemistry and Biochemistry of the Amino Acids*, ed. G. C. Barrett, pp. 399–414. London: Chapman & Hall

Bada, J. L., Deems, L. 1975. Accuracy of dates beyond the C^{14} dating limit using the aspartic acid racemization reaction. *Nature* 255:218–19

Bada, J. L., Finkel, R. 1983. The upper Pleistocene peopling of the New World:

evidence derived from radiocarbon, amino acid racemization and uranium series dating. In *Quaternary Coastlines and Marine Archaeology*, ed. P. M. Masters, N. C. Flemming, pp. 463–79. New York/London: Academic

Bada, J. L., Helfman, P. M. 1975. Amino acid racemization dating of fossil bones. *World Archaeol.* 7:160–73

Bada, J. L., Man, E. H. 1980. Amino acid diagenesis in Deep Sea Drilling Project cores: kinetics and mechanisms of some reactions and their applications in geochronology and in paleotemperature and heat flow determinations. *Earth Sci. Rev.* 16:21–55

Bada, J. L., Masters, P. M. 1982. Evidence for an ~50,000 year antiquity of man in the Americas derived from amino acid racemization of human skeletons. In *Peopling of the New World*, ed. J. E. Ericson, R. E. Taylor, R. Berger, pp. 171–79. Los Altos, Calif: Ballena

Bada, J. L., Protsch, R. 1973. The racemization reaction of aspartic acid and its use in dating fossil bones. *Proc. Natl. Acad. Sci. USA* 70:1331–34

Bada, J. L., Schroeder, R. A. 1975. Amino acid racemization reactions and their geochemical implications. *Naturwissenschaften* 62:71–79

Bada, J. L., Shou, M.-Y. 1980. Kinetics and mechanism of amino acid racemization in aqueous solutions and in bones. See Hare et al 1980, pp. 235–55

Bada, J. L., Kvenvolden, K. A., Peterson, E. 1973. Racemization of amino acids in bones. *Nature* 245:308–10

Bada, J. L., Schroeder, R. A., Carter, G. 1974. New evidence for the antiquity of man in North America deduced from aspartic acid racemization. *Science* 184:791–93

Bada, J. L., Hoopes, E., Darling, D., Dungworth, G., Kessels, H. J., et al. 1979a. Amino acid racemization dating of fossil bones, I. Interlaboratory comparisons of racemization measurements. *Earth Planet. Sci. Lett.* 43:265–68

Bada, J. L., Masters, P. M., Hoopes, E., Darling, D. 1979b. The dating of fossil bones using amino acid racemization. In *Radiocarbon Dating*, ed. R. Berger, H. Suess, pp. 740–56. Los Angeles/Berkeley: Univ. Calif. Press

Bada, J. L., Gillespie, R., Gowlett, J. A. J., Hedges, R. E. M. 1984. Accelerator mass spectrometry based radiocarbon ages of amino acid extracts from Californian paleoindian skeletons. *Nature* 312:442–44

Behrensmeyer, A. K. 1978. Taphonomic and ecologic information from bone weathering. *Paleobiology* 4:150–62

Belluomini, G. 1981. Direct aspartic acid racemization dating of human bones from archaeological sites of central southern Italy. *Archaeometry* 23:125–37

Berger, P., Protsch, R., Reynolds, R., Rozaire, C., Sackett, J. R. 1971. New radiocarbon dates based on bone collagen of California paleoindians. *Contrib. Univ. Calif. Archaeol. Res. Facil. (Berkeley) No. 12*, pp. 43–49

Bernardi, G., Giro, M.-G., Gaillard, C. 1972. Chromatography of polypeptides and proteins on hydroxyapatite columns: some new developments. *Biochim. Biophys. Acta* 278:409–20

Bischoff, J. L., Childers, W. M. 1979. Temperature calibration of amino acid racemization: age implications for the Yuha skeleton. *Earth Planet. Sci. Lett.* 45:172–80

Bischoff, J. L., Rosenbauer, R. J. 1981. Uranium series dating of human skeletal remains from the Del Mar and Sunnyvale sites, California. *Science* 213:1003–5

Butzer, K. W. 1971. *Environment and Archaeology*. Hawthorne, NY: Aldine. 703 pp.

Dungworth, G., Vincken, J. J., Schwartz, A. W. 1974. Compositions of fossil collagens: analysis by gas-liquid chromatography. *Comp. Biochem. Physiol. B* 47:391–99

Engel, M. H., Zumberge, J. E., Nagy, B., Van Devender, T. R. 1978. Variations in aspartic acid racemization in uniformly preserved plants about 11,000 years old. *Phytochemistry* 17:1559–62

Fleming, S. 1976. *Dating in Archaeology*. London: Dent. 272 pp.

Gerow, B. A. 1981. Amino acid dating and early man in the new world: a rebuttal. *Soc. Calif. Archeol. Occas. Pap. Method Theory No. 3*, pp. 1–12

Gorbunoff, M. J. 1984. The interaction of proteins with hydroxyapatite. I–III. *Anal. Biochem.* 136:425–45

Hare, P. E. 1974. Amino acid dating of bone—the influence of water. *Carnegie Inst. Washington Yearb.* 73:576–81

Hare, P. E. 1980. Organic geochemistry of bone and its relation to the survival of bone in the natural environment. In *Fossils in the Making: Vertebrate Taphonomy and Paleoecology*, ed. A. K. Behrensmeyer, A. P. Hill, pp. 208–19. Chicago/London: Univ. Chicago Press

Hare, P. E., Mitterer, R. M. 1966. Nonprotein amino acids in fossil shells. *Carnegie Inst. Washington Yearb.* 65:362–64

Hare, P. E., Turnbull, H. F., Taylor, R. E. 1978. Amino acid dating of Pleistocene fossil materials: Olduvai Gorge, Tanzania. In *Views of the Past: Essays in Old World*

Prehistory and Paleoanthropology, ed. L. G. Freeman, pp. 7–12. The Hague: Mouton

Hare, P. E., Hoering, T. C., King, K., eds. 1980. *Biogeochemistry of the Amino Acids*. New York/Toronto: Wiley. 558 pp.

Harmon, R. S., Mitterer, R. M., Kriausakul, N., Land, L. S., Schwarcz, H. P., et al. 1983. U-series and amino acid racemization geochronology of Bermuda: implications for eustatic sea-level fluctuation over the past 250,000 years. *Palaeogeogr. Palaeoclimatol. Palaeoecol.* 44:41–70

Hay, R. L. 1976. *Geology of the Olduvai Gorge*. Berkeley/Los Angeles: Univ. Calif. Press. 203 pp.

Hedges, R. E. M. 1981. Radiocarbon dating with an accelerator: review and preview. *Archaeometry* 23:3–18

Hedges, R. E. M., Wallace, C. J. A. 1980. The survival of protein in bone. See Hare et al 1980, pp. 35–40

Hoopes, E. A., Peltzer, E. T., Bada, J. L. 1978. Determination of amino acid enantiomeric ratios by gas liquid chromatography of the *N*-trifluoroacetyl-L-prolyl-peptide methyl esters. *J. Chromatogr. Sci.* 16: 556–60

Ike, D., Bada, J. L., Masters, P. M., Kennedy, G., Vogel, J. D. 1979. Aspartic acid racemization and radiocarbon dating of an early milling stone horizon burial in California. *Am. Antiq.* 44:524–30

Karrow, P. F., Bada, J. L. 1980. Amino acid racemization dating of Quaternary raised marine terraces in San Diego County, California. *Geology* 8:200–4

Kessels, H. J., Dungworth, G. 1980. Necessity of reporting amino acid compositions of fossil bones where racemization analyses are used for geochronological applications: inhomogeneities of D/L amino acids in fossil bones. See Hare et al 1980, pp. 527–41

King, K. 1974. Preserved amino acids from silicified protein in fossil radiolaria. *Nature* 252:690–92

King, K., Bada, J. L. 1979. Effect of in situ leaching on amino acid racemization rates in fossil bone. *Nature* 281:135–37

Kriausakul, N., Mitterer, R. M. 1980. Some factors affecting the epimerization of isoleucine in peptides and proteins. See Hare et al 1980, pp. 283–96

Kvenvolden, K. A., Peterson, E., Wehmiller, J., Hare, P. E. 1973. Racemization of amino acids in marine sediments determined by gas chromatography. *Geochim. Cosmochim. Acta* 37:2215–25

Lajoie, K. R., Peterson, E., Gerow, B. A. 1980. Amino acid bone dating: a feasibility study, southern San Francisco Bay region, California. See Hare et al 1980, pp. 477–89

Li, R.-W., Lin, D.-X. 1979. Geochemistry of amino acids of fossil bones from deposits of Peking Man, Lantian Man and Yuanno Man in China. *Sci. Geol. Sin. No. 1*, pp. 56–62 (In Chinese)

Lowenstein, J. M. 1980. Immunospecificity of fossil collagens. See Hare et al 1980, pp. 41–51

Manning, J. M., Moore, S. 1968. Determination of D- and L-amino acids by ion exchange chromatography as L-D and L-L dipeptides. *J. Biol. Chem.* 243:5591–97

Masters, P. M. 1982a. Amino acid racemization in structural proteins. In *Biological Markers of Aging*, ed. M. E. Reff, E. L. Schneider, pp. 120–37. *NIH Publ. No. 82-2221*, US Dep. Health Hum. Serv., Washington DC

Masters, P. M. 1982b. An amino acid racemization chronology for Tabun. In *The Transition from Lower to Middle Paleolithic and the Origin of Modern Man*, ed. A. Ronen, pp. 43–53. Oxford: Br. Archaeol. Rep. (BAR)

Matsu'ura, S., Ueta, N. 1980. Fraction dependent variation of aspartic acid racemization age of fossil bone. *Nature* 286:883–84

Matter, P., Miller, H. W. 1972. The amino acid composition of some Cretaceous fossils. *Comp. Biochem. Physiol. B* 43:55–66

McMenamin, M. A. S., Blunt, D., Kvenvolden, K. A., Miller, S. E., Marcus, L. F., et al. 1982. Amino acid geochemistry of fossil bones from Rancho La Brea asphalt deposits, California. *Quat. Res.* 18:174–83

Miller, E. J., Martin, G. R. 1968. The collagen of bone. *Clin. Orthop. Relat. Res.* 59:195–232

Miller, G. H., Hare, P. E. 1980. Amino acid geochronology: integrity of the carbonate matrix and potential of molluscan fossils. See Hare et al 1980, pp. 415–43

Mitterer, R. M. 1972. Calcified proteins in the sedimentary environment. In *Advances in Organic Geochemistry, 1971*, ed. H. R. Von Gaertner, H. Wehner, pp. 441–51. Oxford: Pergamon

Mitterer, R. M. 1975. Ages and diagenetic temperatures of Pleistocene deposits of Florida based on isoleucine epimerization in *Mercenaria*. *Earth Planet. Sci. Lett.* 28:275–82

Neuberger, A. 1948. Stereochemistry of amino acids. *Adv. Protein Chem.* 4:298–383

Payen, L. A., Rector, C. H., Ritter, E., Taylor, R. E., Ericson, J. E. 1978. Comments on the Pleistocene age assignment and associations of a human burial from the Yuha desert, California. *Am. Antiq.* 43:448–53

Schroeder, R. A. 1974. *Kinetics, mechanism and geochemical applications of amino acid racemization in various fossils.* PhD thesis. Univ. Calif., San Diego. 275 pp.

Schroeder, R. A., Bada, J. L. 1973. Glacial-postglacial temperature difference deduced from aspartic acid racemization in fossil bones. *Science* 182:479–82

Scott, W. E., McCoy, W. D., Shroba, R. S., Rubin, M. 1983. Reinterpretation of the exposed record of the last two cycles of Lake Bonneville, Western United States. *Quat. Res.* 20:261–85

Shou, M.-Y. 1979. *Kinetics and mechanisms of several amino acid diagenetic reactions in aqueous solutions and in fossils.* PhD thesis. Univ. Calif., San Diego. 144 pp.

Skelton, R. R. 1983. *Amino acid racemization dating: a test of its reliability for North American archaeology.* PhD thesis. Univ. Calif., Davis. 343 pp.

Stafford, T. W., Jull, A. J. T., Zabel, T. H., Donahue, D. J., Duhamel, R. C., et al. 1984. Holocene age of the Yuha burial: direct radiocarbon determinations by accelerator mass spectrometry. *Nature* 308:446–47

Steinberg, S. M. 1982. *Part I: The marine organic chemistry of α-keto acids and oxalic acid; Part II: The chemical decomposition reactions of proteins in calcareous fossils and their effects on amino acid based geochronological methods.* PhD thesis. Univ. Calif., San Diego. 397 pp.

Steinberg, S. M., Bada, J. L. 1983. Peptide decomposition in the neutral pH region via the formation of diketopiperazines. *J. Org. Chem.* 48:2295–98

Szabo, B. J. 1980. Results and assessment of uranium series dating of vertebrate fossils from Quaternary alluviums in Colorado. *Arct. Alp. Res.* 12:95–100

Taylor, R. E. 1983. Non-concordance of radiocarbon and amino acid racemization deduced age estimates on human bone. *Radiocarbon* 25:647–54

Taylor, R. E., Payen, L. A. Gerow, B., Donahue, D. J., Zabel, T. H., et al. 1983. Middle Holocene age of the Sunnyvale human skeleton. *Science* 220:1271–73

Taylor, R. E., Payen, L. A., Prior, C. A., Slota, P. J., Gillespie, R., et al. 1985. Major revisions in the Pleistocene age assignments for North American human skeletons by ^{14}C accelerator mass spectrometry: none older than 11,000 ^{14}C years B.P. *Am. Antiq.* In press

Tuross, N., Eyre, D. R., Holtrop, M. E., Glimcher, M. J., Hare, P. E. 1980. Collagen in fossil bones. See Hare et al 1980, pp. 53–63

van Kleef, F. S. M., DeJong, W. W., Hoenders, H. J. 1975. Stepwise degradations and deamidation of eye lens protein α-crystallin during ageing. *Nature* 258: 264–66

Weiner, S., Kustanovich, Z., Gil-Av, E., Taub, W. 1980. Dead Sea Scroll parchments: unfolding of the collagen molecules and the racemization of aspartic acid. *Nature* 287:820–23

White, E. M., Hannus, L. A. 1983. Chemical weathering of bone in archaeological soils. *Am. Antiq.* 48:316–22

Wyckoff, R. W. G. 1972. *The Biochemistry of Animal Fossils.* Bristol, Engl: Scientechnia. 152 pp.

Zumberge, J. E., Engel, M. H., Nagy, B. 1980. Amino acids in bristlecone pine: an evaluation of factors affecting racemization rates and paleothermometry. See Hare et al 1980, pp. 503–25

Ann. Rev. Earth Planet. Sci. 1985. 13 : 269–96

MANTLE METASOMATISM

Michael F. Roden[1] *and V. Rama Murthy*

Department of Geology and Geophysics, University of Minnesota, Minneapolis, Minnesota 55455

INTRODUCTION

The idea of "mantle metasomatism" was explicitly introduced by Bailey (1970, 1972) to explain certain features of the highly alkaline magmatism associated with uplifted and rifted continental regions. Since then, numerous studies have shown that metasomatic changes in mantle rocks are common and may play a key role in determining the chemical and isotopic characteristics of magmas derived from the mantle. This article reviews the current status of this concept and summarizes the evidence concerning the nature and timing of metasomatism in the mantle. We also attempt to demonstrate the relevance of these data to the fundamental problem of alkali basalt genesis and the mass-balance estimates of incompatible elements in the Earth.

In discussing mantle metasomatism, we retain here the classical definition of a process wherein both "infiltration" and "diffusion" transport mobile components from one region to another (e.g. Thompson 1959, Korzhinskii 1965). Since the transport of mobile components is involved, the affected region can either be "enriched" or "depleted" in these components relative to the premetasomatic state, depending on whether these mobile components were added or removed from the region. In the following, we use "mantle metasomatism" to refer only to the *enrichment* process, whereby the abundances of incompatible trace elements [specifically, K, Rb, Sr, Ba, light rare earth elements (LREE), Ti, Nb, Zr, P, U, and Th) are increased relative to the presumed primitive (Sun 1982) or depleted (Jagoutz et al 1979) mantle precursor. It is important to note, however, that there is also evidence for the opposite process—a *depletion* of incompatible trace elements in wall-rock peridotite adjacent to pyroxene-rich dikes (e.g. Boudier & Nicolas 1977).

[1] Current address: Department of Geology, University of Georgia, Athens, Georgia 30602

0084–6597/85/0515–0269$02.00

Although the results of mantle metasomatism (i.e. the increase in abundance of incompatible trace elements in depleted mantle) are well described, the actual nature of the process, as well as its implications for the distribution of trace elements in the Earth, remains uncertain. For example, is the metasomatizing agent a silicate melt or a H_2O–CO_2 fluid? Recent experimental evidence indicates that fluids, especially H_2O-rich fluids, can be effective transport media for incompatible elements at high pressures (Wendlandt & Harrison 1979, Mysen 1983, Schneider & Eggler 1984). A further question of more fundamental importance is whether the metasomatism of depleted peridotite creates the source material of alkalic lavas (e.g. Wass & Rogers 1980, Menzies & Murthy 1980a), or whether metasomatism is a more local phenomenon of relatively small volume that is related to the intrusion and crystallization of basalts in the upper mantle (e.g. Wyllie 1980, Wilshire et al 1980). If the latter model is correct, then metasomatism cannot explain the high incompatible element abundances of alkali basalts. A corollary of the above question is whether or not metasomatized mantle material is a volumetrically important reservoir for these elements, which include the significant heat-producing elements as well as many of their daughter isotopes.

Before proceeding, we first note that several recent reviews dealing with the subject of mantle metasomatism have been published (Bailey 1982, 1984, Dawson 1980, 1984, Frey 1984, Wilshire 1984, Harte 1983, Menzies 1983, Hawkesworth et al 1983).

TERMINOLOGY AND SOME ASSUMPTIONS

A fundamental assumption that underlies the studies of mantle processes and specifically mantle metasomatism is that the chemistry and textures of mantle-derived rocks (i.e. basalts and their ultramafic inclusions, and tectonically emplaced peridotite massifs) can be used to deduce the nature of mantle processes. Typical mantle-derived inclusions that occur in alkalic basalts and kimberlites are oblate bodies of peridotite or pyroxenite less than a meter across (e.g. Frey & Prinz 1978). These inclusions contain minerals indicative of equilibration under pressures and temperatures characteristic of the upper mantle. Because of their rapid transport to the surface and rapid cooling (i.e. chemical quenching), ultramafic inclusions provide the most reliable evidence regarding mantle chemistry. Thus, in the following we emphasize data from inclusions, although where possible we integrate evidence from alkalic basalts and peridotite massifs. (Note that "xenolith" and "nodule" are used interchangeably with "inclusion" in this paper and in the literature.)

We use here the terminology of Dawson (1984) because of its descriptive nature that does not entail any genetic interpretations. "Patent metasomatism" refers to metasomatism that is texturally evident, i.e. replacement of primary phases by metasomatic phases is evident. Note that patent metasomatism is essentially synonymous with the "modal metasomatism" of Harte (1983).

Commonly, patent metasomatism in inclusions is evidenced by the presence of hydrous (e.g. amphibole, phlogopite) and/or anhydrous (e.g. zircon, apatite, rutile, crichtonite-series minerals) phases rich in incompatible elements. (Incompatible elements are those that are relatively concentrated in the melt or fluid phase compared with equilibrium solid phases.) Thus, the incompatible-element-rich nature of the inclusion is an intrinsic feature of its constituent phases (e.g. Frey & Prinz 1978, Stosch & Seck 1980). Moreover, textural as well as compositional data indicate that these metasomatic phases formed in the mantle and are not in equilibrium with the host lava (e.g. Harte 1983, Menzies 1983). Consequently, the enrichment event (i.e. metasomatism) is a mantle process.

In contrast to patent metasomatism, "cryptic metasomatism" entails no textural evidence; however, there is a chemical signature that requires a metasomatic enrichment of incompatible trace elements in peridotite nodules, which on the basis of their major element chemistry show extreme depletion of the "basaltic" component. In other words, there is a decoupling of the trace element chemistry from the major element chemistry.

Korzhinskii (1970) stated that "if the chemical composition of a rock is altered when its minerals are replaced by others, the processes are said to be metasomatic." Consequently, there are two measurable effects of a metasomatic process: (*a*) a change in the number or proportion of phases, and (*b*) a change in the bulk composition of the system. In order to recognize metasomatism, it is necessary to assume a primary mineral assemblage as well as the initial chemical composition of the system. Numerous petrologic studies (summarized in Ringwood 1975) of phase equilibria in model systems, peridotite massifs, and ultramafic inclusions indicate that the upper 200 km of the mantle is comprised predominantly of Mg-rich lherzolite and harzburgite. These rocks are composed of Mg-rich olivine, enstatite, and lesser amounts of diopside and Cr-rich spinel or pyrope. [The nature of the aluminous phase is a function of pressure (e.g. Ringwood 1975).] Thus, we assume that the protolith prior to mantle metasomatism consisted of these minerals in proportions appropriate for spinel or garnet lherzolite, i.e. olivine > enstatite > diopside + spinel/pyrope (at pressures exceeding the upper stability limit, approximately 6–8 kbar, of plagioclase in peridotite).

Although lherzolite and harzburgite are probably the dominant rock type in the upper mantle, evidence from mantle-derived xenoliths as well as peridotite massifs show that this lherzolite/harzburgite "wall rock" is veined by more Fe- and pyroxene-rich rocks (Wilshire & Pike 1975, Irving 1980, Wilshire et al 1980). Some of these pyroxenites contain amphibole or phlogopite (e.g. Francis 1976a, Wilshire et al 1980, Frey & Prinz 1978) and may be intimately associated with metasomatism of the peridotite wall rock (Wilshire et al 1980, Irving 1980, Kempton et al 1984, Roden et al 1984a, Wilshire 1984). In this review, we emphasize metasomatism of peridotite wall rock; we implicitly assume that some pyroxenites may be causal agents of this metasomatism, and that others may predate the metasomatic event and thus may themselves be metasomatized (Menzies 1983).

In addition to textural criteria for metasomatism, there must be chemical and isotopic criteria. In particular, an estimate of the composition of the protolith is required. We assume that the bulk of the upper mantle represents a residue from partial melting of *primitive* mantle material (e.g. McKenzie & O'Nions 1983). Numerous studies of peridotite massifs (e.g. Loubet et al 1975, Frey 1984), ultramafic xenoliths (e.g. Jagoutz et al 1979), and models for the source of mid-ocean ridge basalts (Gast 1968, Schilling 1975) support this contention. Models for primitive mantle composition (Table 1; Wood et al 1979, Ringwood 1979, Sun 1982, Taylor 1982) indicate that it had approximately twice the ordinary chondrite abundances of the refractory incompatible trace elements [e.g. rare earth elements (REE), Sr, Ba, etc] and a flat, chondrite-normalized REE abundance pattern (e.g. Sun 1982). In contrast, the primitive mantle was depleted in the volatile incompatible elements (Rb, K, Pb, etc) relative to chondrites. This primitive mantle composition sets upper limits for the abundances of incompatible elements in the protolith prior to metasomatism, on the assumption that most of the upper mantle represents a residue from prior partial-melting episodes of the primitive mantle (Figure 1).

PHASE ASSEMBLAGES OF METASOMATIZED INCLUSIONS

Only phase assemblages in inclusions exhibiting patent metasomatism are distinctive. Most commonly, amphibole and/or phlogopite are present. In discussing the mineralogy of such inclusions, it is convenient to treat inclusions from alkalic basalts separately from inclusions in kimberlites because the phases present, as well as their compositions, differ between the two suites.

Amphibole, typically a pargasite (Dawson & Smith 1982), is by far the preeminent phase in patently metasomatized peridotite inclusions from

Table 1 Incompatible element abundances (in ppm) in various mantle materials

	Primitive mantle[a]	"Depleted" but fertile mantle[b]	Metasomatized mantle (*n*)[c]		
Rb	0.48–0.86	0.02–0.31	NUN	0.042–2.83	(8)
			STP	0.079–3.36	(12)
Ba	4.9–7.6	2.3–6.1	NUN	3.8–42	(8)
K	180–252	8.1–140	NUN	80–1070	(8)
			STP	63–970	(11)
La	0.50–0.71	0.051–0.51	NUN	0.29–8.3	(8)
			DW	1.6–4.2	(8)
			VIC	0.88–3.4	(3)
			SC	0.98–1.4	(4)
			STP	0.58–8.1	(10)
Sr	16–23	5.9–28	NUN	12–82	(8)
			STP	8.0–43	(10)
Yb	0.34–0.44	0.32–0.50	NUN	0.037–0.40	(8)
			STP	0.088–1.1	(10)
			DW	0.068–0.19	(8)
			VIC	0.024–0.12	(3)
			SC	0.083–0.27	(4)
Rb/Sr	0.030–0.037	0.0034–0.029	NUN	0.0027–0.12	(10)
			STP	0.012–0.16	(7)
			ATQ	0.012–0.064	(5)
Sm/Nd	0.30–0.33	greater than chondrites[d]	NUN	0.12–0.25	(7)
			STP	0.077–0.31	(7)
			DW	0.084–0.20	(8)
La/Yb][e]	1	0.12–0.79	NUN	3.3–29	(8)
			STP	2.4–26	(9)
			DW	5.7–23	(8)
			SC	2.1–7.9	(4)
			VIC	19–25	(3)

[a] Values from Sun (1982), Taylor (1982), and Wood et al (1979).

[b] Range of abundances in six "primitive" lherzolite inclusions from alkalic basalts (Jagoutz et al 1979). These peridotites are relatively rich in clinopyroxene and spinel; consequently, they are fertile with respect to basaltic components. However, these peridotites have also lost a component (small-percent melt or fluid) rich in incompatible elements, and thus they are depleted in these elements with respect to primitive mantle.

[c] Range of abundances in metasomatized mantle-derived peridotites; number of samples indicated in parentheses. NUN = Nunivak Island, Alaska (Menzies & Murthy 1980b, Roden et al 1984a); STP = St. Paul's Rocks, equatorial Atlantic (Frey 1970, Melson et al 1972, M. K. Roden et al 1984); DW = Dreiser Weiher, West Germany (Stosch & Seck 1980); VIC = Victoria, Australia (Frey & Green 1974); SC = San Carlos, Arizona (Frey & Prinz 1978); ATQ = Ataq, South Yemen (Menzies & Murthy 1980b).

[d] Nd abundances interpolated.

[e] Chondrite-normalized (Evensen et al 1978) ratio.

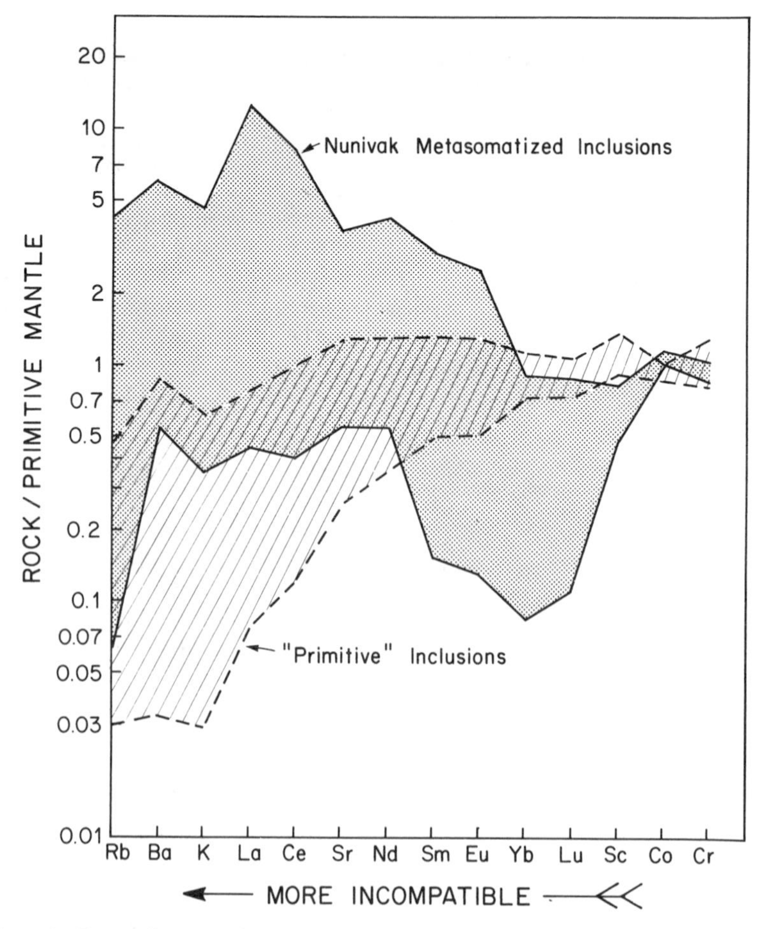

Figure 1 Ranges for trace element abundances in "primitive" spinel peridotite inclusions from alkalic basalts (Jagoutz et al 1979) and in metasomatized spinel peridotite inclusions from Nunivak Island, Alaska (Roden et al 1984a) normalized to primitive mantle abundances (Sun 1982). The Nunivak inclusions are a typical example of metasomatized inclusion suites. Primitive inclusions are clinopyroxene-rich and, consequently, fertile with respect to basaltic components. However, they have lost an incompatible-element-rich component (small-percent melt or fluid) and thus have lower abundances of the highly incompatible elements (e.g. Rb, Ba, K, La) than those of the primitive mantle. These inclusions are probably representative of depleted (but fertile) mantle, an important constituent of the uppermost mantle in oceanic and many continental regions. Although the abundance ranges for the metasomatized inclusions overlap the ranges for the primitive inclusions, the metasomatized inclusions have higher ratios of the highly incompatible to moderately incompatible elements (e.g. K/Yb, La/Yb) than those of either the primitive inclusions or primitive mantle. Moreover, many of the metasomatized inclusions have higher concentrations of the highly incompatible elements than those of primitive mantle.

alkalic basalts (e.g. Varne & Graham 1971, Francis 1976b, Wass & Rogers 1980, Kurat et al 1980). This phase may exhibit obvious replacement textures with respect to primary silicates and oxides (e.g. Francis 1976b, Kempton et al 1984). Phlogopite or biotite is also a common metasomatic phase (Francis 1976b, Ehrenberg 1982a, Frey & Green 1974, Frey & Prinz 1978, Menzies & Murthy 1980b, Reid et al 1975); typically, this mica is Mg- and Ti-rich [$Mg/(Mg+Fe^{2+}) = 0.5$–0.9, TiO_2 greater than 1 wt% (Bachinski & Simpson 1984)]. Apatite occurs in inclusions from a number of localities; this phase is usually associated with amphibole (Wilshire & Trask 1971, Frey & Green 1974, Wass & Rogers 1980, Roden et al 1984a). In addition to solid phases, fluid inclusions within silicate phases are common in metasomatized ultramafic inclusions (e.g. Zindler & Jagoutz 1984); this fluid is presumed to be carbon rich (e.g. Bergman & Dubessy 1984, and references therein). Less commonly reported phases from metasomatized alkalic basalt inclusions include sphene (Lloyd & Bailey 1975), titanomagnetite (Lloyd & Bailey 1975), ilmenite (Ehrenberg 1982a), spinel (Kempton et al 1984), and an unidentified Mn- and Ba-rich phase (Kurat et al 1980). Consequently, a "typical" metasomatic inclusion from an alkalic basalt contains pargasitic amphibole with lesser amounts of Mg- and Ti-rich mica, apatite, a carbon-rich fluid $\pm$ sphene, ilmenite, and titanomagnetite.

By contrast, ultramafic inclusions from kimberlites that exhibit patent metasomatism typically have phlogopite as a modally important hydrous phase. The mica is commonly more Mg-rich and Ti-poor than the micas from alkalic basalt inclusions (Bachinski & Simpson 1984). Amphibole, when present, is usually Al-poor richterite, in contrast to the Al-rich pargasite typical of alkalic basalt inclusions (Dawson & Smith 1982). In general, the phase assemblage tends to be more variable, and a number of rare phases occur that have not been reported from alkalic basalt inclusions. For example, amphibole-phlogopite veins in peridotite inclusions from Bultfontein Mine, South Africa, contain a Ca–Zr–Cr–titanate and a Ba- and Sr-rich rutile (Erlank & Rickard 1977, Jones et al 1982). Two distinct metasomatic events appear to be recorded: an early, relatively low-Ti event that precipitated K-rich, low-Ti mica (0.44–0.86% TiO_2), Ba–Sr–Ca–K–Cr–Zr titanate of the crichtonite series, Cr–spinel, and Mg–ilmenite; and a later high-Ti event that precipitated Ti-rich mica (1.4–4.4% TiO_2) and Zr–Fe^{3+}–Cr–Ba–Sr-rich rutile (Jones et al 1982). A series of crichtonite-structure minerals have also been described from the Bultfontein, Jagersfontein, and De Beers (all South Africa) kimberlites (Haggerty et al 1983, Haggerty 1983). These phases are commonly extremely rich in Ba, Sr, and the LREE (e.g. Jones et al 1982). Moreover, associated with crichtonites in the heavy mineral separates from the

Jagersfontein kimberlite are Zr–Nb–Ba-rich armalcolite and an unidentified Zr-bearing silicate containing 2 wt% REE (Haggerty 1983). These latter minerals are from heavy mineral concentrates, and consequently the link between these minerals and mantle metasomatism is based on their association with crichtonite-series minerals.

In metasomatized inclusions from the Matsoku kimberlite pipe (Lesotho), veins and disseminated grains of sulfides as well as phlogopite, ilmenite, and rutile are present (Harte et al 1975). The sulfides (pyrrhotite, pentlandite, and chalcopyrite) are typically intergrown in single grains. Sulfides also occur in the matrix of polymict breccias from the Bultfontein and De Beers kimberlites (Lawless et al 1979). The clasts include a variety of peridotites, eclogites, and megacrysts cemented by ilmenite, phlogopite, rutile, and sulfides. These remarkable inclusions are interpreted to be samples of old conduit fillings from the mantle (Lawless et al 1979); the mineralogy of the matrix minerals resembles that in other metasomatized inclusions.

Another group of xenoliths possibly related to mantle metasomatism are the mica-amphibole-rutile-ilmenite-diopside $\pm$ zircon-bearing (MARID) inclusions from kimberlites (Dawson & Smith 1977, Dawson 1980, Jones et al 1982). Phlogopite dominates these inclusions and is compositionally distinct from primary-textured phlogopites in peridotites: the MARID micas have higher FeO and lower NiO and Cr_2O_3 contents than the peridotite micas (Dawson & Smith 1977). As for rutiles associated with crichtonite-series minerals, MARID rutiles contain significant amounts of Cr_2O_3 and Fe_2O_3, and they may also contain up to 5 wt% Nb_2O_5 (Dawson & Smith 1977). Rare sphene and an unidentified Ca-titanate also occur (Dawson & Smith 1977). The presence of graded bedding and banding in some MARID inclusions indicates that these inclusions probably formed as crystal segregates, possibly from a kimberlitic magma (Dawson & Smith 1977). Dawson (1980) suggested that a continuum may exist between MARID inclusions and metasomatized peridotites, with the peridotites representing wall rock somewhat removed from the magma chamber in which the MARID inclusions crystallized. Also, Jones et al (1982) favored a possible genetic link between MARID inclusions and veined peridotites from Bultfontein.

An unusual hydrous and possibly metasomatic assemblage occurs in spinel peridotite inclusions from the Green Knobs (New Mexico) and Buell Park (Arizona) ultramafic breccia pipes (Smith & Levy 1976, Smith 1979). Chlorite, amphibole, titanoclinohumite, antigorite, and magnesite are all present and appear to reflect increasing hydration of peridotite at temperatures less than 700°C at 45–60 km depth. Consistent with an origin in the upper mantle for these minerals is the occurrence of similar phases as inclusions in garnet xenocrysts from several other breccia pipes in the same volcanic field (McGetchin & Besancon 1973, Hunter & Smith 1981).

TRACE ELEMENT ABUNDANCES IN METASOMATIZED INCLUSIONS

Abundances of incompatible trace elements in peridotite inclusions from alkalic basalts were particularly instrumental in the formation of the mantle metasomatic model. The most important observations were that (*a*) incompatible trace element abundances could not be predicted from major element abundances (e.g. Frey 1984, Zindler & Jagoutz 1984), and (*b*) incompatible trace elements were commonly greater in abundance in the inclusions than in models for primitive mantle (Figure 1, Table 1; e.g. Frey 1984). A two-component mixing model was developed to explain these observations, in which one of the components is a metasomatic component and the other is residual mantle that has had a partial melt extracted (e.g. Varne & Graham 1971, Frey & Green 1974).

LREE enrichment relative to primitive mantle or chondrite abundances is particularly diagnostic of mantle metasomatism (Figure 1, Table 1; e.g. Wilshire 1984, Frey 1984). The LREE (e.g. La, Ce) are typically more incompatible in magma-peridotite systems than the heavy REE [HREE (e.g. Yb, Lu); Frey et al 1978]. In peridotites the bulk rock CaO content, which is approximately proportional to modal clinopyroxene, is a measure of fertility (defined as the capacity to produce basaltic magmas) with respect to basaltic components. This contention follows from the observation that clinopyroxene is a major contributor to melts of peridotite (e.g. Mysen 1983). Consequently, as larger and larger amounts of melt are extracted, the residue becomes increasingly depleted in CaO. Hence, the observed modal variation in Mg-rich peridotites from lherzolite to harzburgite to chromite-bearing dunite can be interpreted as a continuum of residues from increasing amounts of partial melting. One would expect that incompatible trace elements such as the REE would decrease in abundance from lherzolite to dunite. Moreover, because the LREE are relatively more incompatible than the HREE in garnet or spinel peridotite–melt systems (e.g. Frey et al 1978), the ratio LREE/HREE should also decrease in the same sequence. Peridotites that are residues from partial melting of primitive mantle with chondritic relative REE abundances will be relatively depleted in LREE, i.e. their La/Yb ratio will be less than the La/Yb ratio in chondrites (Figure 1, Table 1).

The moderately incompatible trace elements such as the HREE follow the above predictions: There is a positive correlation between CaO and HREE abundances in peridotite inclusions (e.g. Frey 1984, Zindler &

Jagoutz 1984). However, the highly incompatible LREE elements exhibit either no correlation or an inverse correlation with CaO (e.g. Frey 1984, Zindler & Jagoutz 1984). This observation is based on data for relatively LREE-enriched spinel peridotites from a number of localities: Assab, Ethiopia (Varne & Graham 1971); Victoria, Australia (Frey & Green 1974), Dreiser Weiher, West Germany (Stosch & Seck 1980), Kapfenstein, Austria (Kurat et al 1980), Nunivak Island, Alaska (Menzies & Murthy 1980b, Roden et al 1984a), and Geronimo volcanic field, Arizona (Kempton et al 1984). Moreover, at a number of these localities the relative LREE enrichment correlates with the presence of amphibole. The observed LREE enrichment cannot be rationalized in a simple partial-melting model in which the peridotites are residues. More complex models are required, and in particular the metasomatic model has been proposed. In essence, this model is a mixing model in which one component, a lherzolitic or harzburgitic residue from partial melting, defines the major element and compatible trace element abundances. A second component, the metasomatic fluid or melt, defines the incompatible element abundance (Frey & Green 1974, Frey & Prinz 1978). This model is supported by the presence of phases, such as amphibole, phlogopite, and apatite, which can be attributed to the reaction of peridotite with the metasomatic fluid or melt. Moreover, recent experimental studies (Wendlandt & Harrison 1979, Mysen 1979, 1983) show that CO_2- and especially H_2O-rich fluids are effective carriers of incompatible trace elements at mantle pressures (Schneider & Eggler 1984). As a consequence of these observations, relative LREE enrichment has become accepted as a signature of metasomatism and allows us to identify peridotites that have been subjected to this process.

Abundances of other incompatible elements besides the LREE are affected by mantle metasomatism. Plots of primitive mantle-normalized abundances versus relative incompatibility show that LREE-enriched peridotites have relatively high abundances of Rb, Ba, K, and Sr (Figure 1, Table 1). Presumably other incompatible elements, such as Th, U, Ta, Nb, P, Hf, and Zr, also increase in abundance as a consequence of metasomatism, although sufficient data are lacking. Models for metasomatized source regions inferred from alkalic basalt compositions are enriched relative to primitive mantle in all the incompatible elements (e.g. Clague & Frey 1982). Furthermore, mica pyroxenites and amphibole-apatite inclusions typically have very high abundances of Nb, Zr, U, Th, and P (Lloyd & Bailey 1975, Wass & Rogers 1980). These inclusions are thought to represent veins in metasomatized peridotite or completely replaced ("ultrametasomatized") peridotite.

The observed enrichment in compatible trace elements could conceivably result from contamination along grain boundaries by host magmas. Alkalic basalt or kimberlite hosts are typically rich in the same elements that distinguish metasomatized peridotite. However, numerous studies of mineral separates from metasomatized inclusions show that high abundances of incompatible elements are an intrinsic feature of the constituent minerals of the peridotite. For example, diopsides from relatively LREE-enriched inclusions are invariably LREE enriched even when acid washed to remove grain surface contaminants (e.g. Varne & Graham 1971, Frey & Green 1974, Frey & Prinz 1978, Roden et al 1984a, Zindler & Jagoutz 1984). When amphibole is present, it is also relatively LREE enriched (Varne & Graham 1971, Wass et al 1980, Roden et al 1984a); moreover, amphibole contains significant amounts of K, Rb, Sr, and Ba, and phlogopite is rich in K, Rb, and Ba (Smith et al 1979, Menzies & Murthy 1980a, Basaltic Volcanism Study Project 1981). The incompatible trace elements may be concentrated in other minor phases when present: REE and Sr in apatite (e.g. Wass et al 1980); Ba, K, Rb, and Cs in fluid inclusions (e.g. Zindler & Jagoutz 1984); LREE, Ti, Zr, Ba, Sr, K, and U in crichtonite-series minerals (Jones et al 1982, Haggerty 1983); Ti, Ba, K, and LREE in armalcolite (Haggerty 1983); Ti, Zr, Ba, Nb, and Ta in rutile (Haggerty 1983); Nb and Ta in ilmenite (Harte 1983); and various incompatible elements in rare and as yet unidentified phases (Kurat et al 1980, Haggerty 1983). Nonetheless, when mass-balance calculations are performed based on measured mineral abundances, commonly it is impossible to account for the total bulk rock abundances of some incompatible elements, especially the alkalies and Ba (e.g. Zindler & Jagoutz 1984). Presumably, the unaccounted-for proportion of these elements resides along grain boundaries; this portion of the incompatible element abundances may or may not be related to the metasomatic event.

Not all amphibole-forming events in the upper mantle are associated with a significant enrichment in incompatible element abundances. For example, some amphibole lherzolites from Kapfenstein, Austria, and Itinome-Gata, Japan, are relatively LREE depleted (Kurat et al 1980, Tanaka & Aoki 1981), and there is no significant LREE enrichment associated with the formation of amphibole in lherzolite from Green Knobs, New Mexico (M. F. Roden & D. Smith, unpublished data).

The discussion so far has focused on trace element abundances in spinel lherzolites from alkalic basalts, but some garnet peridotites from kimberlites are also metasomatized. As in the case of the spinel lherzolites, there is a general correlation between composition and REE pattern: The common coarse-grained and Mg-rich peridotites from South African kimberlites have relatively low contents of Al_2O_3 and CaO (typically less than 2 wt%) and tend to be relatively LREE enriched (Shimizu 1975, Nixon et al 1981, Basaltic Volcanism Study Project 1981, Harte 1983). In contrast, however, garnet peridotites more fertile with respect to basaltic components have flatter REE patterns and may even be relatively LREE depleted (e.g. Nixon et al 1981). The decoupling of the LREE from CaO and Al_2O_3 requires a mixing process, i.e. a metasomatic introduction of LREE into CaO- and Al_2O_3-poor inclusions (Nixon et al 1981). However, the model is based primarily on whole-rock data; mineral data are limited, and moreover, the fractionation of the REE between garnet and clinopyroxene complicates the relation of mineral REE patterns to bulk rock REE patterns. Consequently, alternative explanations such as subsolidus exsolution of garnet from a clinopyroxene-rich cumulate have been proposed to explain the LREE enrichment of anhydrous, Mg-rich inclusions (Shimizu 1975). However, evidence from vein minerals at Bultfontein and from heavy mineral separates from Jagersfontein provides support for metasomatism. These minerals are relatively LREE enriched and have high abundances of other incompatible elements. In particular, the Bulfontein vein minerals are hosted by peridotite; these data show that enrichment events similar to those inferred for spinel lherzolite inclusions have occurred in the mantle sampled by kimberlite (Jones et al 1982, Haggerty 1983).

Gurney & Harte (1980) and Harte (1983) suggest that deformed and relatively "hot," Fe–Ti-rich peridotites from South African kimberlites may also be metasomatized, i.e. "Fe–Ti enrichment." These inclusions tend to have relatively flat REE patterns, but they do exhibit some geochemical features such as high Ti *and* high Cr *but* low Ca contents that appear to require a metasomatic or mixing event (Harte 1983). Ehrenberg (1982a) suggests that compositionally similar sheared garnet peridotites from The Thumb minette, Colorado Plateau, may have also suffered a Fe–Ti metasomatic event; this contention is supported by garnet zoning consistent with an influx of TiO_2 (Smith & Ehrenberg 1984).

Garnet peridotites occur rarely in nonkimberlitic hosts, such as the Lashaine Volcano (Tanzania) ankaramite and The Thumb minette (Dawson et al 1970, Ehrenberg 1979). Some inclusions from both localities are LREE enriched and consequently metasomatized (Ridley & Dawson 1975, Ehrenberg 1982b). For example, sample BD738 from Lashaine is LREE enriched, contains phlogopite, and has a relatively low $^{143}Nd/^{144}Nd$ ratio, indicating that the LREE enrichment is relatively old (Ridley & Dawson 1975, Cohen et al 1984). In addition to the evidence for Fe–Ti metasomatism described above, Ehrenberg (1982b) found that peridotites with the lowest Ca and Al contents are relatively LREE enriched. This LREE enrichment was inferred to result from mantle metasomatism, but mass-balance calculations indicate that the LREE are concentrated along grain boundaries. Consequently, the nature and timing of the enrichment event are uncertain.

Ehrenberg (1982a) and Harte (1983) recognized the generality of three types of metasomatism in garnet peridotite inclusions. These are (*a*) relative LREE enrichments in Ca- and Al-poor, Mg-rich "cold" peridotites, (*b*) Fe–Ti enrichment without LREE enrichment in some sheared peridotites, and (*c*) introduction of phlogopite in some patently metasomatized inclusions.

In summary, evidence from spinel lherzolite inclusions from alkalic basalts indicates that mantle metasomatism is characterized by the formation of phases rich in incompatible trace elements. Relative LREE enrichment and anomalously high abundances of other incompatible elements are characteristic features of metasomatized inclusions. The evidence from garnet peridotites is consistent with this picture but more equivocal because of a lack of sufficient mineral data. The most convincing evidence shows that the formation of vein minerals, including exotic crichtonite-series minerals, was associated with an influx of the LREE and other incompatible elements (e.g. Jones et al 1982).

Whether the crystallization of peridotite plus metasomatic component is an open- or closed-system process is unclear. Stosch & Seck (1980) postulated open-system behavior from the observation that many clinopyroxenes from metasomatized inclusions have La/Ce ratios less than those of chondrites (e.g. Frey & Green 1974, Stosch & Seck 1980, Roden et al 1984a). If highly alkaline mafic melts such as kimberlites and nephelinites provide us with reasonable models for the relative REE abundances in the metasomatic component (e.g. Frey & Green 1974), we would expect this component to have a La/Ce ratio greater than chrondrites and, because of its high abundance of LREE, to dominate the LREE budget of the metasomatized peridotite in a closed-system process. Some metasomatized peridotites do conform to this situation and may reflect a nearly closed-system crystallization of peridotite plus metasomatic component. However, the clinopyroxenes and whole-rock metasomatized peridotites that have La/Ce ratios less than those of chondrites must reflect open-system behavior.

ISOTOPIC SYSTEMATICS AND THE TIMING OF MANTLE METASOMATISM

In the previous section we showed that relative LREE enrichment is a diagnostic signature of mantle metasomatism. A consequence of this enrichment is the fractionation of Sm from Nd, i.e. the Sm/Nd ratio typically decreases during metasomatism. This consistent behavior of Sm relative to Nd makes the ^{147}Sm–^{143}Nd isotopic system particularly useful in constraining the age of mantle metasomatism. Moreover, the geochemistry of both elements is relatively well understood: Both elements are chemically similar and are concentrated in identifiable phases such as clinopyroxene and amphibole (Stosch 1982, Zindler & Jagoutz 1984); they are also relatively nonvolatile and consequently were present in chondritic relative abundances in the primitive mantle.

In principle, any isotopic system in which there is a significant fractionation of parent from daughter element during metasomatism could be used, but the other two isotopic systems, Rb–Sr and U–Pb, commonly applied to ultramafic inclusions suffer from disadvantages. Most importantly, the nature of fractionation between the parent and daughter in metasomatic processes is variable in these systems. Thus some phlogopite-bearing metasomatized peridotites have very high Rb/Sr ratios (e.g. Ridley & Dawson 1975, Hawkesworth et al 1983, Roden et al 1984a), but many other metasomatized peridotites have quite low Rb/Sr ratios that are significantly less than that estimated for primitive mantle (Table 1; e.g. Roden et al 1984a, Menzies & Murthy 1980b). Another serious problem results from the observation that much of the Rb may be contained in fluid inclusions or along grain boundaries in phlogopite-free inclusions (e.g. Zindler & Jagoutz 1984). The origin of the Rb that resides in these sites is uncertain, and consequently the relevance of the measured bulk rock Rb/Sr ratio to metasomatism is unclear.

Use of the U–Pb system to constrain the timing of metasomatic events is hampered by poor knowledge of the behavior of U and Pb during metasomatism, as well as by a relative lack of measurements on mantle inclusions. Limited data indicate that as is the case for Rb–Sr, fractionation of U from Pb during metasomatism is variable: Vollmer & Norry (1983) argued for an increase in the U/Pb ratio in the inferred metasomatized source for the Nyiragongo nephelinites (Zaire); however, Lashaine inclusion BD738, a phlogopite-bearing garnet peridotite with a very low $^{143}Nd/^{144}Nd$ ratio, has very unradiogenic Pb (Cohen et al 1984).

Sr and Nd results for metasomatized inclusions can be divided into two groups that correlate in general with mode of occurrence: Metasomatized xenoliths from alkalic basalts tend to have low $^{87}Sr/^{86}Sr$ and high $^{143}Nd/^{144}Nd$ ratios, whereas metasomatized garnet peridotites from kimberlites tend to have high $^{87}Sr/^{86}Sr$ and low $^{143}Nd/^{144}Nd$ ratios. Xenoliths from alkalic basalts commonly have higher $^{143}Nd/^{144}Nd$ and lower Sm/Nd ratios than chondrites (e.g. Menzies & Murthy 1980b). These features indicate that the system has been LREE enriched for a comparatively short period of time. Consideration of the evolution of $^{143}Nd/^{144}Nd$ with time in the inclusion in relation to mantle evolutionary models allows upper age limits to be placed on the event that lowered the Sm/Nd ratio, i.e. the metasomatism (Figure 2). Such modeling invariably indicates that the metasomatism occurred relatively recently: for example, at Nunivak, less than 200 Myr ago (Menzies & Murthy 1980b, Roden et al 1984a), at San Carlos, less than 500 Myr ago (Zindler & Jagoutz 1984); at Dreiser Weiher, less than 200 Myr ago (Stosch et al 1980); and at Geronimo volcanic field, less than 500 Myr ago (Kempton et al 1984). Diffusion modeling of Fe–Mg zoning in garnets from The Thumb minette, Navajo volcanic field, also supports relatively recent metasomatism under the Colorado Plateau (Smith & Ehrenberg 1984). The calculations indicate that this zoning, which

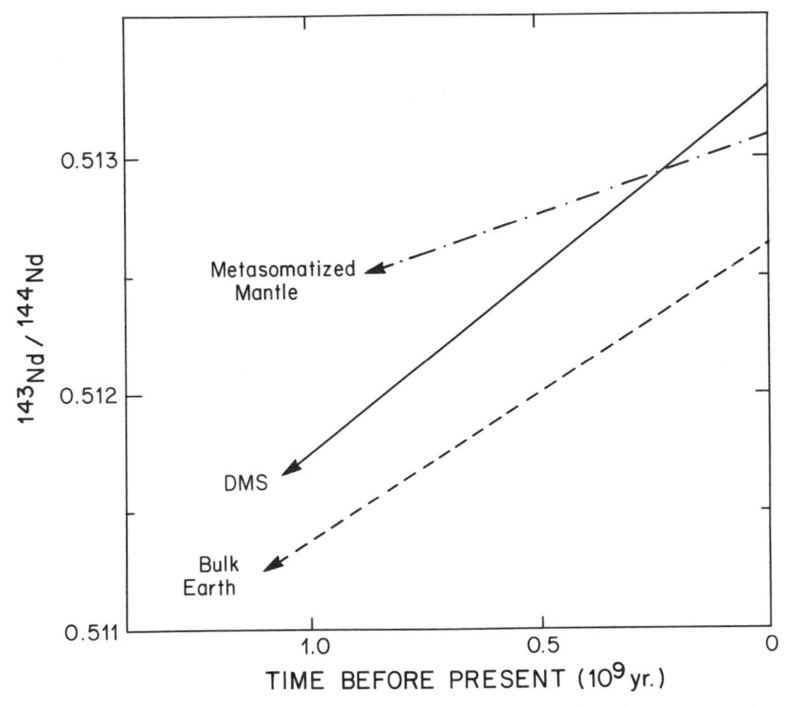

Figure 2 Nd evolution diagram, illustrating how an upper age limit for metasomatism is determined. The growth of $^{143}Nd/^{144}Nd$ in three reservoirs over the past 1 Gyr is indicated by the curves. The bulk Earth curve was calculated assuming that the Earth is essentially chondritic with respect to the REE (present-day $^{143}Nd/^{144}Nd = 0.51264$). The DMS curve is a schematic evolution model for the source of mid-ocean ridge basalt (MORB) in which this reservoir is presumed to have differentiated at some time in the past (>1 Gyr) from a chondritic reservoir. This reservoir is the most radiogenic (with respect to Nd) basalt-producing reservoir in the present-day mantle. Consequently, the DMS curve defines upper limits to permissible $^{143}Nd/^{144}Nd$ ratios in mantle reservoirs that can be melted to form basalts. Thus, the intersection of the growth curve for metasomatized mantle (here modeled as a relatively LREE-enriched peridotite with $^{147}Sm/^{144}Nd = 0.11$, $^{143}Nd/^{144}Nd = 0.5131$) with the DMS curve defines an upper limit to the age of metasomatism, provided metasomatized mantle contributes to present-day basalts.

is a consequence of Fe–Ti metasomatism, could have been preserved for no more than approximately 1000 yr in the mantle. These ages are only upper limits to the age of metasomatism; consequently, the metasomatism may have occurred shortly before eruption. Thus, with respect to time, the metasomatism is intimately associated with the petrogenesis of the host basalts. This time relationship is consistent with models calling for precursory metasomatism in the source regions for alkaline basalts (e.g. Menzies & Murthy 1980a, Wass & Rogers 1980, Boettcher & O'Neil 1980).

In addition to supplying time constraints, the relatively high $^{143}Nd/^{144}Nd$ ratios of metasomatic minerals in nodules from alkalic basalts provide important clues as to the source of metasomatic fluids and insights into the nature of the process. The high $^{143}Nd/^{144}Nd$ isotopic ratios require that the source of the metasomatic component was characterized by long-term "depleted" behavior (Sm/Nd $>$ chondrites). The production of the LREE enrichments from such a depleted source implies that the metasomatic component originated as a very small-percent melt or fluid in equilibrium with a large volume of high $^{143}Nd/^{144}Nd$ depleted mantle (e.g. Frey & Prinz 1978, Mysen 1983, M. K. Roden et al 1984).

In contrast to the high $^{143}Nd/^{144}Nd$, low $^{87}Sr/^{86}Sr$ metasomatized inclusions common to alkalic basalts, metasomatized garnet peridotites from South African kimberlites tend to have high $^{87}Sr/^{86}Sr$ and low $^{143}Nd/^{144}Nd$ ratios (e.g. Menzies & Murthy 1980c, Basu & Tatsumoto 1979, Hawkesworth et al 1983, Cohen et al 1984). If these inclusions have evolved as closed systems since the metasomatic event caused the decrease in the Sm/Nd ratio, then the low $^{143}Nd/^{144}Nd$ ratios require that the metasomatism be very old, i.e. greater than 1 Gyr (e.g. Hawkesworth et al 1983, Cohen et al 1984). Sample BD738 from the Lashaine ankaramite is particularly noteworthy because of its extremely radiogenic $^{87}Sr/^{86}Sr$ ratio and unradiogenic $^{143}Nd/^{144}Nd$ ratio; for example, the separated clinopyroxene has $^{87}Sr/^{86}Sr = 0.83604$ and $^{143}Nd/^{144}Nd = 0.51127$. This inclusion has the least radiogenic Nd and the most radiogenic Sr ratios yet detected from a mantle inclusion, and it records an old (approximately 2 Gyr) metasomatic event (Cohen et al 1984). Moreover, Hawkesworth et al (1983) noted the similar variation in Sr and Nd isotopic ratios between Karoo lavas (South Africa) and diopsides from peridotite inclusions from southern African kimberlites. Both lavas and inclusions may be derived from continental lithosphere where enriched reservoirs have existed for 1.0 to 1.4 Gyr (Hawkesworth et al 1983). The preservation of old metasomatic reservoirs beneath stable, old cratons such as South Africa is consistent with the proposal that old continental lithosphere may contain incompatible-element-enriched regions (e.g. Brooks et al 1976).

As noted earlier, the fractionation in the U–Pb system in metasomatic processes is poorly understood. Assuming that the event that fractionated U from Pb is the same metasomatic event that fractionated Sm from Nd and Rb from Sr, the data from Lashaine sample BD738 indicate that the U/Pb ratio decreases in metasomatism, contrary to the suggestion of Vollmer & Norry (1983). A single bulk rock amphibole-bearing peridotite from Nunivak Island plots near the geochron (Zartman & Tera 1973), but the Nunivak metasomatism is so recent (e.g. Roden et al 1984a) that there has not been sufficient time for isotopic ratios to evolve and reflect parent-daughter fractionation during metasomatism. Clearly, more data are needed from samples of old metasomatized peridotites similar to BD738 in order to obtain evidence regarding the fractionation of U from Pb during metasomatism.

Boettcher & O'Neil (1980) and Kyser et al (1982) measured oxygen isotopic ratios in amphiboles and micas from various mantle-derived materials, including megacrysts from kimberlites and alkalic basalts and minerals from peridotite and pyroxenite inclusions. The relation of these materials to the metasomatism of peridotite wall rock in the mantle is variable, but at least some of the analyzed material is from metasomatized peridotites. The $\delta^{18}O$ values for phlogopites and micas range from 4.68 to 6.01 and are not distinguishable from primary basalt values (Boettcher & O'Neil 1980). The narrow range of both $\delta^{18}O$ and δD values (-58 to -79) suggests a rather uniform H_2O isotopic composition in deep-seated fluids. Based on the similarity of δD values in mantle hydrous minerals with those in the carbonate-free fraction of subducted sediments (Savin & Epstein 1970), Boettcher & O'Neil (1980) and Graham & Harmon (1983) allude to the possibility that a component of mantle H_2O is related to subducted materials.

Rare gas data are very limited. If metasomatism involves the influx of a U- and Th-rich fluid into depleted and degassed mantle, and if the metasomatic component is also ultimately derived from depleted mantle (as some Nd isotopic evidence indicates), then the metasomatic component should have a relatively low $^3He/(U+Th)$ ratio (e.g. Kyser & Rison 1982). With time, the metasomatic peridotite would develop a relatively low $^3He/^4He$ ratio. Four garnet peridotite inclusions from the Matsoku and Jagersfontein pipes have low $^3He/^4He$ ratios compared with those for mid-ocean ridge basalts, and for three of these four peridotite inclusions, the ratios are low compared with that for the atmosphere (Kyser & Rison 1982). Whether all four of these inclusions are metasomatized is unclear from the petrographic description; however, one sample from Matsoku definitely is: It is one of the coarse-grained, LREE-enriched inclusions described in the

Basaltic Volcanism Study Project (1981). One of the Jagersfontein inclusions contains phlogopite, although it is unclear whether the mica is primary or secondary.

In summary, the Sm–Nd system has been most useful in constraining the timing and source characteristics of metasomatism because relative LREE enrichment, i.e. a decrease in the Sm/Nd ratio, appears to be a consistent feature of metasomatism. The data indicate that metasomatism commonly precedes alkalic basalt genesis by a time period shorter than the resolution of the Sm–Nd system; the metasomatism may occur immediately before eruption. Moreover, the isotopic signature of the metasomatic component indicates that it is ultimately derived from material with a long history of relative LREE depletion. Finally, there is the intriguing possibility, based on isotopic evidence from southern African inclusions and the Karoo basalts, that incompatible-element-enriched metasomatized mantle reservoirs may be preserved for a long time under stable continental crust (e.g. Hawkesworth et al 1983).

METASOMATISM AND ALKALIC BASALT GENESIS

Alkalic and undersaturated basalts from oceanic and continental regions are characterized by very high abundances of incompatible elements and extreme LREE enrichment relative to chondrites (e.g. Kay & Gast 1973, Sun & Hanson 1975, Frey et al 1978). However, the abundance of incompatible trace elements in the major minerals of the predominant mantle lithologies (anhydrous lherzolite and harzburgite) are too low to produce alkaline magmas by physically reasonable schemes of partial melting (e.g. Griffin & Murthy 1969, Basu & Murthy 1977, Stosch 1982, Zindler & Jagoutz 1984). Extremely small amounts of melting [e.g. 0.5–2% (Kay & Gast 1973)] of *primitive* mantle can yield melts with LREE enrichment similar to many alkalic basalts and nephelinites. In view of the inferred difficulty in separating a very small amount of melt from a residue, alternative models for alkalic basalt genesis have been developed.

Because of mass-balance problems in generating alkalic basalts from fertile lherzolite, Griffin & Murthy (1969) suggested that amphibole, phlogopite, and intergranular components rich in incompatible elements must play a role in the production of these basalts. Consequently, in their model the source peridotite is metasomatized (as we have defined the word). This metasomatized source model has been adopted by many as a general model for alkalic basalt genesis (e.g. Lloyd & Bailey 1975, Boettcher & O'Neil 1980, Menzies & Murthy 1980a, Wass & Rogers 1980, among others). Moreover, this metasomatized source model is supported by quantitative modeling of trace element abundances in alkalic basalts. Commonly, these models indicate that the source for the alkalic basalts was enriched in incompatible elements, especially LREE, compared with primitive mantle (e.g. Sun & Hanson 1975, Frey et al 1978, Clague & Frey 1982). However, the relatively low $^{87}Sr/^{86}Sr$ and high $^{143}Nd/^{144}Nd$ ratios of many alkalic basalts (e.g. Menzies & Murthy 1980b, Allègre et al 1981, Chauvel & Jahn 1984, Roden et al 1984b) indicate that if the source was metasomatized, then the metasomatism occurred shortly before basalt genesis. The isotopic systematics of alkalic basalts have led to alternative models that do not involve metasomatism, such as complex evolutionary processes in magma chambers (e.g. O'Hara 1977), zone refining of chondritic (with respect to REE) mantle (e.g. Alibert et al 1983), or melting of near-chondritic mantle (e.g. Feigenson et al 1983). Note that the last two models still require an event (metasomatism?) to lower the Sm/Nd ratio from a value greater than that of chondrites in order to explain the relatively high $^{143}Nd/^{144}Nd$ ratios typical of alkalic basalts. Overall, the simplest trace element models for the source region of alkalic basalts are consistent with a metasomatized source, but in view of the nonuniqueness of these models, they are insufficient to prove that the source was metasomatized.

Isotopic and trace element evidence from peridotite inclusions in alkalic basalts and kimberlites typically show that even the most fertile peridotites have lost a component rich in incompatible trace elements (e.g. Jagoutz et al 1979, Menzies & Murthy 1980a, Basaltic Volcanism Study Project 1981). A general picture that emerges is that the lherzolitic mantle reflects an episodic history in which incipient melt formation is a common process. These incipient melts are likely to be rich in H_2O and CO_2 (e.g. Olafsson & Eggler 1983) and will tend to be highly mobile (Spera 1981). Movement of these fluids under tectonic and/or thermal control can locally metasomatize a given region of the mantle by the formation of phases such as amphibole, phlogopite, and other incompatible-element-rich phases. Subsequent melting of the metasomatized domain can generate alkalic magmas with appropriate major and trace element abundances by relatively larger and intuitively more reasonable amounts of partial melting [e.g. 5–20% (Menzies & Murthy 1980a)].

If the metasomatic source model is correct, then the process of metasomatism may be responsible for the onset of alkalic volcanism. Spera (1981) suggested that significant amounts of heat can be transported upward by metasomatic fluids. Consequently, if the metasomatized region was initially near solidus temperatures, the addition of heat during metasomatism could induce magmatism. Moreover, the influx of volatiles into anhydrous peridotite would depress the solidus temperature (e.g.

Olafsson & Eggler 1983). The overlap in isotopic composition between some host or related basalts and metasomatic or vein minerals in peridotite inclusions (e.g. Menzies & Murthy 1980b, Menzies & Wass 1983, Bergman et al 1981, Kempton et al 1984) is consistent with a precursory metasomatic event in the source region acting as a trigger for melting. Furthermore, these alkaline lavas from tectonically active regions commonly have $^{143}Nd/^{144}Nd$ ratios that indicate that the metasomatic enrichment in LREE in their source regions happened shortly before eruption (e.g. Menzies & Murthy 1980b, Kempton et al 1984).

In contrast to the above alkalic basalts, some highly alkaline lavas and hypabyssal intrusives from stable cratonic areas have isotopic ratios indicative of a source metasomatized for a significantly long time (possibly greater than 1 Gyr) before eruption. For example, mica-rich kimberlites from southern Africa and lamproites (Mg- and K-rich lavas and hypabyssal intrusives) from Western Australia have very high incompatible element abundances and very low $^{143}Nd/^{144}Nd$ ratios (Smith 1983, McCulloch et al 1983). The preservation and aging of old metasomatized regions in stable cratonic areas may be consequences of the temperature distribution and the nature of the geotherm in shield areas.

Evidence From Ultramafic Massifs

Data from amphibole-bearing massif peridotites that are texturally and mineralogically analogous to metasomatized peridotite inclusions are rare. This lack of information is unfortunate because such studies could yield information on the structural relations between metasomatized and unmetasomatized lithologies within peridotite bodies. These structural relations could potentially prove important in constraining the nature of the metasomatic process, but they are unavailable from inclusion studies because of small sample size.

Rocks geochemically analogous to metasomatized inclusions exist in ultramafic massifs. For example, relatively LREE-enriched, pargasite-bearing spinel peridotites occur at Zabargad Island, Red Sea, and at St. Paul's Rocks, equatorial Atlantic Ocean (Bonatti et al 1981, Frey 1970, M. K. Roden et al 1984). The St. Paul's Rocks occurrence is important because the peridotites are isotopically similar (Sr, Nd, Pb) to oceanic island basalts; consequently, this occurrence is consistent with models in which the source of alkaline oceanic island basalts is metasomatized prior to melting (e.g. Clague & Frey 1982). To account for the relatively high $^{143}Nd/^{144}Nd$ ratios, M. K. Roden et al (1984) developed an "auto-metasomatic" model in which incompatible elements are scavenged by an H_2O- and CO_2-rich fluid from a large volume of depleted mantle to enrich a smaller volume of depleted mantle. Such a model is consistent with the isotopic compositions of many metasomatized alkalic basalt inclusions, as well as with models that tie metasomatism to the movement of fluids into the amphibole stability field (e.g. Schneider & Eggler 1984).

There is a pressing need for further studies of peridotite massifs in which primary amphibole or phlogopite is present. Particularly useful would be geochemical and structural studies in massifs where volume and structural relations between anhydrous, unmetasomatized peridotite and metasomatized peridotite can be studied.

SUMMARY AND IMPLICATIONS

In the above discussion we have presented evidence supporting the existence of metasomatic processes in the mantle. These processes redistribute incompatible elements and lead to anomalous enrichments of these elements in otherwise depleted mantle. Typically, the metasomatism is evidenced by relative LREE enrichment and the presence of hydrous and other incompatible-element-rich phases. These phases formed by the reaction of the metasomatic fluid (or melt) with anhydrous peridotite. Isotopic evidence from some ultramafic inclusions in kimberlites shows that the process has been active over a sizable portion of the Earth's history, i.e. at least the last 2 Gyr. Furthermore, it appears that the process has also occurred relatively recently and is probably taking place presently in regions of alkaline volcanism, as inferred from the isotopic evidence from ultramafic inclusions in alkalic basalts.

In this section we briefly review syntheses that attempt to integrate the various observations into coherent models for mantle metasomatism. We may distinguish two types of end-member genetic models: (*a*) those in which metasomatism is a local phenomenon and is commonly related to basaltic magmatism, and (*b*) those in which metasomatism is a widespread phenomenon in the lithosphere. In the latter models, metasomatized mantle may be an important reservoir for incompatible elements in the Earth.

The first category of models includes both those in which metasomatism is a consequence of the intrusion and crystallization of a basaltic magma derived externally to the metasomatized peridotite (e.g. Wilshire 1984, Kempton et al 1984) and those in which a melt or fluid derived internally metasomatizes a portion of residual peridotite (e.g. Mysen 1983, M. K. Roden et al 1984, Zindler & Jagoutz 1984). Without extensive field studies of lithologic relationships in metasomatized peridotite massifs, combined with tracer studies utilizing isotopic ratios, it is difficult to distinguish between an internal or external source for the melt or fluid. However, there seems to be no question that at least some examples of mantle metasomatism are local phenomena of small volume related to the intrusion and crystallization of basaltic magma. For example, a number of studies have shown that the occurrence of amphibole in and/or the enrichment in incompatible elements of peridotite wall rock is spatially related to pyroxenite or mica- and amphibole-rich dikes (e.g. Francis 1976a, Wilshire et al 1980, Irving 1980). Moreover, in a number of cases the metasomatic or vein minerals are isotopically similar to host or related basalts (e.g. Menzies & Murthy 1980b, Bergman et al 1981, Kempton et al 1984, Roden et al 1984a). Thus, the metasomatic fluid or melt may be related to surface volcanism. Finally, metasomatism of wall rock adjacent to crystallizing magmas in the peridotite–CO_2–H_2O system is to be expected based on phase equilibria studies (e.g. Olafsson & Eggler 1983): At relatively low pressure (less than approximately 17 kbar) there is no C-bearing solid phase on the solidus (provided that the relatively oxidized system is relevant to magmas in the mantle; see Eggler 1983). Consequently, as a basaltic magma crystallizes in this system, it must become saturated with a fluid phase if it initially contained some CO_2. If this fluid phase migrates into adjacent wall rock of peridotitic composition, it will react with the wall rock and amphibole will precipitate. The fluid will be buffered to more CO_2-rich compositions by the amphibole, and consequently its ability to transport solutes will diminish (Olafsson & Eggler 1983, Mysen 1983, Schneider & Eggler 1984).

The "autometasomatic" model in which an internally derived metasomatic fluid or melt metasomatizes depleted peridotite is attractive because it explains one of the most puzzling aspects of mantle metasomatism: In many examples, the metasomatic component is characterized by a relatively high $^{143}Nd/^{144}Nd$ ratio (see reviews of Menzies 1983, Zindler & Jagoutz 1984). Consequently, the relative LREE enrichment of the metasomatic component is a recent phenomenon. In the autometasomatic model, this enrichment results from either a magma (e.g. Mysen 1983) or a fluid scavenging (e.g. M. K. Roden et al 1984) the incompatible elements from a large volume of depleted and relatively high $^{143}Nd/^{144}Nd$ mantle. The ultimate result is a redistribution and concentration of the incompatible elements in a portion of the depleted mantle.

The "local" metasomatic models described above involve a temperature drop at near-constant pressure. In contrast to these models, a number of other models entail widespread metasomatism of lithosphere, especially the subcontinental lithosphere (e.g. Schneider & Eggler 1984, Hawkesworth et al 1983, Vollmer 1983). In these models, the primary cause of metasomatism is a decrease in pressure accompanied by temperature and composition changes. For example, Schneider & Eggler (1984) suggest that a widespread metasomatic layer may exist coincident with the upper pressure limit to the stability of amphibole. In this model, fluids precipitate amphibole and are buffered to CO_2-rich compositions as they migrate upward and encounter the amphibole stability field. The ability of CO_2-rich fluids to transport incompatible elements is less than that of H_2O-rich fluids (Mysen 1983, Schneider & Eggler 1984), and consequently a layer rich in amphibole and incompatible elements will develop. Other workers have invoked somewhat similar processes, such as the movement of a fluid off a subducted slab resulting in the formation of phlogopite-pyroxenite bodies in the overlying peridotite (e.g. Wyllie & Sekine 1982) or the convective upwelling of CO_2- and H_2O-bearing peridotite through decarbonation or dehydration reactions (e.g. Schilling et al 1980). All these models have a common consequence that metasomatism is a widespread phenomenon in the upper mantle. Hence, the metasomatic reservoir may be quite large and should thus be considered in the mass-balance calculations for the fractionation of the incompatible elements between the crust and mantle. Such calculations typically yield results indicating that depletion of approximately one third to one half of the mantle can account for the abundances of incompatible elements in the crust (e.g. Cohen & O'Nions 1982, and references therein). However, if a sizable portion of the mantle is metasomatized and enriched in incompatible elements, then conceivably more of the mantle may be depleted. Evidence from the Karoo flood basalts and from southern African xenoliths from kimberlites supports the existence of a large volume of mantle geochemically similar to metasomatized mantle. Such reservoirs may be present beneath old cratons, where they would be protected from convective overturn (see review by Hawkesworth et al 1983).

Much work remains to be done in order to ascertain the size of the metasomatic reservoir, to determine the ultimate origin and composition of the fluids, and to understand why the phase assemblages in patently metasomatized inclusions from alkalic basalts appear to differ from the assemblages in patently metasomatized inclusions from kimberlites. For example, are the fluids derived from primitive, undegassed mantle or are they recycled from subducted lithosphere (e.g. Wyllie & Sekine 1982)? Key evidence may come from isotopic studies of recently metasomatized mantle where the fractionation of parent from daughter element during the metasomatic event has not had time to obscure the isotopic signature of the metasomatic component. However, if the fluid has been recycled from old recycled lithosphere, there may be no distinct isotopic signature of the metasomatic component relative to oceanic basalts (e.g. Hofmann & White 1982, Vollmer 1983).

Another important question is whether metasomatic "keels" are common to all old cratons. The evidence from the xenoliths and basalts of southern Africa indicates that a reservoir with relatively high $^{87}Sr/^{86}Sr$ and relatively low $^{143}Nd/^{144}Nd$ ratios may underlie this region (e.g. Hawkesworth et al 1983). The isotopic geochemistry of this reservoir would be similar to that of old metasomatized xenoliths. Consequently, this high $^{87}Sr/^{86}Sr$ reservoir may be identical to metasomatized mantle and may possibly be in the form of a widespread metasomatized layer such as that visualized in some of the above models. If such a reservoir is common to all cratons, then its presence has important ramifications for the interpretation of the isotopic compositions of continental basalts. Consequently, one important direction for future research is to study patently metasomatized ultramafic inclusions from other cratons to determine if they are indicative of the presence of an old enriched keel.

ACKNOWLEDGMENTS

Preparation of this review was supported by NSF grant #EAR 791959 to V. Rama Murthy. We thank our colleagues B.-M. Jahn and J. Mahoney for their helpful criticisms of a draft of this paper. This article is publication number 1081 of the School of Earth Sciences, Department of Geology and Geophysics, University of Minnesota, Minneapolis, Minnesota 55455.

Literature Cited

Alibert, C., Michard, A., Albarede, F. 1983. The transition from alkali basalts to kimberlites: isotope and trace element evidence from melilitites. *Contrib. Mineral. Petrol.* 82:176–86

Allègre, C. J., Dupre, B., Lambret, B., Richard, P. 1981. The subcontinental versus suboceanic debate, I. Lead-neodymium-strontium isotopes in primary alkali basalts from a shield area: the Ahaggar volcanic suite. *Earth Planet. Sci. Lett.* 52:85–92

Bachinski, S. W., Simpson, E. L. 1984. Ti-phlogopites of the Shaw's Cove minette: a comparison with micas of other lamprophyres, potassic rocks, kimberlites, and mantle xenoliths. *Am. Mineral.* 69:41–56

Bailey, D. K. 1970. Volatile flux, heat focusing and generation of magma. *Geol. J. Spec. Issue* 2:177–86

Bailey, D. K. 1972. Uplift, rifting and magmatism in continental plates. *J. Earth Sci. (Leeds)* 8:225–39

Bailey, D. K. 1982. Mantle metasomatism—continuing chemical change within the earth. *Nature* 296:525–30

Bailey, D. K. 1984. Kimberlite: "The Mantle Sample" formed by ultrametasomatism. In *Kimberlites I: Kimberlites and Related Rocks*, ed. J. Kornprobst, pp. 323–34. Amsterdam: Elsevier. 466 pp.

Basaltic Volcanism Study Project. 1981. *Basaltic Volcanism on the Terrestrial Planets*. New York: Pergamon. 1286 pp.

Basu, A. R., Murthy, V. R. 1977. Ancient lithospheric lherzolite xenolith in alkali basalt from Baja California. *Earth Planet. Sci. Lett.* 35:246–57

Basu, A., Tatsumoto, M. 1979. Samarium-neodymium systematics in kimberlites and in the minerals of garnet lherzolite inclusions. *Science* 205:398–401

Bergman, S. C., Dubessy, J. 1984. CO_2–CO fluid inclusions in a composite peridotite xenolith: implications for upper mantle oxygen fugacity. *Contrib. Mineral. Petrol.* 85:1–13

Bergman, S. C., Foland, K. A., Spera, F. J. 1981. On the origin of an amphibole-rich vein in a peridotite inclusion from the Lunar Crater Volcanic Field, Nevada, U.S.A. *Earth Planet. Sci. Lett.* 56:343–61

Boettcher, A. L., O'Neil, J. R. 1980. Stable isotope, chemical, and petrographic studies of high-pressure amphiboles and micas: evidence for metasomatism in the mantle source regions of alkali basalts and kimberlites. *Am. J. Sci.* 280-A:594–621

Bonatti, E., Hamlyn, P., Ottonello, G. 1981. Upper mantle beneath a young oceanic rift: peridotites from the island of Zabargad (Red Sea). *Geology* 9:474–79

Boudier, F., Nicolas, A. 1977. Structural controls on partial melting in the peridotites. *Oreg. Dep. Geol. Mineral Ind. Bull.* 96:63–78

Brooks, C., James, D. E., Hart, S. R. 1976. Ancient lithosphere: its role in young continental volcanism. *Science* 193:1086–94

Chauvel, C., Jahn, B.-M. 1984. Nd–Sr isotope and REE geochemistry of alkali basalts from the Massif Central, France. *Geochim. Cosmochim. Acta* 48:93–110

Clague, D. A., Frey, F. A. 1982. Petrology and trace element geochemistry of the Honolulu Volcanics, Oahu: implications for the oceanic mantle below Hawaii. *J. Petrol.* 23:447–504

Cohen, R. S., O'Nions, R. K. 1982. The lead, neodymium and strontium isotopic structure of ocean ridge basalts. *J. Petrol.* 23:299–324

Cohen, R. S., O'Nions, R. K., Dawson, J. B. 1984. Isotope geochemistry of xenoliths from East Africa: implications for development of mantle reservoirs and their interaction. *Earth Planet. Sci. Lett.* 68:209–20

Dawson, J. B. 1980. *Kimberlites and Their Xenoliths.* Berlin: Springer-Verlag. 252 pp.

Dawson, J. B. 1984. Contrasting types of upper-mantle metasomatism? In *Kimberlites II: The Mantle and Crust-Mantle Relationships*, ed. J. Kornprobst, pp. 289–94. Amsterdam: Elsevier. 393 pp.

Dawson, J. B., Smith, J. V. 1977. The MARID (mica-amphibole-rutile-ilmenite-diopside) suite of xenoliths in kimberlite. *Geochim. Cosmochim. Acta* 41:309–24

Dawson, J. B., Smith, J. V. 1982. Upper-mantle amphiboles: a review. *Mineral Mag.* 45:35–46

Dawson, J. B., Powell, D. G., Reid, A. M. 1970. Ultrabasic xenoliths and lava from the Lashaine Volcano, northern Tanzania. *J. Petrol.* 11:519–31

Eggler, D. H. 1983. Upper mantle oxidation: evidence from olivine-orthopyroxene-ilmenite assemblages. *Geophys. Res. Lett.* 10:365–68

Ehrenberg, S. N. 1979. Garnetiferous ultramafic inclusions in minette from the Navajo Volcanic Field. In *The Mantle Sample: Inclusions in Kimberlites and Other Volcanics*, ed. F. R. Boyd, H. O. A. Meyer, pp. 330–44. Washington DC: Am. Geophys. Union. 423 pp.

Ehrenberg, S. N. 1982a. Petrogenesis of garnet lherzolite and megacrystalline nodules from The Thumb, Navajo Volcanic Field. *J. Petrol.* 23:507–47

Ehrenberg, S. N. 1982b. Rare earth element geochemistry of garnet lherzolite and megacrystalline nodules from minette of the Colorado Plateau province. *Earth Planet. Sci. Lett.* 57:191–210

Erlank, A. J., Rickard, R. S. 1977. Potassic richterite bearing peridotites from kimberlite and the evidence they provide for upper mantle metasomatism. *Extended Abstr., Int. Kimberlite Conf., 2nd, Santa Fe, N. Mex*

Evensen, N. M., Hamilton, P. J., O'Nions, R. K. 1978. Rare-earth abundances in chondritic meteorites. *Geochim. Cosmochim. Acta* 42:1199–1213

Feigenson, M. D, Hofmann, A. W., Spera, F. J. 1983. Case studies on the origin of basalt. 2. The transition from tholeiitic to alkalic volcanism on Kohala volcano, Hawaii. *Contrib. Mineral. Petrol.* 84:390–405

Francis, D. M. 1976a. Amphibole pyroxenite xenoliths: cumulate or replacement phenomena from the upper mantle, Nunivak Island, Alaska. *Contrib. Mineral. Petrol.* 58:51–61

Francis, D. M. 1976b. The origin of amphibole in lherzolite xenoliths from Nunivak Island, Alaska. *J. Petrol.* 17:357–78

Frey, F. A. 1970. Rare earth and potassium abundances in St. Paul's Rocks. *Earth Planet. Sci. Lett.* 7:351–60

Frey, F. A. 1984. Rare earth element abundances in upper mantle rocks. In *Rare Earth Element Geochemistry, Developments in Geochemistry 2*, ed. P. Henderson, pp. 153–203. Amsterdam: Elsevier

Frey, F. A., Green, D. H. 1974. The mineralogy, geochemistry and origin of lherzolite inclusions in Victorian basanites. *Geochim. Cosmochim. Acta* 38:1023–59

Frey, F. A., Prinz, M. 1978. Ultramafic inclusions from San Carlos, Arizona: petrologic and geochemical data bearing on their petrogenesis. *Earth Planet. Sci. Lett.* 38:129–76

Frey, F. A., Green, D. H., Roy, S. D. 1978. Integrated models of basalt petrogenesis: a study of quartz tholeiites to olivine melilitites from South Eastern Australia utilizing geochemical and experimental petrological data. *J. Petrol.* 19:463–513

Gast, P. W. 1968. Trace element fractionation and the origin of tholeiitic and alkaline magma types. *Geochim. Cosmochim. Acta* 32:1057–86

Graham, C. M., Harmon, R. S. 1983. Stable isotope evidence on the nature of crust-mantle interactions. In *Continental Basalts and Mantle Xenoliths*, ed. C. J. Hawkesworth, M. J. Norry, pp. 20–45. Cheshire, Engl: Shiva. 272 pp.

Griffin, W. L., Murthy, V. R. 1969. Distribution of K, Rb, Sr and Ba in minerals relevant to basalt genesis. *Geochim. Cosmochim. Acta* 33:1389–1414

Gurney, J. J., Harte, B. 1980. Chemical variations in upper mantle nodules from southern African kimberlites. *Philos. Trans. R. Soc. London Ser. A* 297:273–93

Haggerty, S. E. 1983. The mineral chemistry of new titanates from the Jagersfontein kimberlite, South Africa: implications for metasomatism in the upper mantle. *Geochim. Cosmochim. Acta* 47:1833–55

Haggerty, S. E., Smyth, J. R., Erlank, A. J., Rickard, R. S., Danchina, R. V. 1983. Lindsleyite (Ba) and mathiasite (K): two new chromium-titanates in the crichtonite series from the upper mantle. *Am. Mineral.* 68:494–505

Harte, B. 1983. Mantle peridotites and processes—the kimberlite sample. In *Continental Basalts and Mantle Xenoliths*, ed. C. J. Hawkesworth, M. J. Norry, pp. 46–91. Cheshire, Engl: Shiva. 272 pp.

Harte, B., Cox, K. G., Gurney, J. J. 1975. Petrography and geological history of upper mantle xenoliths from the Matsoku kimberlite pipe. *Phys. Chem. Earth* 9:477–506

Hawkesworth, C. J., Erlank, A. J., Marsh, J. S., Menzies, M. A., Van Calsteren, P. 1983. Evolution of the continental lithosphere: evidence from volcanics and xenoliths in southern Africa. In *Continental Basalts and Mantle Xenoliths*, ed. C. J. Hawkesworth, M. J. Norry, pp. 111–38. Cheshire, Engl: Shiva. 272 pp.

Hofmann, A. W., White, W. M. 1982. Mantle plumes from oceanic crust. *Earth Planet. Sci. Lett.* 57:421–36

Hunter, W. C., Smith, D. 1981. Garnet peridotite from Colorado Plateau ultramafic diatremes: hydrates, carbonates, and comparative geothermometry. *Contrib. Mineral. Petrol.* 76:312–20

Irving, A. J. 1980. Petrology and geochemistry of composite ultramafic xenoliths in alkalic basalts and implications for magmatic processes within the mantle. *Am. J. Sci.* 280-A:389–426

Jagoutz, E., Palme, H., Baddenhausen, H., Blum, K., Cendules, M., et al. 1979. The abundances of major, minor, and trace elements in the earth's mantle as derived from primitive ultramafic nodules. *Proc. Lunar Planet. Sci. Conf., 10th*, pp. 2031–50

Jones, A. P., Smith, J. V., Dawson, J. B. 1982. Mantle metasomatism in 14 veined peridotites from Bultfontein Mine, South Africa. *J. Geol.* 90:435–53

Kay, R. W., Gast, P. W. 1973. The rare earth content and origin of alkali-rich basalts. *J. Geol.* 81:653–82

Kempton, P. D., Menzies, M. A., Dungan, M. A. 1984. Petrography, petrology and geochemistry of xenoliths and megacrysts from the Geronimo volcanic field, southeastern Arizona. In *Kimberlites II: The Mantle and Crust-Mantle Relationships*, ed. J. Kornprobst, pp. 71–84. Amsterdam: Elsevier. 393 pp.

Korzhinskii, D. S. 1965. The theory of systems with perfectly mobile components and processes of mineral formation. *Am. J. Sci.* 263:193–205

Korzhinskii, D. S. 1970. *Theory of Metasomatic Zoning*. Oxford: Clarendon. 162 pp. Trans. A. Augrell (From Russian)

Kurat, G., Palme, H., Spettel, B., Baddenhausen, H., Hofmeister, H., et al. 1980. Geochemistry of ultramafic xenoliths from Kapfenstein, Austria: evidence for a variety of upper mantle processes. *Geochim. Cosmochim. Acta* 44:45–60

Kyser, T. K., Rison, W. 1982. Systematics of rare gas isotopes in basic lavas and ultramafic xenoliths. *J. Geophys. Res.* 87:5611–30

Kyser, T. K., O'Neil, J. R., Carmichael, I. S. E. 1982. Genetic relations among basic lavas and ultramafic nodules: evidence from oxygen isotope compositions. *Contrib. Mineral. Petrol.* 81:88–102

Lawless, P. J., Gurney, J. J., Dawson, J. B. 1979. Polymict peridotites from the Bultfontein and De Beers Mines, Kimberley, South Africa. In *The Mantle Sample: Inclusions in Kimberlites and Other Volcanics*, ed. F. R. Boyd, H. O. A. Meyer, pp. 145–55. Washington DC: Am. Geophys. Union. 423 pp.

Lloyd, F. E., Bailey, D. K. 1975. Light element metasomatism of the continental mantle: the evidence and the consequences. *Phys. Chem. Earth* 9:381–416

Loubet, M., Shimizu, N., Allègre, C. J. 1975. Rare earth elements in alpine peridotites. *Contrib. Mineral. Petrol.* 53:1–12

McCulloch, M. T., Jaques, A. L., Nelson, D. R., Lewis, J. D. 1983. Nd and Sr isotopes in kimberlites and lamproites from Western Australia: an enriched mantle origin. *Nature* 302:400–3

McGetchin, T. R., Besancon, J. R. 1973. Carbonate inclusions in mantle-derived pyropes. *Earth Planet. Sci. Lett.* 18:408–10

McKenzie, D., O'Nions, R. K. 1983. Mantle reservoirs and ocean island basalts. *Nature* 301:229–31

Melson, W. G., Hart, S. R., Thompson, G. 1972. St. Paul's Rocks, equatorial Atlantic: petrogenesis, radiometric ages and implications on sea-floor spreading. *Geol. Soc. Am. Mem.* 132:241–72

Menzies, M. A. 1983. Mantle ultramafic xenoliths in alkaline magmas: evidence for mantle heterogeneity modified by magmatic activity. In *Continental Basalts and*

Mantle Xenoliths, ed. C. J. Hawkesworth, M. J. Norry, pp. 92–110. Cheshire, Engl: Shiva. 272 pp.

Menzies, M. A., Murthy, V. R. 1980a. Mantle metasomatism as a precursor to the genesis of alkaline magmas—isotopic evidence. *Am. J. Sci.* 280-A: 622–38

Menzies, M. A., Murthy, V. R. 1980b. Nd and Sr isotope geochemistry of hydrous mantle nodules and their host alkali basalts: implications for local heterogeneities in metasomatically veined mantle. *Earth Planet. Sci. Lett.* 46: 323–34

Menzies, M. A., Murthy, V. R. 1980c. Enriched mantle: Nd and Sr isotopes in diopsides from kimberlite nodules? *Nature* 283: 634–36

Menzies, M. A., Wass, S. Y. 1983. CO_2- and LREE-rich mantle below eastern Australia: a REE and isotopic study of alkaline magmas and apatite-rich mantle xenoliths from Southern Highlands Province, Australia. *Earth Planet. Sci. Lett.* 65: 287–302

Mysen, B. O. 1979. Trace element partitioning between garnet peridotite minerals and water-rich vapor: experimental data from 5 to 30 kbar. *Am. Mineral.* 64: 274–87

Mysen, B. O. 1983. Rare earth element partitioning between (H_2O+CO_2) vapor and upper mantle minerals: experimental data bearing on the conditions of formation of alkali basalt and kimberlite. *Neues Jahrb. Mineral. Abh.* 146: 41–65

Nixon, P. H., Rogers, N. W., Gibson, I. L., Grey, A. 1981. Depleted and fertile mantle xenoliths from southern African kimberlites. *Ann. Rev. Earth Planet. Sci.* 9: 285–309

O'Hara, M. J. 1977. Geochemical evolution during fractional crystallization of a periodically refilled magma chamber. *Nature* 266: 503–7

Olafsson, M., Eggler, D. H. 1983. Phase relations of amphibole, amphibole-carbonate, and phlogopite-carbonate peridotite: petrologic constraints on the asthenosphere. *Earth Planet. Sci. Lett.* 64: 305–15

Reid, A. M., Donaldson, C., Brown, R. W., Ridley, W. I., Dawson, J. B. 1975. Mineral chemistry of peridotite inclusions from Lashaine Volcano, North Tanzania. *Phys. Chem. Earth* 9: 525–37

Ridley, W. I., Dawson, J. B. 1975. Lithophile and trace element data bearing on origin of peridotite inclusions from the Lashaine Volcano, Tanzania. *Phys. Chem. Earth* 9: 545–57

Ringwood, A. E. 1975. *Composition and Petrology of the Earth's Mantle.* New York: McGraw-Hill. 618 pp.

Ringwood, A. E. 1979. *Origin of the Earth and Moon.* Berlin: Springer-Verlag. 295 pp.

Roden, M. F., Frey, F. A., Francis, D. M. 1984a. An example of consequent metasomatism in peridotite inclusions from Nunivak Island, Alaska. *J. Petrol.* 25: 546–77

Roden, M. F., Frey, F. A., Clague, D. A. 1984b. Geochemistry of tholeiitic and alkalic lavas from the Koolan Range, Oahu, Hawaii: implications for Hawaiian volcanism. *Earth Planet. Sci. Lett.* 69: 141–58

Roden, M. K., Hart, S. R., Frey, F. A., Melson, W. G. 1984. Sr, Nd and Pb isotopic and REE geochemistry of St. Paul's Rocks: the metamorphic and metasomatic development of an alkali basalt mantle source. *Contrib. Mineral. Petrol.* 85: 376–90

Savin, S. M., Epstein, S. 1970. The oxygen and hydrogen isotope geochemistry of ocean sediments and shales. *Geochim. Cosmochim. Acta* 34: 43–64

Schilling, J.-G. 1975. Rare earth variations across "normal segments" of the Reykjanes Ridge, 60–53°N, Mid-Atlantic Ridge, 29°S, and East Pacific Rise, 2–19°S, and evidence on the composition of the underlying low-velocity layer. *J. Geophys. Res.* 80: 1459–73

Schilling, J.-G., Bergeron, M. B., Evans, R. 1980. Halogens in the mantle beneath the North Atlantic. *Philos. Trans. R. Soc. London Ser. A* 297: 147–78

Schneider, M. E., Eggler, D. H. 1984. Compositions of fluids in equilibrium with peridotite: implications for alkaline magmatism-metasomatism. In *Kimberlites I: Kimberlites and Related Rocks*, ed. J. Kornprobst, pp. 383–94. Amsterdam: Elsevier. 466 pp.

Shimizu, N. 1975. Rare earth elements in garnets and clinopyroxenes from garnet lherzolite nodules in kimberlites. *Earth Planet. Sci. Lett.* 25: 26–31

Smith, C. B. 1983. Pb, Sr and Nd isotopic evidence for sources of southern African Cretaceous kimberlites. *Nature* 304: 51–54

Smith, D. 1979. Hydrous minerals and carbonates in peridotite inclusions from the Green Knobs and Buell Park kimberlitic diatremes on the Colorado Plateau. In *The Mantle Sample: Inclusions in Kimberlites and Other Volcanics*, ed. F. R. Boyd, H. O. A. Meyer, pp. 345–56. Washington DC: Am. Geophys. Union. 423 pp.

Smith, D., Ehrenberg, S. N. 1984. Zoned minerals in garnet peridotite nodules from the Colorado Plateau: implications for mantle metasomatism and kinetics. *Contrib. Mineral. Petrol.* 86: 274–85

Smith, D., Levy, S. 1976. Petrology of the Green Knobs diatreme and implications for the upper mantle beneath the Colorado Plateau. *Earth Planet. Sci. Lett.* 29:107–25

Smith, J. V., Hewig, R. L., Ackermand, D., Dawson, J. B. 1979. K, Rb, and Ba in micas from kimberlite and peridotitic xenoliths and implications for origin of basaltic rocks. In *Kimberlites, Diatremes, and Diamonds: Their Geology, Petrology and Geochemistry*, ed. F. R. Boyd, H. O. A. Meyer, pp. 241–55. Washington DC: Am. Geophys. Union. 399 pp.

Spera, F. J. 1981. Carbon dioxide in igneous petrogenesis: II. Fluid dynamics of mantle metasomatism. *Contrib. Mineral. Petrol.* 77:56–65

Stosch, H. G. 1982. Rare earth element partitioning between minerals from anhydrous spinel peridotite xenoliths. *Geochim. Cosmochim. Acta* 46:793–811

Stosch, H. G., Seck, H. A. 1980. Geochemistry and mineralogy of two spinel peridotite suites from Dreiser Weiher, West Germany. *Geochim. Cosmochim. Acta* 44:457–70

Stosch, H. G., Carlson, R. W., Lugmair, G. W. 1980. Episodic mantle differentiation: Nd and Sr isotopic evidence. *Earth Planet. Sci. Lett.* 47:263–71

Sun, S.-S. 1982. Chemical composition and origin of the Earth's mantle. *Geochim. Cosmochim. Acta* 46:179–93

Sun, S.-S., Hanson, G. N. 1975. Origin of Ross Island basanitoids and limitations upon the heterogeneity of mantle sources for alkali basalts and nephelinites. *Contrib. Mineral. Petrol.* 52:77–106

Tanaka, T., Aoki, K.-I. 1981. Petrogenetic implications of REE and Ba data on mafic and ultramafic inclusions from Itinome-Gata, Japan. *J. Geol.* 89:369–90

Taylor, S. R. 1982. Lunar and terrestrial crusts: a contrast in origin and evolution. *Earth Planet. Sci. Lett.* 29:233–41

Thompson, J. B. Jr. 1959. Local equilibrium in metasomatic processes. In *Research in Geochemistry*, ed. P. H. Abelson, 1:427–57. New York: Wiley. 511 pp.

Varne, R., Graham, A. L. 1971. Rare earth abundances in hornblende and clinopyroxene of a hornblende lherzolite xenolith: implications for upper mantle fractionation processes. *Earth Planet. Sci. Lett.* 13:11–18

Vollmer, R. 1983. Earth degassing, mantle metasomatism and isotopic evolution of the mantle. *Geology* 11:452–54

Vollmer, R., Norry, N. J. 1983. Unusual isotopic variations in Nyiragongo nephelinites. *Nature* 301:141–43

Wass, S. Y., Rogers, N. W. 1980. Mantle metasomatism—precursor to continental alkaline volcanism. *Geochim. Cosmochim. Acta* 44:1811–24

Wass, S. Y., Henderson, P., Elliott, C. J. 1980. Chemical heterogeneity and metasomatism in the upper mantle—evidence from rare earth and other elements in apatite-rich xenoliths in basaltic rocks from eastern Australia. *Philos. Trans. R. Soc. London Ser. A* 297:333–46

Wendlandt, R. F., Harrison, W. J. 1979. Rare earth partitioning between immiscible carbonate and silicate liquids and CO_2 vapor: results and implications for the formation of light rare earth-enriched rocks. *Contrib. Mineral. Petrol.* 69:409–19

Wilshire, H. G. 1984. Mantle metasomatism: the REE story. *Geology* 12:395–98

Wilshire, H. G., Pike, J. E. N. 1975. Upper-mantle diapirism: evidence from analogous features in alpine peridotite and ultramafic inclusions in basalt. *Geology* 3:467–70

Wilshire, H. G., Trask, N. J. 1971. Structural and textural relationships of amphibole and phlogopite in peridotite inclusions, Dish Hill, California. *Am. Mineral.* 56:240–55

Wilshire, H. G., Pike, J. E. N., Meyer, C. E., Schwarzman, E. C. 1980. Amphibole-rich veins in lherzolite xenoliths, Dish Hill and Deadman Lake, California. *Am. J. Sci.* 280-A:576–93

Wood, D. A., Joron, J.-L., Treuil, M., Norry, M., Tarney, J. 1979. Elemental and Sr isotope variations in basic lavas from Iceland and the surrounding ocean floor: nature of mantle source inhomogeneties. *Contrib. Mineral. Petrol.* 70:319–39

Wyllie, P. J. 1980. The origin of kimberlite. *J. Geophys. Res.* 85:6902–10

Wyllie, P. J., Sekine, T. 1982. The formation of mantle phlogopite in subduction zone hybridization. *Contrib. Mineral. Petrol.* 79:375–80

Zartman, R. E., Tera, F. 1973. Lead concentration and isotopic composition in five peridotite inclusions of probable mantle origin. *Earth Planet. Sci. Lett.* 20:54–66

Zindler, A., Jagoutz, E. 1984. Trace element and Nd and Sr isotope systematics of peridotite nodules from Peridot Mesa, San Carlos, Arizona. *Geochim. Cosmochim. Acta.* In press

Ann. Rev. Earth Planet. Sci. 1985. 13:297–314

THE GEOLOGICAL RECORD OF INSECTS

F. M. Carpenter and L. Burnham

Department of Entomology, Museum of Comparative Zoology, Harvard University, Cambridge, Massachusetts 02138

INTRODUCTION

Insects constitute the largest class of the phylum Arthropoda. Like other members of the phylum, they have an exoskeleton and are metamerically segmented. The insect head consists of six fused segments; the thorax three segments, each with a pair of segmented legs; and the abdomen eleven segments, which may be reduced or modified in some specialized species. Most insects have two pairs of wings attached to the second and third thoracic segments, but some have only one pair, and others have secondarily lost both pairs.

Wings are important in the classification of insects at both the ordinal and familial levels, and especially so for fossil specimens. The cuticular nature of the wings helps their preservation as fossils, even in situations where the rest of the insect has been decomposed or otherwise destroyed.

Insects are basically terrestrial, and they respire by a series of tracheal tubes that open laterally on the thorax and abdomen. Nevertheless, many orders of insects have become secondarily adapted to freshwater habitats, either as immature stages or as adults, or both. In these instances, the tracheae terminate in gills that allow oxygen in solution in the water to diffuse into the insect's respiratory system.

Most insects become fossils by falling into a body of water, where they may be entombed in silts or other fine sediments and eventually be preserved, if not destroyed by various scavengers and detritivores. Chitin and protein, primary constituents of the insect exoskeleton, are rapidly destroyed by a variety of microorganisms, and insects must generally undergo immediate burial under anaerobic conditions for fossilization to take place. Hence, most insects are fossilized under only very precise

0084-6597/85/0515-0297$02.00

(usually catastrophic) environmental conditions; under normal conditions, preservation is an unlikely event. Thus it must be stressed that there is much bias in the fossil record of the insects, and that the absence of an order or family from a fossil-bearing deposit does not necessarily mean that the taxon was not in existence at the time.

The origin of insects is not clear. Several diverse theories have been advanced, suggesting their origin from trilobites, crustaceans, myriopods, or the hexapod class Diplura. For all of these proposals, however, the evidence is very weak and speculative. Since the insects and the members of these other groups have evolved along very different lines for at least 350 Myr, the absence of convincing evidence among existing arthropods for such a relationship is not surprising. The fossil record, as presently known, contributes nothing to our understanding of the actual origin of the insects. The earliest known insects, although belonging to the more generalized part of the insect hierarchy, show no morphological features indicative of their ancestral stock. On the other hand, there is a bewildering array of fragmentary remains of arthropods from the Cambrian to the Lower Carboniferous that do not fall within any of the known higher arthropod categories, existing or extinct. It is probable that the ancestral stock of the insects will eventually be found among such material. Until then, we can only speculate on their nature.

Although differing views on insect evolution have been presented in the literature (Martynov 1937, Rohdendorf & Rasnitsyn 1980), the concept currently accepted by most entomologists is that there were four main stages in the history of the insects. The first stage consisted of primitively wingless species comprising the subclass Apterygota, represented now by bristletails (order Archaeognatha) and silverfish (order Zygentoma). In nearly all respects, both morphological and physiological, the Apterygota are the most generalized insects known.

The second stage began with the origin of the winged insects, or Pterygota. This was probably the most important step in the history of the insects, and it is almost certainly more responsible for the success of the group than any other factor. Flight not only provided an effective means of dispersal but also offered a means of escape from predators and greater flexibility in obtaining food. It is not surprising that over 99% of all extant insects belong to the subclass Pterygota. Nevertheless, the origin of wings in insects is by no means clear; several theories have been proposed, but none of them has received general acceptance by entomologists concerned with insect evolution (Kukalová-Peck 1978, Rasnitsyn 1981, Kingsolver 1984). The most primitive of the winged insects, termed the paleopterous Pterygota, have a simple, hinged articulation of the wings to the thorax. They are represented in the present fauna by only two orders: the dragonflies (order Odonata) and the mayflies (order Ephemeroptera) (Martynov 1925). Although many of the Paleoptera are good fliers, the nature of their wing articulation prevents them from flexing their wings back over the abdomen when they are at rest [Figure 1 (*top*)]. These paleopterous Pterygota make up less than 1% of the species of insects in existence today.

The third stage was marked by the development of a more complicated wing articulation (resulting from a modification of the third axillary sclerite) that allowed the wings to be placed back over the abdomen when the insect was at rest (Snodgrass 1935). This enabled these insects, termed the neopterous Pterygota, to exploit a diversity of new habitats. No longer hindered by outstretched wings, they were able to crawl into dense foliage or seek refuge under bark or stones or in soil [Figure 1 (*bottom*)]. Many of them have modified their wings in various ways. In several orders, such as the fleas (Siphonaptera) and lice (Anoplura), the wings have been secondarily lost as an adaptation to ectoparasitism. In others, the fore wings are present but have been modified as either leathery or thickened protective covers, as in the cockroaches (order Blattodea), grasshoppers (order Orthoptera), true bugs (order Hemiptera), and several other orders. The most primitive of these neopterous Pterygota undergo direct postembryonic development: the immature stages, usually termed nymphs, pass gradually through a series of molts to the adult form. Since the wings develop externally, increasing in size each nymphal instar, these insects are termed exopterygote Neoptera. Nymphs typically resemble the adults, feed on the same food, and live in the same habitat. Fifteen existing orders, including the three orders mentioned above, are part of this group, which accounts for about 12% of the living species of insects.

The fourth, and presently final, stage of the insects' history occurred when some of the Neoptera evolved a more complicated or indirect type of postembryonic development. In these insects, known as the endopterygote Neoptera, wing buds develop in the immatures, or larvae, as invaginations of the body wall. The wings are not evaginated until the insect reaches the pupal stage—a quiescent, nonfeeding state, during which marked reorganization of larval tissues and organs take place. The larvae show little morphological resemblance to the adults, feed on different food, and live in different habitats. Nine existing orders are assigned to the endopterygote Neoptera. These are the largest and most familiar of the insect orders and include the beetles (order Coleoptera), true flies (order Diptera), ants, bees, and wasps (order Hymenoptera), etc. Approximately 85% of all living insects belong to this category.

The geological record of these four stages in the evolution of the insects shows several surprises (Figure 2). The earliest known insects are from

Figure 1 (*top*) *Dunbaria* (Palaeodictyoptera, Permian, Kansas, Division Paleoptera). (*bottom*) *Gerarus* (Protorthoptera, Upper Carboniferous, Illinois, Division Neoptera).

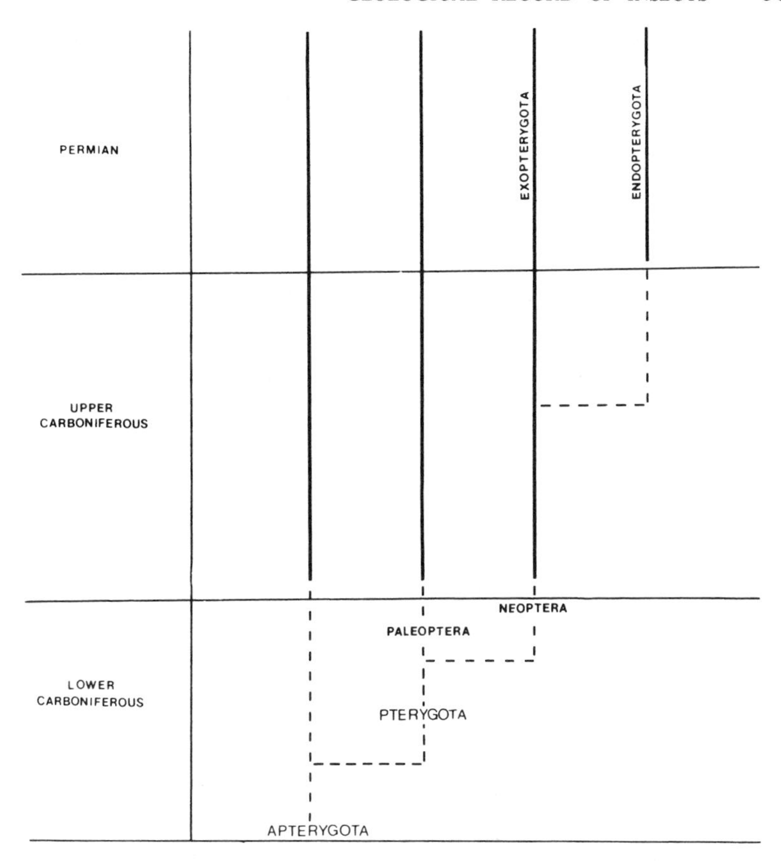

Figure 2 Geological record of main stages in insect evolution. Broken lines indicate absence of fossil record.

Upper Carboniferous deposits, and it is obvious that the insects had already reached three of the four stages in their evolution by this time. The endopterygote Neoptera first appear as early as the Lower Permian, where they are represented by two living orders, the scorpion flies (order Mecoptera) and lacewings (order Neuroptera). Since both these orders show considerable diversity, we can infer that the endopterygotes undoubtedly arose in the Upper Carboniferous. Presumably, also, wings and wing-flexing arose earlier than the Upper Carboniferous, probably in the Lower Carboniferous. This means that these four stages in insect evolution had taken place before the end of the Carboniferous, and that most probably the first three stages were attained even earlier. Unfortunately, we have no fossil evidence for this assumption at present.

Table 1 Geological ranges of the orders of insects

Classification	Geologic range
Subclass Apterygota	
Order Archaeognatha	U. Carb.–Recent
Order Zygentoma	Tert.–Recent
Subclass Pterygota	
Division Paleoptera	
Order Ephemeroptera	U. Carb.–Recent
Order Palaeodictyoptera[a]	U. Carb.–Perm.
Order Megasecoptera[a]	U. Carb.–Perm.
Order Diaphanopterodea[a]	U. Carb.–Perm.
Order Protodonata[a]	U. Carb.–Perm.
Order Odonata	Perm.–Recent
Division Neoptera	
Section Exopterygota	
Order Plecoptera	Perm.–Recent
Order Protorthoptera[a]	U. Carb.–Perm.
Order Grylloblattodea	Recent
Order Blattodea	U. Carb.–Recent
Order Dermaptera	Jur.–Recent
Order Caloneurodea[a]	U. Carb.–Perm.
Order Manteodea	Tert.–Recent
Order Isoptera	L. Cret.–Recent
Order Zoraptera	Recent
Order Protelytroptera[a]	Perm.
Order Orthoptera	U. Carb.–Recent
Order Phasmida	Tri.–Recent
Order Embioptera	Tert.–Recent
Order Miomoptera[a]	U. Carb.–Perm.
Order Psocoptera	Perm.–Recent
Order Mallophaga	Recent
Order Anoplura	Recent
Order Thysanoptera	Perm.–Recent
Order Hemiptera	Perm.–Recent
Section Endopterygota	
Order Mecoptera	Perm.–Recent
Order Neuroptera	Perm.–Recent
Order Glosselytrodea[a]	Perm.–Jur.
Order Trichoptera	Perm.–Recent
Order Lepidoptera	L. Cret.–Recent
Order Coleoptera	Perm.–Recent
Order Strepsiptera	Tert.–Recent
Order Diptera	Tri.–Recent
Order Siphonaptera	Tert.–Recent
Order Hymenoptera	Tri.–Recent

[a] Extinct orders.

The discussion that follows traces the geological history of the insects (Table 1). During the past century, over 50 extinct orders of insects have been named and described, but most of these were based on fragmentary material and have subsequently been synonymized. At present, only 9 extinct orders are sufficiently well known to justify their recognition; 4 belong to the Paleoptera, 4 to the exopterygote Neoptera, and 1 is considered to belong to the endopterygote Neoptera (Brues et al 1954, Carpenter 1976, Wootton 1981).

PALEOZOIC ERA

Upper Carboniferous

Eleven orders of insects belonging to the Apterygota, Paleoptera, and exopterygote Neoptera are recorded from the Upper Carboniferous. Seven of them are extinct. Insects occur only rarely in Upper Carboniferous deposits, and the majority of them have been found at only three well-known localities: Commentry, France; Illinois, USA; and Kuznetzk Basin, USSR. Most of the specimens are from Westphalian and Stephanian beds, although the few specimens from the Namurian belong to neopterous as well as paleopterous orders and therefore suggest a much earlier origin for the Paleoptera.

APTERYGOTA The existing order Archaeognatha is represented in Upper Carboniferous beds of France and Illinois by species belonging to an extinct suborder, Monura. In most respects, Monura are considered more generalized than the extant members of the Archaeognatha (Sharov 1957b, Carpenter 1985a,b, Carpenter & Richardson 1985). They lacked a pair of terminal processes, or cerci, characteristic of all living Apterygota, but they did have the median process well developed. The absence of cerci has generally been considered a derived condition, but their absence in the Monura suggests that this may in fact represent an ancestral condition. Although only a few species of Archaeognatha are known from the Upper Carboniferous, they show considerable diversity in structure (Carpenter & Richardson 1985).

PALEOPTEROUS PTERYGOTA The existing order Ephemeroptera (mayflies) is represented in the Upper Carboniferous of France by a single adult specimen. Since the order is generally recognized as the least specialized of the Pterygota, its presence in the Upper Carboniferous is not unexpected. Unlike existing species, in which the hind wings are much reduced in size, the fore and hind wings in this Carboniferous species are nearly identical. The abdomen bears a pair of cerci and a median process, as in existing mayflies (Carpenter 1963).

The extinct order Palaeodictyoptera was one of the largest in the Upper Carboniferous, having a total of 17 families. Most of the species were of moderate size, but a few had wingspans as great as 40 cm (Kukalová 1969a,b, 1970). Typically, fore and hind wings were similar, but in more specialized species the hind wings are distinctly broader than the front wings. The prothorax bore a pair of small, winglike lobes, and the abdomen terminated in a pair of long cerci. The most unusual feature of these insects was the nature of the mouthparts, which formed a sucking beak containing five slender stylets. Such a beak is characteristic of insects that feed on liquid food, and it is presumed that the Palaeodictyoptera used their beaks to pierce plants and thus obtain nutrients in the form of phloem. Several nymphs belonging to this order have been found, and all of them have divergent wing buds and beaks like the adults. Since these nymphs show no structural modifications for an aquatic life, they are presumed to have been terrestrial.

Members of the extinct order Megasecoptera resembled the Palaeodictyoptera and had similar mouthparts, but they lacked the small winglike lobes on the prothorax, and their wings had a more specialized venation and shape. The nymphs were like those of the Palaeodictyoptera and are also presumed to have been terrestrial. This was a much smaller order in the Carboniferous than the Palaeodictyoptera, however, with only about eight families.

The extinct order Diaphanopterodea was related to the Palaeodictyoptera and Megasecoptera and had similar mouthparts, but differed from them in having a more reduced venation. Most distinctive about the Diaphanopterodea was their ability to fold their wings back over the abdomen at rest. Examination of the wing articulation has shown that this wing position was not accomplished by modification of the third axillary sclerite, as in the Neoptera, but by a different mechanism (Kukalová-Peck 1974). Consequently, the Diaphanopterodea are considered highly specialized Paleoptera that evolved a wing-flexing mechanism independently of the true Neoptera. The order Diaphanopterodea was small in the Carboniferous, with only four families known.

The extinct order Protodonata resembled dragonflies, with some venational differences. They had strong mandibles and very spiny fore legs. All were large insects, and some species, with a wing expanse of 70 cm, were the largest insects ever to have existed. The structure of their mouthparts and legs strongly indicates that they were predators, catching prey in flight, like modern dragonflies. Although they were widely distributed in the Carboniferous, they did not have much diversity, and only two families are known from that period (Carpenter 1943).

EXOPTERYGOTE NEOPTERA Five orders in this division are known from the Upper Carboniferous. Three of them, the Protorthoptera, Caloneurodea, and Miomoptera, are extinct. The other two, the Blattodea and Orthoptera, although closely related to the others, are both living. The development of wing flexing accounts for much of the adaptive radiation present in these insects. For example the thickening of the fore wings, which occurred many times independently, results in their diminished function in flight and in an increased role as protective covers for the insect's thorax and abdomen. As part of this same process, the hind wings were correspondingly widened, and lines of folding developed over the anal region, allowing that part of the wing to fold up like a fan. The Blattodea and Protorthoptera are examples of this phase of evolution. Subsequently, however, in several exopterygote orders the coriaceous nature of the fore wing became less marked, and the anal area was reduced to nearly its original size. Two of the extinct orders, the Caloneurodea and Miomoptera, lack an expanded area of the hind wing, and the fore wing is not coriaceous. The phylogenetic position of these orders is therefore unclear.

The insects placed in the order Protorthoptera have several conspicuous features in common, including coriaceous fore wings, at least slightly expanded anal areas of the hind wings, chewing mouthparts, cursorial hind legs, and prominent but not long cerci. In other respects, however, these insects showed extraordinary diversity of body structures. In some of the Carboniferous species, the prothorax is very elongate and armed with long spines [see Figure 1 (*bottom*)]; in others, one set of mouthparts, the maxillary palpi, are extraordinarily long; and in others the fore legs are elongate and raptorial. Attempts to divide this order into suborders and even into orders have proven unsatisfactory (Burnham 1983). As the order Protorthoptera is now understood, it is the largest of all extinct orders, and over thirty protorthopteran families have been recorded from the Upper Carboniferous alone.

The existing order Blattodea (cockroaches) is well represented in the Upper Carboniferous. Unfortunately, the classification of fossil cockroaches is difficult because morphological characters employed in the systematics of living species are not preserved in the fossils. In fact, the fore wings are usually the only parts of cockroaches preserved, and they lack the characters necessary for family- or generic-level determination. Although hundreds of species of Upper Carboniferous cockroaches have been described, they are generally placed in only four or five families. However, specimens of cockroaches are usually the most numerous of all insects in the Upper Carboniferous, comprising approximately 80% of all insects collected in this period. They are commonly found in coal shale, associated with such plants as horsetails, club mosses, and seed ferns. They probably lived in swamps and other wet regions, which provided ideal conditions for their preservation as fossils.

In general, the Upper Carboniferous cockroaches were similar in appearance to those now living, the differences in wing and body structure being slight. However, one distinctive feature of Carboniferous cockroaches was that the females had long external ovipositors. These do not occur in living species, and their presence in Carboniferous cockroaches suggests that females may have inserted their eggs in specific, perhaps concealed, places (Schneider 1977, 1978a,b).

The existing order Orthoptera (grasshoppers, katydids, crickets) includes species with the characteristics of the Protorthoptera but differs from them in having the hind legs modified for jumping. Only one family is known from the Upper Carboniferous.

The extinct order Caloneurodea included insects whose fore and hind wings were similar, but with the hind pair lacking the expanded anal area. Little is known about their body structure, except that the antennae and legs were long and slender and that short cerci were present. The order has usually been placed in the exopterygote Neoptera, but Sharov believed them to have been endopterygotes. Only two families are known from the Carboniferous (Sharov 1966).

The extinct order Miomoptera consisted of small to very small insects whose fore and hind wings were similar, without an enlarged anal area on the hind pair.

Permian

The Permian insect fauna is better known than that of any other pre-Tertiary fauna. Extensive collections from the Soviet Union, Czechoslovakia, the United States, and Australia have provided the basis for much of this knowledge. Twenty-two orders, including the 9 extinct ones, are in the Permian record. These are distributed as follows: Apterygota, 1 order; Paleoptera, 6 orders; exopterygote Neoptera, 10 orders; and endopterygote Neoptera, 5 orders. With the combination of 9 extinct and 13 extant orders, Permian insects apparently comprised the most morphologically diverse insect fauna that is known.

All of the orders found in the Upper Carboniferous continued into the Permian, some with significant changes in diversity. The Permian Archaeognatha, however, were very similar to those of the previous period and belonged not only to the same suborder (Monura) but also to the same family. The Monura were apparently already on the decline, and they are not known beyond the end of the Permian. The Ephemeroptera, on the other hand, were very diverse in the Permian. They were represented by 5 families, including many genera, and they are preserved as both nymphs and adults (Carpenter 1979). The adults had well-developed mandibles, in contrast to the existing mayflies, in which the mouthparts are atrophied and nonfunctional. The nymphs were similar to those of the living species, bearing lateral, tracheal gills. The Palaeodictyoptera were much less diversified than in the Carboniferous, with only 6 families represented. In some families there was a marked reduction and even complete loss of the hind wings (Sharov & Sinitshenkova 1977, Sinitshenkova 1980). The Megasecoptera, in contrast, were much more diverse in the Permian (Carpenter 1947), with their 12 families showing striking modifications of body and wings. The Diaphanopterodea, having 6 families, were likewise more diverse in the Permian and are considered to have been the most highly specialized of the Paleoptera. The Protodonata, although well represented in the Permian, were similar in size and venation to those of the Carboniferous and belonged to the same families. The Protorthoptera apparently reached the peak of development in the Permian, represented there by a total of at least 40 families (Carpenter 1950, 1966). The Blattodea were much less abundant in the Permian than in the Carboniferous. The Orthoptera, with 5 families, also showed much greater diversity than in the Carboniferous (Sharov 1971), as did the Caloneurodea, represented by 8 families. The Miomoptera are known from the Permian only by 3 families; one was a Carboniferous family (Palaeomanteidae), the members of which were extremely abundant in the Lower Permian and which are represented by nymphs as well as adults (Sharov 1957a).

The following orders occurred for the first time in the Permian:

PALEOPTERA The existing order Odonata (dragonflies and damselflies) was represented by six families belonging to three suborders; two of the suborders are extinct, the third (Zygoptera) is extant. All of the Permian Odonata were small and delicate (Carpenter 1931, 1947). Their nymphs are unknown.

EXOPTERYGOTE NEOPTERA The existing order Plecoptera (stone flies) is known in the Permian by several families that are closely related to the more generalized extant families, and they have been shown to possess tracheal gills and other modifications for an aquatic environment.

The extinct order Protelytroptera is known only in the Permian and by species that closely resemble the beetles (Coleoptera). The fore wings were modified in most species to form wing covers, or elytra, and the hind wings were able to fold up under the elytra at rest. However, the nature of the hind wings and the presence of cerci indicate that they were probably related to the earwigs (Dermaptera) and Protorthoptera, rather than to the Coleoptera (Carpenter & Kukalová 1964, Kukalová 1965).

The existing order Psocoptera (bark lice), which are very small insects related to the thrips (Thysanoptera), were represented by several families that were remarkably similar to some of the Hemiptera (true bugs) that occurred in the Permian (Becker-Migdisova & Vishniakova 1962).

The existing order Thysanoptera (thrips) is first recorded from Permian deposits, in which it is represented by two families and many genera. The wing venation of these Permian species is very similar in many ways to that of the Psocoptera and the Hemiptera (Homoptera); several of the genera now assigned to the Thysanoptera have previously been placed in both of those orders (Vishniakova 1981).

The existing order Hemiptera (true bugs), which are characterized by having the mouthparts modified to form a sucking beak, is known in the Permian by a very large number of families belonging to the suborder Homoptera. Four extant superfamilies and one living family (Cixiidae) have been identified (Becker-Migdisova 1960).

ENDOPTERYGOTE NEOPTERA The existing order Mecoptera (scorpion flies) first appeared in the Lower Permian, and by the end of that period it was represented by 11 families, nearly twice as many as exist now. Included are several species that apparently belonged to a living family (Nannochoristidae). In some other Permian families, the venational patterns grade into those of certain Permian Trichoptera (caddis flies) (Martynova 1962b).

The existing order Neuroptera (alderflies, snake flies, and lacewings) also made its first appearance in the Lower Permian, and by the end of the period six families belonging to all three of the extant suborders (Sialoidea, Raphidioidea, and Planipennia) were represented (Ponomarenko 1977b).

The existing order Trichoptera (caddis flies) is also known in the Lower Permian, with a total of four families for the period. The immature stages are unknown (Martynova 1962c, Sukatscheva 1976).

The members of the extinct order Glosselytrodea, known only from the Permian to the Jurassic, resemble certain Neuroptera in venation and in other details of wing structure. These features suggest that the order was endopterygote, but the evidence is by no means conclusive. Only three families are known (Martynova 1952, Sharov 1966).

The existing order Coleoptera (beetles) was represented in the Permian by six families, all of them belonging to the suborder Archostemata, the most generalized of all Coleoptera. At the present time there are only two families in that suborder (Ponomarenko 1969, 1977a).

MESOZOIC ERA

Triassic

The insect fauna of the Triassic is less known than that of any other period since the first appearance of the insects in the Upper Carboniferous. This is probably a result of the scarcity of freshwater deposits. The most striking aspect of the Triassic fauna is that all but one (Glosselytrodea) of the extinct orders that were present in the Permian are absent. In other respects, the Triassic record offers few surprises. Previously existing groups continued to diversify during this period, especially the Ephemeroptera, Odonata, and most of the exopterygote Neoptera (Pritykina 1981). The Archaeognatha were becoming increasingly modern, as evinced by the presence of a species belonging to the extant family Machilidae (Paclt 1972). Three living orders, one in the Exopterygota and two in the Endopterygota, also appeared for the first time in the Triassic:

EXOPTERYGOTE NEOPTERA The order Phasmida (walking sticks), related to the Orthoptera, is known from the Triassic of Australia and is represented by a single family (Martynov 1928, Martynova 1962a).

ENDOPTERYGOTE NEOPTERA The order Diptera (true flies) is represented in the Triassic by species assigned to eight different families, all apparently belonging to the suborder Nematocera, which contains the least specialized families of the order. These Triassic fossils have a curiously specialized wing venation, however, and may eventually turn out to belong to an extinct suborder (Rohdendorf 1964).

Representatives of the order Hymenoptera (sawflies, ants, bees, wasps) first appear in the Triassic and belong to a total of 59 genera in two families, both of which are extant and belong to the relatively unspecialized suborder Symphyta (Rasnitsyn 1964, 1969, 1975).

Jurassic

The insect fauna of the Jurassic period is better known than that of either the Triassic or Cretaceous, primarily because of the extensive insect-bearing localities discovered in the Soviet Union. Many existing families appeared during this period (Popov 1971, Rasnitsyn 1976, 1982, Vishniakova 1976), although extinct families still generally predominated. The Jurassic Diptera, for example, consisted of 29 families, of which all but 4 are now extinct; and the Hymenoptera included 17 families, of which all but 5 are now extinct. In marked contrast, however, three fourths of the Jurassic families of Coleoptera are still existing.

EXOPTERYGOTE NEOPTERA The order Dermaptera (earwigs) is the only new order to occur in the Jurassic. The fossils resembled living species in general appearance but differed in having cerci that are long and multisegmented, with only a suggestion in some of the coalescence of segments that eventually formed the forceps that are characteristic of all living Dermaptera (Vishniakova 1980).

Cretaceous

Few insect-bearing sediments from the Cretaceous are known, and our knowledge of the Cretaceous insect fauna has therefore been limited. However, relatively recent discoveries of Cretaceous ambers with insect inclusions have added greatly to the known fauna. Furthermore, the excellent preservation of most of the specimens has enabled detailed comparisons with living material. It now appears that the great majority of Cretaceous insects belong to existing families and represent an essentially modern fauna. This is, in part, owing to the origin and radiation of flowering plants that took place during the Cretaceous. Many of the Diptera and Hymenoptera preserved in amber from this period belong to families that are important pollinators. The Cretaceous is also a very important period with respect to the social insects. Ants first appeared at this time and have been found in amber from New Jersey and Siberia. Both workers and reproductive forms have been described, and it is clear that at least some form of social organization had already been achieved within the Hymenoptera (Burnham 1978). Two orders that make their first appearance in the Cretaceous are the Isoptera and the Lepidoptera:

EXOPTERYGOTE NEOPTERA The Isoptera (termites) have been found in Cretaceous sediments from both Canada and England (Jarzembowski 1981) and have been assigned to the existing family Hodotermitidae. Since all living members of this family are fully social, it is virtually certain that the Cretaceous species were also social (Jarzembowski 1981).

ENDOPTERYGOTE NEOPTERA The Lepidoptera (moths and butterflies) occur for the first time in the Cretaceous (Skalski 1979), and their appearance undoubtedly correlates with the origin of the angiosperms. Several species belonging to the extant family Micropterygidae, which is considered ancestral to the rest of the Lepidoptera, have been described.

CENOZOIC ERA

Tertiary

Since insect-bearing deposits of Tertiary age are numerous and accessible, our knowledge of the insects of this period is extensive. As early as the beginning of the Tertiary, the insect fauna was like that of the present in most respects. Nearly all of the families in the early Tertiary exist today, and by the middle of the period roughly three fourths of all genera are extant. In addition, five new orders occurred in the Tertiary:

APTERYGOTA The order Zygentoma (silverfish), although probably in existence at least as far back as the Triassic, first appears in the Lower Oligocene (Silvestri 1912).

EXOPTERYGOTE NEOPTERA The order Manteodea (mantids) is represented in the fossil record by species that have been included in extant genera belonging to two different families.

The order Embioptera (web spinners) is represented in the Tertiary by species assigned to three different families, one of which is extinct (Davis 1940).

ENDOPTERYGOTE NEOPTERA The order Siphonaptera (fleas) is known by two species belonging to a single extant genus (Peus 1968).

The Strepsiptera are entomophagous parasites and are known only from Baltic and Dominican amber. They are represented by two species belonging to two extant genera (Kinzelbach 1979).

Quaternary

Insects collected from this period occur primarily in peat and tar deposits, and (with only rare exceptions) have been assigned to existing species. These have been studied mainly in connection with the effects of Pleistocene glaciation on the geographical distribution of insects.

SUMMARY

The record of insects in the Upper Carboniferous shows that by the end of that period they had passed through all four major stages of their evolutionary history. The first three of these had apparently been completed in Lower Carboniferous time or earlier, before the fossil record of the insects actually begins. The Permian was a period of very rapid expansion of the orders. All 11 previously existing orders continued from the Carboniferous, and 11 others appeared before the end of the period. Some of the paleopterous orders had begun to dwindle, but others expanded proportionally. However, it was the exopterygote Neoptera that changed the most, resulting in a confusing combination of characters suggestive of several orders (as in the Protorthoptera), and similar radiations occurred among the Permian endopterygote Neoptera. Most of these lines died out early in the Mesozoic. Five orders had haustellate mouthparts, presumably for feeding on plant juices. It is probable that by the end of the Permian the insects had achieved as complicated a relationship with the flora of the time as our contemporary insects have with the flowering plants, which developed rapidly during the Cretaceous. The interval of time between the development of the flowering plants and the present was about the same (100 Myr) as that from the beginning of the Upper Carboniferous to the end of the Permian.

The transition from the Permian to the Mesozoic was abrupt, with only one of the nine extinct orders surviving beyond the Permian. The Diptera and the Hymenoptera, two of our largest orders at present, appeared in the Triassic and by the Jurassic included a substantial number of existing families. The appearance of the Lepidoptera in the Cretaceous, and especially of the social insects (such as the termites and ants), set the stage for the Tertiary. By mid-Tertiary, on the average, about 60% of the genera of insects are extant. A few species, apparently morphologically inseparable from some now living, have turned up in several of the Tertiary ambers.

ACKNOWLEDGMENT

This research was supported by National Science Foundation Grant DEB 8205398, F. M. Carpenter, Principal Investigator.

Literature Cited

Becker-Migdisova, Ye. E. 1960. New Permian Homoptera from European USSR. *Tr. Paleontol. Inst. Akad. Nauk SSSR* 76:1–112 (In Russian)

Becker-Migdisova, Ye. E., Vishniakova, V. N. 1962. Order Psocoptera. In *Principles of Paleontology*, ed. B. B. Rohdendorf, pp. 226–35. Moscow: Akad. Nauk SSSR (In Russian)

Brues, C. T., Melander, A. L., Carpenter, F. M. 1954. Classification of insects. *Bull. Mus. Comp. Zool., Harv. Univ.* 108:1–91

Burnham, L. 1978. Survey of social insects in the fossil record. *Psyche* 85:85–133

Burnham, L. 1983. Studies on Upper Carboniferous insects: 1. The Geraridae (Order Protorthoptera). *Psyche* 90:1–57

Carpenter, F. M. 1931. The Lower Permian insects of Kansas. Pt. 2. The orders Palaeodictyoptera, Protodonata, and Odonata. *Am. J. Sci.* 21(5):97–139

Carpenter, F. M. 1943. Studies on the Carboniferous insects from Commentry, France. Pt. 1. Introduction and families Protagriidae, Meganeuridae, and Campylopteridae. *Bull. Geol. Soc. Am.* 54:527–54

Carpenter, F. M. 1947. Lower Permian insects from Oklahoma. Pt. 1. Introduction and the orders Megasecoptera, Protodonata, and Odonata. *Proc. Am. Acad. Arts Sci.* 76:25–54

Carpenter, F. M. 1950. The Lower Permian insects of Kansas. Pt. 10. The order Protorthoptera: family Liomopteridae and its relatives. *Proc. Am. Acad. Arts Sci.* 78:185–219

Carpenter, F. M. 1963. Studies on the Carboniferous insects from Commentry, France. Pt. IV. The genus *Triplosoba*. *Psyche* 70:120–28

Carpenter, F. M. 1966. The Lower Permian insects of Kansas. Pt. 11. The orders Protorthoptera (continued) and Orthoptera, with a discussion of the Orthopteroid Complex. *Psyche* 73:46–88

Carpenter, F. M. 1976. Geological history and evolution of the insects. *Proc. Int. Congr. Entomol., 15th, Washington DC*, pp. 63–70

Carpenter, F. M. 1979. Lower Permian insects from Oklahoma. Pt. 2. Orders Ephemeroptera and Palaeodictyoptera. *Psyche* 86:261–90

Carpenter, F. M. 1985a. Studies on Paleozoic Archaeognatha (Insecta) from France, the

Soviet Union, and the United States. *Psyche*. In press

Carpenter, F. M. 1985b. Hexapoda: Classes Collembola, Protura, Diplura, Insecta. In *Treatise on Invertebrate Paleontology*. Lawrence, Kans: Geol. Soc. Am. In press

Carpenter, F. M., Kukalová, J. 1964. The structure of the Protelytroptera, with description of a new genus from Permian strata of Moravia. *Psyche* 71:183–97

Carpenter, F. M., Richardson, E. S. Jr. 1985. Archaeognatha (Insecta) from the Upper Carboniferous of Illinois. *Psyche*. In press

Davis, C. 1940. Taxonomic notes on the order Embioptera, XVIII. *Proc. Linn. Soc. NSW*. 65:362–87

Jarzembowski, E. A. 1981. An early Cretaceous termite from southern England (Isoptera: Hodotermitidae). *Syst. Entomol.* 6:91–96

Kingsolver, J. G. 1984. Aerodynamics, thermoregulation, and the evolution of insect wings: differential scaling and evolutionary change. *Evolution*. In press

Kinzelbach, R. K. 1979. Das erste neotropische Fossil der Fächerflügler (Stuttgarter Bernsteinsammlung: Insecta, Strepsiptera). *Stuttg. Beitr. Naturkd.* 52(B):1–14

Kukalová, J. 1965. Permian Protelytroptera, Coleoptera, and Protorthoptera of Moravia. *Sb. Geol. Ved. Paleontol.* 6:61–98

Kukalová, J. 1969a. Revisional study of the order Palaeodictyoptera in the Upper Carboniferous shales of Commentry, France, Pt. I. *Psyche* 76:163–215

Kukalová, J. 1969b. Revisional study of the order Palaeodictyoptera in the Upper Carboniferous shales of Commentry, France. Pt. II. *Psyche* 76:439–86

Kukalová, J. 1970. Revisional study of the order Palaeodictyoptera in the Upper Carboniferous shales of Commentry, France. Pt. III. *Psyche* 77:1–44

Kukalová-Peck, J. 1974. Wing-folding in the Paleozoic insect order Diaphanopterodea, with a description of a new representative of the family Elmoidae. *Psyche* 81:315–33

Kukalová-Peck, J. 1978. Origin and evolution of insect wings and their relation to metamorphosis, as documented by the fossil record. *J. Morphol.* 156:53–125

Martynov, A. V. 1925. Über zwei Grundtypen der Flügel bei den Insecten und ihre Evolution. *Z. Morphol. Oekol. Tiere* 4:465–501

Martynov, A. V. 1928. A new fossil form of Phasmatodea from Galkino (Turkestan) and on Mesozoic phasmids in general. *Ann. Mag. Nat. Hist.* 1(10):319–28 (In Russian)

Martynov, A. V. 1937. Studies on the geological history and phylogeny of the insects (Pterygota), Pt. 1. *Tr. Paleontol. Inst. Akad. Nauk SSSR* 7:1–148 (In Russian)

Martynova, O. M. 1952. The order Glosselytrodea in Permian beds of the Kamerovsky region. *Tr. Paleontol. Inst. Akad. Nauk SSSR* 40:187–96 (In Russian)

Martynova, O. M. 1962a. Order Phasmatodea. In *Principles of Paleontology*, ed. B. B. Rohdendorf, pp. 151–60. Moscow: Akad. Nauk SSSR (In Russian)

Martynova, O. M. 1962b. Mecoptera. In *Principles of Paleontology*, ed. B. B. Rohdendorf, pp. 283–92. Moscow: Akad. Nauk SSSR (In Russian)

Martynova, O. M. 1962c. Trichoptera. In *Principles of Paleontology*, ed. B. B. Rohdendorf, pp. 294–99. Moscow: Akad. Nauk SSSR (In Russian)

Paclt, J. 1972. Grundsätzliches zur Chorologie und Systematik der Felsenspringer. *Zool. Anz.* 188:422–29

Peus, F. 1968. Über die beiden Bernstein-Flöhe (Insecta, Siphonaptera). *Paläontol. Z.* 42:62–72

Ponomarenko, A. G. 1969. Historical development of the Coleoptera-Archostemata. *Tr. Paleontol. Inst. Akad. Nauk SSSR* 125:1–240 (In Russian)

Ponomarenko, A. G. 1977a. Mesozoic Coleoptera. *Tr. Paleontol. Inst. Akad. Nauk SSSR* 161:1–119 (In Russian)

Ponomarenko, A. G. 1977b. Paleozoic members of the Megaloptera (Insecta). *Paleontol. Zh.* 1977(1):78–86 (In Russian)

Popov, Y. A. 1971. Historical development of the hemipterous infra-order Nepomorpha. *Tr. Paleontol. Inst. Akad. Nauk SSSR* 129:1–227 (In Russian)

Pritykina, L. N. 1981. New Triassic Odonata from Central Asia. *Tr. Paleontol. Inst. Akad. Nauk SSSR* 183:5–42

Rasnitsyn, A. P. 1964. New Triassic Hymenoptera from Middle Asia. *Paleontol. Zh.* 1964(1):88–96 (In Russian)

Rasnitsyn, A. P. 1969. The origin and evolution of the Hymenoptera. *Tr. Paleontol. Inst. Akad. Nauk SSSR* 123:1–196 (In Russian)

Rasnitsyn, A. P. 1975. Hymenoptera Apocrita of the Mesozoic. *Tr. Paleontol. Inst. Akad. Nauk SSSR* 147:1–132 (In Russian)

Rasnitsyn, A. P. 1976. Living representatives of the order Protoblattodea. *Dokl. Akad. Nauk SSSR* 228(2):502–4 (In Russian)

Rasnitsyn, A. P. 1981. A modified paranotal theory of insect wing origin. *J. Morphol.* 168:331–38

Rasnitsyn, A. P. 1982. Triassic and Jurassic insects of the genus *Shurabia* (Grylloblattida, Geinitziidae). *Paleontol. Zh.* 1982(3):78–87 (In Russian)

Rohdendorf, B. B. 1964. Historical development of the Diptera. *Tr. Paleontol. Inst. Akad. Nauk SSSR* 100:1–311 (In Russian)
Rohdendorf, B. B., Rasnitsyn, A. P., eds. 1980. Historical development of insects. *Tr. Paleontol. Inst. Akad. Nauk SSSR* 178:1–268 (In Russian)
Schneider, J. 1977. Zur Variabilität der Flügel paläozoischer Blattodea (Insecta), Teil I. *Freiberg. Forschungsh. C* 326:87–105
Schneider, J. 1978a. Variabilität der Flügel paläozoischer Blattodea (Insecta), Teil II. *Freiberg. Forschungsh. C* 334:21–39
Schneider, J. 1978b. Zur Taxonomie und Biostratigraphie de Blattodea (Insecta) des Karbon und Perm der DDR. *Freiberg. Forschungsh. C* 340:1–152
Sharov, A. G. 1957a. Nymphs of Miomoptera from Permian deposits of the Kuznetsk Basin. *Dokl. Akad. Nauk SSSR* 112: 1106–8 (In Russian)
Sharov, A. G. 1957b. Distinctive Paleozoic wingless insects of a new order Monura. *Dokl. Akad. Nauk SSSR* 115:795–98 (In Russian)
Sharov, A. G. 1966. The position of the orders Glosselytrodea and Caloneurodea in the system of insects. *Paleontol. Zh.* 1966(3):84–93 (In Russian)
Sharov, A. G. 1971. Phylogeny of the Orthopteroidea. *Tr. Paleontol. Inst. Akad. Nauk SSSR* 118:1–212 (In Russian)
Sharov, A. G., Sinitshenkova, N. D. 1977. New Palaeodictyoptera from the USSR. *Paleontol. Zh.* 1977(1):48–63 (In Russian)
Silvestri, F. 1912. Die Thysanuren des baltischen Bernstein. *Schr. Phys.-Oekon. Ges. Königsberg Preuss.* 53:42–66
Sinitshenkova, N. D. 1980. A revision of the order Permothemistida (Insecta). *Paleontol. Zh.* 1980(4):91–106 (In Russian)
Skalski, A. W. 1979. A new member of the family Micropterygidae (Lepidoptera) from the Lower Cretaceous of Transbaikalia. *Paleontol. Zh.* 1979(2):90–97
Snodgrass, R. E. 1935. *Principles of Insect Morphology*. New York: McGraw-Hill. 667 pp.
Sukatscheva, I. D. 1976. Caddis-flies of the suborder Permotrichoptera. *Paleontol. Zh.* 1976(2):94–105 (In Russian)
Vishniakova, V. N. 1976. Relict Archipsyllidae (Insecta, Psocoptera) in the Mesozoic fauna. *Paleontol. Zh.* 1976(2): 76–84 (In Russian)
Vishniakova, V. N. 1980. Earwigs (Insecta, Forficulida) from the Upper Jurassic of the Karatau Range. *Paleontol. Zh.* 1980(1): 78–94 (In Russian)
Vishniakova, V. N. 1981. New Paleozoic and Mesozoic Lophioneuridae (Thripida, Lophioneuridae). *Tr. Paleontol. Inst. Akad. Nauk SSSR* 183:43–61 (In Russian)
Wootton, R. J. 1981. Palaeozoic insects. *Ann. Rev. Entomol.* 26:319–44

Ann. Rev. Earth Planet. Sci. 1985. 13 : 315–44

DOWNHOLE GEOPHYSICAL LOGGING

A. Timur

Chevron Oil Field Research Company, La Habra, California 90631

M. N. Toksöz

Earth Resources Laboratory, Department of Earth, Atmospheric, and Planetary Sciences, Massachusetts Institute of Technology, Cambridge, Massachusetts 02139

INTRODUCTION

Downhole geophysical logging is the method of making geophysical measurements in drill holes and interpreting these measurements for evaluation of geological properties of the subsurface. The downhole measurements can be made with great spatial resolution with a single sonde, containing both source and receivers, which are separated by 10 cm to tens of meters. Measurements can be repeated every few centimeters as the sonde moves upward or downward in the borehole. This technique is called well logging. It is also possible to make downhole measurements with the sensor in the borehole at depth and the source at the surface. When the sensor package is moved to a different depth, the source signal is repeated. When these measurements are made with seismic waves, the process is called vertical seismic profiling (VSP). In VSP the surface source can be any appropriate system that would generate a seismic pulse (dynamite, air gun) or a sinusoidal wave (vibroseis), and the downhole sensor can be a hydrophone or borehole geophone. Vertical seismic profiling is a rapidly expanding technique, and some of its applications are described in a series of recent books (Gal'perin 1974, Hardage 1983, Balch & Lee 1984, Toksöz & Stewart 1984). Another method of downhole measurement is possible when closely spaced drill holes are available, with the source placed in one drill hole and the detector (receiver) in another. This technique, called cross-

0084–6597/85/0515–0315$02.00

hole measurement, is used extensively in geotechnical applications but has limited applications in the Earth sciences because of the necessity of closely spaced, deep drill holes. In this paper we concentrate only on the well-logging method, since it covers the broadest spectrum of downhole geophysical measurements.

A large number of wells (more than 50,000 in the United States alone) are drilled each year for scientific studies, for resource exploration and production purposes, and for waste disposal. Drilling alone provides limited information about the geologic formations at depth. Recovery and analysis of cores (samples) are the primary means for determining petrological and petrophysical properties. Core analyses, however, have some drawbacks. Cores represent a very small sample of the formation in a heterogeneous geologic environment. Material properties change as a result of drilling, coring, recovery, and laboratory procedures, and measured core properties may be different than in situ properties. In some instances (poorly consolidated materials, fractured rocks), core recovery is a major problem. In well logging a wide range of physical properties (electrical, nuclear, acoustic) are measured in the borehole as a function of depth. These data are then interpreted in terms of geologic properties, generally with the aid of laboratory data and theoretical models. Borehole measurements involve downhole sensors (sondes), a means of transmitting data to the surface, and recording and processing facilities at the surface. The amount of data depends on the type of measurement, varying from less than 10^3 bits m^{-1} of hole for some to more than 10^7 bits m^{-1} for others (such as full-waveform acoustic logs). A schematic diagram of a logging system is shown in Figure 1.

Since its introduction in 1928 by the Schlumberger brothers for resistivity measurements, well logging has expanded rapidly to include the measurements of many other properties in the borehole. The primary application of well logging has been in the petroleum industry. As a result, the most emphasis has been placed on the characterization of sedimentary rocks and their pore fluids—water or hydrocarbon. The application of well logging to other areas such as mining, geotechnical engineering, water resources, and scientific studies is expanding rapidly. Today there are more than 40 separate physical measurements that can be made in a borehole. A listing of the more commonly used measurements is given in Tables 1–4, classified according to measured physical property. The three primary properties are electrical (Table 1), nuclear (Table 2), and acoustic (Table 3). In addition there is a spectrum of other measurements with specific applications (Table 4). Detailed descriptions of individual logs are available in the literature (Jennings & Timur 1973, Evans & Gouilloud 1979, Segesman 1980, Hilchie

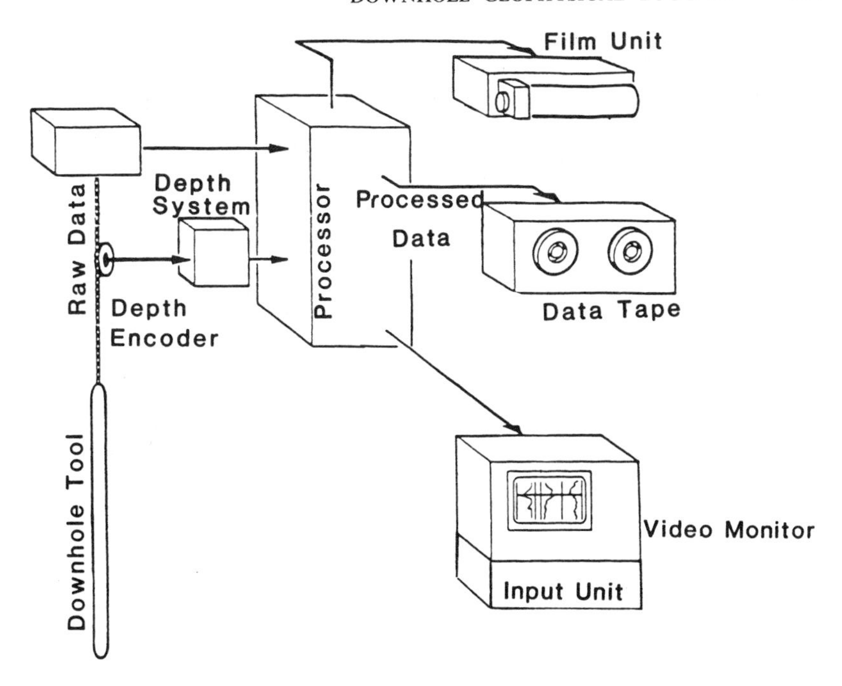

Figure 1 Schematic diagram of a logging system with a computerized data acquisition and interpretation unit.

1978, Serra et al 1980, Timur 1982). In this paper we cover the basic principles and more recent developments in well-logging methods.

ELECTRICAL METHODS

Electrical properties are measured in boreholes over a broad frequency range, from direct current to gigahertz. Numerous tools and techniques (Table 1) have been developed to characterize the electrical properties of the borehole environment, to determine the true resistivity of the formation, and to interpret these properties in terms of rock matrix, pore, and pore-fluid properties.

Except for spontaneous potential (SP) and dielectric logs (EPT), the electrical logs listed in Table 1 measure resistivity or conductivity. Logging tool design, electrode spacings, and frequencies are arranged to focus the current to a zone of investigation either close to the borehole or deep in the formation. Drilling fluid has finite resistivity, and when it diffuses into the

Table 1 Electrical and electromagnetic logs

Method	Measured property	Applications
Spontaneous potential (SP)	Electrochemical and electrokinetic potentials	Shales and shaliness formation water resistivity (R_w), bed boundaries
Nonfocused electric log	Resistivity	(*A*) Water and gas/oil saturation (*B*) Porosity of water zones (*C*) R_w in zones of known porosity (R_0) (*D*) True resistivity of formation (R_t) (*E*) Resistivity of invaded zone (R_{x0})
Focused induction conductivity log	Resistivity	*A*, *B*, *C*, *D*, very good for estimating R_t in either freshwater or oil-base mud
Focused resistivity logs	Resistivity	*A*, *B*, *C*, *D*, especially good for determining R_t of thin beds, depth of invasion
Focused and nonfocused microsensitivity logs	Resistivity	Resistivity of invaded zone (R_{x0})
Dielectric logs (EPT)	Dielectric permittivity	Lithology, water and gas/oil saturations independent of salinity

formation it creates an "invaded zone" (Figure 2). Many quantities need to be measured directly or determined as part of the resistivity log interpretation. These include the resistivities of the mud filtrate (R_{mf}), the formation water (R_w), the invaded zone (R_{x0}), and the diameter of the invaded zone (d_i); these are then used to determine the resistivity of the formation (R_t).

In most sedimentary rocks, the porosity and salinity of the saturating fluid determine the formation resistivity. The relationship is expressed by using the formation factor F, where[1]

$$F = \frac{R_0}{R_w}. \tag{1}$$

[1] In shaly formations, Equation (1) would have a correction factor based on the effective concentration of clay exchange cations. This correction can be determined by laboratory measurement of cation exchange capacity (Waxman & Smits 1968).

Table 2 Nuclear radiation logs

Method	Measured property	Applications
Gamma ray	Total gamma rays due to natural radioactivity	Shales and shaliness
Spectral gamma ray	Gamma-ray spectra of natural radioactivity	Lithologic identification
Gamma-gamma (Compton)	Electron density	Porosity, lithology
Gamma-gamma (photoelectric)	Photoelectric absorption cross section	Lithology
Neutron-gamma	Hydrogen content	Porosity
Neutron–thermal neutron	Hydrogen content	Porosity—gas vs liquid
Neutron–epithermal neutron	Hydrogen content	Porosity—gas vs liquid
Pulsed neutron capture	Decay rate of thermal neutrons	Water and gas/oil saturations behind casing
Induced gamma ray	Gamma-ray spectra induced by neutron interactions	Elemental composition of the formation rock and fluids

R_0 is equivalent to R_t when the rock is saturated with water only. Empirical observations indicate a power-law dependence of F on porosity φ, i.e.

$$F = A\varphi^{-m}, \tag{2}$$

where A and m are constants, determined by measurements on core samples. Generally A is close to 1 and m is about 2. [Equation (2) is usually referred to as Archie's Law.] In an extensive study of 1800 sandstone samples from 15 oil fields, Timur et al (1972) obtained values of $A = 1.13$

Table 3 Acoustic logs

Method	Measured property	Applications
Conventional acoustic	Compressional wave velocity	Seismic velocity, porosity
Full-waveform acoustic	Microseismograms (waveforms)	Compressional and shear velocities, attenuations, lithology, porosity, permeability fractures
Bond logging	High-frequency microseismograms	Cement bond quality

Table 4 Other widely used logs

Method	Measured property	Applications
Caliper	Borehole diameter	Location of mud cake, washout
Deviation log	Azimuth and inclination of borehole	Borehole position
Dipmeter	Four orthogonal resistivities	Dip and strike of beds, sedimentary features
Gravity meter	Gravity field	Formation density
Mud logging	Mud, drill cuttings, drilling variables; evaluation, drilling control	Real time formation properties
Nuclear magnetic resonance	Free hydrogen, relaxation rate of hydrogen	Effective porosity and permeability
Televiewer	Ultrasonic reflections from borehole surface	Fractures, orientation of fracture planes, bed boundaries, break-out, in situ stress
Wireline formation testing	Pressure; fluid gradients; water, oil, and gas sampling	Reservoir pressure and indication of permeability; fluid gradients; water oil, and gas samples

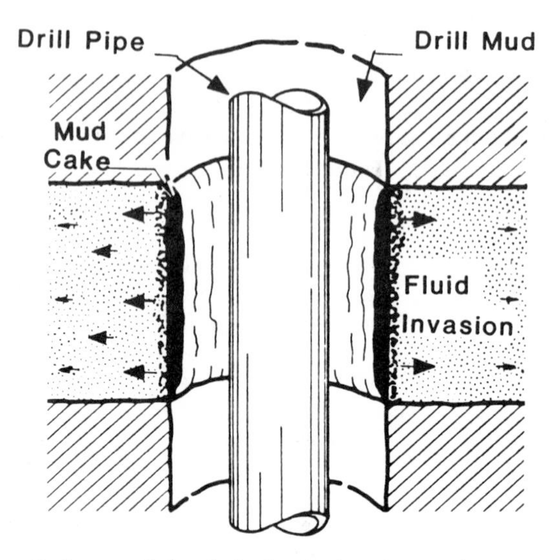

Figure 2 Schematic diagram of a borehole, showing invasion zone in a permeable formation.

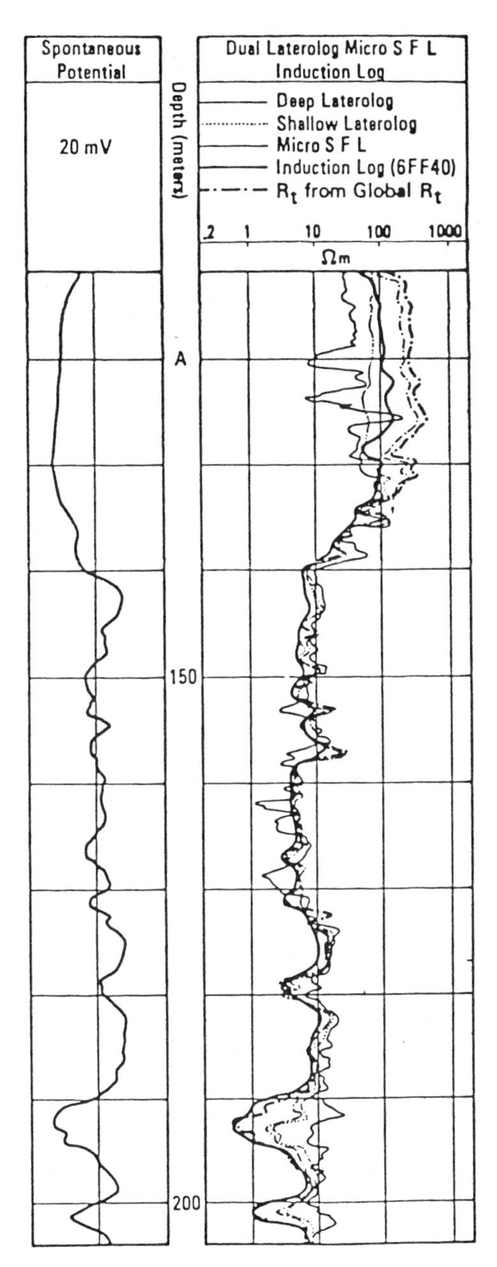

Figure 3 An example of SP and resistivity logs. Listing of logs given in Table 1. True formation resistivity profile R_0 is calculated from observed logs [data from Schlumberger Ltd. (see Ancel et al 1974)].

and $m = 1.73$. It is important to mention that clay minerals could have a significant effect on formation resistivity and on the constants A and m, especially in sandstones.

Resistivity Logs

An example of various borehole electrical resistivity measurements is illustrated in Figure 3 with logs from the Illizi basin in Algeria. Curves shown are spontaneous potential (SP), the dual (deep and shallow) laterolog, focused microresistivity (MSFL), and the induction log. The SP log is interpreted as a measure of shaliness, with low values indicating relatively "clean" sands. The two lower sands at 192 and 201 m depth are water bearing (low resistivity), and the upper zone above 34 m is hydrocarbon bearing (high resistivity).

To calculate the true resistivity profile, a computer program utilizes all available resistivity logs to determine R_t as well as R_{x0} and d_i (Mayer & Sibbit 1980). The evaluation is based on a step-resistivity model of the borehole, the invaded zone, and the unaltered formation. Tool response characteristics are included in the calculations. The inversion or the fitting of the observed log data is an iterative process. This iteration is continued until the values for the three unknown parameters (R_t, R_{x0}, d_i) are optimized at each level to a predetermined statistical significance.

The logs in Figure 3 were processed with this method until the resistivity profile R_t shown in the figure was obtained. The high R_t value in the upper sand is the key to determining hydrocarbon (instead of water) saturation.

Dielectric Logs

The resistivity and conductivity methods listed in Table 1 depend on the salinity of the water in the formation to evaluate water saturation and/or hydrocarbon saturation. As the salinity is reduced, it becomes more difficult to differentiate hydrocarbons from water, and the accuracy of saturation determination becomes unreliable in formations containing freshwater or water of unknown salinity. Most of the electrical tools operate within the frequency range of 35 Hz (focused resistivity) to 20 kHz (focused conductivity).

Recent developments in electrical logging have concentrated on very high frequencies and dielectric properties. The electromagnetic propagation tool (EPT) operates at frequencies in the gigahertz range and measures the phase shift and attenuation of an electromagnetic wave traveling in the formation (Calvert et al 1977). The phase shift is converted to propagation time, which is related to the dielectric constant. The relative dielectric constant ε_r' for water is significantly higher than that for gas, oil, and matrix materials, and it is almost independent of the salinity of the water (Table 5).

Table 5 Relative dielectric constants and propagation time for various Earth materials (data from Segesman 1980)

Mineral	$\varepsilon_r' = \varepsilon'/\varepsilon_0$	t_{pl} (ns m^{-1})
Sandstone	4.65	7.2
Dolomite	6.8	8.7
Limestone	7.5–9.2	9.1–10.2
Anhydrite	6.35	8.4
Halite[a]	5.6–6.35	7.9–8.4
Gypsum[a]	4.16	6.8
Petroleum	2.0–2.4	4.7–5.2
Shale	5–25	7.45–16.6
Freshwater at 25°C	78.3	29.5

[a] Average value from publications.

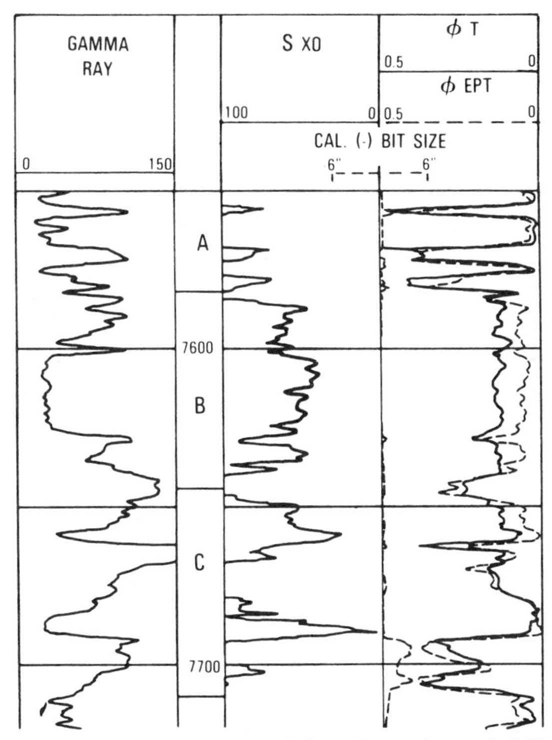

Figure 4 Porisity and saturation from dielectric logs. Gamma log on the left is for reference. S is water saturated; φ_{EPT} is water-filled porosity measurements based on dielectric logs. φ_T is total porosity obtained from density and neutron log combination. Note that in zone B, φ_{EPT} is lower than φ_T, indicating hydrocarbon presence (Wharton et al 1980).

Propagation time can be used to determine water-filled porosity if the lithology is known. In conjunction with the other porosity tools, the EPT evaluates hydrocarbon saturation.

An example of the use of a dielectric log to obtain water saturation is shown in Figure 4. The matrix travel time t_{pm} is used to calculate the porosity φ_{EPT}, plotted in track 3 of Figure 4. The water saturation S_{x0} is calculated and plotted in track 2. Also plotted in this figure for purposes of comparison is the total porosity φ_T obtained from density and neutron log combination (described in the next section). In zone B, φ_{EPT} is lower than φ_T, thereby indicating the presence of hydrocarbons.

NUCLEAR METHODS

The first borehole measurements of the nuclear properties of Earth formations were conducted in 1939 to determine the level of natural radioactivity. The technology has evolved to include a broad spectrum of measurements for use in both open and cased holes (Table 3). Two reprint volumes (Hoyer et al 1976, Lawson et al 1978) present a more detailed description of these methods. In this section we emphasize the most recent developments.

Natural Gamma-Ray Logs

Potassium (^{40}K), uranium, and thorium are three principal radioactive elements in the Earth's crust that emit gamma rays of characteristic energies (Figure 5). By counting gamma rays over a spectrum of energy bands, we can compute the abundance of these three elements near the borehole. The total flux of radiation can be used to identify potassium-rich rocks, shales among sedimentary rocks, and granitic composition among crystalline rocks.

Spectral gamma-ray logging is carried out in practice by counting gamma rays in a number of predetermined energy windows. Relative abundances are determined by the inversion of these data. However, the inversion process is complicated by several factors. Uranium and thorium decay with a series of gamma emissions, associated with each of their daughter products, before an eventual disintegration to lead. In order to accumulate statistically significant count rates, it would be useful to count daughter product emissions as well. Unlike the laboratory, a borehole is not a practical location for long-time counting. Thus, one is forced to assume that equilibrium exists between daughter and parent elements. In addition, all gamma rays in this energy band may be subject to Compton scattering, causing significant blurring of the recorded spectra. The magnitude of this effect is largely controlled by formation properties.

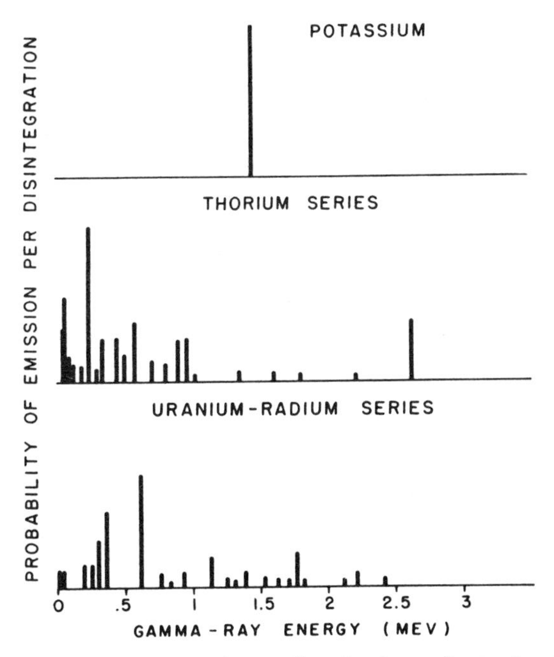

Figure 5 Gamma-ray energy spectra from radioactive decay of potassium (^{40}K), thorium, and uranium series.

Natural gamma-ray spectroscopy can be used for lithology identification, for detailed stratigraphic correlation, for recognition of different facies, for determination of reservoir shaliness and source-rock potential, and for mineral evaluation, such as for coal, uranium, and potash (Fertl 1979). The example in Figure 6 illustrates the response in an organic-rich (source-rock) shale in contrast to a typical shale. Relatively clean limestone (4122–4218 ft) is characterized by low concentrations of potassium, uranium, and thorium. Below the limestone, the typical shale response is illustrated by high potassium and thorium and moderate uranium concentrations. Above the limestone, the organic-rich (source-rock) shale is characterized by the low thorium but high uranium concentrations.

Gamma-Gamma Logs

In addition to use in computing levels of natural radioactivity, artificial gamma-ray sources are also utilized to determine formation density. A cesium-137 source placed downhole emits gamma rays at a known energy of 550 keV. Compton scattering is the dominant interaction in this energy range, but the photoelectric effect becomes important at energies <75 keV.

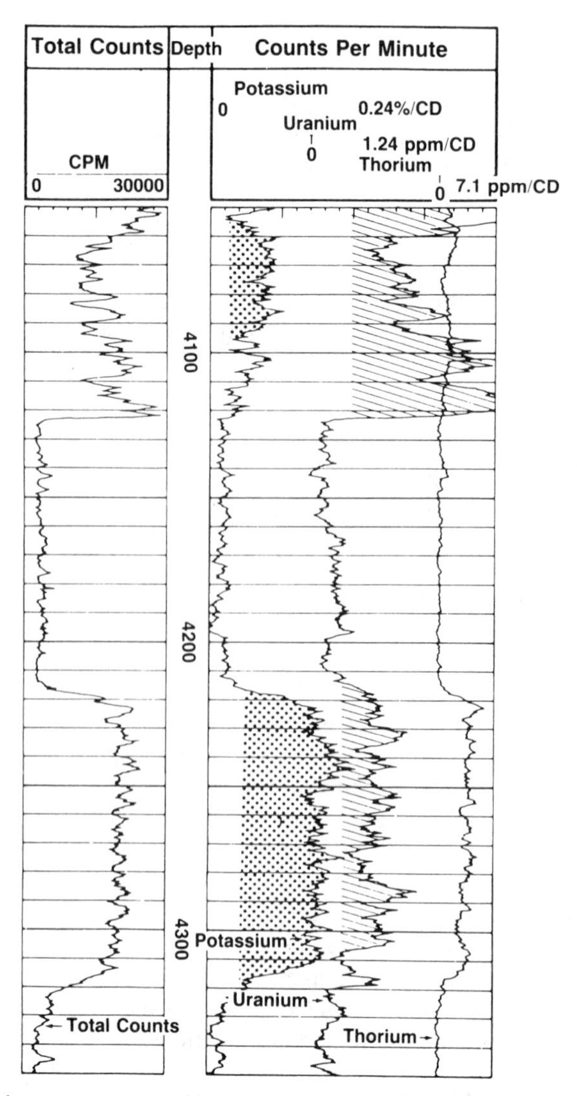

Figure 6 Natural gamma-ray spectral log. Total gamma counts are on the left; potassium, thorium, and uranium abundances based on gamma-ray spectral decomposition are on the right. Stippled zones on potassium curve indicate shales, and shaded zones of the uranium curve denote areas of organic-material abundance. Top shale is a good candidate for a source rock (Fertl 1979).

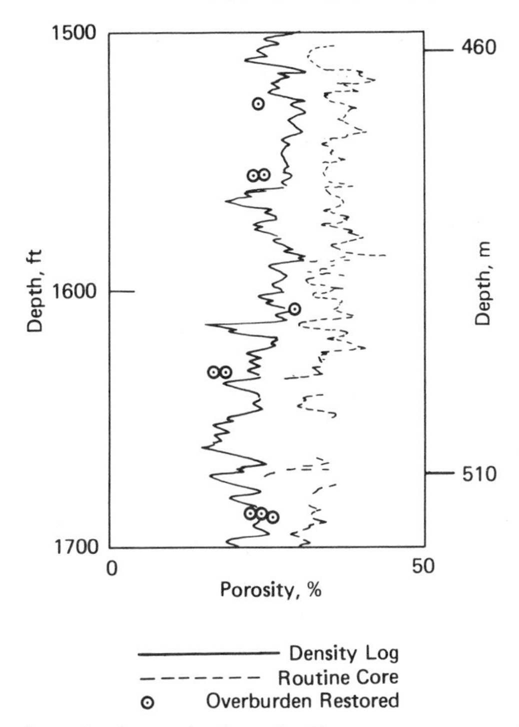

Figure 7 Comparison of rock porosity determined by core measurement (dashed line) and from neutron density log (solid line). "Routine core" measurement is at atmospheric pressure. Points are measured core porosity at confining pressure, equal to the overburden pressure (Neuman 1978).

If we consider only Compton-scattered gamma rays, the number of gamma rays backscattered to a detector at a fixed distance is related to the overall electron density of the formation. Electron density is then converted to bulk density by an empirical calibration process.

In practice, a gamma-ray source and two detectors are mounted on a sidewall skid, which is applied to the wall of the borehole (Wahl et al 1964). The radioactive source emits gamma rays continuously that interact with the formation. Gamma rays are counted within the energy band where the Compton scattering is predominant. The bulk density ρ_b measured by the tool is the volumetric average of all the constituents of the rock matrix. For a sedimentary rock composed of matrix and water-filled pores, if the matrix density ρ_m and the fluid density ρ_f are known, the porosity φ can then be calculated from the volumetric average equation

$$\rho_b = (\rho_f - \rho_m)\varphi + \rho_m. \tag{3}$$

An example of a density log comparison with core analysis porosity is shown in Figure 7 for a poorly consolidated sandstone. Porosities from density logs are significantly less than those from routine core analysis measurements at atmospheric pressure. However, when the core porosities are measured under simulated subsurface pressure conditions, core and log porosities become much closer (Neuman 1978).

A recent development in the gamma-gamma method is the downhole measurement of a photoelectric absorption cross section (Felder & Boyeldieu 1979). In the energy band observed by the detector in the borehole, the higher part of the gamma-ray energy is governed by the electron density. The lower-energy part of the spectrum is governed by photoelectric absorption effects. Since photoelectric absorption is strongly dependent on the atomic number Z^4, the low-energy spectrum reflects the lithology effect much more strongly. In the "lithodensity" log, the electron density effect is eliminated by taking the ratio of the count rates in the low- and high-energy windows, and the resulting ratio is used to obtain the photoelectric absorption cross section P_e. Typical values of P_e (b atom^{-1}) are listed in Table 6 for various Earth materials (Edmonson & Raymer 1979). Figure 8 illustrates the response of this measurement in different lithologies. The P_e curve is a good lithology indicator. It is affected very little by the nature of the fluids in the pores. In combination with density and

Table 6 Photoelectric absorption cross sections (data from Edmonson & Raymer 1979)

Material	P_e (b atom^{-1})	Specific gravity
Quartz	1.81	2.65
Calcite	5.08	2.71
Dolomite	3.14	2.87
Anhydrite	5.05	2.96
Halite	4.65	2.71
Siderite	14.7	3.94
Pyrite	17.0	5.00
Barite	267	4.48
Kaolinite	1.83	2.42
Chlorite	6.30	2.77
Illite	3.45	2.53
Montmorillonite	2.04	2.12
Water (fresh)	0.358	1.00
Water (10^5 ppm NaCl)	0.734	1.06
Water (2×10^5 ppm NaCl)	1.12	1.12
Oil [n (CH_2)]	0.119	ρ_{oil}
Gas (CH_4)	0.095	ρ_{gas}

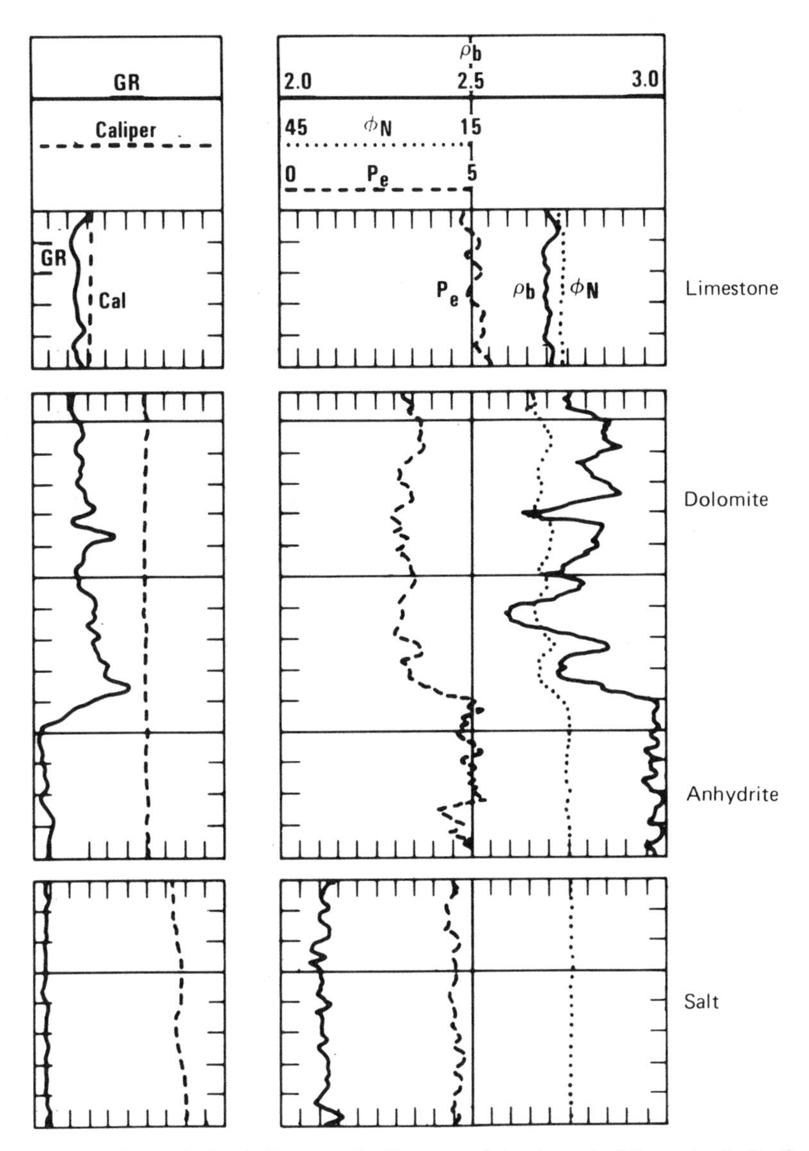

Figure 8 Lithodensity log in low-porosity limestone, dolomite, anhydrite, and salt. On the left are gamma and caliper logs. ρ_b = bulk density, φ_N = neutron porosity, P_e = photoelectric absorption cross section (b atom^{-1}). Note the high P_e values for limestone and anhydrite, as shown in Table 6 (Felder & Boyeldieu 1979).

neutron logs, it can be used to analyze complex lithologies (Gardner & Dumanoir 1980).

Neutron Logs

In neutron logging, a steady-state neutron source is used to determine porosity indirectly through measurements of hydrogen number density. Fast neutrons (energy greater than 3 MeV) traveling through matter generally lose their energy by scattering (elastic collisions with other atoms). Among the common elements, hydrogen has the highest scattering cross section (20 b), i.e. it has the highest probability of colliding with a

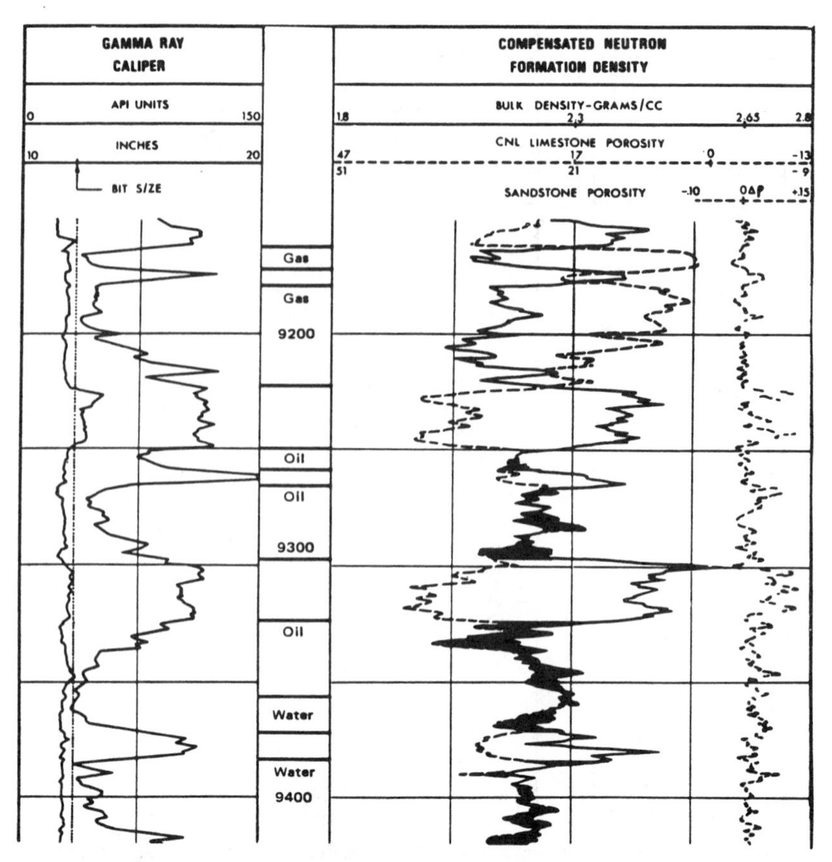

Figure 9 Bulk density and compensated thermal neutron (CNL) porosity log responses in water, oil, and gas zones. Two porosity scales are given for CNL (limestone and sandstone), since the calibration depends on rock type. Note that CNL "porosity" is lower in gas zones because of fewer hydrogen atoms. Hydrated clays produce the opposite behavior. For water, both porosities are about the same (Ancel et al 1974).

neutron. Since hydrogen has about the same mass as a neutron, the average energy loss per collision is also highest for hydrogen. The slowing down of high-energy neutrons depends primarily on the amount of hydrogen present. In rocks, hydrogen is usually found as water or hydrocarbon. By restricting a detector to count only thermal (<0.25 eV) neutrons that have lost their energy by scattering, one can determine the amount of hydrogen and, hence, the apparent porosity. Hydrogen that is present as part of the rock matrix (e.g. hydrated clays) may cause errors in the estimate of porosity.

An example of a simultaneous recording of bulk density measured by gamma-gamma logs and of porosity measured by the compensated thermal neutron log is given in Figure 9. Bulk density can be converted to porosity by using Equation (3). This combination gives porosity and is useful for identifying pore fluids (gas, water, oil). Because of the calibration process, both measurements of porosity are in agreement only in clean, water-filled limestones. In gas zones, "neutron porosity" is lower, since gas has fewer hydrogen atoms than water per unit pore volume. In shales, neutron porosity is too high because of hydrogen bound to the surface of clays.

After a neutron has been moderated to thermal energy levels, it may be absorbed by a nucleus in the formation. Conservation of energy necessitates the emission of a "gamma ray of capture" at a wavelength that is characteristic of the element involved. Absorption cross sections of the common elements vary from 0.0002 b for oxygen to 0.3 b for hydrogen and 31.6 b for chlorine. Early neutron logs were designed to count the capture gamma rays. More recent tools utilize dual detectors to count thermal neutrons and so are able to compensate for borehole effects such as salinity, mud cake, etc.

Pulsed Neutron Logs

Pulsed neutron logs utilize a source emitting high-energy neutron pulses (14 MeV) at high repetition rates. These neutrons lose most of their energy through inelastic scattering during the first few microseconds to become epithermal. Then, through elastic collisions, they slow down to thermal energy and finally are captured by some nuclei, emitting a capture gamma ray.

The most common pulsed neutron logging technique, pulsed neutron capture logging, involves monitoring the rate of decay of the capture gamma-ray flux. Since the number of capture events in a medium is proportional to the number of neutrons present, this technique yields the macroscopic capture cross section. Present logging technology utilizes two detectors and various time gates to measure the formation capture cross section Σ_b. For a saturated rock this includes contributions from the rock matrix, the interstitial water, and the interstitial hydrocarbons:

$$\Sigma_b = \Sigma_m(1-\varphi)+\Sigma_w S_w \varphi + \Sigma_{hc}(1-S_w)\varphi, \tag{4}$$

where Σ_m, Σ_w, and Σ_{hc} are the neutron absorption cross sections for rock matrix, water, and hydrocarbon, respectively, and S_w is the water saturation in fraction of pore volume.

The neutron absorption cross section is very sensitive to chlorine. Since chlorine is primarily in brine ($NaCl + H_2O$), pulsed neutron logs have a response similar to resistivity logs. However, unlike a resistivity log, they can be run in cased holes. Primary applications have been in locating hydrocarbon-bearing formations; monitoring gas/oil, gas/water, and oil/water contacts; and determining water saturation behind casing. One of the more useful applications of pulsed neutron logs is in determining residual oil saturation before instituting an enhanced oil recovery project. Examples of many of these applications are described in papers compiled in the reprint volumes cited earlier.

Another pulsed neutron method utilizes the induced gamma-ray spectra resulting from both fast neutron interactions and neutron capture. Again, a pulsed neutron source is used to generate 14-MeV neutrons. These interact with formation nuclei, which emit gamma rays of well-defined energies for each specific nuclear reaction. Therefore, elements in a formation can be identified by their gamma-ray emission lines, and concentrations can be determined by the relative intensity of these lines. This is equivalent to neutron activation analysis in a borehole.

These principles were used in developing the carbon/oxygen ratio logs for determining hydrocarbon saturation independent of formation water salinity, and in developing gamma-ray spectroscopy (GST) logs for determining the elemental abundances. The elements C, Ca, Fe, O, S, and Si are determined from the fast neutron interactions, and the elements Ca, Cl, Fe, H, S, and Si from the neutron capture. The results are presented as ratios, as shown in examples given in Table 7.

Table 7 Elemental yield ratios for induced gamma-ray spectroscopy (modified from Westaway et al 1980)

Yield ratio	Interaction	Application
C/O	Inelastic	Carbon-oxygen ratio
Cl/H	Capture	Salinity indicator
H/(Si + Ca)	Capture	Porosity indicator
Fe/(Si + Ca)	Capture	Lithology (iron indicator)
Si/(Si + Ca)	Capture and inelastic	Lithology

ACOUSTIC LOGGING METHOD

Acoustic logs have become an integral part of well logging since the first downhole measurements of velocities were conducted in 1927 to obtain time-depth data for seismic interpretation. For conventional acoustic logs, the sonde hangs in the center of a borehole. Seismic waves are generated by a source transducer at one end of the sonde and are recorded by one or more receivers at the other end. The travel times of refracted waves propagating through the formation allow the calculation of formation velocity.

Historically, the greatest use of the acoustic logs has been for the determination of compressional wave velocities V_p as a function of depth. The compressional wave data have been used both for interpreting seismic reflection sections and for determining formation porosity.

For this purpose, the linear relationship

$$\frac{1}{V_p} = c_1\varphi + c_2 \tag{5}$$

is used for estimating porosity φ from the compressional wave travel time $1/V_p$ by using empirical coefficients c_1 and c_2 (which are determined by measurements on core samples). This method generally works well for clean sandstones and for carbonates. As in other conventional porosity logs, however, variations in lithology make the porosity estimates from compressional wave transit times alone unreliable. In these instances, conventional acoustic logs are used in conjunction with density and/or neutron logs.

FULL-WAVEFORM ACOUSTIC LOGS

Full-waveform acoustic logging is rapidly emerging as an important borehole technique for formation evaluation and for seismic exploration. Borehole measurements of compressional and shear wave velocities and attenuations define formation properties and are used for the modeling and interpretation of seismic reflection and vertical seismic profiling (VSP) data.

In conventional acoustic logging, a tool may contain one or two sources and a pair of receivers. Only the time delay (or moveout) of the compressional headwave is recorded and the slowness (inverse of velocity) calculated. With the full-waveform acoustic logging tools, the entire microseismogram at each receiver is recorded digitally. These new digital tools generally have a longer (3–20 m) source/receiver separation than is commonly used in conventional borehole-compensated sonic logging. With a combination of multiple sources and receivers, more than 60 records of data can now be obtained at each depth. Thus, a large volume of data is generated for even a shallow well. A typical logging tool is shown in Figure 10.

In Figures 11*a* and 11*b*, two examples of full-waveform data are shown as a function of depth. Figure 11*a* is from a section of evaporites, carbonates, and shales. Different formations stand out. The section shows that there are different wave types propagating with different velocities and having different amplitudes. Figure 11*b* is an example of a log in a carbonate sequence. In addition to the section, individual wave traces are shown at selected depths. Different wave types are also indicated.

For the simple case of a fluid-filled borehole in a formation where both P- and S-wave velocities are higher than the fluid velocity, there are essentially four types of elastic waves that propagate: two headwaves and two guided waves. The well-known P-wave begins as a compressional wave in the borehole fluid, is critically refracted into the formation as a P-wave, and then is refracted back into the fluid as a compressional wave. The so-called S-wave begins as a compressional wave in the borehole fluid, is critically refracted into the formation as an S-wave, and then is refracted back into the fluid as a compressional wave. Between the P and S headwave arrivals there exists a ringy packet called the leaky or PL mode (Cheng & Toksöz 1980). The leaky mode amplitude, and hence the appearance of the P-wave train, varies strongly with the Poisson ratio.

In this review we do not go into the mathematical treatment of wave propagation in a borehole. This has been done by a number of investigators for boreholes of varying degrees of complexity (Biot 1952, White &

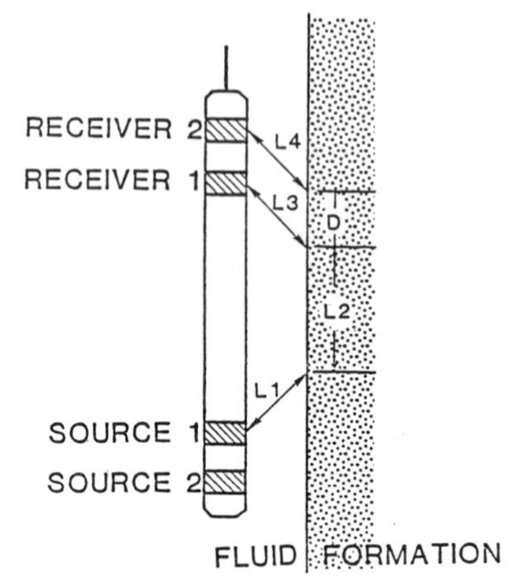

Figure 10 Schematic of a full-waveform acoustic sonde with two transmitters and two receivers. P-wave refraction paths through the formation are shown.

Zechman 1968, Peterson 1974, Tsang & Rader 1979, Cheng & Toksöz 1981, Schoenberg et al 1981, Paillet & White 1982, Baker 1984, Tubman et al 1984).

Information From Full-Waveform Logs

Vast amounts of information about the formation and the borehole are contained in full-waveform logs. Compressional and shear wave velocities

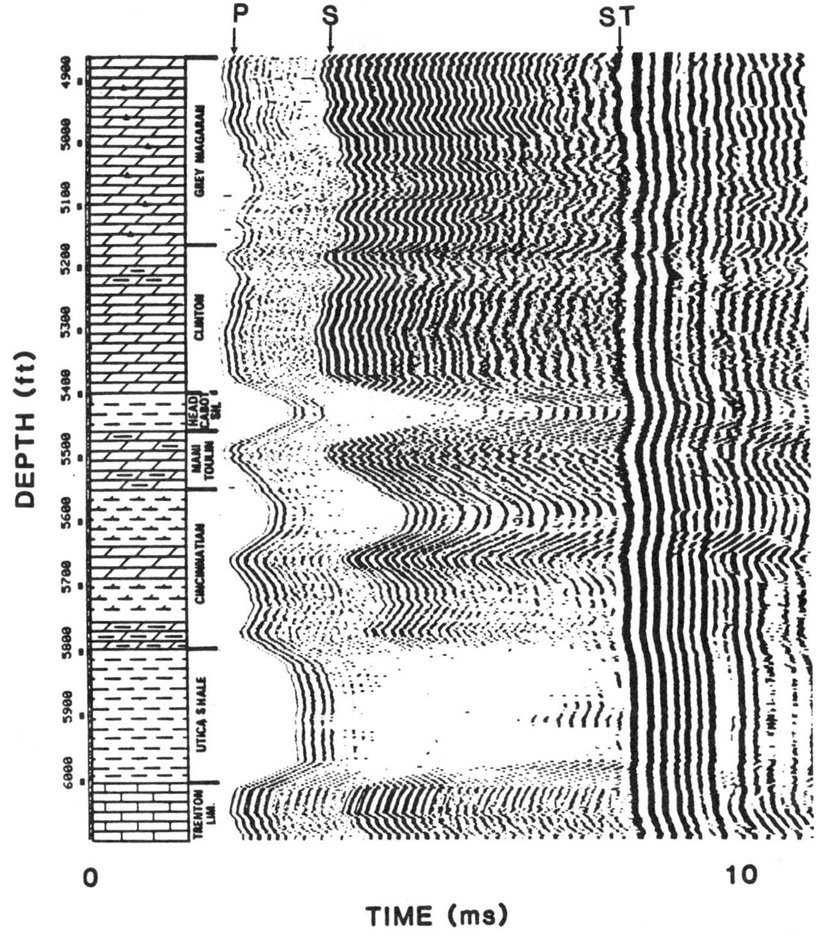

Figure 11a A full-waveform acoustic log section through carbonates, shales, and evaporites recorded by Elf Aquitaine sonde EVA. Waveform (microseismogram) is plotted at each depth. Source-receiver spacing is 10 m. P indicates refracted P-waves, and S refracted shear waves. ST refers to Stoneley or tube waves guided by the borehole fluid-solid interface. Waves between S and ST are another guided wave train.

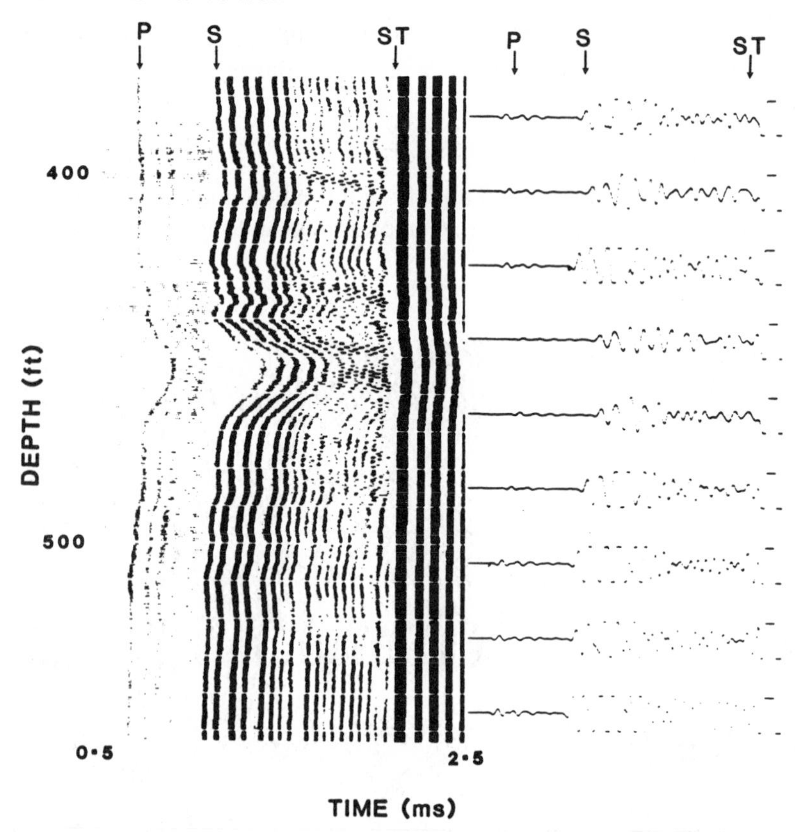

Figure 11b A full-waveform acoustic seismogram section recorded by Welex sonde. Individual seismograms are shown to illustrate the waveform. P, S, and ST are as defined in Figure 11*a*.

of the formation are among the most desired measurements. These can be obtained from direct measurements of travel-time differences between receivers. A detailed description of such measurement techniques is given by Willis & Toksöz (1983). In Figure 12 the P- and S-wave velocities from full-waveform logs and the corresponding core measurements are shown. Note that the core data, although in general agreement with log-derived velocities, differ at individual points. This difference arises because of two factors. First, log velocities are average values over 60-cm spacing, while core measurements are made on 10-cm-long samples. Second, cores are subject to alteration due to cutting, removal from borehole, and repressurization in the laboratory.

The importance of shear wave velocities V_s for identifying lithology is

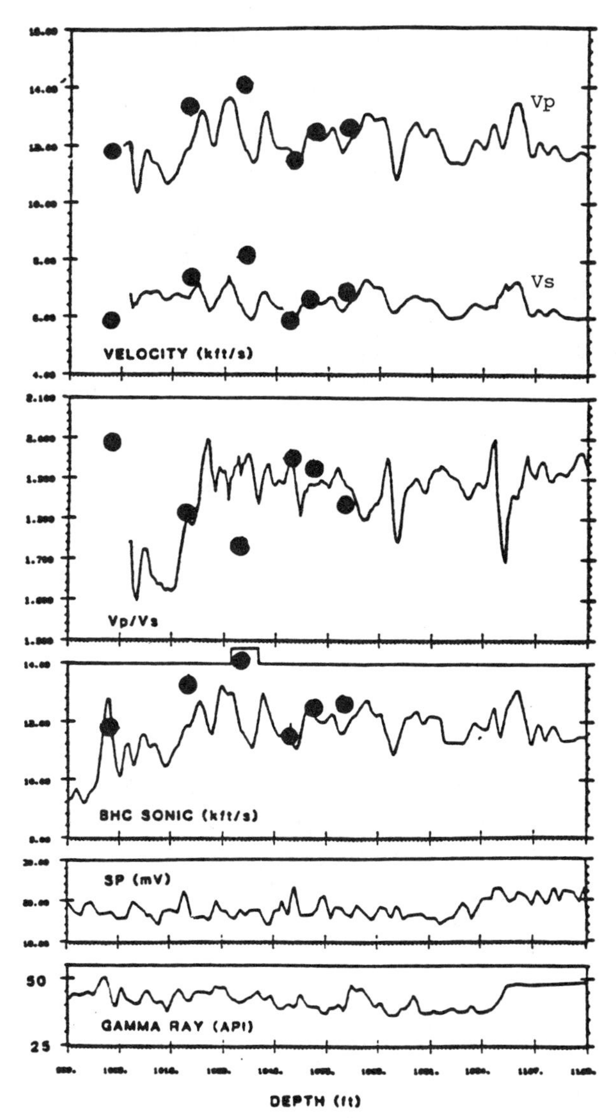

Figure 12 Comparison of compressional and shear wave velocities measured in the core samples in the laboratory (●) with full-waveform sonic log velocities (*top*) and V_p/V_s velocity ratios (*middle*). Bottom velocity curve shows P-wave velocities obtained by standard sonic log. SP and gamma logs are given for purposes of comparison.

illustrated in Figure 13, where V_p/V_s is plotted versus V_p for three rock types (sandstones, limestones, and dolomites). Compressional wave velocities (V_p) vary over a wide and overlapping range for these rock types, depending on porosity and lithification. However, the V_p/V_s ratio acts as an identifier. If lithology is known, V_p/V_s ratios can determine gas or water saturation.

In addition to velocities and attenuations, full-waveform acoustic logs can be used to detect and characterize fractures intersecting a borehole. Figure 14 shows the scattering and attenuations of full waveforms by a hydrofracture. The attenuation of the waves resulting from induced fluid flow between the borehole and formation can be used to determine the formation permeability (Rosenbaum 1974, White 1983, Hsui & Toksöz 1984).

Full-waveform acoustic logging technology and applications are expanding rapidly into acoustic imaging and characterization of the rock

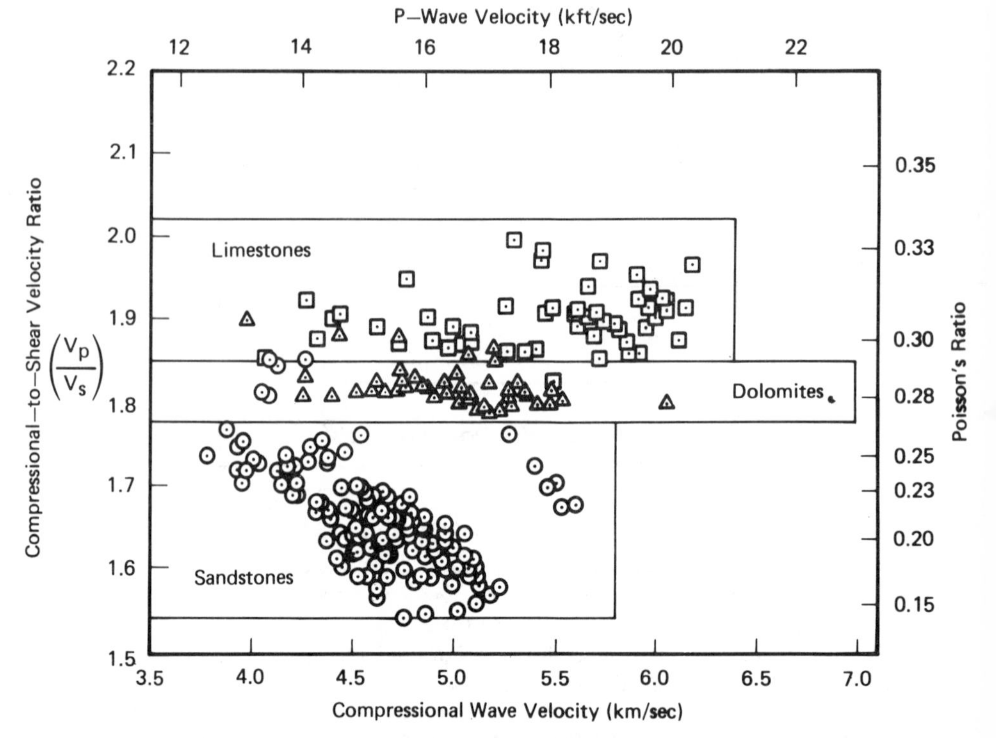

Figure 13 Compressional-to-shear velocity ratios V_p/V_s versus compressional wave velocity plot for different rock types. Data are from core measurements. Note that although there is considerable overlap on compressional velocities, the V_p/V_s ratio separates the lithologies.

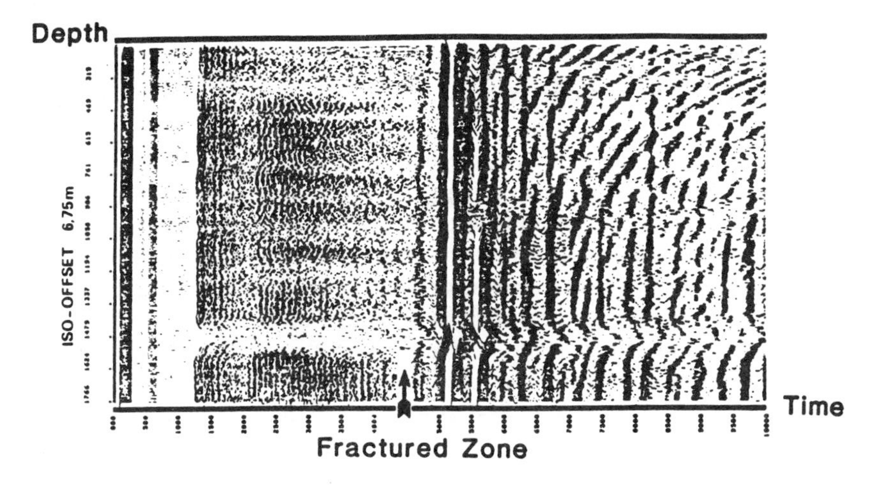

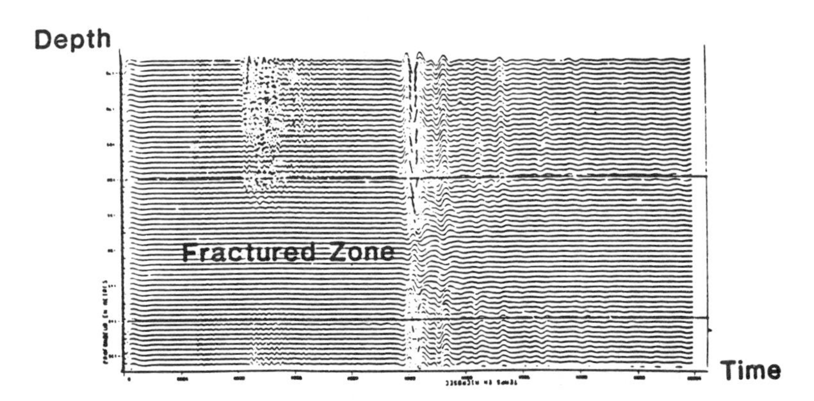

Figure 14 Scattering and attenuation of full-waveform acoustic waves by a fracture (hydrofracture) intersecting the borehole. The full-waveform section is similar to that shown in Figure 11*a*, except for the characteristic signature of the waveform. Bottom figure is an expanded version of the top, showing the details of the waveform at the fracture.

around a borehole. The technique works well in soft sediments, sedimentary rocks, and crystalline rocks.

OTHER WELL-LOGGING METHODS

A number of well-logging technology methods not included in the previous three groups are listed in Table 4. Four of these are briefly described below to indicate the broad scope of the overall well-logging technology.

Borehole Geometry Logs

A knowledge of borehole geometry is essential for correcting and interpreting borehole geophysical measurements. A cross section of the hole is measured by caliper logs with mechanical arms expanding against the borehole. New caliper logs using acoustic pulses have also become available. Caliper logs and hole cross sections are important for making hole size corrections for other logs, for calculating hole volume to determine cement requirements, for measuring mud-cake thicknesses that would indicate permeable zones, and for determining in situ stress.

In some applications, such as directional drilling, a knowledge of the complete geometry of the hole, including the azimuth and deviation, is required. Deviation logs provide two independent caliper curves: a continuous measurement (*a*) of the hole axis relative to vertical and (*b*) of the orientation of the calipers with respect to magnetic north. More recently, inertial guidance units are being applied to hole deviation measurements. These devices can measure deviation of borehole axis relative to the Earth's axis of rotation. Since they do not depend on magnetic field measurements, they can be used in cased holes.

Televiewer

High-frequency acoustic logs can be used to image the borehole wall. A single transducer rotates at constant speed, emitting acoustic pulses in the

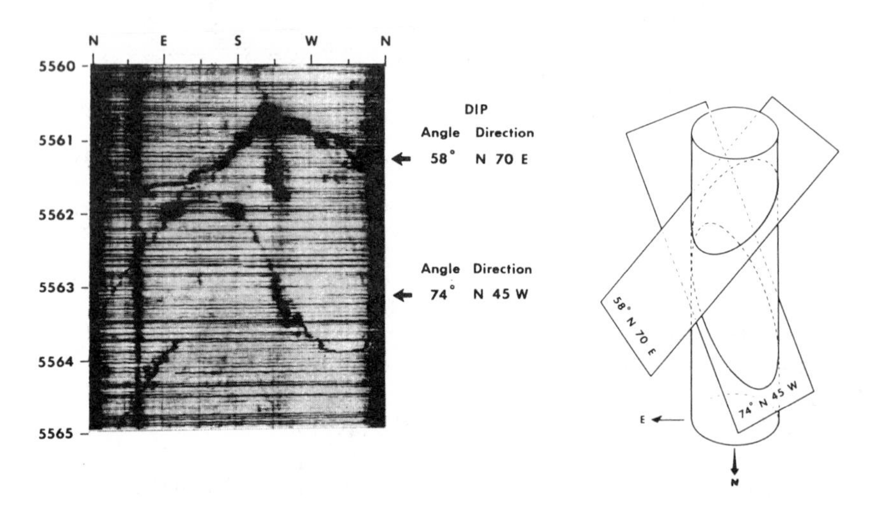

Figure 15 Borehole televiewer image of two fractures intersecting a borehole. The attitudes of fracture planes are illustrated on the right (Seismograph Service Corporation 1981).

megahertz range and receiving their echoes from the borehole face. The azimuth of the beam is also recorded. As with the transmission method, both travel times and amplitudes are used. The borehole televiewer produces a continuous acoustic picture of a borehole wall (Figure 15). It is used in an open hole to delineate fracture location and orientation, bedding planes, and vuggy porosity. In a cased hole, it is used to determine the size and distribution of perforations and to inspect casing failures.

In fractured reservoirs, the borehole televiewer is an essential formation evaluation tool, providing data that no other combination of logs can produce. The usefulness of this technology has been further increased with recent improvements in data-recording and in image-enhancement techniques to improve delineation of fractures.

Dipmeter Logs

Dipmeters are used to obtain a continuous record of formation dips for structural, stratigraphic, and sedimentary studies. Resistivities are recorded by electrodes on four orthogonal pads pressing against the wall of the borehole. The pads are held in the same plane, perpendicular to the axis of the hole. In addition to the resistivities, also recorded are the two independent measurements of the hole diameter, the orientation of the tool, and the hole deviation. These data from the high-resolution, four-arm dipmeter are then processed to display large numbers of dip calculations for geological interpretations (Cox 1970; Figure 16).

Applications of the dipmeter for exploration and development geology are numerous and are not described here. Recent interpretations utilize pattern recognition to correlate four dipmeter resistivity curves (Vincent et al 1977) and artificial intelligence concepts to interpret the dips in terms of geological models (Davis et al 1981).

Nuclear Magnetism Log

Downhole nuclear magnetism measurements in the Earth's magnetic field yield log responses unique to the formation fluids but not to the rock matrix. The measurements conducted by this technique, the free fluid index (FFI) and the spin-lattice relaxation time (T_1), reflect pore size distributions in sandstones, which can then be used to estimate permeability (Timur 1969).

CONCLUSIONS

Well logging is a broad science that measures petrophysical properties in the borehole. Current technology provides a wide range of electrical, nuclear, and acoustic parameters. The ability to measure these properties in

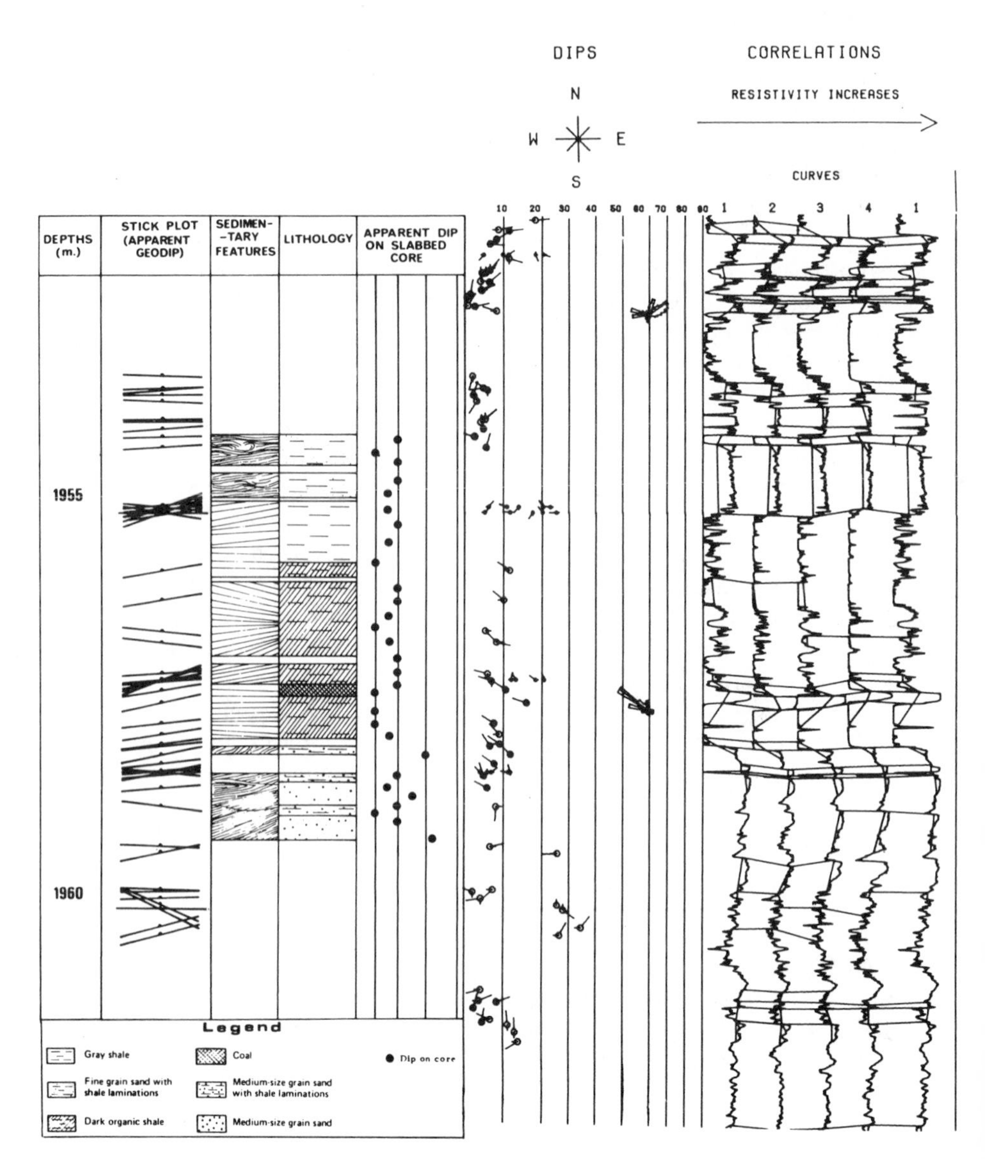

Figure 16 Dipmeter log data and interpretation in light of core data. Four curves on the right are the raw dipmeter records. Tad poles at the center are the calculated dips. Sedimentary interpretation is given on the left (Vincent et al 1977).

situ, and with fine spatial resolution, represents a very powerful tool for subsurface geophysical measurement. Several tens of physical properties may now be measured simultaneously during one trip into a borehole. Data acquisition rates are now so high that computers are required both downhole and at the surface to process the data.

The potential of well-logging methods was realized early by the petroleum industry, but the methods are rapidly growing in the engineering and scientific communities. The integration of logging data with traditional subsurface measurements is one of the greatest challenges facing Earth scientists today.

ACKNOWLEDGMENTS

This research was sponsored in part by the Full Waveform Acoustic Logging Consortium at the Earth Resources Laboratory of the Massachusetts Institute of Technology. The authors would like to thank James Mendelson for his critical reading of the manuscript.

Literature Cited

Ancel, C., Couve de Murville, E., Dadrian, C., Deines, D., Goetz, J., et al. 1974. *Schlumberger Well Evaluation—Nigeria.* Paris: Schlumberger Afr. 2nd ed.

Baker, L. J. 1984. The effect of the invaded zone on full wave train acoustic logging. *Geophysics* 49: 796–809

Balch, A. H., Lee, M. W., eds. 1984. *Vertical Seismic Profiling: Technique, Applications, and Case Histories.* Boston, Mass: Int. Hum. Resour. Dev. Corp. 488 pp. 1st ed.

Biot, M. A. 1952. Propagation of elastic waves in a cylindrical bore containing a fluid. *J. Appl. Phys.* 23: 977–1005

Calvert, R. J., Rau, R. N., Wells, L. E. 1977. *Electromagnetic propagation—a new dimension in logging.* Presented at Ann. Calif. Reg. Meet. Soc. Pet. Eng. of AIME, 47th (Pap. 6542)

Cheng, C. H., Toksöz, M. N. 1980. Modelling of full waveform acoustic logs. *Trans. Soc. Prof. Well Log Anal.* 21: J1–12

Cheng, C. H., Toksöz, M. N. 1981. Elastic wave propagation in a fluid-filled borehole and synthetic acoustic logs. *Geophysics* 46: 1042–53

Cox, J. W. 1970. The high resolution dipmeter reveals dip-related borehole and formation techniques. *Trans. Soc. Prof. Well Log Anal.* 11: D1–26

Davis, R. R., Hall, J. E., Boutemy, Y. L. 1981. *A dual porosity CNL logging system.* Presented at Ann. Fall Meet. Soc. Pet. Eng. of AIME, 56th, San Antonio, Tex. (Pap. 10296)

Edmonson, H., Raymer, L. L. 1979. Radioactive logging parameters for common minerals. *Log Anal.* 20: 38–47

Evans, C. B., Gouilloud, M. 1979. The changing role of well logging in reservoir evaluation—a challenge of the 1980's. *World Pet. Congr., 10th, Bucharest* 3: 181–89

Felder, B., Boyeldieu, C. 1979. The lithodensity log. *Trans. Eur. Form. Eval. Symp., Soc. Prof. Well Log Anal., London,* Pap. O

Fertl, W. H. 1979. Gamma ray spectral data assists in complex formation evaluation. *Log Anal.* 20: 3–37

Gal'perin, E. I. 1974. *Vertical Seismic Profiling,* Moscow Nedra, 1971. English Trans. Soc. Explor. Geophys. Spec. Publ. No. 12, Tulsa

Gardner, J. S., Dumanoir, J. L. 1980. Litho-density log interpretation. *Trans. Soc. Prop. Well Log Anal.* 21: N1–23

Hardage, B. A., ed. 1983. *Vertical Seismic Profiling: Part A: Principles.* London/Amsterdam: Geophys. Press. 450 pp. 1st ed.

Hilchie, D. W. 1978. *Applied Open Hole Log Interpretation.* Golden, Colo: D. W. Hilchie Inc.

Hoyer, W. A., Hilchie, D. W., Jordan, J. R., Mills, W. R., Tittman, J., Wichmann, P. A. 1976. *SPWLA Reprint Volume—Pulsed Neutron Logging,* Soc. Prof. Well Log Anal.

Hsui, A., Toksöz, M. N. 1984. A model for

tube wave attenuation: its application to the determination of in situ formation permeability. *Geophysics*. In press

Jennings, H. Y., Timur, A. 1973. Significant contributions in formation evaluation and well testing. *J. Pet. Technol.* 25:1432–46

Lawson, B. L., Hoyer, W. A., Pickett, G. R. 1978. *SPWLA Reprint Volume—Gamma Ray, Neutron and Density Logging*, Soc. Prof. Well Log Anal.

Mayer, C., Sibbit, A. 1980. *Global, a new approach to computer-processed log interpretation.* Presented at Ann. Calif. Reg. Meet. Soc. Pet. Eng. of AIME, 47th (Pap. 9341)

Neuman, C. H. 1978. *Log and core measurements of oil in place, San Joaquin Valley, California.* Presented at Ann. Calif. Reg. Meet. Soc. Pet. Eng. of AIME (Pap. 7146)

Paillet, F., White, J. E. 1982. Acoustic models of propagation in the borehole and their relationship to rock properties. *Geophysics* 47:1215–28

Peterson, E. W. 1974. Acoustic wave propagation along a fluid-filled cylinder. *J. Appl. Phys.* 45:3340–50

Rosenbaum, J. H. 1974. Synthetic microseismograms: logging in porous formations. *Geophysics* 39:14–32

Schoenberg, M., Marzetta, T., Aron, J., Porter, R. 1981. Space-time dependence of acoustic waves in a borehole. *J. Acoust. Soc. Am.* 70:1496–1507

Segesman, F. F. 1980. Well-logging method. *Geophysics* 45:1667–84

Seismograph Service Corporation. 1981. *Technical Specifications on Seisviewer Logging.* Birdwell Div.

Serra, O., Baldwin, J., Quirein, J. 1980. Theory, interpretation and practical applications of natural gamma ray spectroscopy. *Trans. Soc. Prof. Well Log Anal.* 21:Q1–30

Timur, A. 1969. Pulsed nuclear magnetic resonance studies of porosity, movable fluid and permeability of sandstones. *J. Pet. Technol.* 21:775–86

Timur, A. 1982. Open hole well logging. *Proc. Int. Meet. Pet. Eng., Beijing*, pp. 639–74 (Soc. Pet. Eng. Pap. 10037)

Timur, A., Hempkins, W. B., Worthington, A. E. 1972. *Porosity and pressure dependence of formation resistivity factor for sandstones.* Presented at Form. Eval. Symp. Can. Well Log. Soc., 4th, Calgary (Pap. D)

Toksöz, M. N., Stewart, R. R., eds. 1984. *Vertical Seismic Profiling: Part B: Advanced Concepts.* London/Amsterdam: Geophys. Press. 419 pp. 1st ed.

Tsang, L., Rader, D. 1979. Numerical evaluation of transient acoustic waveform due to a point source in a fluid-filled borehole. *Geophysics* 44:1706–20

Tubman, K. M., Cheng, C. H., Toksöz, M. N. 1984. Synthetic full waveform acoustic logs in cased boreholes. *Geophysics* 49:1051–59

Vincent, Ph., Gartner, J. E., Attali, G. 1977. *GEODIP, an approach to detailed dip determination using correlation by pattern recognition.* Presented at Ann. Meet. Soc. Pet. Eng. of AIME, 52nd, Denver, Colo. (Pap. 6823)

Wahl, J. S., Tittman, J., Johnstone, C. W., Alger, R. P. 1964. The dual spacing formation density log. *J. Pet. Technol.* 16:1411–16

Waxman, M. H., Smits, L. J. M. 1968. Electrical conductivities in oil-bearing shaly sands. *Soc. Pet. Eng. J.* 8:107–22

Westaway, P., Hertzog, R., Plusek, R. E. 1980. *The Gamma spectrometer tool inelastic and capture gamma ray spectroscopy for reservoir analysis.* Presented at Ann. Fall Meet. Soc. Pet. Eng. of AIME, 55th, Dallas, Tex. (Pap. 9461)

Wharton, R. P., Hazen, G. A., Rau, R. N., Best, D. L. 1980. *Electromagnetic propagation logging: advances in technique and interpretation.* Presented at Ann. Fall Meet. Soc. Pet. Eng. of AIME, 55th, Dallas, Tex. (Pap. 9267)

White, J. E. 1983. *Underground Sound: Application of Seismic Waves.* Amsterdam/Oxford/New York: Elsevier. 253 pp. 1st ed.

White, J. E., Zechman, R. E. 1968. Computed response of an acoustic logging tool. *Geophysics* 33:302–10

Willis, M. E., Toksöz, M. N. 1983. Automatic P and S velocity determination from full waveform digital acoustic logs. *Geophysics* 48:1631–44

Ann. Rev. Earth Planet. Sci. 1985. 13 : 345–83

THE MIDCONTINENT RIFT SYSTEM

W. R. Van Schmus

Department of Geology, University of Kansas, Lawrence, Kansas 66045

W. J. Hinze

Department of Geosciences, Purdue University, West Lafayette, Indiana 47907

1. INTRODUCTION

One of the most prominent features on gravity and aeromagnetic maps of the United States is a series of major, generally linear anomalies that extend from central Kansas to Lake Superior and then turn southward into central Michigan (Figures 1, 2). This system of anomalies clearly reflects major geological features, and as such it must represent a very significant episode in the history of the North American continental lithosphere. The geologic features associated with these anomalies, particularly axial basins filled with basalt and immature clastic rocks along with evidence of crustal extension, indicate that the episode was probably one of incipient rifting.

Terminology

The features discussed in this review have been referred to by a variety of names, both descriptive and genetic. The major feature, which extends from central Kansas to Lake Superior (Figure 1), was originally recognized about 40 years ago as a major positive gravity anomaly and has been referred to as the "midcontinent gravity high" or the "midcontinent gravity anomaly." Subsequent aeromagnetic studies have shown that the designation "midcontinent geophysical anomaly" (or MGA) is more appropriate, and we follow this usage here, with the restriction that it only applies to the segment from Kansas to Lake Superior. Other segments or proposed extensions are referred to individually (e.g. Lake Superior basin,

0084–6597/85/0515–0345$02.00

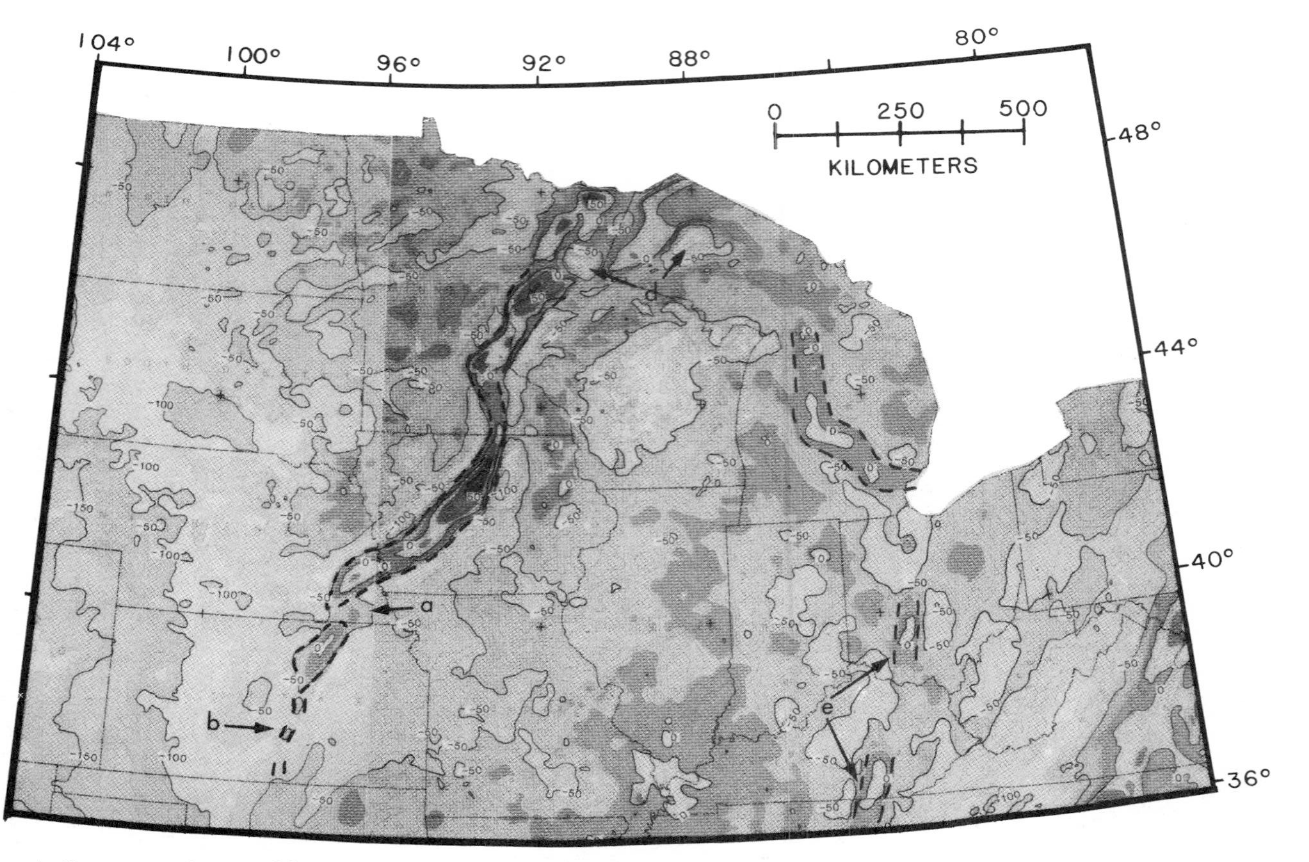

Figure 1 Bouguer gravity map of the north-central portion of the United States, showing the major anomalies associated with the midcontinent rift system (outlined with dashed lines). Letters refer to specific features mentioned in the text. Modified from Society of Exploration Geophysicists (1982). Heavy contours at 50-mgal intervals.

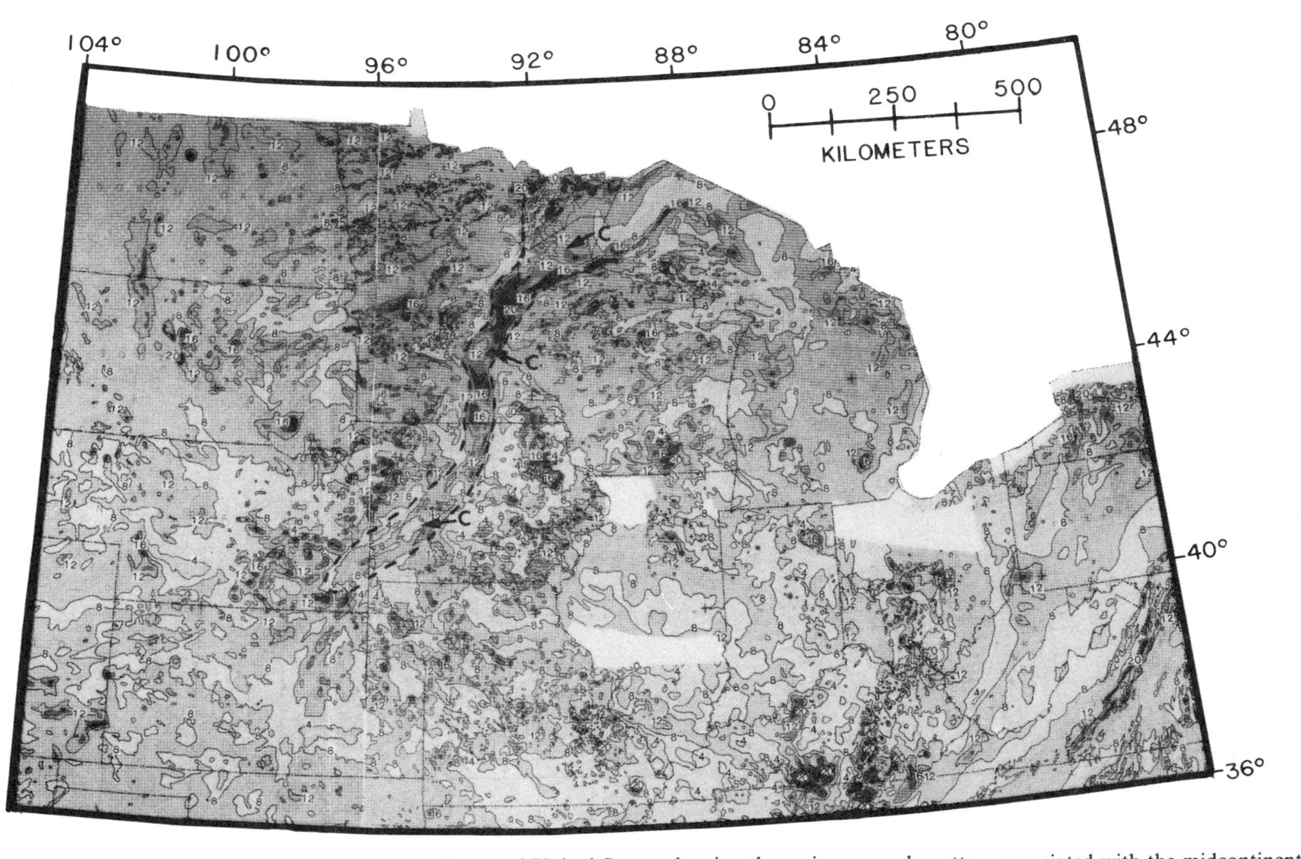

Figure 2 Aeromagnetic anomaly map of the north-central United States, showing the major anomaly pattern associated with the midcontinent geophysical anomaly (Figure 1; dashed lines). Modified after US Geological Survey (1982). Letters refer to features mentioned in the text. Heavy contours at 400-gamma intervals.

mid-Michigan geophysical anomaly, east-continent geophysical anomalies, etc.).

Rocks associated with the MGA are exposed in the Lake Superior region and comprise the classical association of bimodal volcanic, plutonic, and clastic sedimentary units of the 1.0–1.2-Gyr-old "Keweenawan" suite (cf Morey & Green 1982). Genetic interpretation of the MGA and associated features as due to an incipient rift system goes back many years (e.g. Lyons 1959, King & Zietz 1971), and these features have been referred to as the "central North American rift system" (Ocola & Meyer 1973) or, more simply, the "midcontinent rift system" (Wold & Hinze 1982). In this review we follow the latter usage in genetic contexts.

Regional Geologic Setting

A major feature of the rift system is that it cuts across several Precambrian basement terranes of quite different age, structure, and composition in the craton that existed prior to 1.2 Gyr ago (Figure 3). The northern portion, including the Lake Superior basin, occurs in the Archean Superior Province of the Canadian Shield. In this region the Archean rocks are generally divided into a northern granite-greenstone terrane that has remained relatively stable since it formed 2.6–2.8 Gyr ago and a southern gneiss-migmatite terrane that formed 2.6–3.6 Gyr ago but was subjected to extensive deformation during Early Proterozoic orogenies (Morey & Sims 1976). The Lake Superior basin occurs within the granite-greenstone terrane, just to the north of the boundary (Great Lakes Tectonic Zone; Sims et al 1980) between the terranes. The rift system broadens within the Lake Superior basin, it becomes much wider, and the regional trend changes from NE-SW (MGA) to NW-SE (mid-Michigan segment). Klasner et al (1982) have suggested that this change in character and orientation may have been affected by differences between the two Archean terranes.

South of the Lake Superior basin, the southeastern and southwestern segments are narrower and occur for the most part in early Proterozoic crust, although the northern part of the MGA segment occurs in the Archean gneiss-migmatite terrane. Archean crust is not found south of central Wisconsin and southern Minnesota (Figure 3; Van Schmus & Bickford 1981, Nelson & DePaolo 1985). The Proterozoic crustal basement rocks in the midcontinent region formed 1300–1900 Myr ago and consist of 1600–1900-Myr-old new crust [which probably formed as orogenic continental margin assemblages (Van Schmus & Bickford 1981)] and 1300–1500-Myr-old anorogenic plutonic and volcanic rocks that formed by remelting of the older Proterozoic crust (Thomas et al 1984, Nelson & DePaolo 1985, Anderson 1983). According to Klasner et al (1982), the orientation of various segments of the rift system may be controlled by

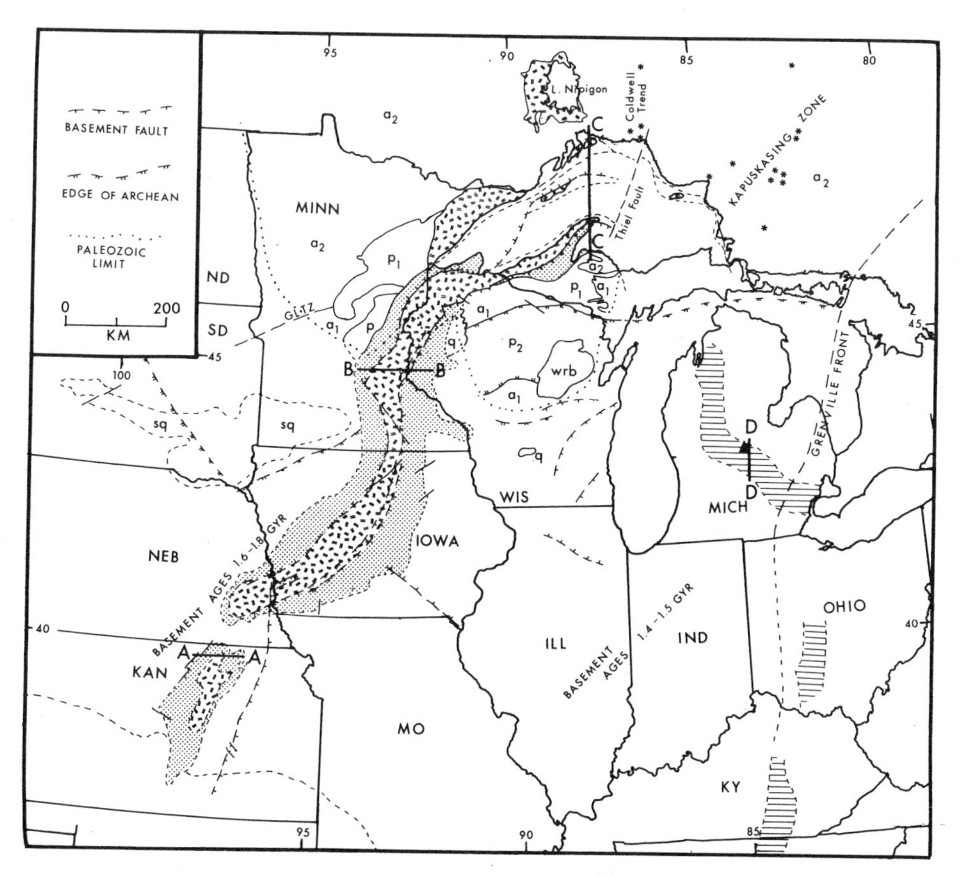

Figure 3 Generalized geologic map showing major features of the prerift Precambrian basement and principal geologic units associated with the midcontinent rift system. Random "=" represent central rift igneous rocks and associated sedimentary rocks, including Oronto Group; stippled pattern represents late Upper Keweenawan clastic rocks of the posttectonic phase (Bayfield Group, Jacobsville Sandstone, etc) in basins flanking the main rift sequence; horizontal-ruled pattern represents rift basin fill whose character is unknown in detail in the eastern region; asterisks north and northeast of Lake Superior represent alkaline complexes; dashed line labeled GLTZ is the Great Lakes Tectonic Zone. Abbreviations for older geologic units are as follows: a_1, older Archean gneiss-migmatite terrane; a_2, Archean granite-greenstone terrane; p_1, early Proterozoic cratonic cover; p_2, Early Proterozoic igneous complex; sq, Sioux Quartzite; q, other Early Proterozoic quartzites; wrb, 1500-Myr-old Wolf River Batholith. Compiled by the authors from various sources. Section lines AA, BB, CC, DD refer to Figure 7.

structures within the Early Proterozoic basement. However, knowledge of structures in the basement is very limited, and other factors could be dominant.

Lithologic units associated with the midcontinent rift system are exposed only in the Lake Superior region and comprise the "Keweenawan" suite of rocks (Morey & Green 1982). They can be divided roughly into two groups: rift-basin igneous rocks (Duluth Gabbro, Keweenawan Volcanics) with associated red clastic rocks, and later crustal-depression-filling clastic rocks. These rocks are summarized in more detail below. Equivalent rocks are known from subsurface samples and indicate that there was extensive igneous activity and sedimentation along the midcontinent rift system at about 1100 Myr ago. This appears to be the youngest Precambrian igneous activity in the midcontinent region.

To the east of the midcontinent region is the Grenville Province, which represents a major crustal event. The Grenville Province is large and complex, but the combination of high-grade metamorphism, reworked continental crust, compressional structures, and abundant sialic crustal material strongly suggests that it represents a continental collision (e.g. Young 1980). Geochronologic studies of Grenville rocks show that much of the igneous and metamorphic activity of the Grenville event occurred 1150–900 Myr ago, with an earlier peak of activity 1100 ± 50 Myr ago and a second peak of activity 950 ± 50 Myr ago (e.g. Baer 1981). The proximity in time and space of the Grenville Province and the midcontinent rift system has led to suggestions that the two are genetically related, as is discussed below.

2. EXTENT AND GEOPHYSICAL EXPRESSION

Attention was first drawn to the MGA in Woollard's (1943) transcontinental gravity profile, which crossed the anomaly in eastern Kansas. As additional gravity data were observed and compiled (e.g. Lyons 1950, Black 1955, Thiel 1956, Coons et al 1967, Craddock et al 1969, Society of Exploration Geophysicists 1982), the gravity signature of the feature and its extent began to take shape. The anomaly, as first recognized, extends from Kansas to the western end of Lake Superior and is a major feature of the gravity anomaly field of the United States (Figure 1). A large amplitude range of up to 160 mgal over its central maximum and bordering minima, a cross-cutting relationship to the regional anomaly pattern, and a width of approximately 150 km combine to produce a characteristic and obvious anomaly. The MGA is also readily apparent in the magnetic anomaly field (Figure 2) as first recognized by King & Zietz (1971), although it is less prominent than in the gravity field.

Since the delineation of the MGA, additional segments have been identified suggesting that the feature extends south into Oklahoma as well as in an arcuate pattern through Lake Superior and then southerly across Michigan and Ohio as far south as Tennessee. Additional minor components may extend northward from Lake Superior into Canada.

Midcontinent Geophysical Anomaly (MGA)

The MGA consists of several generally southwesterly trending coincident gravity and magnetic positive anomalies flanked by minima, and it extends from western Lake Superior in Minnesota and Wisconsin through central Iowa and southeastern Nebraska into Kansas. It strikes southwest from Lake Superior, makes a sharp turn in south-central Minnesota, and continues southeastward to the Iowa border, where it resumes its southwesterly course into southeastern Nebraska. At this point the anomaly shifts abruptly to the east by about 60 km, then continues south-southwesterly from the Kansas-Nebraska border into central Kansas. Data from subsurface samples in southeast Nebraska (Lidiak 1972, Treves & Low 1984) indicate that the anomaly marked "a" to the northeast of the Kansas segment in Figure 1 is not represented by Keweenawan units at the Precambrian surface and may not be part of the MGA. The gravity anomalies associated with the MGA north of central Kansas are measured in tens of milligals and thus are easily observed. The Minnesota and Iowa segments have the greatest amplitudes, with values ranging from greater than +70 mgal in Minnesota to less than −110 mgal in the flanking minimum in north-central Iowa.

In most discussions the MGA is terminated in central Kansas, but as pointed out by Lyons (1959), a subtle gravity anomaly component continues on strike with the major anomaly into Oklahoma ("b" in Figure 1) and perhaps as far south as southern Oklahoma. More recently, Yarger (1983) has shown that a system of linear magnetic anomalies that are particularly prominent in a second vertical derivative map extend south-southwesterly from the major magnetic anomaly in northern Kansas to at least the Oklahoma border. These lines of evidence strongly support continuation of the MGA to at least Oklahoma, but the anomalies, both gravity and magnetic, are greatly attenuated and indicate that the character of their source is different south of central Kansas, and the underlying disturbance of the crust is greatly reduced.

The gravity and magnetic anomalies of the MGA are correlative in a general sense—that is, major maxima and minima are spatially coincident (Figures 1, 2); however, in detail the anomalies are commonly not correlated. Gravity anomalies are smoothly varying over the central positive anomaly as well as the flanking minima. In contrast, the magnetic signature over the central gravity high consists of linear anomalies that parallel the strike of the feature and may range in length from a few kilometers to more than a hundred kilometers. In addition, prominent but featureless magnetic minima may occur within the confines of the central positive gravity anomalies (King & Zietz 1971). Broad, flanking magnetic "quiet" zones indicate increased depth to magnetic sources adjacent to the central anomaly. The differences in the gravity and magnetic anomalies are caused by the fundamental potential-field relationships associated with monopolar sources (gravity) versus dipolar sources (magnetics) and, more importantly, by geologic variations that differentially affect the controlling physical properties (density and magnetic polarization) of the sources.

Most of the MGA is overlain by Phanerozoic rocks that prevent direct study of the anomaly source. Drill holes to basement are limited in number and poorly distributed, so investigation of the source of the MGA has been largely by indirect geophysical methods. However, outcrops of the Precambrian basement at the western end of Lake Superior can be extrapolated into the subsurface by anomalous geophysical fields, principally gravity and magnetic anomalies. Thus, Thiel (1956) showed that the central positive gravity anomaly is associated with Keweenawan basalt deposited in an elongate depression, and that the flanking minima correlate with prisms of clastic rocks that thin away from the central gravity high. The central basin of basalt has been disrupted along high-angle reverse faults and brought into juxtaposition with younger sedimentary rocks. The latter were deposited in later, broader basins that originally completely covered the basalts, and the faults generally parallel the axis of the central, basalt-filled basin. This simple structural interpretation is the basis of current interpretations, although subsequent studies have shown that deeper crustal manifestations have a marked, and in some cases the predominant, effect upon geophysical anomalies.

As another example, King & Zietz (1971) have shown that mafic volcanic rocks are not only denser but also more magnetic than the surrounding basement country rocks, and that the relatively less-dense clastic rocks are essentially nonmagnetic. Furthermore, the mafic lavas retain a remanent magnetization acquired at the time of crystallization that may be three to five times as great as the magnetization induced by the Earth's present magnetic field and that differs radically in direction from the present field. This remanent magnetization complicates the magnetic anomaly pattern of the MGA, especially where lavas have been structurally deformed subsequent to their extrusion.

The situation is even more complicated, because the Earth's magnetic field reversed itself at least twice (N to R, R to N) during the period of development of the source of the MGA. As a result, reversely polarized lavas, dikes, sills, and plutons occur near the base of the igneous section (Green 1982), which result in strong negative (rather than positive) magnetic anomalies for some Keweenawan mafic units. King & Zietz (1971) have shown that the effect of remanent magnetization explains the relatively low magnetic values along the axis of the gravity anomaly, the persistent low along the western edge, and the marked positive along the eastern margin of the MGA. The long linear anomalies are probably related to upturned or faulted volcanic units that parallel the length of the anomaly. The broad, featureless magnetic minima along the axis of the gravity high, such as those observed in northwestern Wisconsin, over the Twin Cities basin at roughly 45°N, and along the length of the MGA in Iowa ("c" in Figure 2), are interpreted to be remnants of clastic basin fill that originally blanketed volcanic-filled troughs (King & Zietz 1971). The low-density clastic rocks of these basins contrast with the surrounding high-density lavas, producing subtle gravity minima in residual anomaly maps (Coons et al 1967, Hildenbrand et al 1982).

Lake Superior Basin

Both surface geologic data and geophysical anomalies establish the direct connection between the MGA and the Lake Superior basin. The evidence for this connection and the geology and geophysics of the Lake Superior region are summarized in a recent series of papers edited by Wold & Hinze (1982). The anomaly pattern of the MGA broadens and becomes much more complex as it passes into the basin near the western end of Lake Superior. Positive gravity and magnetic anomalies that occur over exposed and buried mafic volcanic rock delineate the limbs of an elongate basin, showing that the structural basin generally is outlined by the present shorelines of the lake (Hinze et al 1966, 1982). However, the anomaly pattern is considerably more complex than suggested by this simple interpretation. The gravity anomaly pattern is perturbed by relatively low-density clastic sedimentary rocks that overlie thick mafic volcanic rocks. Variations in thickness and density of the clastic rocks impose a variable negative effect upon positive anomalies associated with density contrasts of the mafic volcanic and intrusive rocks. Stripping off gravitational effects of the mass deficiency due to the low-density Upper Keweenawan clastic rocks shows that the entire lake is a positive gravity anomaly surrounded by negative anomalies. Particularly intense negative anomalies occur over the Bayfield Peninsula of Wisconsin and Keweenaw Bay of Michigan ("d" in Figure 1), which suggests that volcanic rocks of the basin are thin or absent in these areas (Hinze et al 1982). Magnetic anomalies of the Lake Superior basin are complicated by the strong remanent magnetization of the Keweenawan igneous rocks, which is aligned at a high angle to the

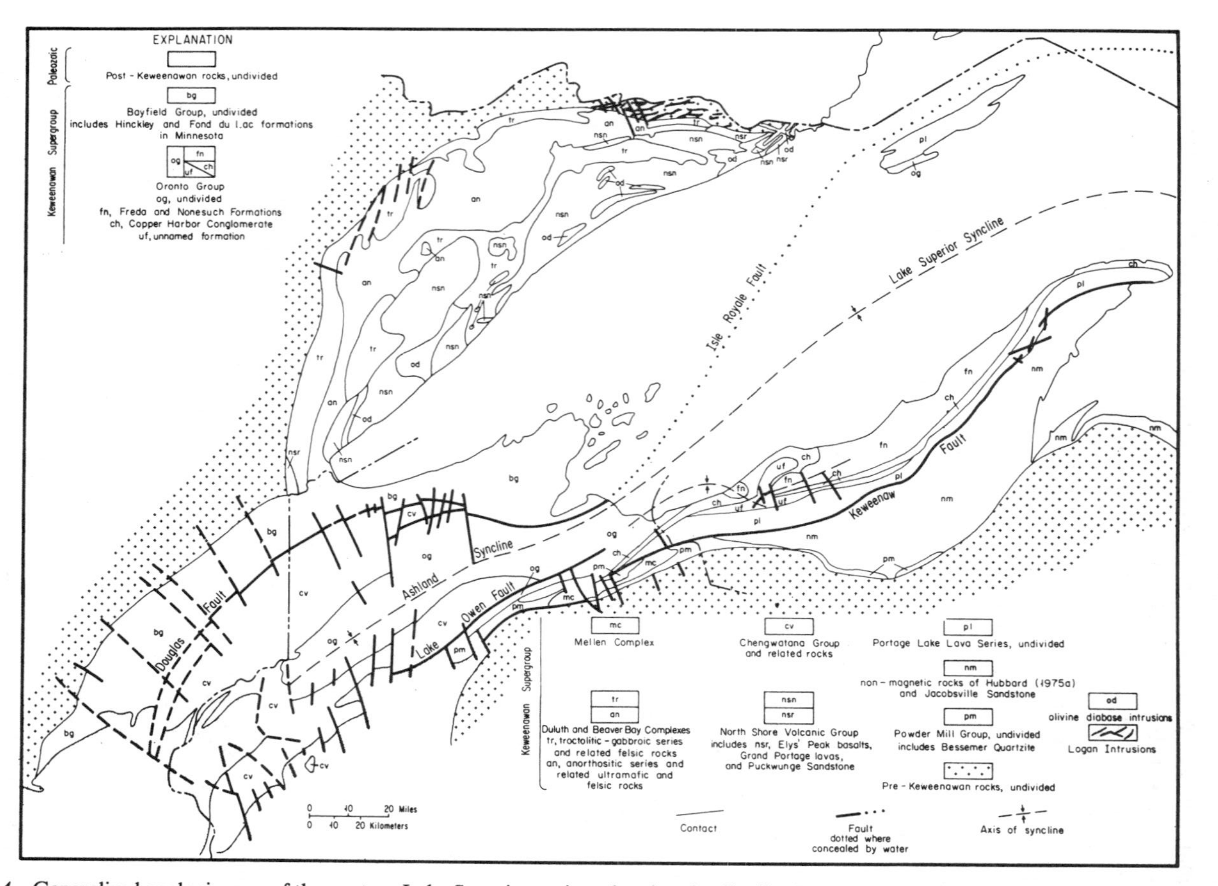

Figure 4 Generalized geologic map of the western Lake Superior region, showing the distribution and inferred stratigraphy of Keweenawan rocks (Morey & Green 1982). Stratigraphic units are not necessarily shown in sequence.

present geomagnetic field. However, the axis of the structural basin is defined by a magnetic minimum (Figures 1, 4), and numerous faults paralleling and transecting the basin are indicated by discontinuities in the magnetic anomaly pattern (Hinze et al 1982).

Mid-Michigan Geophysical Anomaly

The mid-Michigan anomaly, which transects the Michigan basin from the northern margin to the southeast corner (Figure 1; Hinze 1963), is strikingly similar to the midcontinent geophysical anomaly (Hinze & Merritt 1969). Oray et al (1973) demonstrated the direct connection of this anomaly to the anomalies at the eastern end of the Lake Superior basin, and Hinze et al (1975) interpreted the anomaly as a rift feature that is a continuation of the structure associated with the MGA and Lake Superior basin. This interpretation was supported by results from a deep drill hole located in central Michigan on the maximum of the mid-Michigan geophysical anomaly (Figure 3; Sleep & Sloss 1978). At a depth of 3715 m the hole penetrated Keweenawan red clastic rocks (Catacosinos 1981), which underlie the basal Cambrian formations; it bottomed at a depth of 5324 m in metamorphosed mafic volcanic rocks, which were first encountered from 4970 to 4998 m and again from 5252 m to the bottom of the hole (McCallister et al 1978). Subsequently, deep seismic reflection profiling by the Consortium for Continental Reflection Profiling (COCORP) identified a structural trough containing layered formations beneath the Phanerozoic sedimentary rocks and coinciding with the mid-Michigan anomaly (see Section 4). The gravity anomaly of the mid-Michigan geophysical anomaly is continuous across the state of Michigan (Figure 1). However, the associated magnetic anomaly only occurs along segments of the feature (Figure 2), which suggests that volcanic rocks either are not present or are thin in portions of the rift, and that the positive gravity anomaly is related either to low-magnetization mafic intrusives at depth or to volcanic rocks that have been altered to a nonmagnetic form (Hinze et al 1982).

East-Continent Geophysical Anomalies

The southern limit of the eastern arm of the midcontinent rift system is generally placed in southeastern Michigan, either at the termination of the magnetic anomaly of the mid-Michigan feature or near the US-Canadian border slightly farther east, where the gravity anomaly terminates (Figure 1). However, more speculatively, a series of rectangular, N-S trending gravity maxima that extend southward across western Ohio into central Kentucky and Tennessee ("e" in Figure 1) have been interpreted as a southerly continuation of this arm of the rift (Lyons 1970, Halls 1978, Keller et al 1982). This interpretation is based on the form of the gravity anomalies and related positive magnetic anomalies, as well as on limited crustal seismic refraction and basement lithologic data. The lithologic data were all acquired from deep drill holes that penetrated through the Phanerozoic sedimentary rocks into the crystalline basement, and they are interpreted by Keller et al (1982) as indicating a strong correlation between the positive anomalies and bimodal igneous suites. These rocks are believed to have originated in the same rifting event that caused the MGA, but they were subsequently metamorphosed where they occur east of the Grenville Front (Lidiak et al 1984). Additional, but more speculative, rift blocks that may correlate with the MGA have been identified by Keller et al (1983) in the eastern midcontinent.

3. PETROLOGIC AND GEOCHEMICAL ASPECTS OF THE RIFT SYSTEM

Lithologic units associated with the midcontinent rift system can be divided into two major suites. The first includes units directly associated with the rifting: gabbro, bimodal volcanics, and clastic sedimentary rocks controlled by the rift structures and present as interflow sediments. The second suite consists of clastic sediments that formed after the main rifting phase and were deposited in large crustal depressions along the rift system. These units are related to each other and to older cratonic basement by a series of well-defined structures. Most data come from exposures in the Lake Superior region, complemented by fragmentary data from sparse subsurface samples, drill holes to basement, and regional geophysical interpretation. Figure 3 summarizes the geology of the midcontinent rift system, and various aspects are outlined below.

Lithologic Suites

As mentioned above, there are two major suites of rocks generally regarded as "Keweenawan": an igneous-sedimentary unit that occurs as primary rift-basin fill, and a later, entirely sedimentary one that occurs as late-stage fill of basins overlying the initial rock suite. In addition, a third, essentially pre-Keweenawan suite occurs locally and may have been deposited in structural depressions later occupied by the rift. The general stratigraphic relationships of Keweenawan and related units are complex and have recently been reviewed by Morey & Green (1982). Weiblen & Morey (1980), Green (1982), and Weiblen (1982) have reviewed the geology and geochemistry of the igneous units. A series of papers by Ojakangas & Morey (1982), Merk & Jirsa (1982), Daniels (1982), Morey & Ojakangas (1982), and Kalliokoski (1982), along with recent papers by Elmore (1984) and Catacosinos (1981), provide good summaries of the geology of the sedimentary units. Van Schmus et al (1982) have summarized the geochronology of Keweenawan rocks in the Lake Superior region, and Halls & Pesonen (1982) have reviewed the paleomagnetic aspects. Table 1 briefly outlines the key stratigraphic relationships in the Lake Superior region, and Figure 4 gives the geology as exposed in the western part of Lake Superior.

VOLCANIC UNITS Volcanic rocks constitute a major part of the units that fill the Lake Superior basin, with aggregate thicknesses up to 10 km in some localities (Green 1982). The volcanic units were extruded as fissure-fed continental flood basalts, and White (1972) and Green (1982) interpret the distribution of volcanic rocks as due to eruption from approximately eight major centers in the Lake Superior region (Table 2). Thus, individual flow units are not found everywhere but locally thicken and thin. This aspect of the volcanism complicates stratigraphic correlation, but the major reversals of the Earth's magnetic field that occurred during Keweenawan volcanism have been used extensively for regional correlation (Morey & Green 1982). Most of the volcanism occurred during an intermediate period of reversed field and after return to normal polarity; only limited volcanism occurred in the Lake Superior region during the "lower normal" polarity period of time.

The volcanic units were erupted into a broad, subsiding basin in the

Table 1 Stratigraphic summary for the Lake Superior region

Interval			Lithologic units	Approximate age (Myr)
			Post-Keweenawan, pre-Cambrian (poorly defined)	
Keweenawan Supergroup	Upper	Normal	Upper: mostly cratonic detritus (Bayfield Group and equivalent)	<1100
Keweenawan Supergroup	Upper	Normal	Lower: mostly Keweenawan detritus (Oronto Group and equivalent)	<1100
Keweenawan Supergroup	Middle	Normal	Interflow sediments	ca. 1100
Keweenawan Supergroup	Middle / Lower	Normal / Rev.	Later volcanic flows, sills, dikes; gabbro, anorthosite	
Keweenawan Supergroup	Lower	Rev.	Early volcanic flows, sills, dikes	1100 to 1200
Keweenawan Supergroup	Lower	Normal	Prevolcanic sediments	
		Normal	Pre-Keweenawan basinal clastic rocks (Sibley Gp.)	ca. 1340

Lake Superior region, rather than into a narrow, fault-bounded rift valley. The axis of the basin at the top of the volcanic rocks generally follows the centerline of the lake, but it is shifted slightly to the south due to the asymmetric character of the basin (Figure 4). Geologic and geophysical evidence indicates that in general the dip of the volcanic rocks on the southern margin is steeper than for those on the northern side of the lake, with the dip of the southern margin rocks locally approaching vertical. Individual volcanic flow units probably do not universally thicken toward the center of the basin, although they probably thicken as a whole. Merk & Jirsa (1982) present evidence that volcanic units originally extended beyond the erosional limits of the present basin and were fed from fissures that we now observe as dike swarms in the adjacent older rocks (Green 1982). Gravity anomalies of the eastern half of Lake Superior are muted in comparison with those in the western half. This feature may in part be related to the depth of burial of the mafic rocks, but modeling by Hinze et al (1982) suggests the cause is a thinner volcanic layer in the east that has been much less drastically deformed than in the west.

As indicated above, detailed information about the distribution of volcanic rocks and their structure is much more limited for the buried southeastern and southwestern extensions of the rift system. However, the limited drill hole data that do exist (Craddock 1972, Lidiak 1972, Sleep & Sloss 1978, Yaghubpur 1979, Bickford et al 1979, Treves & Low 1984) indicate that volcanic units are present over most, if not all, of the length of the rift system (Figure 3). However, there is also evidence that the volcanic rocks are locally interbedded with or overlain by sedimentary units, much as in the Lake Superior region, and that the volcanic units are probably not continuous. The multiple extrusive center model proposed for the Lake Superior region probably is also reasonable for the rest of the rift system.

Table 2 Keweenawan volcanic centers (after Green 1982)

Volcanic plateaus	Polarity	Thickness (km)
Keweenaw Point–Isle Royale	N	2.5–5.2
Chengwatana	N	6
North Shore (normal)	N	3.7–7.1
Mamainse-Michipicoten	N	4.1–7.1
Mamainse (reversed)	R	2.5
Osler	R	2.8
Ironwood–Grand Portage–Nopeming	R	3.0–6.3
Siemens Creek	N	0.1

Although basaltic compositions predominate, the volcanic rocks of the Lake Superior basin span the range from olivine tholeiite to rhyolite (Figure 5), and Green (1982) estimates that the population is bimodal, with the rhyolitic maximum significantly lower than the basaltic maximum. The basalts typically have high Al contents, and the transitional group has alkali contents that overlap the alkali-tholeiitic basalt fields. Green (1982) regards the Keweenawan volcanics as similar to other plateau basalts. However, attempts to model magma evolution have been only partly successful, and it is not yet clear how the various groups are related to one another or to Keweenawan plutonic units such as the Duluth Complex (see below). In any case, it appears from Sr, Pb, and Nd isotopic data (Leeman 1977, Dosso & Murthy 1982) that crustal contamination was significant in many areas for rhyolitic components, but that primitive mafic compositions have been derived directly from the mantle. One of the major questions is the degree to which the rhyolitic magmas were derived by fractional crystallization of basaltic magmas or by partial melting of older crustal rocks.

Basaltic rocks have also been encountered in the subsurface along the MGA and in a deep drill hole in central Michigan (McCallister et al 1982, Lidiak 1972). However, there are few chemical or isotopic data on these occurrences that can be used to define their origin.

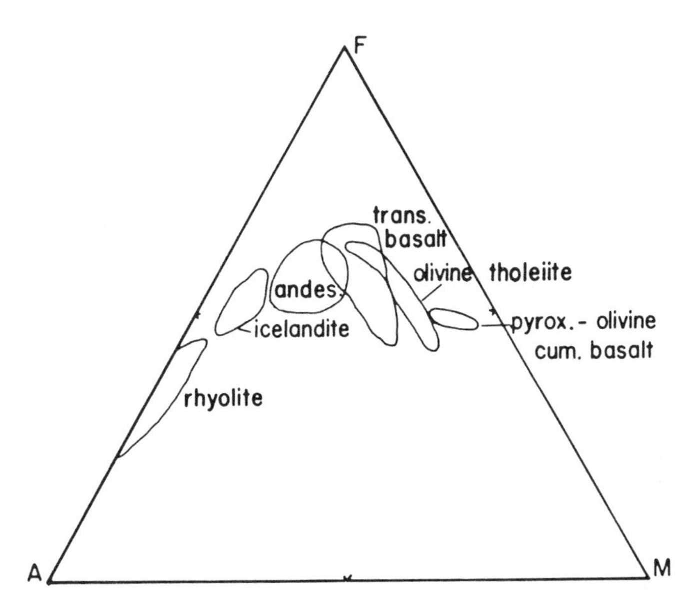

Figure 5 AFM diagram for Keweenawan lavas of the North Shore Volcanic Group, western shore of Lake Superior (Green 1982).

PLUTONIC UNITS The geology, petrology, and geochemistry of plutonic units associated with the midcontinent rift system in the Lake Superior region have been reviewed recently by Weiblen (1982). The plutonic units can be divided into three principal suites: large gabbroic bodies, basaltic sills and dikes, and alkaline complexes. The first two are intimately associated with Keweenawan volcanic rocks, and their association with the rift system is obvious. The third group, a series of alkaline intrusions north and northeast of Lake Superior (Figure 3), is closely associated with the rift system in space and time, and many models of rift system development include the alkaline suite as an integral part of their history (Weiblen 1982).

Probably the best known of the large mafic intrusive units are the Duluth Complex sequence in Minnesota (Weiblen & Morey 1980), west of Lake Superior (Figure 4), and the Mellen Complex in northwestern Wisconsin. Both of these units have many features of typical layered gabbro sequences, although the stratigraphies and petrology are somewhat more complex than typical models. The various units are not related to one another by a simple model of cumulates or magmatic differentiation; instead, they indicate a complicated intrusive history that includes cross-cutting relationships and internal intrusive relationships. Early models of the Duluth Complex treated it as a lopolith (Grout 1918), but later workers have argued against this model because it does not fit the geophysical data, particularly the gravity data; for example, White (1966) argues that there is no evidence for the central feeder needed for the lopolith model. More recently, Weiblen & Morey (1980) have called on a "half-graben" model, in which the igneous units of the Duluth Complex were emplaced into space developed by subsiding crustal blocks during extensional phases of the rift system.

Although the general compositions of the igneous rocks in the mafic complexes are tholeiitic, they do not follow simple differentiation trends, and Weiblen & Morey (1980) present arguments for the existence of two main igenous sequences in the Duluth Complex. The earlier phase involved extensive and efficient melt and crystal segregation to produce rock types with very contrasting mineralogies, such as peridotite, anorthosite, and felsic differentiates. The later phase, referred to as the troctolitic phase, involved intrusion of magma more continuously, with less crystal segregation but with nonuniform fractionation trends. Weiblen & Morey (1980) have suggested that the magma responsible for the earlier phase can be related to that which provided the earlier, reversely polarized lavas of the North Shore Volcanic Group in Minnesota, and that the magma associated with the troctolitic phase can be related to the later, normally polarized lavas of these volcanics.

Mafic intrusive rocks are also extensively represented in smaller plutonic bodies, primarily as sills and dikes throughout the Lake Superior region, including Lake Nipigon, north of Lake Superior (Figure 3). These have also been recently reviewed by Weiblen (1982), who concludes that two distinct magma types can be recognized. The earlier dikes and sills tend to have quartz tholeiite compositions, whereas the later ones tend to have olivine tholeiite compositions. These compositional differences also correlate with the magnetic polarity of the dikes and sills; the earlier quartz tholeiite ones are generally reversely polarized, while the later ones are normally polarized. Weiblen & Morey (1980) and Weiblen (1982) have also concluded that it is reasonable to regard all the mafic igneous rocks of the Lake Superior basin as genetically related, with an overall subdivision into two compositional suites. The earlier suite includes the cumulate units of the Duluth Complex, quartz tholeiite sills and dikes, and Lower Keweenawan volcanic units, most of which are reversely polarized but appear to be related to one another in a rather complex way petrogenetically (Weiblen 1982). The second, later suite consists of the troctolite series of the Duluth Complex and equivalent units in other intrusions, olivine tholeiite sills and dikes, and the later, normally polarized volcanic units that may represent a simpler system of fractionating magma chambers, feeder dikes, and overlying volcanic flows (Weiblen & Morey 1980).

The other major plutonic suite of general Keweenawan age in the Lake Superior region is a series of alkaline complexes that occur to the north of Lake Superior as the Coldwell alkaline province and to the northeast of Lake Superior and as nepheline-carbonatite complexes in the Kapuskasing structural zone (Currie 1976, Weiblen 1982; Figure 3). The alkaline complexes are not tied to the Keweenawan igneous activity by any direct cross-cutting, intrusive, or depositional relationships, but they are approximately the same age (1000–1200 Myr; see below), and both reversed and normal magnetic polarities occur. Therefore, it is tempting to try to relate these units to an early stage in the tectonic evolution of the Lake Superior region (Burke & Dewey 1973, Weiblen 1982). However, although the alkaline and carbonatite complexes may be about the same age as the Keweenawan rifting and igneous activity, the Kapuskasing structural zone itself is a Late Archean to Early Proterozoic feature (Watson 1980, Percival & Card 1983).

At present there is very little direct evidence of plutonic units along the rest of the midcontinent rift system of the magnitude and variety found in the Lake Superior region. However, there are many gravity and magnetic anomalies along the system that could easily be related to such units, and some drill holes have penetrated mafic and alkaline bodies in the subsurface of Iowa and Nebraska near the MGA. These could be approximately the same age as the rift system (Yaghubpur 1979, Brookins et al 1975).

Unfortunately, there have been no good age determinations on these units. Nonetheless, it is not unreasonable to expect that some features similar to those in the Lake Superior region occur along the rest of the midcontinent rift system; many of the anomalies in the immediate vicinity of the main anomaly system are potential targets of study in the search for such units.

SEDIMENTARY UNITS There are basically four sequences of sedimentary rocks associated with the rift system in the Lake Superior basin. In all cases the rocks are clastic units of varying grain size and texture, from conglomerates to shales. The older (pre-Keweenawan) suite is quite limited in occurrence and may not have a direct relationship to rift development. The three younger suites are directly associated with development of the rift system and can be divided according to whether they represent deposition during active rift volcanism (synvolcanic), whether they were deposited in a central trough that remained in the rift system after cessation of igneous activity (postvolcanic), or whether they were deposited in a broad sag basin that developed as a result of regional subsidence of the rift complex (sag basin).

Pre-Keweenawan sedimentary rocks The principal sedimentary unit that can be delegated to pre-Keweenawan status is the Sibley Group that occurs north of Lake Superior. It is about 1340 Myr old (see below) and underlies volcanic rocks of the Osler Group in the Thunder Bay–Nipigon region. Recent summaries of these rocks include those of Franklin et al (1980) and Ojakangas & Morey (1982). Franklin et al (1980) have suggested that the Sibley Group was deposited in a basin that formed as a third arm of the rift system, but these rocks are apparently considerably older than other early-stage Keweenawan rocks, and the location may either be fortuitous or represent a basin developed along structural weaknesses that later helped to control the rift system.

Synvolcanic sedimentary rocks Sedimentary units that are interbedded with volcanic rocks are somewhat limited in extent and have been reviewed recently by Merk & Jirsa (1982). They are typically coarse, immature, polymictic, red-bed clastic rocks that were deposited by streams flowing over the surfaces of Keweenawan volcanic flows. We have included in this category older units that occur as prevolcanic sedimentary rocks in the western Lake Superior region (Ojakangas & Morey 1982) but that are apparently close in age to the onset of Keweenawan igneous activity, since they appear to fit as part of the tectonic suite. The oldest Keweenawan lavas have normal polarity, and sedimentary units of this suite have both normal and reversed polarity, which indicates that some of them were deposited after the onset of volcanism. The beds of these formations are generally mature, quartz-rich sandstones that were deposited in shallow basins (Ojakangas & Morey 1982); these features suggest that rift development was not very advanced and, further, that volcanism began very early in the development of the rift.

Postvolcanic rift basin units The units included in this suite are postvolcanic rocks that show close association with the main rift tectonic activity and consist primarily of clastic debris derived from the volcanic rocks of the rift system. These units are generally included in the Oronto Group (Daniels 1982) and Solor Church Formation (Morey & Ojakangas 1982) and mark the beginning of Late Keweenawan time (Morey & Green 1982). The principal formations are the Copper Harbor Conglomerate (Elmore 1984), the Nonesuch Shale, and the Freda Sandstone. The Copper Harbor Conglomerate represents alluvial deposition contemporaneous with and following Late Keweenawan volcanism in depressions of the rift system. The Nonesuch Shale has been interpreted as representing anoxic lacustrine environments that flanked the rift system and were formed by the combined action of alluvial, volcanic, and tectonic processes (Daniels 1982). It is locally interbedded with units of the Copper Harbor Conglomerate. The Freda Sandstone overlies the other units and marks the transition upward into fluvial sedimentation more characteristic of a stable craton.

Younger sedimentary units Morey & Ojakangas (1982) have divided the postvolcanic sedimentary rocks of the Lake Superior region into those that consist of detritus derived primarily from the rift system itself (largely volcanic sources, discussed above) and those that consist of detritus derived primarily from the surrounding older Precambrian craton. The latter suite typically overlies the former suite and may be regarded as postdating active rifting. Instead, the depositional basins were formed largely by vertical processes, perhaps as crustal subsidence due to the increased mass of the crust along the rift system. The principal units that represent this phase of sedimentation include the Bayfield Group in Wisconsin, the Fond du Lac Formation and Hinckley Sandstone in Minnesota (Morey & Ojakangas 1982), and the Jacobsville Sandstone (Kalliokoski 1982) in the eastern Lake Superior region.

Formations with equivalent lithologic and structural relationships occur along almost the entire length of the midcontinent rift system and have been encountered in numerous drill holes to basement. Key examples are the Rice Formation in the subsurface of Kansas (Scott 1966), rocks similar to (and mapped as) the Bayfield Group in Iowa (Yaghubpur 1979), and red clastic rocks encountered in a deep drill hole in the Michigan basin, over the mid-Michigan anomaly (Catacosinos 1981, Fowler & Kuenzi 1978). These units have a much wider distribution than the rocks of the main rift system, and they approach thicknesses of 2–4 km.

Geochronology

Although there have been numerous geochronologic investigations of units of the Keweenawan suite, they have been of relatively little use in establishing the stratigraphic relationships among the Keweenawan units or between these units and others in the region of comparable age. There are two principal reasons for this. First, it appears that much of the development of the rift system took place over a relatively short period of time. In particular, Silver & Green (1972) have shown through U–Pb analyses of zircons that most of the igneous activity in the Lake Superior region took place over a very short period of time, about 1110 ± 10 Myr ago, so that resolution of relative ages during this event is probably only possible using the U–Pb method on zircons. Unfortunately, however, zircons are not common in most of the igneous units present in the Lake Superior region but are primarily restricted to felsic phases. Other common dating techniques that have been or could be applied to igneous rocks of the region either have poorer precision (and hence poorer resolution) or are more susceptible to later disturbances. As a result, there are many reported ages from 900 to 1200 Myr ago (Van Schmus et al 1982), but the range in ages is probably more apparent than real. Also, attempts to obtain meaningful ages on sedimentary units have been only partially successful, with many of the results uncertain because of later loss of radiogenic daughter product or because of incorporation of inherited daughter product in the detrital components.

One aspect of the history of the rocks that has helped establish relative as well as approximate absolute chronologies has been the paleomagnetic history. As mentioned above, there were at least two major reversals of the Earth's field during Keweenawan times (N to R, R to N) that can be used to help establish relative ages. Also, the paleopole position migrated rapidly during Keweenawan times, so that poles from units relatively close together in age are spatially resolved on the polar wander path for the interval involved (Figure 6; Van Schmus et al 1982, Halls & Pesonen 1982). The chronology that is summarized below is thus a result of both radiometric and paleomagnetic dating, along with traditional stratigraphic data.

Pre-Keweenawan sedimentary units, notably the Sibley Group, apparently formed about 1340 Myr ago (Franklin et al 1980) and probably predate the onset of Keweenawan structural and igneous development by at least 100 Myr. The earliest igneous event that might be related to the onset of rifting and igneous activity is the emplacement of basaltic dikes in the Sudbury, Ontario, region about 1225 Myr ago (Van Schmus 1975) during the early period of normal polarity. Although no genetic link has been demonstrated, structurally or petrochemically, these dikes may correspond in age to the volcanic units formed during the early normal polarity time. Age studies on the carbonatitic and alkaline complexes north and northeast of Lake Superior are not sufficiently precise or accurate to determine whether these bodies are older than, contemporaneous with, or younger than the main Keweenawan activity, but Weiblen (1982) has placed them in the older category.

The vast majority of igneous activity, in terms of both volume and stratigraphic units, apparently occurred over a very short time span about 1110 Myr ago (Silver & Green 1972). This includes much of the igneous activity during the reversed polarity interval as well as that occurring after the return to normal polarity. In fact, the dated units bracket the magnetic reversal from reverse to normal, thus dating it rather precisely at 1110 Myr. The only dates that are younger than 1100 Myr are either Rb–Sr whole-rock or K–Ar by various techniques. In both cases, there is considerable uncertainty because of possible loss of radiogenic daughter product, so that the existing data should be interpreted to indicate the possibility of younger units, but that there may in fact be very little igneous activity associated with the rift system after 1100 Myr. This is not unreasonable, since major igneous activity in more recent rift systems (e.g. Oslo graben, Rio Grande rift, Kenya rift) took place over intervals of only 20–30 Myr (Williams 1982). Rb–Sr ages from postvolcanic sedimentary units, particularly the None-

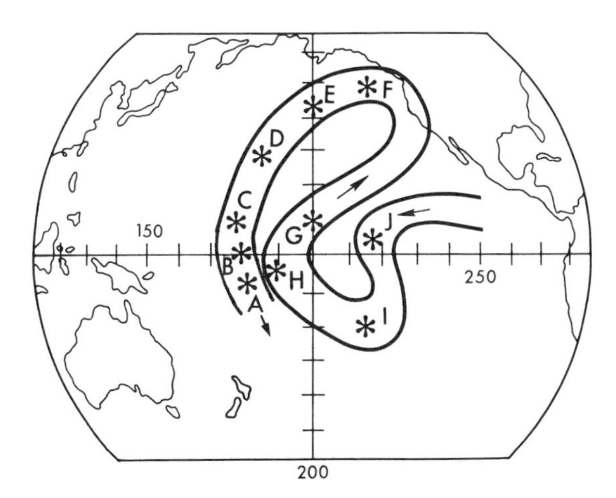

Figure 6 Sketch of apparent polar wander path for Lake Superior region units 1000–1500 Myr old (as listed in Table 3). After Van Schmus et al (1982) and Halls & Pesonen (1982).

such Shale (1023±46 Myr; Chaudhuri & Faure 1967), yield ages close to 1100 Myr, which further indicates that volcanic activity could not have continued to much later than 1100 Myr ago.

The ages of the posttectonic sedimentary units are even more uncertain because they are not amenable to normal radiometric techniques. However, paleomagnetic data on younger units such as the Freda Sandstone and Jacobsville Sandstone show that the paleopole positions of these units are only slightly farther along the Keweenawan polar wander curve (Figure 6) than the Nonesuch Shale or the younger volcanic units (Halls & Pesonen 1982). If the rapid pole position movement that characterizes the earlier part of the curve continued through deposition of the Freda and Jacobsville sandstones, then they are probably not much younger than 1100–1000 Myr.

One of the more interesting aspects of the chronologic data is that the paleopole position changed very rapidly during the main period of igneous activity and, presumably, associated rifting (Table 3, Figure 6). This suggests that the craton was actively moving at the time and was also undergoing significant changes in direction that could have induced tensional forces (Halls & Pesonen 1982).

Initial isotopic data for radiometric systems in Keweenawan rocks are still somewhat limited. Van Schmus et al (1982) summarized Rb–Sr data that existed to that time and concluded that the mean initial Sr-87/Sr-86 ratio for Keweenawan igneous rocks was 0.7046±0.0005 on a regional basis, too close to the uncertainties in a mantle growth curve for significant interpretation. On a more detailed basis, Leeman (1977) reported a significant range for initial Sr and initial Pb composition from volcanic rocks in Minnesota, and Dosso & Murthy (1982) found significant

Table 3 Paleomagnetic poles for Lake Superior region[a] (Figure 6)

Unit	Age (Myr)	Polarity	Pole position (long)	Pole position (lat)
A. Jacobsville Sandstone	?	N	183	−9
B. Freda Sandstone	?	N	180	1
C. Nonesuch Shale	1023	N	177	10
D. Middle Keweenawan	1110	N	183	29
E. Middle Keweenawan	1110	R	203	42
F. Logan Sills	1150	R	220	49
G. Lower Keweenawan	?	N	200	10
H. Sudbury Dikes	1225	N	189	−3
I. Sibley Group	1340	N	214	−20
J. Croker Island Complex	1475	N	217	5

[a] After Van Schmus et al (1982) and Halls & Pesonen (1982).

variations in initial Sr and Nd isotopic ratios. In general the isotopic data can be interpreted to indicate the involvement of older crustal rocks in the evolution of the felsic Keweenawan magmas. However, the basaltic magmas were apparently derived from enriched, heterogeneous mantle without significant crustal contamination.

Economic Deposits

Economic deposits commonly accompany rift systems, and the midcontinent rift system is no exception (Norman 1978, Weiblen 1982). There are at least four known or possible types of deposits associated with the rift system. The first of these is the classical native copper deposits of the Keweenaw Peninsula in Michigan (White 1968). These are basically stratabound deposits that formed as a result of hydrothermal solutions acting on copper-bearing lavas and volcaniclastic debris of the region (White 1971). Similar, but much smaller, examples occur in the Mamainse Point region along the east shore of Lake Superior. So far no occurrences of such deposits have been reported from the buried part of the rift system, but it is a distinct possibility that they exist. In fact, McCallister et al (1982) reported trace amounts of native copper from altered basaltic rocks encountered in the deep hole in the Michigan basin.

The second type of deposit is Cu–Ni sulfide mineralization associated with the lower phases of the Duluth Gabbro Complex in Minnesota (Weiblen 1982) as well as other plutons, sills, and dikes. Most of these are subeconomic, but the potential exists that igneous intrusions associated with the midcontinent rift system in places other than the Lake Superior region may have significant mineralization. Some mafic intrusions of possible Keweenawan age have been explored in Iowa, so far without success (Yaghubpur 1979).

The third type of occurrence is also primarily Cu mineralization, in this case associated with the 1100–1200-Myr-old alkaline and carbonatite complexes north and east of Lake Superior (Currie 1976, Norman 1978). As with the other types of mineralization, similar occurrences are not known from the buried portions of the rift system, but if the Lake Superior complexes are genetically related to the rifting, such features may occur elsewhere along the rift system. One possibility, in fact, is the Elk Creek carbonatite-bearing complex associated with the Elk Creek anomaly in southeast Nebraska (Irons 1979). It is reported to underlie basal Paleozoic sedimentary rocks (Brookins et al 1975), and a nearby gabbro plug in the subsurface yields an Ar–Ar age of about 1200 Myr (Treves 1981).

The existence of the fourth type of economic occurrence is still speculative. The Nonesuch Shale is locally organic rich, and Lee & Kerr (1984) and Dickas (1984) have recently summarized the hydrocarbon potential of similar types of units within the central sequence of the rift system along the MGA. Several test wells are currently underway or planned, and we shall have to await the outcome of this exploration. Even if the results from such drilling and related geophysical surveys are negative, they should eventually help in developing a better understanding of the stratigraphic sections and geological relationships within the buried rift basins.

Sawkins (1982) has summarized the metallogenesis associated with rift systems in general. The wide variety of mineral deposits, particularly strata-bound copper deposits, found associated with continental rift systems indicates that there may be significant economic potential, both metallic and nonmetallic, along the buried portions of the midcontinent rift system.

4. STRUCTURAL RELATIONSHIPS

Surface and Near-Surface Structure

The structure of the midcontinent rift system remains poorly known despite many years of investigation because outcrops are limited where the feature is at the surface in the Lake Superior region and because there is poor distribution of deep drill holes along the rest of its length. The lack of direct geologic information on the rift has focused attention on the interpretation of geophysical data, especially gravity and magnetic anomalies and seismic studies. However, gravity and magnetic studies have poor horizontal and vertical resolution, and seismic data have until recently been confined to regional crustal seismic measurements and shallow seismic investigations in Lake Superior. Two recent COCORP reflection profiles have helped in interpretation of the buried rift structure in Kansas (Serpa et al 1984) and central Michigan (Brown et al 1982), but in general structural detail is poorly constrained. Figure 7 summarizes interpretations of the rift structure at four places along the rift system.

Figure 7 Four interpretations of cross sections of the midcontinent rift system (Figure 3). (*a*) Based on COCORP data across the northern part of the Kansas segment (Serpa et al 1984), showing a basin formed by normal faults; stippled pattern = crystalline basement. (*b*) Based on interpretation of gravity, magnetic, and borehole data for the Twin Cities region of Minnesota (King & Zietz 1971), showing a central horst with flanking clastic basins. Note absence of deep root structure. (*c*) Based on surface geology in the Lake Superior region (Green 1982), showing inferred relationships of volcanic units (OS = Osler Series, PLV = Portage Lake volcanics, PM = Powder Mill Group) to Upper Keweenawan sedimentary rocks (UKS) and Lower to Middle Precambrian basement (LMPC). Note absence of deep root structure. (*d*) Based on COCORP survey across mid-Michigan anomaly (Brown et al 1982), showing major reflectors. Note absence of any significant vertical offsets in Precambrian units, suggesting a relatively smooth basin (Serpa et al 1984).

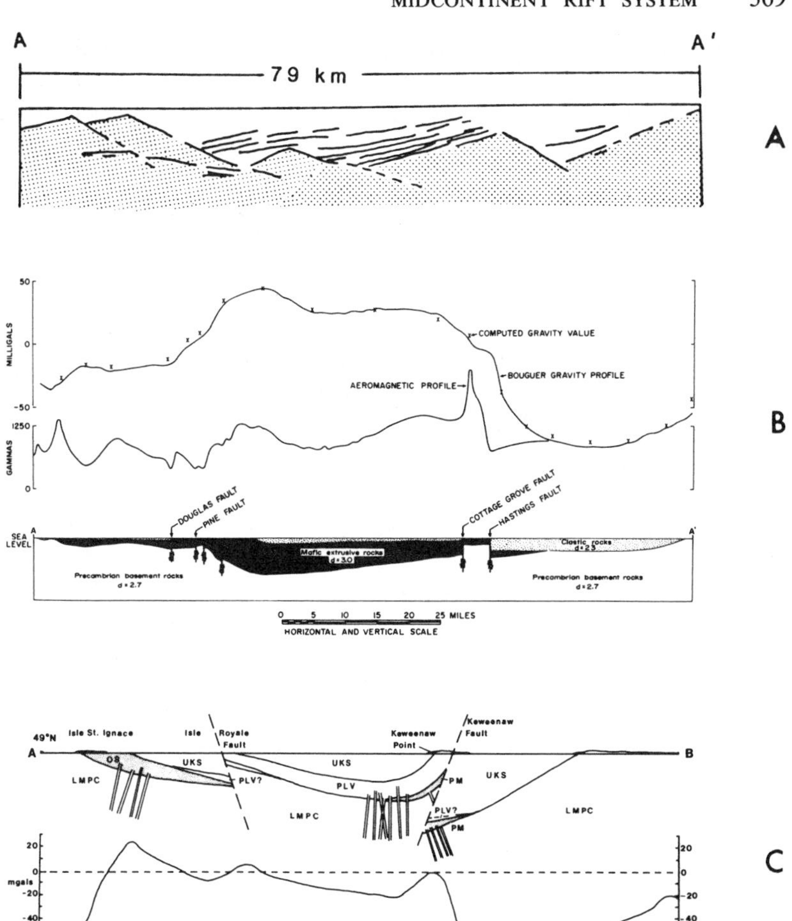
A
A'
79 km
A
50
MILLIGALS
0
-50
1250
GAMMAS
0
COMPUTED GRAVITY VALUE
BOUGUER GRAVITY PROFILE
AEROMAGNETIC PROFILE
B
DOUGLAS FAULT
PINE FAULT
COTTAGE GROVE FAULT
HASTINGS FAULT
SEA LEVEL
Mafic extrusive rocks d=3.0
Clastic rocks d=2.5
Precambrian basement rocks d=2.7
Precambrian basement rocks d=2.7
0 5 10 15 20 25 MILES
HORIZONTAL AND VERTICAL SCALE
49°N
Isle St. Ignace
Isle Royale Fault
Keweenaw Point
Keweenaw Fault
A
B
OS
UKS
LMPC
PLV?
UKS
PLV
PM
UKS
LMPC
PLV?
PM
LMPC
20
0
mgals
-20
-40
-60
-80
C

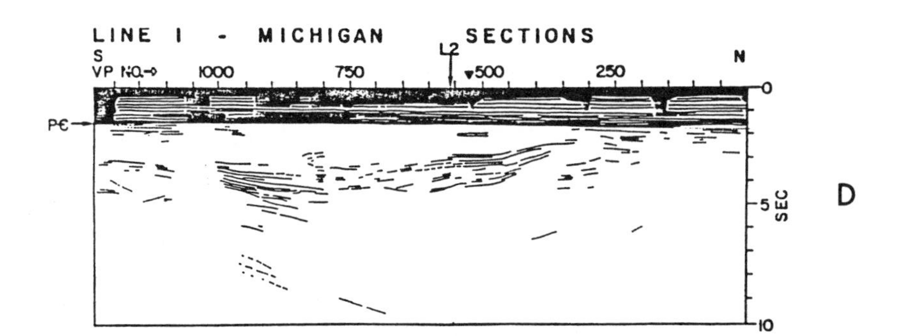
LINE I - MICHIGAN SECTIONS
S
N
VP NO.
1000
750
L2
500
250
PЄ
0
5 SEC
10
D

Geologic observations and geophysical interpretations show that faulting is characteristic of rift system units. The most striking faults observed in the midcontinent rift system are high-angle reverse faults, which formed along the margins of the central basins before deposition of Cambrian clastic sediments (Figure 7*b*). These faults cannot be interpreted as typical graben faults because, as White (1966) has pointed out, the patterns of sedimentation related to the faults and their attitudes prohibit such an interpretation. In fact, no major normal faults occur in the Lake Superior basin, and no evidence exists for an axial graben buried beneath the surficial rocks of the basin (Hinze et al 1982). The major marginal reverse faults of the basin (the Keweenaw, Douglas, and Isle Royale faults; see Figure 4) are particularly important structural features that have led to the development of prominent horsts. These horsts, which bring older volcanic rocks into juxtaposition with flanking clastic sedimentary rocks of a late-stage basin, are only observed in the western half of the lake and the extension to the south. This high-angle reverse-faulting event, which appears anomalous in an extensional (rift) domain, is a commonly observed phenomenon of continental rifts where subsequent compression has modified the structural regime (Milanovsky 1981). Therefore, if the faults bounding the central horst blocks are reverse faults, there must have been later compression.

The second major type of faulting in the Lake Superior basin and in areas of volcanic rock outcrop is represented by transverse faults that cut the older marginal thrust faults (Figure 4). Major strike-slip faults are not observed in the basin. Well-developed folds are commonly observed in the volcanic and overlying sedimentary rocks of the Lake Superior basin; they are believed to be pene- and postcontemporaneous with basin formation, while the faulting postdates folding (Davidson 1982).

A regional structural fabric that either predated or formed at the same time as the rifting is shown by the alignment of mafic dike swarms in the Lake Superior region (Figure 3). These dike swarms tend to parallel the northwest, northeast, and southern shores of Lake Superior and may indicate the orientation of extensional forces during volcanism.

The upper crustal manifestations of the rift system associated with the MGA have until recently been determined primarily by modeling of gravity and magnetic anomalies, utilizing geologic information extrapolated from outcrops in the Lake Superior region and drill hole data. Most models are similar to those proposed by Thiel (1956) on the basis of observed gravity anomalies. A typical example of a derived geologic model (King & Zietz 1971) is shown in Figure 7*b*. The detailed gravity anomaly profile modeled in this illustration was collected by Craddock et al (1963) across the Twin Cities basin in Minnesota and western Wisconsin. The gravity anomaly maximum is due to high-density mafic volcanic rocks that are preserved in an axial horst. The negative gravity anomaly on the eastern margin of the profile is caused by a wedge of low-density Upper Keweenawan clastic rocks. These rocks are younger than the volcanic rocks, so that the volcanic rocks have been relatively uplifted into juxtaposition with the clastic rocks.

A recent COCORP seismic reflection profile over the MGA in northern Kansas has provided new insight into the structure of the axial trough of the midcontinent rift (Serpa et al 1984). Their interpretation of the reflection data (Figures 7*a*, 8) shows an asymmetric basin consisting of two stratigraphic units. The lower unit, composed of high-amplitude reflections between 1 and 3 s, is interpreted as a 5-km-thick segment of interbedded basalts and clastic rocks. An overlying seismically transparent unit is interpreted to be made up of predominantly clastic rocks varying from 1 to 3 km in thickness. The nature of these rocks is supported by drill hole data (Bickford et al 1979). East-dipping reflectors (Figure 8) that truncate the inferred volcanic sequence across the entire width of the trough are interpreted as faults and cause the structural asymmetry. An earlier COCORP survey across the mid-Michigan anomaly in central Michigan (Figure 7*d*) indicated similar reflection sequences, and these are also interpreted as basal volcanic units with overlying clastic units (Brown et al 1982). A deep hole that had earlier been drilled near the site of the COCORP survey penetrated 1315 m of Precambrian red clastic rocks before bottoming in a metamorphosed mafic volcanic unit at a depth of 5324 m. Brown et al (1982) found only indirect evidence of faulting in the trough in the form of relatively steep dips, sharp flexures, and structural benches.

Interpretation of the seismic reflection profiles across the mid-Michigan and midcontinent geophysical anomalies suggests that the principal shallow manifestation of the midcontinent rift system is a trough that initially filled with interbedded volcanic and clastic rocks that were later covered by clastic sediments. Contrary to classical views of surface deformation associated with rifts, a central graben with marginal, steeply deepening normal faults has not been observed for the midcontinent rift system. Instead, faulting is limited to tilted fault blocks, with extension presumably taken up by rotation of fault blocks (Serpa et al 1984); Weiblen & Morey (1980) have called on similar processes for the extension that accompanied emplacement of the Duluth Complex. The interpretations from the COCORP surveys also indicate that later high-angle reverse faulting, with associated horsts, is not a universal feature of the midcontinent rift system and may be limited to the western Lake Superior region and the northern part of the MGA (Figures 7*b,c*).

The principal differences between the Lake Superior basin and the rest of the rift system are that it is broader and, apparently, deeper. Serpa et al

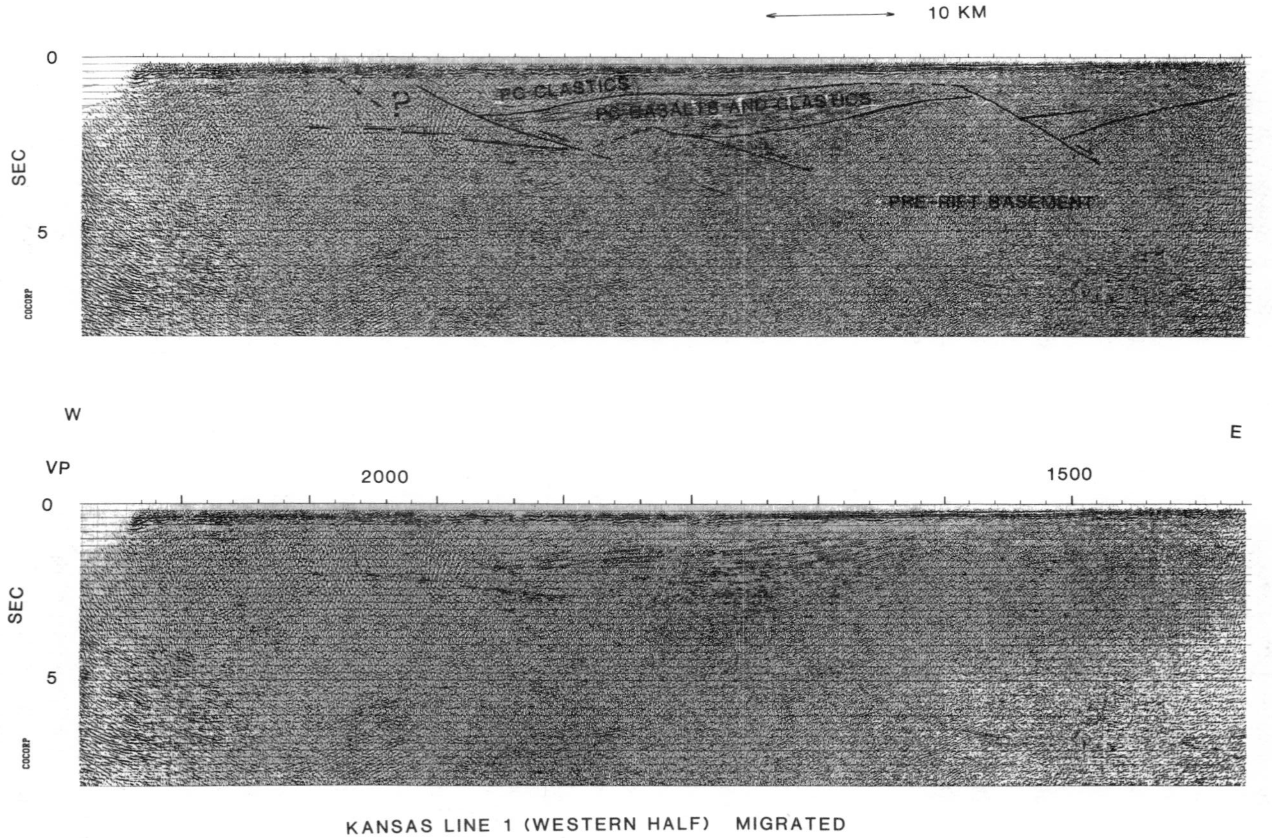

Figure 8 COCORP seismic reflection data for part of the survey across the MGA in northern Kansas (Serpa et al 1984), showing in more detail the rift basin model of Figure 7*a*.

(1984) suggest a maximum depth of 8 km for the trough in northern Kansas, while geological data (Halls 1966), gravity data (Hinze et al 1982), and seismic refraction studies (Leutgert & Meyer 1982) indicate depths in excess of 10 km in the Lake Superior basin.

In addition to the vertical aspects of the structure of the rift system, several distinct lateral features can be noted. In particular, there are several places where the trend of the rift system is disrupted, as noted above. The largest offset occurs in southern Minnesota, where the MGA swings to the southeast and then back to the southwest. Some representations of this offset have shown it as a distinct break and have proposed a major transverse or transform fault (King & Zietz 1971, Craddock 1972, Chase & Gilmer 1973). However, the gravity map of the region clearly shows that the major gravity high is continuous, though narrower, throughout this segment. Thus, we consider it more probable that the rift system was deflected along preexisting structures, although the possibility of later faulting along the same structures cannot be discounted. The second major offset along the MGA occurs in southeastern Nebraska, at the Kansas border (Figures 1, 3). In this case there appears to be a definite discontinuity, and either the rift system was offset by later faulting or the rifting jumped along a preexisting structure. It is interesting to note that this offset coincides with extension of gravity and magnetic lineaments that trend SE-NW across Missouri and probably mark major structures in the older Proterozoic basement (Kisvarsanyi 1984, Arvidson et al 1984, Hildenbrand et al 1982). Interpretation of these structures has not progressed to the point where it can be stated whether they predate or postdate the midcontinent rift system. Other discontinuities occur at the southeast end of the mid-Michigan geophysical anomaly and between the east-continent anomalies. The significance of these are not known.

Deep Structure

Characteristically, the entire crust is disturbed in a rifting event, and expressions of this disturbance remain in paleorifts as a vestige of the thermal-tectonic processes (Ramberg & Morgan 1984). Seismic and gravity studies indicate that such is the case in the midcontinent rift system. Seismic refraction studies in the Lake Superior basin and along the MGA (e.g. Smith et al 1966, Cohen & Meyer 1966, Ocola & Meyer 1973, Leutgert & Meyer 1982) indicate higher crustal seismic velocities and a thicker crust. Halls (1982) has used crustal time terms derived from seismic experiments in the Lake Superior region to map crustal thickness. His results show a crust in central Lake Superior that reaches thicknesses in excess of 50 km, with arms of thickened crust extending southward from the ends of the lake along the MGA and mid-Michigan anomaly. COCORP seismic profiling in northern Kansas, across the MGA (Serpa et al 1984), indicates a disturbed crust. Serpa et al infer the presence of intrusions beneath the rift trough from seismic velocities, reflection character, and modeling of the gravity anomaly data. Similarly, Keller et al (1982) interpret the seismic results of Warren (1968) in the vicinity of the east-continent anomalies as evidence of disturbed crust.

Gravity modeling of the midcontinent rift system prior to the early 1970s was generally successful in explaining the entire anomaly (axial high with the marginal minima) with high-density volcanic material in an axial trough between adjacent clastic wedges of lower density rocks (Figures 7*b*,*c*). However, with the increasing availability of deep crustal seismic information and the realization that the entire crust remains disturbed as a result of the rifting event, it became apparent that a significant portion of the gravity high associated with the rift was due instead to increased density in the lower crust. Ocola & Meyer (1973), Hinze et al (1982), and Chandler et al (1982) ascribe the positive gravity anomalies in varying degrees to an increased crustal density, presumably caused by multiple mantle intrusions. These intrusions may have been initiated with the weakening of the crust under the extensional forces associated with the Keweenawan rifting event, and they probably served as feeders to the surface volcanic rocks. A typical example of gravity modeling that includes higher density lower crust is shown in Figure 9. The block of 3.0 g cm^{-3} density material shown is not intended to represent a homogenous unit of intruded material that is now anomalously denser and of higher velocity than the adjacent crust; instead, it should be taken to represent the average properties of a region of high-density intrusions. The shape of the high-density block is open to considerable modification because of ambiguities in modeling gravity data.

The gravity model across the western Lake Superior basin shown in Figure 9 incorporates the thickness of the crust beneath the rift zone that has been indicated by seismic studies. The thickened crust produces a regional gravity minimum that, when summed with the gravity high associated with the near-surface volcanic rocks filling the trough, results in negative anomalies bounding the central maximum. These marginal minima are represented in Figure 9 and are characteristically found along the linear gravity maximum of the midcontinent rift system. Commonly, these minima are related to the marginal clastic-rock-filled basins. However, White (1966) argues that in several areas of the Lake Superior basin and the MGA, the clastic rocks are inadequate to explain fully the observed minima. Thus, it follows that the minima at least partly reflect the broad gravity anomaly caused by an increased depth to Moho along the axis of the rift system.

Another aspect of the deep structure that has been discussed recently is

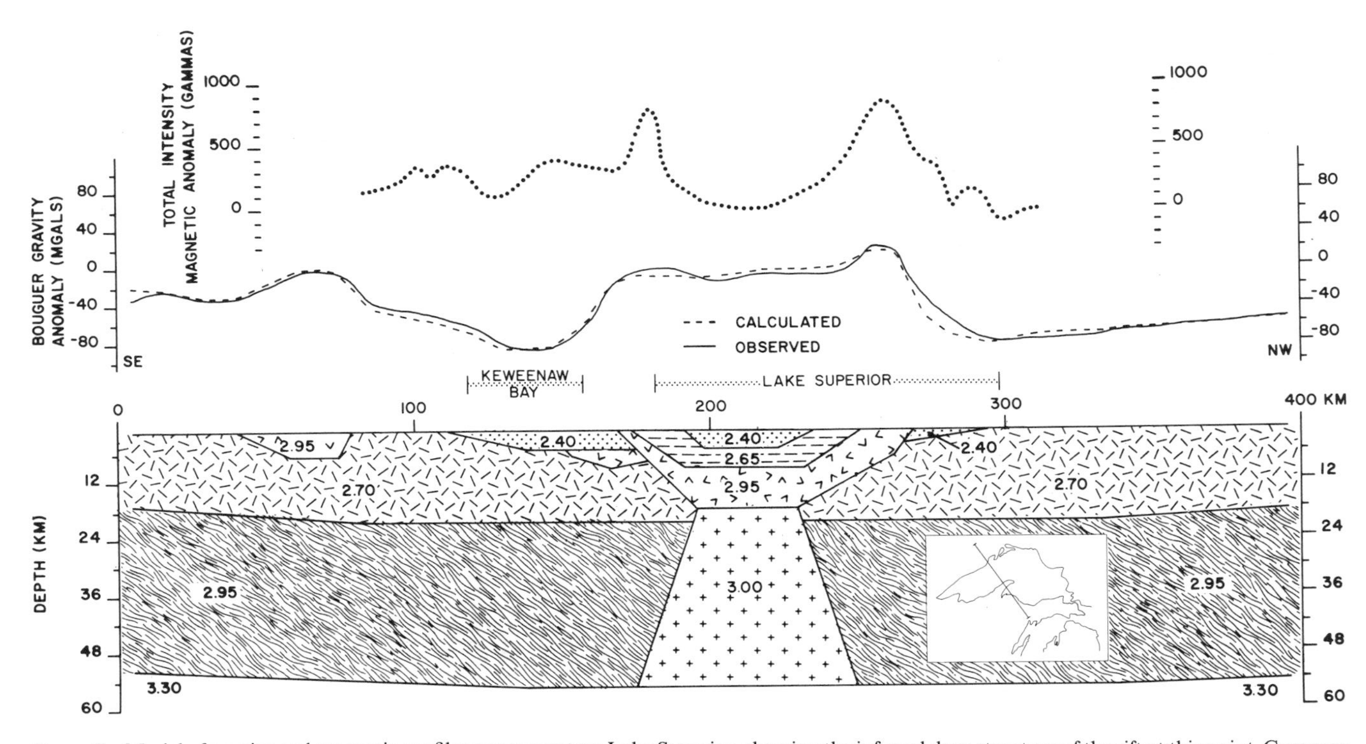

Figure 9 Model of gravity and magnetic profiles across western Lake Superior, showing the inferred deep structure of the rift at this point. Compare with cross sections in Figures 7*b*,*c*. Numbers indicated for geologic units are assumed densities in grams per cubic centimeter. Lithologies are identified by densities as follows: 2.70 = upper crust, 2.95 = lower crust and mafic extrusives, 3.3 = upper mantle, 2.65 = Oronto Group; 2.40 = Jacobsville Sandstone and Bayfield Group, and 3.0–3.05 = "hybrid" introduced crust. After Hinze et al (1982).

the question of how far the older crust has separated during the rifting event. It is apparent that at the southern end (in Kansas), crustal separation must become negligible. Chandler (1983) has recently analyzed offsets of magnetic anomalies in the western Lake Superior region and concludes that crustal separation was about 60 km, consistent with earlier estimates of 65–75 km (Chase & Gilmer 1973) and 70 km (Sims et al 1980). The width of the anomaly in these areas is comparable with or slightly wider than the rest of the rift system (excluding the Lake Superior basin), which indicates that the rift never opened more than about 60–75 km.

Although many older rift systems show evidence of moderate to extensive reactivation, the midcontinent rift system shows little evidence of such behavior. There are a few minor structures along the rift system (Coons et al 1967, Hinze et al 1975) but no major faults that cut the Phanerozoic cover. In fact, there is very little evidence of a concentration of seismic activity today in the region occupied by the MGA or any other segment of the system (see, for example, York & Oliver 1976).

5. ORIGIN AND EVOLUTION OF THE RIFT SYSTEM

Although the overwhelming consensus is that the MGA, Lake Superior Basin, mid-Michigan anomaly, etc, represent continental rifting, there is much less agreement about the underlying causes for the rifting. In general, a rift can be classified as an active type, in which thermal perturbation of the mantle causes disturbance of the overlying crust, or a passive type, in which plate interaction causes internal tension (Baker & Morgan 1981). Because of the generally large volumes of basaltic magma associated with the midcontinent rift system, an active-type system seems more probable. However, because of the proximity of the Grenville Province, it is possible that extensional forces behind a continental collisional zone (Tapponnier & Molnar 1976, Tapponnier et al 1982) may have contributed significantly to the development of the rift.

In the context of an active model for the formation of the midcontinent rift system, several authors have proposed that the Lake Superior basin is situated over a former hotspot, that a rifting (rrr) triple junction formed, but that only two of the arms (the MGA and mid-Michigan segments) developed extensively. The third, failed arm in this model has not been uniquely identified, but suggestions include the Kapuskasing structural zone northeast of Lake Superior (Burke & Dewey 1973), the Coldwell Alkaline Complex trend north of Lake Superior (Mitchell & Platt 1978), and the Nipigon basin north of Lake Superior (Franklin et al 1980). All three of these possibilities have problems, however. As mentioned above, the Kapuskasing structural zone is an Archean crustal feature (Watson 1980, Percival & Card 1983). It is bounded on the southeast by an inferred northwest-dipping thrust fault that brought deep-seated Archean rocks to the surface. The time of uplift is uncertain; Watson (1980) suggested an Archean age, but geochronologic data summarized by Percival & Card (1983) suggest that a more probable time of uplift is Early Proterozoic because the structure is cut by roughly 1800-Myr-old alkaline-carbonatite complexes. In any case, it appears that the Kapuskasing structural zone predates development of the rift system by several hundred million years. Furthermore, since the Kapuskasing structure was apparently formed by thrusting (Percival & Card 1983), it is not a likely candidate for the failed third arm of the rift.

Igneous activity in the Nipigon region and alkaline complex development in general north of Lake Superior could be related to regional tension associated with development of the rift system, but it does not seem to be localized well enough to define a failed third arm of a rift system developing over a hotspot. Mitchell & Platt (1978) pointed out that mafic dike swarms in the Lake Superior region tend to fall into three groups, paralleling the northeast, northwest, and south shores of the lake. They suggested that the dikes may have been emplaced into zones of weakness developed during the start of rifting, and that the three orientations were consistent with the geometry of rifting over a hotspot. Thus, the presence of a well-defined third arm may not be a prerequisite for favoring the former existence of a hotspot in the Lake Superior region.

Models of passive origin, calling on intraplate tensional forces, have been proposed by Donaldson & Irving (1972), Gordon & Hempton (1983), Weiblen (1982), and Baer (1981). These are generally related to the stress field established in the North American plate in conjunction with the development of the Grenville Province, although details vary considerably depending on whether authors accept a plate collision model for the Grenville Province (e.g. Young 1980, Gordon & Hempton 1983) or call on some other mechanisms (e.g. Baer 1981, Weiblen 1982). One of the factors complicating attempts to fit models of this type into development of the Grenville Province is that a detailed understanding of the Grenville Province is still far from complete, and there is still considerable dispute as to whether there is an identifiable suture or what the actual processes were. Therefore, full understanding of the midcontinent rift system may have to await more complete understanding of the Grenville Province.

The midcontinent rift system is a very large crustal feature, comparable in scale with the current East African rift system (Figure 10). The African system is also very complex, with large segments having extensive igneous activity while others have very little igneous activity (Williams 1982), and

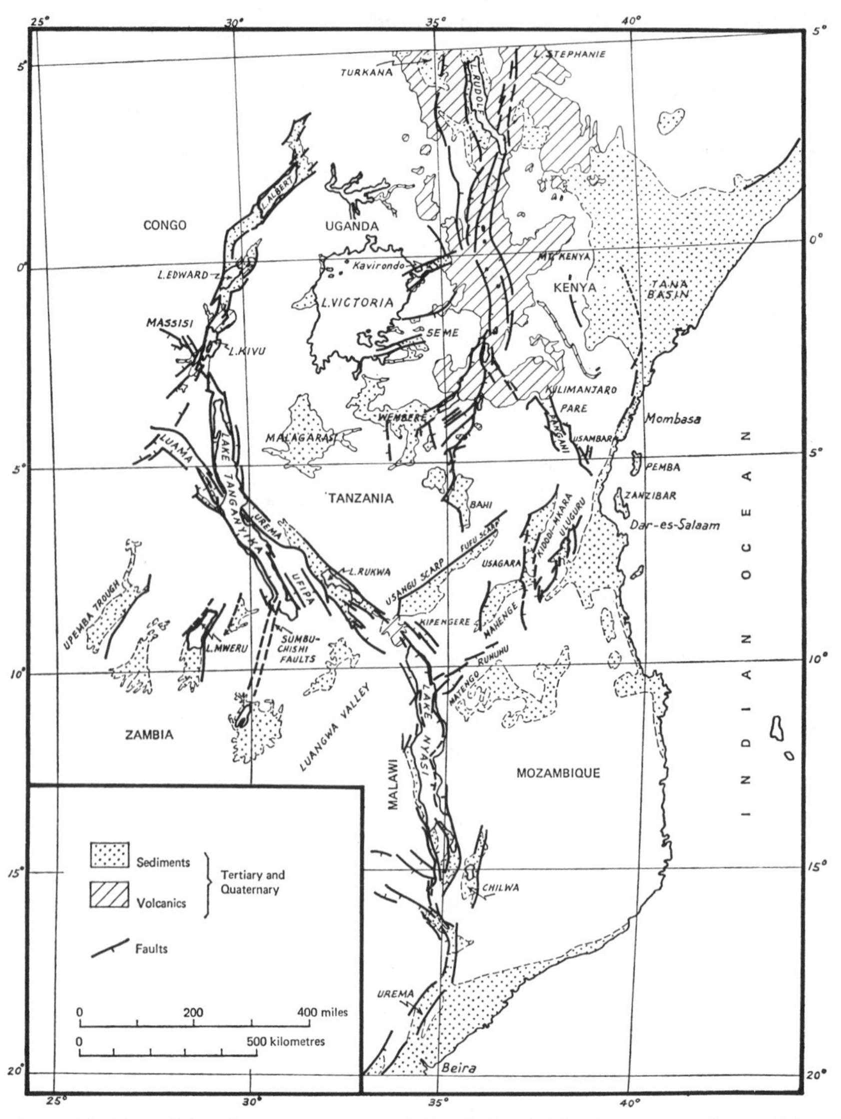

Figure 10 Map of the rift system in eastern Africa (Baker 1971). Note the complexity of the rift system, with numerous branches and satellite rifts. Compare with the midcontinent rift system for scale and complexity.

with many branches and satellite rifts. There appears to be little doubt that active rifting is a major component of development of the East African system, and we believe that it is also reasonable to attribute much of the development of the midcontinent rift system to active, rather than passive, forces. In this context, several authors have pointed out that the development of the midcontinent rift system occurred at a time during which rifting was prevalent throughout the world (Stewart 1976, Sawkins 1976, Halls 1978), and thus it may be related to a major extensional stress field that developed at this time in continental lithosphere, perhaps as a result of widespread plate movement and breakup.

The proximity in time of the rifting and development of the Grenville system and others like it also suggests strongly that these tectonic regimes must be related to one another in a broad sense. Although there may be no direct connection between formation of the midcontinent rift system and development of the Grenville orogen, there appears to be strong evidence for a connection between later events in the rift system (such as later compression) and in the Grenville Province; full understanding of one probably cannot be accomplished without full understanding of the other.

In summary, the midcontinent rift system is a major feature of the North American continental lithosphere. It began about 1200 Myr ago, probably as an active rift system in conjunction with a period of continental breakup on a global scale. Interpretation of geophysical data suggests that much, if not most, of the rift system contains mafic intrusive rocks in a central root zone, with variable amounts of shallower intrusive mafic rocks, mafic volcanic rocks, and clastic sedimentary rocks. The latter rocks formed both in axial rift basins and (later) in regional basins that formed as the crust subsided along the rift system. Later structural modification of the rift system may have been influenced strongly by development of the nearby Grenville orogen. The rift system has been shown to contain valuable mineral deposits, and exploration along the buried portions of the rift may reveal other occurrences of nonferrous metals. In addition, the rift system is currently being examined by the petroleum industry as a possible source of oil and gas.

ACKNOWLEDGMENTS

This paper was made possible through past and present support from the NSF, the NRC, and other federal, state, or private sources, including current NSF Grant EAR 82-19137 to WRVS. We wish to thank M. E. Bickford, C. Craddock, P. Morgan, and L. Serpa for comments on the manuscript and L. Serpa for use of illustrations.

Literature Cited

Arvidson, R. E., Bindschadler, D., Bowring, S., Eddy, M., Guinness, E., Leff, C. 1984. Bouguer images of the North American craton and its structural evolution. *Nature* 311:241–43

Anderson, J. L. 1983. Proterozoic anorogenic granite plutonism of North America. *Geol. Soc. Am. Mem.* 161:133–54

Baer, A. J. 1981. A Grenvillian model of Proterozoic plate tectonics. In *Precambrian Plate Tectonics*, ed. A. Kröner, pp. 353–85. Amsterdam: Elsevier

Baker, B. H. 1971. Explanatory note on the structure of the southern part of the African rift system. In *Tectonics of Africa*, pp. 543–48. Paris: UNESCO

Baker, B. N., Morgan, P. 1981. Continental rifting: progress and outlook. *Eos, Trans. Am. Geophys. Union* 62:585–86

Bickford, M. E., Harrower, K. L., Nusbaum, R. L., Thomas, J. J., Nelson, G. E. 1979. Preliminary geologic map of the Precambrian basement rocks of Kansas. *Kansas Geol. Surv. Map M-9*, scale 1:500,000, with accompanying notes. 9 pp.

Black, W. A. 1955. Study of the marked positive gravity anomaly in the northern midcontinent region of the United States. *Geol. Soc. Am. Bull.* 66:1531 (Abstr.)

Brookins, D. G., Treves, S. B., Bolivar, S. L. 1975. Elk Creek, Nebraska, carbonatite: strontium geochemistry. *Earth Planet. Sci. Lett.* 28:79–82

Brown, L., Jensen, L., Oliver, J., Kaufman, S., Steiner, D. 1982. Rift structure beneath the Michigan Basin from COCORP profiling. *Geology* 10:645–49

Burke, K., Dewey, J. L. 1973. Plume generated triple junctions: key indicators in applying plate tectonics to old rocks. *J. Geol.* 81:406–33

Catacosinos, P. A. 1981. Origin and stratigraphic assessment of pre–Mt. Simon clastics (Precambrian) of Michigan Basin. *Am. Assoc. Pet. Geol. Bull.* 65:1617–20

Chandler, V. W. 1983. Correlation of magnetic anomalies in east-central Minnesota and northwestern Wisconsin: constraints on magnitude and direction of Keweenawan rifting. *Geology* 11:174–76

Chandler, V. W., Bowman, P. L., Hinze, W. J., O'Hara, N. W. 1982. Long wavelength gravity and magnetic anomalies of the Lake Superior Basin structure. See Wold & Hinze 1982, pp. 223–38

Chase, C. G., Gilmer, T. H. 1973. Precambrian plate tectonics: the midcontinent gravity high. *Earth Planet. Sci. Lett.* 21:70–78

Chaudhuri, S., Faure, G. 1967. Geochronology of the Keweenawan rocks, White Pine, Michigan. *Econ. Geol.* 62:1011–33

Cohen, T. J., Meyer, R. P. 1966. The Midcontinent Gravity High: gross crustal structure. In *The Earth Beneath the Continents, Am. Geophys. Union Geophys. Monogr.*, ed. J. S. Steinhart, T. J. Smith, 10:141–65

Coons, R. L., Woollard, G. P., Hershey, G. 1967. Structural significance and analysis of mid-continent gravity high. *Am. Assoc. Pet. Geol. Bull.* 51:2381–99

Craddock, C. 1972. Keweenawan geology of east-central and southeastern Minnesota. In *Geology of Minnesota: A Centennial Volume*, ed. P. K. Sims, G. B. Morey, pp. 416–24. Minneapolis: Minn. Geol. Surv.

Craddock, C., Thiel, E. C., Gross, B. 1963. A gravity investigation of the Precambrian of southeastern Minnesota and western Wisconsin. *J. Geophys. Res.* 68:6015–32

Craddock, C., Mooney, H. M., Kolehmainen, V. 1969. Simple Bouguer gravity map of Minnesota and northwestern Wisconsin. *Minn. Geol. Surv. Misc. Map Ser., Map M-10*, scale 1:1,000,000, with discussion. 14 pp.

Currie, K. L. 1976. The alkaline rocks of Canada. *Geol. Surv. Can. Bull. 239*. 229 pp.

Daniels, P. A. Jr. 1982. Upper Precambrian sedimentary rocks: Oronto Group, Michigan-Wisconsin. See Wold & Hinze 1982, pp. 107–33

Davidson, D. M. Jr. 1982. Geological evidence relating to interpretation of the Lake Superior Basin structure. See Wold & Hinze 1982, pp. 5–14

Dickas, A. B. 1984. Midcontinent rift system: Precambrian hydrocarbon target. *Oil Gas J.* 82(Oct. 15):151–59

Donaldson, J. A., Irving, E. 1972. Grenville Front and rifting of the Canadian Shield. *Nature* 237:139–40

Dosso, L., Murthy, V. R. 1982. Keweenawan volcanism of the north shore of Lake Superior: implications for continental mantle evolution. *Eos, Trans. Am. Geophys. Union* 63:461 (Abstr.)

Elmore, R. D. 1984. The Copper Harbor Conglomerate: a late Precambrian fining-upward alluvial fan sequence in northern Michigan. *Geol. Soc. Am. Bull.* 95:610–17

Fowler, J. H., Kuenzi, W. D. 1978. Keweenawan turbidites in Michigan (deep borehole red beds): a foundered basin sequence developed during evolution of a Proterozoic rift system. *J. Geophys. Res.* 83:5833–43

Franklin, J. M., McIlwaine, W. H., Poulsen, H. K., Wanless, R. K. 1980. Stratigraphy and depositional setting of the Sibley

Group, Thunder Bay district, Ontario, Canada. *Can. J. Earth Sci.* 17:633–51

Gordon, M. B., Hempton, M. R. 1983. The Keweenawan rift: a collision-induced rift caused by the Grenville Orogeny? *Eos, Trans. Am. Geophys. Union* 64:852 (Abstr.)

Green, J. C. 1982. Geology of the Keweenawan extrusive rocks. See Wold & Hinze 1982, pp. 47–56

Grout, F. F. 1918. The lopolith, an igneous form exemplified by the Duluth gabbro. *Am. J. Sci. Ser. 4* 46:516–22

Halls, H. C. 1966. A review of the Keweenawan geology of the Lake Superior region. In *The Earth Beneath the Continents, Am. Geophys. Union Geophys. Monogr.*, ed. J. S. Steinhart, T. J. Smith, 10:3–27

Halls, H. C. 1978. The late Precambrian central North America rift system—A survey of recent geological and geophysical investigations. In *Tectonics and Geophysics of Continental Rifts, NATO Adv. Study Inst., Ser. C.*, ed. E. R. Neumann, I. B. Ramberg, 37:111–23. Boston: Reidel

Halls, H. C. 1982. Crustal thickness in the Lake Superior region. See Wold & Hinze 1982, pp. 239–44

Halls, H. C., Pesonen, L. J. 1982. Paleomagnetism of Keweenawan rocks. See Wold & Hinze 1982, pp. 173–202

Hildenbrand, T. G., Simpson, R. W., Godson, R. H., Kane, M. F. 1982. Digital colored residual and regional Bouguer gravity maps of the conterminous United States with cut-off wavelengths of 250 km and 1000 km. *US Geol. Surv. Geophys. Inv. Map GP-953-A*, scale 1:7,500,000. 2 sheets

Hinze, W. J. 1963. Regional gravity and magnetic anomaly maps of the southern peninsula of Michigan. *Mich. Geol. Surv. Rep. Invest. No. 1.* 26 pp.

Hinze, W. J., Merritt, D. W. 1969. Basement rocks of the southern peninsula of Michigan. In *Studies of the Precambrian of the Michigan Basin, Mich. Basin Geol. Soc. Guideb.*, ed. H. B. Stonehouse, pp. 28–59

Hinze, W. J., O'Hara, N. W., Trow, J. W., Secor, G. B. 1966. Aeromagnetic studies of eastern Lake Superior. In *The Earth Beneath the Continents, Am. Geophys. Union Geophys. Monogr.*, ed. J. S. Steinhart, T. J. Smith, 10:95–110

Hinze, W. J., Kellogg, R. L., O'Hara, N. W. 1975. Geophysical studies of basement geology of southern peninsula of Michigan. *Am. Assoc. Pet. Geol. Bull.* 59:1562–84

Hinze, W. J., Wold, R. J., O'Hara, N. W. 1982. Gravity and magnetic studies of Lake Superior. See Wold & Hinze 1982, pp. 203–21

Irons, L. A. 1979. *A gravity survey of the Humboldt Fault and related structures in southeastern Nebraska.* MS thesis. Univ. Nebr., Lincoln. 67 pp.

Kalliokoski, J. 1982. Jacobsville Sandstone. See Wold & Hinze 1982, pp. 147–55

Keller, G. R., Bland, A. E., Greenberg, J. K. 1982. Evidence for a major late Precambrian tectonic event (rifting?) in the eastern midcontinent region, United States. *Tectonics* 1:213–22

Keller, G. R., Lidiak, E. G., Hinze, W. J., Braile, L. W. 1983. The role of rifting in the tectonic development of the midcontinent, U.S.A. *Tectonophysics* 94:391–412

King, E. R., Zietz, I. 1971. Aeromagnetic study of the midcontinent gravity high of central United States. *Geol. Soc. Am. Bull.* 82:2187–2208

Kisvarsanyi, E. B. 1984. The Precambrian tectonic framework of Missouri as interpreted from the magnetic anomaly map. *Mo. Dep. Natl. Resour. Contrib. Precambrian Geol. No. 14.* 19 pp.

Klasner, J. S., Cannon, W. F., Van Schmus, W. R. 1982. The pre-Keweenawan tectonic history of southern Canadian Shield and its influence in the formation of the Midcontinent Rift. See Wold & Hinze 1982, pp. 27–46

Lee, C. K., Kerr, S. D. 1984. The Midcontinent Rift—A frontier hydrocarbon province. *Oil Gas J.* 82(Aug. 13):144–50

Leeman, W. P. 1977. Pb and Sr isotopic study of Keweenawan lavas and inferred 4 b.y. old lithosphere beneath part of Minnesota. *Geol. Soc. Am. Abstr. with Programs* 9: 1068

Leutgert, J. H., Meyer, R. P. 1982. Structure of the western basin of Lake Superior from cross structure refraction profiles. See Wold & Hinze 1982, pp. 245–55

Lidiak, E. G. 1972. Precambrian rocks in the subsurface of Nebraska. *Nebr. Geol. Surv. Bull. No. 26.* 41 pp.

Lidiak, E. G., Hinze, W. J., Keller, G. R., Reed, J. E., Braile, L. W., Johnson, R. W. 1984. Geologic significance of regional gravity and magnetic anomalies in the east-central midcontinent. In *The Utility of Regional Gravity and Magnetic Anomalies.* Tulsa, Okla: Soc. Explor. Geophys. In press

Lyons, P. L. 1950. A gravity map of the United States. *Tulsa Geol. Soc. Dig.* 18:33–43

Lyons, P. L. 1959. The Greenleaf anomaly, a significant gravity feature. In *Symp. Geophys. Kans., Kansas State Geol. Surv. Bull.*, ed. W. M. Hambleton, 137:105–20

Lyons, P. L. 1970. Continental and oceanic geophysics. In *The Megatectonics of Continents and Oceans*, ed. H. Johnson, B. L.

Smith, pp. 147–66. New Brunswick, NJ: Rutgers Univ. Press

McCallister, R. H., Boctor, N. Z., Hinze, W. J. 1982. Petrology of the spilitic rocks from the Michigan Basin drill hole, 1978. *J. Geophys. Res.* 83:5825–31

Merk, G. P., Jirsa, M. A. 1982. Provenance and tectonic significance of the Keweenawan interflow sedimentary rocks. See Wold & Hinze 1982, pp. 97–106

Milanovsky, E. E. 1981. Aulacogens of ancient platforms: problems of their origin and tectonic development. *Tectonophysics* 73:213–48

Mitchell, R. H., Platt, R. G. 1978. Mafic mineralogy of ferroaugite syenite from the Coldwell Alkaline Complex, Ontario, Canada. *J. Petrol.* 19:627–51

Morey, G. B., Green, J. C. 1982. Status of the Keweenawan as a stratigraphic unit in the Lake Superior region. See Wold & Hinze 1982, pp. 15–25

Morey, G. B., Ojakangas, R. W. 1982. Keweenawan sedimentary rocks of eastern Minnesota and northwestern Wisconsin. See Wold & Hinze 1982, pp. 135–46

Morey, G. B., Sims, P. K. 1976. Boundary between two Precambrian W terranes in Minnesota and its geologic significance. *Geol. Soc. Am. Bull.* 87:141–52

Nelson, B. K., DePaolo, D. J. 1985. Rapid production of continental crust 1.7–1.9 b.y. ago: Nd isotopic evidence from the basement of the North American midcontinent. *Geol. Soc. Am. Bull.* In press

Norman, D. I. 1978. Ore deposits related to the Keweenawan rift. In *Petrology and Geochemistry of Continental Rifts*, ed. E. R. Neumann, I. B. Ramberg, pp. 245–53. Dordrecht, Neth: Reidel

Ocola, L. C., Meyer, R. P. 1973. Central North American rift system: 1. Structure of the axial zone from seismic and gravimetric data. *J. Geophys. Res.* 78:5173–94

Ojakangas, R. W., Morey, G. B. 1982. Keweenawan pre-volcanic quartz sandstones and related rocks of the Lake Superior Region. See Wold & Hinze 1982, pp. 85–96

Oray, E., Hinze, W. J., O'Hara, N. W. 1973. Gravity and magnetic evidence for the eastern termination of the Lake Superior syncline. *Geol. Soc. Am. Bull.* 84:2763–80

Percival, J. A., Card, K. D. 1983. Archean crust as revealed in the Kapuskasing uplift, Superior Province, Canada. *Geology* 11:323–26

Ramberg, I. B., Morgan, P. 1984. Physical characteristics and evolutionary trends of continental rifts. *Proc. Int. Geol. Congr., 27th, Moscow.* Utrecht, Neth: VNU Sci. Press. In press

Sawkins, F. J. 1976. Widespread continental rifting: some considerations of timing and mechanism. *Geology* 4:427–30

Sawkins, F. J. 1982. Metallogenesis in relation to rifting. In *Continental and Oceanic Rifts, Am. Geophys. Union Geodyn. Ser.*, ed. G. Palmason, 8:259–69

Scott, R. W. 1966. New Precambrian(?) formation in Kansas. *Am. Assoc. Pet. Geol. Bull.* 50:380–84

Serpa, L., Setzer, T., Farmer, H., Brown, L., Oliver, J., et al. 1984. Structure of the southern Keweenawan Rift from COCORP surveys across the Midcontinent Geophysical Anomaly in northeastern Kansas. *Tectonics* 3:367–84

Silver, L. T., Green, J. C. 1972. Time constants for Keweenawan igneous activity. *Geol. Soc. Am. Abstr. with Programs* 4:665–66

Sims, P. K., Card, K. D., Morey, G. B., Peterman, Z. E. 1980. The Great Lakes tectonic zone—A major crustal structure in central North America. *Geol. Soc. Am. Bull.* 91:690–98

Sleep, N. H., Sloss, L. L. 1978. A deep borehole in the Michigan Basin. *J. Geophys. Res.* 83:5815–19

Smith, T. J., Steinhart, J. S., Aldrich, L. T. 1966. Crustal structure under Lake Superior. In *The Earth Beneath the Continents, Am. Geophys. Union Geophys. Monogr.*, ed. J. S. Steinhart, T. J. Smith, 10:181–97

Society of Exploration Geophysicists. 1982. *Gravity anomaly map of the United States (exclusive of Alaska and Hawaii).* Tulsa, Okla: Soc. Explor. Geophys. 2 sheets, scale 1:2,500,000

Stewart, J. H. 1976. Late Precambrian evolution of North America: plate tectonics implication. *Geology* 4:11–15

Tapponnier, P., Molnar, P. 1976. Slip-line field theory and large scale continental tectonics. *Nature* 264:319–24

Tapponnier, P., Peltzer, G., LeDain, A. Y., Armijo, R., Cobbold, P. 1982. Propagating extrusion tectonics in Asia: new insights from simple experiments with plasticine. *Geology* 10:611–16

Thiel, E. C. 1956. Correlation of gravity anomalies with the Keweenawan geology of Wisconsin and Minnesota. *Geol. Soc. Am. Bull.* 67:1079–1100

Thomas, J. J., Shuster, R. D., Bickford, M. E. 1984. A terrane of 1350–1400 m.y. old silicic volcanic and plutonic rocks in the buried Proterozoic of the midcontinent and in the Wet Mountains, Colorado. *Geol. Soc. Am. Bull.* 95:1150–57

Treves, S. B. 1981. Some Precambrian gabbroic rocks from southeast Nebraska. In *Regional Tectonics and Seismicity of*

Eastern Nebraska, NUREG Rep. CR-2411, ed. R. R. Burchett, pp. 45–54. Lincoln: Nebr. Geol. Surv.

Treves, S. B., Low, D. J. 1984. The Precambrian geology of Nebraska. *Geol. Assoc. Can. Program with Abstr.* 9:112

US Geological Survey. 1982. Composite magnetic anomaly map of the United States, Part A—Conterminous United States. *US Geol. Surv. Map GP954A*, 2 sheets, scale 1:2,500,000

Van Schmus, W. R. 1975. On the age of the Sudbury dike swarm. *Can. J. Earth Sci.* 86:907–14

Van Schmus, W. R., Bickford, M. E. 1981. Proterozoic chronology and evolution of the midcontinent region, North America. In *Precambrian Plate Tectonics*, ed. A. Kröner, pp. 261–96. Amsterdam: Elsevier

Van Schmus, W. R., Green, J. C., Halls, H. C. 1982. Geochronology of Keweenawan rocks of the Lake Superior region: a summary. See Wold & Hinze 1982, pp. 165–71

Warren, D. H. 1968. Transcontinental geophysical survey (35°–39°N), seismic refraction profiles of the crust from 74° to 87°W longitude. *US Geol. Surv. Map I-535-D*

Watson, J. 1980. The origin and history of the Kapuskasing structural zone, Ontario, Canada. *Can. J. Earth Sci.* 17:866–76

Weiblen, P. W. 1982. Keweenawan intrusive rocks. See Wold & Hinze 1982, pp. 57–82

Weiblen, P. W., Morey, G. B. 1980. A summary of the stratigraphy, petrology and structure of the Duluth Complex. *Am. J. Sci.* 280A:88–133

White, W. S. 1966. Geologic evidence for crustal structure in the western Lake Superior Basin. In *The Earth Beneath the Continents, Am. Geophys. Union Geophys. Monogr.*, ed. J. S. Steinhart, T. J. Smith, 10:28–41

White, W. S. 1968. The native-copper deposits of northern Michigan. In *Ore Deposits of the United States, 1933–1967*, ed. J. D. Ridge, pp. 303–25. New York: Am. Inst. Min. Metall. Pet. Eng.

White, W. S. 1971. A paleohydrologic model for mineralization of the White Pine Copper Deposit, northern Michigan. *Econ. Geol.* 66:1–13

White, W. S. 1972. Keweenawan flood basalts and continental rifting. *Geol. Soc. Am. Abstr. with Programs* 4:532–34

Williams, L. A. J. 1982. Physical aspects of magmatism in continental rifts. In *Continental and Oceanic Rifts, Am. Geophys. Union Geodyn. Ser.*, ed. G. Palmason, 8:193–222

Wold, R. J., Hinze, W. J., eds. 1982. *Geology and Tectonics of the Lake Superior Basin, Geol. Soc. Am. Mem. 156.* 280 pp.

Woollard, G. P. 1943. Transcontinental gravitational and magnetic profile of North America and its relation to geologic structure. *Geol. Soc. Am. Bull.* 54:747–90

Yaghubpur, A. 1979. *Preliminary geological appraisal and economic aspects of the Precambrian basement of Iowa.* PhD thesis. Univ. Iowa, Iowa City. 294 pp.

Yarger, H. L. 1983. Regional interpretation of Kansas aeromagnetic data. *Kansas Geol. Surv. Geophys. Ser. No. 1.* 35 pp.

York, J. E., Oliver, J. E. 1976. Cretaceous and Cenozoic faulting in eastern North America. *Geol. Soc. Am. Bull.* 87:1105–14

Young, G. M. 1980. The Grenville orogenic belt in the North Atlantic continents. *Earth Sci. Rev.* 16:277–88

Ann. Rev. Earth Planet. Sci. 1985. 13: 385–425

PALEOGEOGRAPHIC INTERPRETATION: With an Example From the Mid-Cretaceous

A. M. Ziegler, David B. Rowley, Ann L. Lottes, Dork L. Sahagian, Michael L. Hulver, and Theresa C. Gierlowski

Department of Geophysical Sciences, University of Chicago, Chicago, Illinois 60637

INTRODUCTION

Progress in global paleogeography in the 1970s was related to several key developments in the 1960s, among them the plate tectonics concept, the growth of the paleomagnetic data base, and the ability to determine relative water depths using fossil communities and sedimentary structures. As a result, generalized paleogeographic maps showing continental relationships and environmental interpretations are now available for all Phanerozoic periods (Ziegler et al 1979, 1982b). Progress in the 1980s will result from the recent recognition that continental margins can be stretched more than twofold during rifting (Montadert et al 1979, LePichon & Sibuet 1981, Dewey 1982), and that such stretched regions can be subsequently telescoped by hundreds of kilometers during collisions (Dewey 1982, Bally 1981). Indeed, the western margin of North America was stretched during the late Precambrian to earliest Cambrian, probably by a factor of 2 or more (Bond & Kominz 1984), compressed in the late Jurassic through early Tertiary Sevier through Laramide orogenies by a factor of 2 to 3 (Price 1981), and subsequently stretched in the Great Basin by at least a factor of 2 (Hamilton & Myers 1966, Coney 1980). Such mobility of the continental crust must be accounted for in palinspastic base maps before general paleogeographic relationships can be refined.

0084–6597/85/0515–0385$02.00

In addition, much recent work within orogenic belts has focused attention on the widespread occurrence of relatively small crustal fragments that have been variously named exotic, allochthonous, or suspect terranes. These terranes record conspicuously different geologic histories from adjacent regions, including distinct lithologic, tectonic, magmatic, paleomagnetic, and biogeographic patterns. Terranes are separated from one another by sutures, marked by ophiolites and accretionary prisms, or by strike-slip faults, often of very large displacement. The Wrangellia terrane of western North America (Jones et al 1977) has travelled at least 30° and more likely 90° northward since the Triassic (Stone et al 1982), and during this transit it has passed from the coral reef zone into the temperate clastic zone (Monger et al 1982). It is at least locally separated from inner terranes by the Bridge River and Hozameen ophiolitic and accretionary prism assemblages of British Columbia and Washington State (Davis et al 1978). It is therefore important to establish the sequence and timing of terrane assembly and motion before more realistic global paleogeographic maps are reconstructed. Additional information on the motions of the oceanic plates can be gleaned from such studies (Page & Engebretson 1984).

This paper demonstrates how continent-scale palinspastic and paleogeographic maps are prepared, and how basin development and lithofacies patterns are related to extensional, compressive, and shear tectonic regimes. We illustrate this by a mid-Cretaceous (Cenomanian Stage) map centered on the United States and Mesoamerica, and by the geography and lithofacies of the Recent for comparison. The maps are segments of world-wide maps for 16 stages of the Mesozoic and Cenozoic eras being prepared by the Paleogeographic Atlas Project at the University of Chicago.

PALINSPASTIC RESTORATION

Most of the paleogeographic maps published every year do not employ tectonically restored base maps, and this practice can lead to a significant misrepresentation of relationships in areas that have experienced subsequent deformation. A map that shows regions restored to their undeformed state is called *palinspastic* [from Greek, meaning "stretched back" (Kay 1937)]. Although early paleogeographers (including C. Schuchert, E. Argand, and M. Kay, among many others) appreciated the problems of using unrestored bases (see Kay 1945), it was not until relatively recently that the severity of this problem was widely recognized. The realization of the extreme mobility of continental and oceanic lithosphere within the context of plate tectonics is due to the advent and increasing availability of various geophysical techniques, including multichannel seismic reflection profiles and cryogenically determined paleomagnetic poles, as well as to better geologic and structural maps of most parts of the world. For example, multichannel seismic reflection profiles of the southern Appalachians (Cook et al 1979) and eastern Great Basin (Allmendinger et al 1983) have suggested a minimum of 260 km of shortening and 30–60 km of extension, respectively, in these two regions alone. Paleomagnetic pole data have been particularly important in demonstrating large-scale latitudinal motions of various exotic terranes, such as the Wrangellia and Peninsular terranes, as well as large-scale rotations of terranes of the western Cordillera of North America. Structural techniques have also improved and now allow increasingly reliable determinations of rock strain and sense of shear along zones of high strain (Ramsay & Huber 1983, Simpson & Schmid 1983). These data taken together with improving global stratigraphic and geochronologic data provide much of the information necessary for making palinspastic restorations. In the following sections, we discuss the conceptual framework being used by the Paleogeographic Atlas Project to make first-order palinspastic restorations.

The primary information that is required to make palinspastic restorations is (*a*) the nature of deformations (i.e. do they reflect increases or decreases in surface area), (*b*) the amount of displacement associated with the various structures, and (*c*) the time interval during which displacements occurred. There is now a surprising amount of available information, albeit sometimes controversial, and thus first-order reconstructions are now possible. These reconstructions are required if we are to begin to fully appreciate the complexity of geological history and to recognize the next generation of important regional geological problems.

The primary goal of this paper is to outline how paleogeographic interpretations are made and depicted. As a part of this we present an interpretive map of Mesozoic and Cenozoic tectonic elements of the United States, Mesoamerica, and adjacent oceanic realms of the eastern Pacific, central Atlantic, and Gulf of Mexico (Figure 1; see color insert). This map provides much of the basic information required for making palinspastic restorations and has been used as the basis for the Cenomanian reconstruction depicted in Figure 6 (color insert). Before discussing our maps, we outline both the temporal and tectonic framework employed on them.

Tectonic Time Divisions of the Mesozoic and Cenozoic

Our approach to the temporal subdivisions of the Mesozoic and Cenozoic is based on the assumption that changes in global plate motions result in (or alternatively, result from) approximately synchronous, globally recognizable tectonic events. Such events include major continental collisions, periods of intracontinental rifting and/or ocean opening, and changes in plate velocities. This assumption follows from basic principles of plate kinematics on an Earth with a limited number of plates (some of which represent significant portions of the Earth's surface) that require that a change in relative motion between any two or more of the major plates must in turn result in changes of motion between all neighboring plates until a new quasi-steady state is reached, or until a new perturbing force operates on the system.

Five major changes can be detected during Mesozoic and Cenozoic times, which serve to define six intervals ranging in duration from 35 to 50 m.y. The interval lengths and boundaries have been established by choosing the beginnings or endings of major orogenic events and/or times of marked change in global plate motions. Thus, the sharp bend in the Hawaiian-Emperor and other Pacific hotspot tracks at about 43 m.y. is chosen as the beginning of the latest tectonic interval. This change in Pacific-hotspot relative motion in the middle Eocene corresponds quite well to the initiation of suturing of India with Asia, and since convergence continues today in this area, this interval is synonymous with the Himalayan Orogony. The opening of the Cayman Trough and the eastward motion of the Caribbean plate also occurred during this time (Pindell & Dewey 1982).

The time interval preceding this one corresponds temporally with the Laramide Orogeny, which extends from approximately the mid-late Cretaceous (Coniacian) to the middle Eocene in the type region. In North America, important "Laramide" effects include the reactivation of the Rocky Mountains, the closure of marine troughs in Mexico, and the collision of the Greater Antilles with the Bahamas platform (Pindell & Dewey 1982). Ocean-opening phases between India and Madagascar, Australia and Antarctica, New Zealand and Australia, New Zealand and Antarctica, and Greenland and North America, and northward Pacific-hotspot relative motion (Coney 1978, Henderson et al 1984) all occurred within this interval.

The opening of the South Atlantic near the beginning of the Cretaceous must represent another important change in plate motions, and it serves to define the start of an interval that corresponds roughly in time with the Sevier Orogeny of the western United States overthrust belt (Armstrong 1968). It should be mentioned that foreland thrusting along this belt continued for a long time and spanned the Nevadan, Sevier, and Laramide orogenic intervals. Subduction along the western margin of North America was also active during these intervals, but it was at about the mid-Cretaceous that suturing of Wrangellia occurred in the Pacific Northwest (Davis et al 1978).

Perhaps the most important upset in plate relationships happened at about the mid-Jurassic with the opening of the central Atlantic Ocean.

Extension along the eastern margin of North America coincided with the transformation of the western margin into an Andean system signaled by the onset of the Nevadan and Columbian orogenies of the United States and Canada, respectively. The transtensional opening of the Somali and Mozambique oceanic basins between east and west Gondwana was also an important event of this interval (Rabinowitz et al 1983).

Still earlier episodes of the Mesozoic must be defined on the basis of events far afield of North America. The collision of South and North China at about the mid-late Triassic (Norian) to begin the Indosinian Orogeny (Wang 1980) provides a convenient datum, and it coincides in time with the beginning of stretching along the Atlantic and Gulf margin of the United States. The Cape Orogeny of South Africa terminated at about the Norian (Dingle et al 1983) and may be used to define an interval that includes latest Paleozoic through mid-late Triassic times.

In conclusion, the Cape, Indosinian, Nevadan, Sevier, Laramide, and Himalayan orogenies in succession provide convenient terms for relatively restricted tectonic episodes of the Mesozoic and Cenozoic, and structures portrayed on our tectonic map (Figure 1) are correlated with these time intervals. An important point is that when intervals are defined in this way, the interval boundaries need not, and often do not, correspond with traditionally defined period or even stage boundaries.

In the following sections are discussed the major tectonic environments that we recognize: ocean floor, stretched continental crust, fold-thrust belts, subduction-accretion prisms, and strike-slip systems. The basic styles of deformation associated with each of these environments, the ways these deformations are treated palinspastically, and the methods useful in dating them are described. Specific examples of our treatment of these structures have been chosen from our Cenomanian palinspastic restoration (Figure 6).

Ocean Floor

Oceanic lithosphere is generated by seafloor spreading in a number of tectonic settings, including mid-ocean ridges, back-arc basins, and as the end product of pull-apart basin formation. The limits and age of ocean floor are for the most part readily determined from its typical geophysical signature (Emiliani 1983). Controversies generally arise over the location of continent-ocean boundaries, and over correlations of magnetic anomalies in some areas. The Gulf of Mexico is a good example of an oceanic realm for which both the extent of oceanic crust and the age of ocean floor have proved extremely controversial. The Gulf is characterized by low-amplitude magnetic anomalies that are difficult to correlate with a magnetic reversal sequence, by thick, extensive Jurassic evaporites that impede subsalt seismic reflections, and by thick sedimentary wedges along the continental margins. Comparison of maps depicting the limits and age of ocean floor within the Gulf (e.g. Buffler et al 1980, Cebull & Shurbet 1980, Martin 1978) shows that virtually all of them are different because they are based on different assumptions concerning the relationship of salt to crustal type, as well as on different prejudices concerning the opening history of the Gulf.

We depict the oceanic crust of the Gulf to be of middle-to-late Jurassic age, based on the regional constraints imposed by late Paleozoic stratigraphic and tectonic histories of circum-Gulf continental realms (Pindell & Dewey 1982), on the ages of the synrift Eagle Mills clastics and basaltic volcanics and syn- to postrift Louann salts, on the absence of correlatable magnetic anomalies due to seafloor spreading during the Jurassic quiet zone, and on the relative motion of Gondwana and Mesoamerica with respect to North America (Anderson & Schmidt 1983). Recent International Program of Ocean Drilling/Deep Sea Drilling Project (IPOD/DSDP) drillings in the Florida Straits bottomed in late Precambrian continental crust that is several tens of kilometers seaward of the 3000-m isobath (Schlager et al 1984); these results forcefully indicate the problems of interpreting the circum-Gulf region.

In the past the recognition of linear magnetic anomalies has been used to support arguments concerning the presence of oceanic basement in several areas [Labrador Sea (Srivastava 1978), South Africa (Rabinowitz & La Brecque 1979), Red Sea and Gulf of Aden (Cochran, 1981, 1983a)], but recent results derived primarily from multichannel seismic reflection profiling have shown, at least in the case of South Africa, that the innermost region of identified linear anomalies overlies stretched continental crust and not ocean floor (Austin & Uchupi 1982). With increasing appreciation of the magnitudes and frequent occurrences of stretched continental crust, there will no doubt be many more reinterpretations of this type along continental margins. As a final note, it should be pointed out that there are still important controversies concerning the correlation of linear magnetic anomalies with the reversal time scale. A good example can be seen in the west Philippine Sea, where Weissel (1981) and Shih (1980) have quite different interpretations of the anomaly correlations.

Stretched Continental Crust

Continental stretching involves an increase in surface area without the generation of new crust. The basic geometry is shown in Figure 2. In Figure 2*A* the ruled area of continental crust and subcrustal lithosphere is extended, leading to the development of rotated fault blocks at upper, brittle crustal levels and ductile stretching in the lower crust and subcrustal lithosphere (Figure 2*B*). The hallmark of stretching at surface levels is the

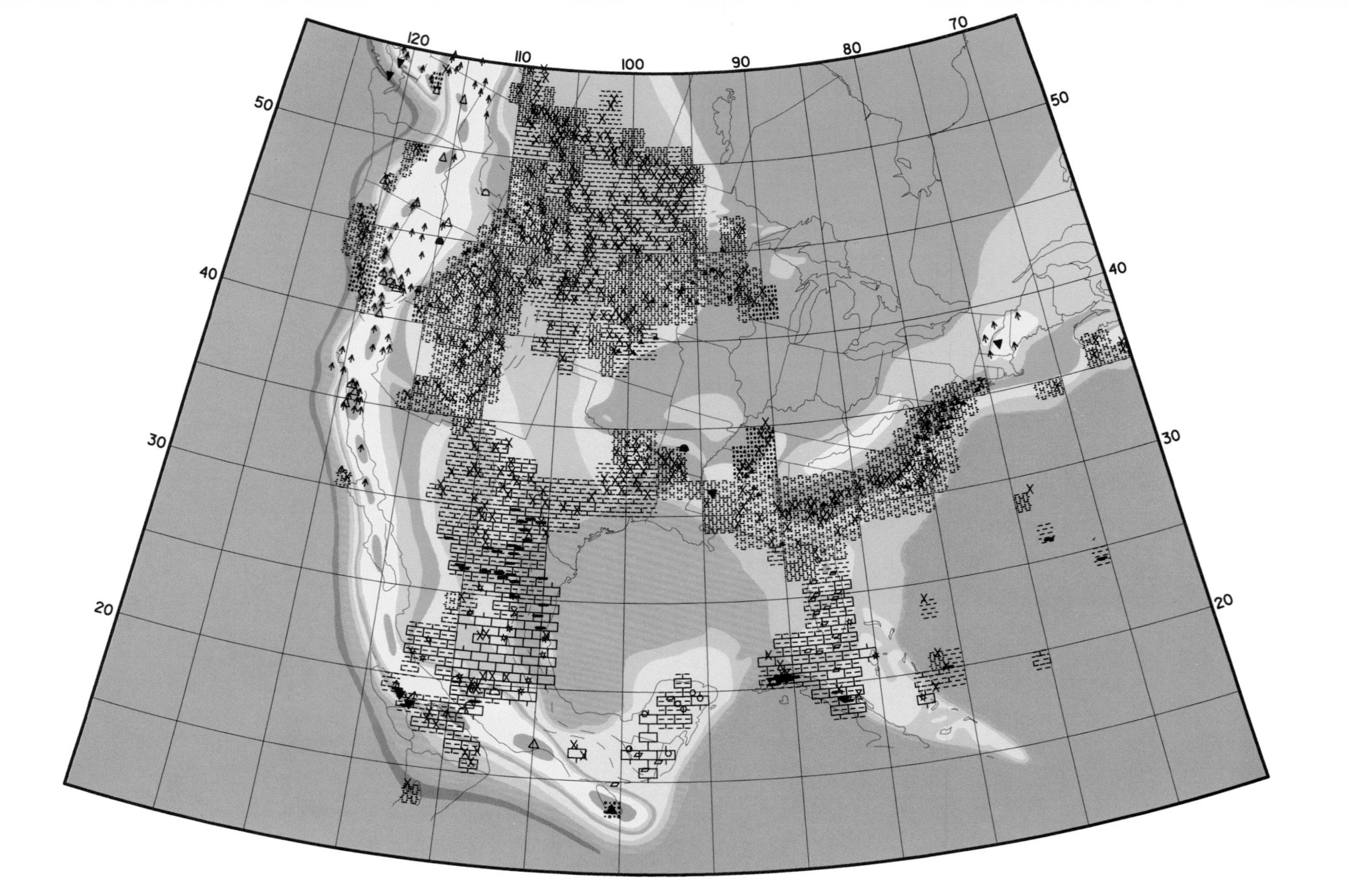

120
110
100
90
80
70
50
40
30
20
50
40
30
20

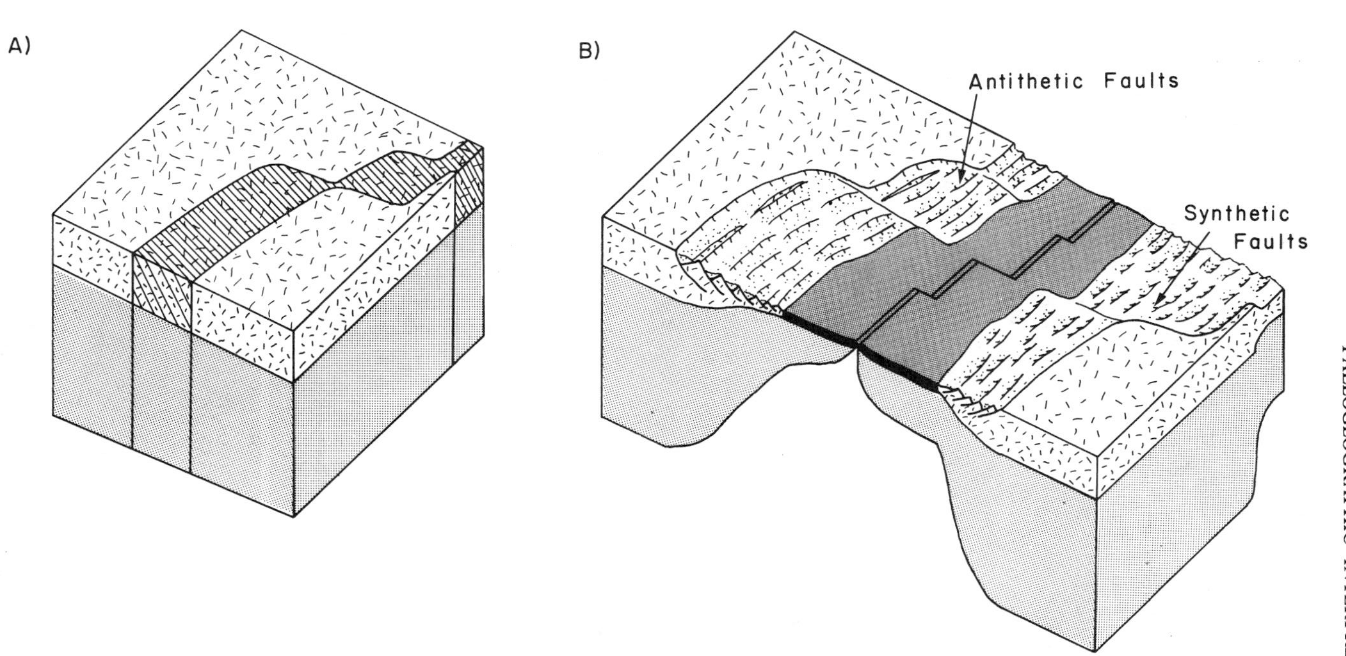

Figure 2 Lithospheric stretching and Atlantic-type margin development. *A* Initial condition of continental crust (random dashes) and subcrustal lithosphere (fine stipple) prior to stretching. Diagonal rules delimit the region to be stretched. *B* Continental crust thinned and stretched by an average factor of 2.5, resulting in pervasive normal faulting in the upper crust (with faults either synthetic or antithetic to the margin) and ductile flow in the lower crust and mantle. Crustal stretching ceases with initiation of seafloor spreading and generation of oceanic crust (black). The ridge axis need not be symmetrically situated within the stretched region. Reconstruction only of continent-ocean boundaries without palinspastically restoring the stretched continental crust to its initial state (diagonally ruled region in *A*) results in significant underfitting of the continents.

development of normal faults and associated basins that can occur as either synthetically or antithetically developed sets across the margin.

Until relatively recently, the geological and geophysical consequences of large-scale continental stretching were not fully appreciated. McKenzie (1978) presented a relatively simple model involving uniform, instantaneous stretching, which he showed to have the following consequences: (*a*) rapid synstretching subsidence where the initial ratio of crustal to lithospheric thickness (Cz/Lz) is greater than ~ 0.2, (*b*) surface uplift where Cz/Lz is less than ~ 0.2, (*c*) elevated geothermal gradients due to concomitant thinning of the thermal lithosphere, and (*d*) poststretching long-term thermal subsidence due to time cooling of the lithosphere. Subsequent work has examined the consequences of stretching over finite time periods (Jarvis & McKenzie 1980, Beaumont et al 1982, Cochran 1983a) and of nonuniform stretching (Royden & Keen 1980, Beaumont et al 1982), as well as the possible geologic and petrologic consequences of lithospheric stretching (Le Pichon & Sibuet 1981, Dewey 1982, Steckler & Watts 1981, Cochran 1981, 1983b, Vink 1982, Wernicke & Burchfiel 1982), but the basic relationships of McKenzie's initial model hold true.

Crustal stretching has affected areas along the East Coast of the United States, the circum-Gulf of Mexico, and in the Basin and Range province (Figure 1). The amounts of stretching in these regions are difficult to determine precisely, but present estimates of average stretching factors range from $\beta = 2$ (100%) to $\beta = 3$ (150%) in all of them [East Coast (Keen 1981), Gulf Coast (J. L. Pindell, written communication), Basin and Range (Hamilton & Myers 1966, Hamilton 1978, Proffett 1977, Wernicke et al 1982)].

Palinspastically and paleogeographically, it is very important to delimit regions affected by crustal stretching. Along Atlantic-type margins, such as the East and Gulf coasts, stretching was succeeded by seafloor spreading. In the past, the basic premise of most continental reconstructions (Bullard et al 1965, Dietz & Holden 1970, Le Pichon & Fox 1971, Pitman & Talwani 1972) was that the best fit was one in which continental crust of conjugate margins did not overlap. Refitting of the continent-ocean boundaries shown on Figure 1 would not yield the original continental positions, because stretching has not been accounted for. Where regions affected by stretching are very wide (e.g. 100–300 km), this systematic underfitting can have a significant impact on reconstructions in other contiguous and noncontiguous regions. For example, misfitting North America, South America, and Africa in the Triassic has significant implications for the latitudinal position of Australia. In order to restore palinspastically regions that have been stretched, we defined the limits of stretched crust and compiled information on the thicknesses of the stretched and unstretched crust. The region that has been stretched can then be restored until its thickness equals that of the adjacent unstretched crust. The limits of stretched continental crust shown on the tectonic map define quite substantial areas that must be palinspastically restored. In regions with good multichannel seismic reflection data, the methods of Le Pichon & Sibuet (1981) and Wernicke & Burchfiel (1982) can be applied to determine stretching factors.

Basin and Range stretching postdates the Cenomanian, and therefore it has to be palinspastically restored for our Cenomanian reconstruction. Although no detailed fault-by-fault reconstructions of this region have yet been made, the general consensus is that about 100% of the essentially east-west extension has occurred in this region since the end of the Eocene (Hamilton & Myers 1966, Hamilton 1978, Wernicke et al 1982, Coney 1980). Although the thickness of the crust prior to Basin and Range extension is unknown, a twofold thinning of the crust seems likely (Wernicke et al 1982). The absence of significant rotation of the Sierra Nevada (Hannah & Verosub 1980) suggests that the amount of stretching in northern and southern parts of the Great Basin has been roughly the same. The extension directions derived from structural analysis of lineations developed in the regional detachment surfaces support an average east-west direction as well [e.g. ~S70°W in southwestern Arizona and California (Davis et al 1980), N70°W in the Snake Range (Miller et al 1983)]. We have therefore decreased the width of the present Basin and Range by approximately one half for our Cenomanian palinspastic restoration.

Fold-Thrust Belts

Fold-thrust belts as used in the tectonic map (Figure 1) refer specifically to those regions characterized by folding and/or thrusting that affect rocks derived from, or deposited on top of, a preexisting continental basement. Fold-thrust belts are thereby distinguished from other folded and thrust rocks that typify accretionary prisms and that were derived from, or deposited on top of, oceanic basement. Under this definition, fold-thrust belts generally form within collisional environments, such as the Himalayas, Appalachians, and Brooks Range, and within noncollisional back-arc regions of compressional ("Andean-type") arcs (Dewey 1980, Molnar & Atwater 1978, Dickinson 1979), including the sub-Andean ranges (Allmendinger et al 1983, Burchfiel 1980), and the Laramide and Sevier belts (Burchfiel & Davis 1975, Burchfiel 1980). Some fold-thrust belts, such as the Canadian Rockies (Monger et al 1982, Monger & Price 1979), result from a progression from collisional to back-arc environments. At present it is not clear what role the subduction of anomalously thick,

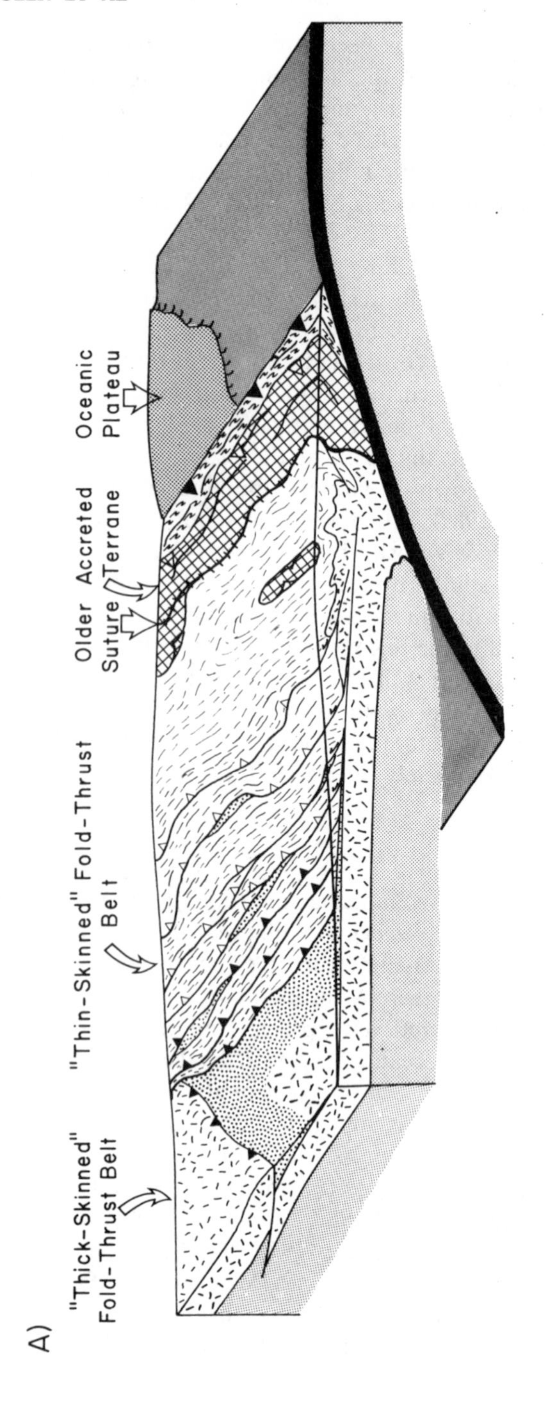
A)
"Thick-Skinned" Fold-Thrust Belt
"Thin-Skinned" Fold-Thrust Belt
Older Suture
Accreted Terrane
Oceanic Plateau

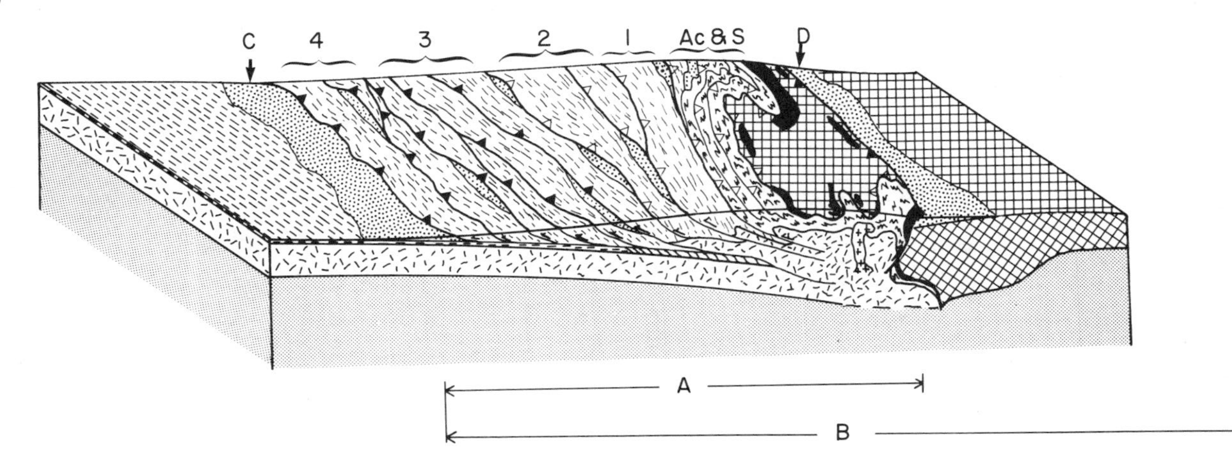

Figure 3 Fold-thrust belts. *A* Geometry of an Andean-type subduction zone with well-developed back-arc fold-thrust belts of "thin-skinned" and "thick-skinned" character. Shallow dip of Benioff Zone probably reflects subduction of either young oceanic lithosphere or an oceanic plateau. Thrust-loaded basins (coarse stipple) overlie depressed continental crust (random dashes). The continental crust is increasingly remobilized in the arcward direction. "Thin-skinned" thrust belt developed by progressive stepping out of thrust front. Previously accreted terrane (crosshatch) and accretionary prism also shown. Arc-related plutons and volcanics not shown. *B* Geometry of a collision-related fold-thrust belt. Collision of a continent (continental crust = random dashes, undeformed cover = dashed lines, deformed cover = wavy dashes) with an arc [arc and forearc basement = crosshatch, ophiolites = black, accretionary wedge = squiggly lines (Ac & S)]. Thrust-loaded basins (coarse stipple) develop in front of fold-thrust belt (C) and near the suture associated with back thrusting (retro-charriage) (D). Fold-thrust belt develops by outward stepping of thrust front from 1 to 4 with time. Sediments within fold-thrust belt change facies from onshore (4) to offshore (1). Crustal thickening within suture is associated with remobilization of basement and anatectic melting to produce syncollisional granites (pluses). Palinspastic restoration of continent side involves increase in width from present (A) to former width (B).

buoyant crust (such as oceanic plateaus and aseismic ridges) plays, but at least some reconstructions of the Laramide Orogeny relate it to subduction of the Farallon plate equivalent of the Hess Rise (Livacarri et al 1981, Henderson et al 1984).

The geometries within fold-thrust belts of different origins are comparable, but as yet only collisional fold-thrust belts have been shown to be associated with very large-scale (i.e. order of several hundred kilometers) crustal shortening (see, for example, Cook et al 1979, LeFort 1975).

The general geometry and structural development of fold-thrust belts of noncollisional and collisional origin are shown in Figures 3*A* and 3*B*, respectively. Fold-thrust belts are generally characterized by the following components in a hinterland direction: (*a*) A "foreland" or thrust-loaded basin, filled with terrestrial and/or marine sediments derived from the adjacent fold-thrust belt that results from flexure of the lithosphere adjacent to a thrust-related load. Evolution of these basins reflects the amount of loading, the flexural properties of the foreland lithosphere, and the rate of sedimentation versus sea-level rise or fall, as discussed by Beaumont (1981), Jordan (1981), and Dewey (1982), among others. (*b*) A frontal detachment in which previously deposited "foreland" basin deposits are commonly observed to be involved in the deformation. (*c*) The fold-thrust belt proper, which is characterized by (1) a temporal sequence of thrusts, such that structurally higher, more hinterland thrusts moved earlier and structurally lower, more foreland thrusts moved later; (2) a change in facies of the involved strata from generally platformal in the foreland to marginal in the hinterland; and (3) the increasing involvement and deformation of preexisting continental basement within the fold-thrust belt toward the hinterland. Laramide-style fold-thrust systems involve less shortening and generally bring basement to the surface immediately adjacent to the frontal belt. COCORP data suggest that the thrusts penetrate and offset the Moho and are not markedly listric (Smithson et al 1979). This contrasts quite strongly with typical "thin-skinned" fold-thrust belts.

The timing of thrust belt development is generally best constrained by ages of synorogenic flysch and molasse deposits. In addition, the ages of the youngest rocks involved in the deformation, as well as the isotopic ages of metamorphic minerals and syn- to posttectonic anatectic granites, also provide data on timing.

The methods generally used to restore palinspastically fold-thrust belts were outlined in classic papers by Dahlstrom (1969, 1970), in which the concept of "balanced cross sections" was introduced. The structural geometry of fold-thrust belts has been the focus of much recent study (Bally 1981, Boyer & Elliott 1982, Suppe 1983), in part owing to increasing interest in the hydrocarbon potential of overthrust environments. Figure 3*B* illustrates the importance of palinspastically restoring a fold-thrust belt. The present width of the fold-thrust belt is marked by A, whereas a width B is suggested for the predeformation state, based on approximate area balancing of the continental crust shown on the end section. Width A represents approximately 65% of the total shortening that occurred during two time intervals. The generation of intermediate palinspastic restorations is quite straightforward, since balancing involves undoing the deformation associated with each fold and thrust, starting in the foreland and progressing to the hinterland. Balanced (or retro-deformable; see Suppe 1983) cross sections of many North American orogenic belts have now been published, including the Canadian Rockies (Bally et al 1966, Price 1981), the Idaho-Wyoming fold-thrust belt (Dixon 1982), and the Valley and Ridge province of the Appalachians (Cook et al 1979). These, among others, provide critical tie-points for palinspastic restorations of these fold-thrust belts.

We have restored palinspastically the Laramide (i.e. Coniacian to middle Eocene) folding and thrusting from southern Mexico to the Canadian border, as well as a portion of the Sevier deformation in the same area. Laramide age deformation can be divided into two components: a generally eastern belt characterized by "thick-skinned" deformation, and a western "thin-skinned" deformation. The eastern belt of the Front Range Rockies, Black Hills, Big Horn Mountains, and Wind River Range appear to be associated with 10–30 km of east-west shortening and probably an equivalent amount of north-south-directed strike-slip motion (Hamilton 1978). We have moved regions affected by thick-skinned deformation back by about 40 km in a southwesterly direction. The western, thin-skinned belt is affected by apparently continuous activity from late Jurassic to middle Eocene times, and it is more difficult to determine what component reflects Cenomanian and younger deformation. Detailed studies in the Idaho-Wyoming fold-thrust belt (Wiltschko & Dorr 1983, Dixon 1982) suggest approximately 40 km of east-west motion. Comparable amounts of displacement appear to have occurred to the south, and therefore we have moved westward by approximately 40 km all regions to the west of the thin-skinned belt.

Other post-Cenomanian fold-thrust-related deformations are recorded in the Pacific Northwest, where imbrication along the Shuksan and Church Mountain thrusts record more than 60 km of east-west shortening (Misch 1966, Vance et al 1980, Monger 1977). In addition, approximately 50 km of northeast-southwest-directed shortening in South Florida and the Bahamas associated with the middle Eocene–Oligocene collision of Cuba and Hispaniola (Gealey 1980, Pindell & Dewey 1982) has been restored. This places the carbonate bank assemblages now observed on the northern margins of these islands back as a southward extension of the present Florida-Bahama platform.

Subduction-Accretion Prisms

Subduction of oceanic crust is often associated with the development of accretionary prisms (Karig 1974, Dickinson & Seely 1979), which, where well developed, represent substantial increases in the area of "continental" crust (Figure 4). Accretionary prisms are characterized by complex associations of trench-fill turbidites, upper and lower slope basin sediments, oceanic pelagic sediments usually of siliceous character, and various igneous and metamorphic rocks of ophiolitic affinities. These assemblages are primarily products of offscraping of rocks of trench or underthrust plate origin that accumulate along the leading edges of many subducting margins. Accretionary prisms are not ubiquitous elements of subduction zones, as was previously assumed, but they are now recognized as the result of interplay between several components, including the rate of sediment influx into the trench from the underriding and overriding plates, the presence or absence of seamounts or other topographic features on the underriding plate, and the age and relative motions of the overriding plate with respect to the subduction zone hinge line.

The rate of sedimentation within oceanic realms is controlled by biogenic productivity, which in turn relates to oceanic circulation patterns and/or proximity and access to continental sources, such as must have produced the Zodiak fan of the North Pacific. An important, but often ignored, factor that strongly controls the rate of sediment influx from the overriding plate is climate (in particular, rainfall). Ziegler et al (1981) pointed out that the absence of sediment fill within the Peru-Chile Trench closely parallels regions of the Andes characterized by low annual rainfall and thus by low net rate of sediment influx to the trench. Regions characterized by thick accretionary prisms form adjacent to river systems draining high mountainous terrain along those portions of continents that lie within areas of relatively high annual rainfall. This is certainly the case along the Oregon-Washington coast, the Chugach and Prince William terranes of southern Alaska, Barbados, Makhran, and Sumatra. Tectonic controls also play a role and have been discussed by Dewey (1980).

The ages of the accretionary prisms shown in Figure 1 indicate when the rocks were accreted; they generally do not reflect the ages of the rocks within the accretionary wedge. The dating of the time of accretion is often problematic. The best and most straightforward way is to date the trench-fill turbidites in each successive package, since the residence time of these sediments in the trench is generally quite short (<3 m.y.). Unfortunately, trench-fill turbidites are often poorly fossiliferous, they can occur within

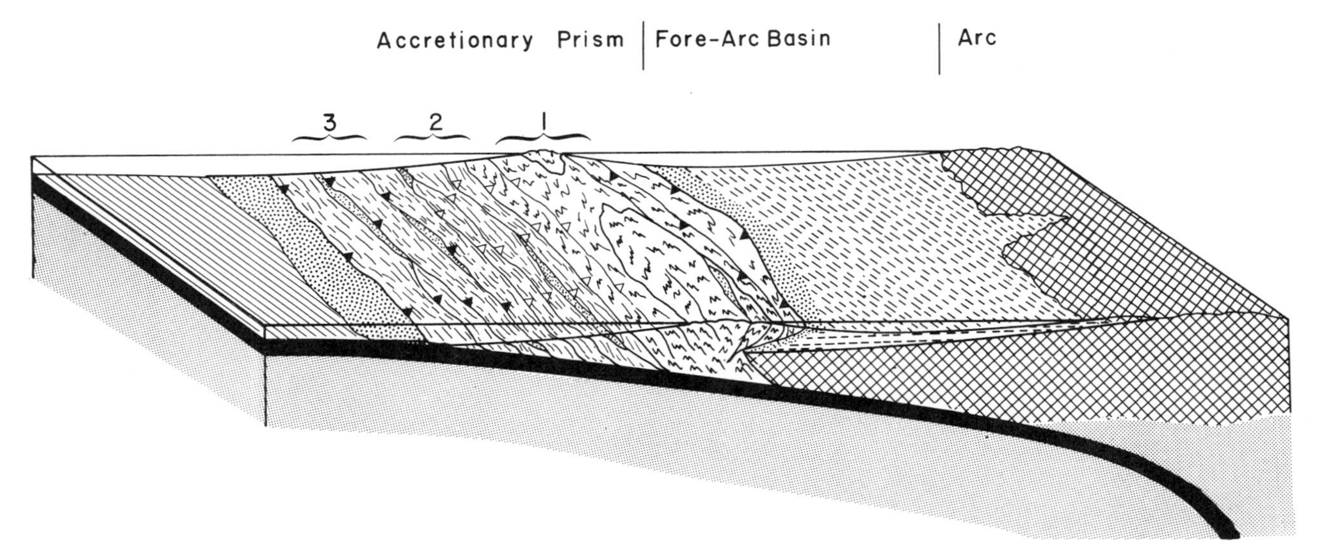

Figure 4 Accretionary prisms and Pacific-type margin development. Development of an accretionary prism (wavy and squiggly lines) by offscraping of trench-fill turbidites (coarse stipple), of oceanic pelagic deposits (ruled), and locally of fragments of ophiolite and seamount from the downgoing slab (oceanic crust = black, mantle = fine regular stipple). Growth of the accretionary prism occurs by progressive oceanward addition of wedge-shaped packages of offscraped material (numbers 1–3), schematically represented by oldest (1 = open triangles, squiggly lines), younger (2 = open triangles, wavy dashes), and youngest (3 = black triangles, wavy dashes). Upper and lower slope basins (fine irregular stipple) record continued motion on thrusts within the accretionary prism. In thick prisms, arcward thrusts (black triangles) emplace older accretionary prism rocks over ponded turbidites (stipple) of mixed forearc ridge and arc derivation and over forearc basin sediments (dashed lines); this process results in a raising of the forearc ridge. Crosshatch is arc and forearc basin basement.

completely disrupted mélange terranes where fossil ages are of uncertain meaning, and they cannot always be differentiated from turbidites within slope basins that postdate accretion. Where ages from the trench-fill turbidites are not available, alternative methods include (*a*) dating of the youngest oceanic pelagic sediments and/or sediments associated with accreted seamount complexes providing maximum ages, (*b*) isotopic dating of metamorphic rocks, such as blueschists, within the prism providing minimum ages, (*c*) isotopic age dating from near-trench igneous rocks providing minimum ages, and (*d*) extrapolation from ages of volcanism within adjacent arc terranes. None of these methods yield unequivocal ages, and whenever possible they have been used in conjunction to determine times of accretion.

Accretionary prisms are relatively easy to account for palinspastically, since the major components of large accretionary wedges are synaccretion trench-fill turbidites. In such cases the areas presently underlain by accretionary wedges that postdate the time of a given reconstruction are simply removed from the map. Thus, on the Cenomanian reconstruction we have removed the areas underlain by later Cretaceous and Tertiary accretion along the west coast of the United States, the Baja Peninsula, and the Middle America Trench. These sequences vary considerably along strike, ranging from regions characterized by thick basalts and associated reef carbonates, such as the Olympic Peninsula and farther south along the Oregon-Washington coast (Snavely et al 1968, Duncan 1982), to turbidite-dominated sequences of the Franciscan (McLaughlin et al 1982). DSDP/IPOD drilling results show that accretion along the Middle American Trench is quite limited (Moore et al 1982). In addition, the accretionary prisms exposed along the north coasts of Cuba and Hispaniola have also been removed. It is possible to make some inferences concerning the age of ocean floor being subducted at the time of the reconstruction if ages are available from offscraped oceanic pelagic sequences. This approach may be the only way to constrain ages of ocean floor in early Mesozoic and older times. More detailed palinspastic restorations are probably inappropriate at this time, since the nature of the subduction process generally leaves a sparse, but exceedingly complex, record of itself.

Strike-Slip Systems

Strictly speaking, strike-slip systems are associated with neither increase nor decrease of surface area. They are therefore the most straightforward tectonic elements to restore palinspastically. Some strike-slip systems, such as the San Andreas, define plate boundaries and generally parallel relative plate motion vectors. Other strike-slip faults define schollen (a crustal fragment bounded by faults) boundaries (Dewey & Sengör 1979) that accommodate various portions of relative motion, such as the faults associated with collision (McKenzie 1972, Dewey & Sengör 1979, Molnar & Tapponnier 1975), subduction (Fitch 1972, Dewey 1980), and transpressional strike-slip (Davis & Burchfiel 1973); these faults do not parallel relative plate motions.

Strike-slip faults commonly have segments that are oblique to the general fault trend. Such segments were described by Harland (1971) as transpressional and transtensional for compressional and extensional segments of strike-slip faults (Figure 5). Subsequently, Lowell (1972) and Harding & Lowell (1979) have shown that transpressional systems are commonly associated with upward-branching fault patterns that they term flower structures. Transtensional segments, on the other hand, are characterized by pull-apart basins, which are associated with crustal stretching (Mann et al 1983). When displacements are large, pull-apart basins can evolve into small ocean basins, such as the Gulf of California and the Andaman Sea.

Several important strike-slip systems are recognized within the area of Figure 1. These include the late Tertiary faults of the San Andreas system—the San Andreas (310 km), San Jacinto (24 km), and Elsinore (40 km) faults—and the Hosgri–San Gregorio (110 km) fault system of the western United States (Dickinson 1983). Earlier Tertiary and late Cretaceous faults include the proto–San Andreas (195 km; Dickinson 1983), as well as the Straight Creek–Fraser River system (150 km; Davis et al 1978) of the Pacific Northwest. The Ross Lake (160 km; Davis et al 1978) and Nacimiento (560 km; Dickinson 1983) faults represent earlier Cretaceous faults. Farther south, in Mexico and Central America, major strike-slip motion is associated with the Jurassic Mohave-Sonora megashear (~800 km)

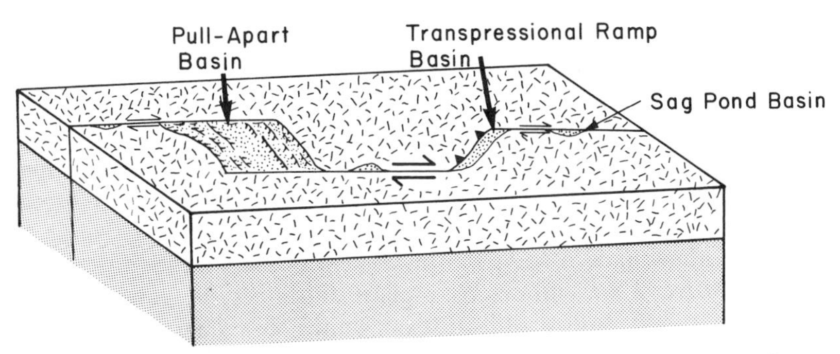

Figure 5 Strike-slip systems. Schematic representation of strike-slip system, with pull-apart basin, transpressional thrust and thrust-loaded basin, and local sag pond basins along the strike-slip fault.

trans-Mexican volcanic belt (~300 km), with latest Cretaceous(?) and younger motion along the Acapulco-Guatemala (~1300 km) fault (Anderson & Schmidt 1983) and its continuation in the Motagua system, and with offshore extension in the Swan transform (Pindell & Dewey 1982). Reconstruction of the Acapulco-Guatemala fault places the Chortis block back against southwestern Mexico and requires that the Middle America Trench evolve during migration of a trench-trench-transform system. Jurassic motion also occurred along the South Florida transform fault (Klitgord et al 1984). Motions on all of these faults, except the South Florida, Mohave-Sonora megashear, and Ross Lake faults, are reversed to produce the Cenomanian palinspastic restoration.

PALEOGEOGRAPHIC METHODOLOGY

Operationally, paleogeography consists basically of (*a*) dating rocks, (*b*) characterizing them environmentally, and (*c*) placing them in a paleolatitudinal and paleolongitudinal context. The first operation is concerned with paleontology, geochronology, and seismic and magnetic reversal stratigraphy; the second with paleoecology, sedimentology, and tectonics; and the third with paleomagnetism, paleoclimatology, and paleobiogeography. Operations (*a*) and (*c*), the "when" and "where" of paleogeography, are well known to most Earth scientists (if not, see Payton 1977, Harland et al 1982, McElhinny & Valencio 1981); thus, the emphasis in the following sections is on operation (*b*), the questions of "what" was deposited and "what" does it mean in terms of paleobathymetry and paleotopography. This discussion parallels and expands on that in Ziegler et al (1979, pp. 480–82).

Lithologic Data

There are 25 types of lithologies that are either common or significant in terms of depositional, climatic, or tectonic regimes and so are appropriate for regional paleogeographic studies (Table 1). For the Paleogeographic Atlas Project, the lithologies of approximately 38,000 localities world-wide have been assessed, and some 750 are displayed on our Cenomanian map (Figure 6) and 1100 on our Recent map (Figure 7). In our standard format, the five most common lithologies are recorded and listed in order of decreasing abundance. A computer printout of the data for Figures 6 and 7 is available from the authors on request and gives the country, state, latitude and longitude to the nearest tenth of a degree, lithologies, environment, formation, reliability of correlation at the stage level, and reference in terms of author, date, page, and locality number. A second printout is also available of the 234 references that were employed in compiling this data base.

The clastic-carbonate sediments (Table 1) are of course pervasive and are herein referred to as the "background lithologies." A rudimentary but

Table 1 Basic lithologic types appropriate for regional paleogeography[a]

Code	Lithology
Clastic-carbonate sediments	
(C)	Conglomerate
(S)	Sandstone
(M)	Mudstone, shale
(L)	Carbonate
Climatically significant sediments	
(T)	Tillite and glacio-marine beds
(P)	Peat, coal
(D)	Dolomite
(G)	Gypsum, anhydrite
(H)	Halite and bittern salts
(E)	Evaporites (G and H above)
(R)	Reefs
Oceanographically significant sediments	
(Q)	Bedded chert, radiolarite, diatomite
(V)	Phosphorite
(W)	Ferromanganese nodules and concretions
(X)	Limonite, goethite, or hematite
(Y)	Chamosite
(Z)	Glauconite
Soils	
(N)	Nonmarine, nondeposition
Acid-basic volcanic sequence	
(K)	Rhyolite, rhyodacite, trachyte, latite
(A)	Andesite, basaltic andesite, dacite
(B)	Basalt, phonolites, basanites, dolerite dikes
Acid-basic intrusive sequence	
(J)	Granite, monozonite, adamellite, alkali granite
(I)	Granodiorite, diorite, albitic granite, tonalite
(F)	Foidite, foyaite, essexite, theralite, etc
"Cooling ages" on intrusive and metamorphic rocks	
(U)	Uplift and unroofing

[a] Letters in parentheses are standard Paleogeographic Atlas Project codes.

effective clastic-carbonate index has been constructed from the relative abundance data of the background lithologies:

1. Conglomerate dominant
2. Sandstone with shale
3. Shale with sandstone
4. Shale
5. Clastics with some carbonate
6. Carbonate with some clastics
7. Pure carbonates.

It is this regrouping of the background lithologies that is portrayed on the maps in the form of computer-generated lithofacies patterns.

The computer has been programed to search each 1° latitude-longitude square for instances of the background lithology number code. When it finds a locality, it plots the appropriate lithology symbol, or averages the lithology if two or more localities are found in the same square. In either case, it centers the symbol on the 1° square, and in this way the symbols connect to form patterns where localities are abundant. The computer has also interpolated shading symbols into "empty" squares if there is data on background lithology in three or more of the eight surrounding squares.

The climatically significant accessory sediments are represented on the maps by discrete symbols, as are all the remaining lithologies. For simplicity, only the most abundant of these accessory lithologies is shown for each locality. Tillite is of course associated with mid- to high-latitude settings, but it is not shown in Figures 6 and 7 because there are no known Cretaceous glacial deposits, and because none of the 40 present glaciers in the United States and Mexico are greater than 5 km^2. It is doubtful that such deposits would survive in the geologic record, so they have not been plotted on our map. Coal and evaporites indicate opposite extremes of precipitation, or more accurately evapotranspiration (which combines precipitation and temperature). An analysis of these deposits through the Mesozoic and Cenozoic (Parrish et al 1982) shows that the coals consistently represent the equatorial and temperate rainy zones, and the evaporites the intermediate subtropical dry zones throughout this time span. Both are found in coastal and tectonically or glacially blocked drainage systems, and so they are useful as environmental and climatic indicators. The coals and peats in Figures 6 and 7 are north temperate coals, while the evaporites represent the north subtropical zone. Carbonates in general, and reefs in particular, were limited throughout the Mesozoic and Cenozoic to 40°N and 40°S as they are today; this zonation is thought to be due to the geometrical effect of light penetration to the seafloor as a function of latitude (Ziegler et al 1984). Deep-water reefs are known at present, but most reefs encountered in the geologic record appear to represent shallow water, based on the diversity of coral species and their association with algal-derived sediments, oolites, and other indicators of shallow conditions.

The oceanographically significant sediments, with the exception of chert, are classed as authigenic minerals and generally occur in trace amounts. It is important to note that the authigenic minerals are detectable mainly in areas of low background sedimentation (Emery 1969), particularly adjacent to desert belts, where there is low clastic influx, and in depths too great for carbonate production. Such conditions were enhanced during periods with rapid transgressions, such as the Cambrian and Cretaceous, when glauconite and phosphorite-rich sediments were generated over large regions. Phosphorite and chert are commonly linked with organic-rich sediments in outer-shelf settings and are likely to have been generated by oceanic upwelling systems of Ekman type (Parrish & Curtis 1982). The term "bioproductite" has been proposed for this compositionally diverse, but genetically linked, triumverate (Ziegler et al 1982a). In such settings it is usually reasonable to infer that water depths exceeded 50 m, because Ekman transport is limited to such conditions. However, dynamic upwelling, wherein deeper waters are forced over shoal areas by surface currents, can result in high surface productivity in shallow areas, so depth assignments based on the occurrence of bioproductites must be made with caution. Ferromanganese nodules occur over large tracts of the deep ocean today, characteristically in areas remote from terrigenous and biogenic source areas and also in lake deposits (Figure 7). Even so, they are rarely found in the rock record, largely because of its bias toward shelf settings. The iron minerals limonite, chamosite, and glauconite represent an onshore-offshore spectrum linked to the oxidation potential of these environments (Berner 1971). The presence of iron minerals, as with other authigenic minerals, is indicative of low background sedimentation, but it may also indicate complete chemical breakdown of soils in the source areas in specific climatic and topographic situations. The paleogeographic distribution of the iron minerals has yet to be analyzed in detail.

Soil types have been taken to indicate paleoclimatic conditions, but they are difficult to date precisely and so cannot be incorporated into paleogeographic data bases on a wide scale. Occasionally, soils can be dated by stratigraphic superposition and provide useful limitations on the position of the shoreline.

The igneous rocks are not as easy to classify as sedimentary rocks, because igneous processes are less efficient in compositional differentiation than surface processes. As a result of variation in nomenclatural usage in the world literature, we have grouped related igneous rock types into felsic, intermediate, and mafic classes (Table 1) based on the relative proportions of quartz, alkali feldspar, plagioclase, and feldspathoids (foids) (Streckeisen 1979). These groupings of the volcanic and intrusive rocks yield satisfying map patterns indicative of general plate tectonic environments. The intermediate types generally occur in linear belts representing the subduction-related igneous rocks. Landward, in the zone of crustal thickening and melting, there is generally a diffuse zone of felsic and some intermediate types. Basaltic compositions are found in most igneous environments, but they predominate, of course, in oceanic, hotspot, or continental rifting situations. The intrusive equivalents, the "foidites," are rare but do occur associated with continental rifts.

Large data bases of radiometric ages of igneous and metamorphic rocks are available in the published record, but for paleogeographic purposes most of these must be interpreted as "cooling ages." As York & Farquhar (1972) have pointed out, "When an area is involved in a prolonged cooling period, the various radioactive clocks will commence recording at widely separated times as the temperatures slowly fall from one critical blocking temperature to the next. Thus the Rb–Sr whole rock and U–Pb zircon clocks will begin to record the lapse of time before any other systems." We have used the Rb–Sr and U–Pb clocks to date the igneous intrusions, but we class the K–Ar ages as cooling ages, with the implication that this cooling takes place during uplift and unroofing of the orogen and therefore is indicative of mountainous terrains.

It is critical in paleogeography, as in other fields, to separate data from interpretation. In the foregoing paragraphs, lithologic types have been treated as the basic data of paleogeography, and comments on environmental interpretation have been restricted to lithologies that are confined to narrow environmental categories. Lithologies constitute data to the extent that the descriptions for a particular time interval are balanced and comprehensive (which, in the world literature, they often are not), and to the extent that the time limits of the interval are well established in the rock section (a more severe problem). Some rock types, such as sandstone and shale, are unrestrictive to environment, although sources may use terms such as greywacke, flysch, turbidite, or trench deposit that imply varying levels of environmental connotation. In this case, sandstone and shale constitute the "data," and the environment the "interpretation," as is explained below.

Environmental Interpretation

On our maps (Figures 6, 7) we employ a strict bathymetric/topographic contour representation of environments (Table 2). This is hazardous in view of the fact that the shoreline is the only contour directly interpretable from the geologic record, so it must be stressed at the outset that the other contours, as represented by the meter scale, should be treated as idealized guidelines that approximate the original contours. We feel that the contour system has the advantage of representational simplicity, and that the tectonic context is given by the paleotectonic maps and implied by the

Table 2 Elevation ranges of environments selected for regional paleogeographic maps

Code	Elevation (m)	Environments	Geologic recognition
	+10,000		
9		Collisional mountains	High-T, high-P metamorphics
	+4,000		
8		Andean-type peaks	Andesites/granodiorites in continental setting
	+2,000		
7		Island-arc peaks Rift shoulders	Andesites/granodiorites in marine setting Adjacent fanglomerates
	+1,000		
6		Inland plains Rift valleys Some forearc ridges	Between environments 5 and 7 Basalts, lake deposits in graben Tectonic mélanges
	+200		
5		Coastal plains Lower river systems Delta tops	Alluvial complexes Major floodplain complexes Swamps and channel sands
	0		
4		Inner shelves Reef-dammed shelves Delta fronts	Heterogeneous marine sediments Carbonates of Bahamian type Topset silts and sands
	−50		
3		Outer shelves Some epeiric basins Pro-deltas	Fine sediments, most "bioproductites" Fine clastics or carbonates Foreset silts and proximal turbidites
	−200		
2		Continental slope/rise Mid-ocean ridges Pro-delta fans	Slump/contourite facies Ocean crust less than 20 m.y. old Bottomset clays and distal turbidites
	−4,000		
1		Ocean floors	Pelagic sequences on ocean crust
	−6,000		
0		Ocean trenches	Turbidites on pelagic sequences
	−12,000		

contour configurations on the paleogeographic maps. A number of comprehensive texts relating sediments to environments are available (Davis 1983, Friedman & Sanders 1978, Reading 1978, Selley 1970), and thus the following discussion is restricted to the problems of environmental interpretation encountered in regional paleogeographic work.

The general elevation ranges of the environments selected for inclusion in Table 2 were determined simply by perusing a geographic map of the world. The ocean-floor categories (codes 0, 1, and mid-ocean ridges of 2) are reasonably straightforward. The well-known subsidence-time relationship of oceanic crust allows a depth assignment of 4 km for the ocean floor 20 m.y. after it was generated. Since the age of the ocean floor is now relatively well known (Sclater et al 1981), the widths of former ocean ridges can be inferred (Thiede 1982) in areas where the crust has not yet been subducted. Present-day trenches are reasonably well defined by the 6-km contour, although some older portions of the Pacific and South Atlantic ocean floor lie beneath this contour. Along subducting margins, the occurrence of turbidites can be used to define the trench contour, although it must be mentioned that turbidites cover large areas of shallower ocean floor adjacent to passive margins (North Atlantic) or to shear margins (western Canada). Oceanic trenches typically confine turbidity currents to movement parallel to the continental margin, so current-direction studies can be used to test the presence of a trench in the rock record. Locally, however, where the trench is filled with sediments (as along present-day coastal Oregon and Washington), this relationship may not hold true. By contrast, in areas of low sediment yield, such as Peru and Chile, the trench may be devoid of sediments. In summary, the best guide to the presence of a trench is the regional perspective of a subducting margin, rather than the presence or absence of turbidite deposits, which might in any case be subducted or overridden in the event of a continent-continent collision.

The continental slope and rise category is a distinctive but areally restricted and tectonically vulnerable environment. Cretaceous examples are known in the form of extensive reef talus deposits in Mexico (Enos 1983), clastic slump facies in Texas (Siemers 1978), and reef talus and mixed hemipelagic and carbonate turbidites in northern Cuba (Pardo 1975). The delta subdivision is shown in codes 2 through 5, but it can, of course, be restricted to the very shallow categories depending on basin morphology and crustal substructure.

The shelf environments (codes 3 and 4) provide the most abundant data sets for paleogeographic studies, and the depth order (and even the depth range) of environments can be determined from paleoecological and sedimentological studies. [See Kauffman (1967) for Cretaceous examples.] Indeed, the subdivision of shelf environments now possible is too fine to be portrayed adequately on regional scale maps. In general, inner-shelf environments are subject to wave and current activity, light penetration, and temperature and salinity variations, and these combine to yield a diversity of sedimentary rock types and biologic assemblages that contrasts sharply with the uniformity of the outer-shelf environments.

Epeiric basins constitute a special problem. These are created in a variety of ways, including (*a*) sea-level rise rapid enough to drown carbonate environments or create estuaries that trap clastic influx; (*b*) foreland thrusting that locally depresses the continental crust, as in the case of the west margin of the Western Interior Basin (Jordan 1981); (*c*) continental stretching and subsequent thermal subsidence of the lithosphere, seen today in the Gulf of Thailand; and (*d*) glacial scouring, as in the case of the present Hudson Bay. Depth interpretations of ancient epeiric basins are difficult because most environmental models are based on the marginal shelf seas characteristic of the present, and because the diversity of oceanographic parameters operating at these depths would vary in ways difficult to predict in enclosed basins. The thicknesses of prograding delta wedges (Asquith 1970), if corrected for stratigraphic and structural loading as well as compaction, could be used in the case of the Western Interior Seaway. Unfortunately, the loading corrections have yet to be applied, and published estimates of 600 m for the late Cretaceous of Wyoming are in our view excessively high.

The reconstruction of elevations in the terrestrial regions creates the biggest problem for the paleogeographer. The coastal plains and lower river valleys of the past do have a rich legacy in the form of swamp and alluvial plain complexes (Figure 6), and the fact that they have been transgressed from time to time bears testimony to their low elevations. The main problem comes with areas from 200 to 1000 m above sea level, an interval that today constitutes about 65% of the land surface (Cogley 1984, Figure 10). Much of this could be classified as inland plains, and such areas are generally not capable of leaving a geologic record of any sort. Many rift valleys, in their early stages, also lie at this elevation, as do the higher portions of forearc ridges. The Jurassic and Cretaceous Franciscan complex of coastal California provides an excellent example of a forearc ridge. It can be dated by K–Ar techniques on blueschist metamorphics (Armstrong & Suppe 1973), but its elevation, which increased through time, has been determined by its effects on sedimentation patterns in the forearc basin (Ingersoll 1979).

The mountainous categories (codes 7, 8, and 9) are recognized chiefly on the basis of igneous and metamorphic associations. In the case of the Andean and island-arc regions, the elevations refer to the andesite peaks and not to the associated mountain terrains that would generally be in the next lower elevation categories. The tectonic and sedimentary context of the andesites can be used to distinguish the Andean- from the island-arc-type mountains. The 4000-m contour was chosen to divide category 8 from 9 simply because most recent collision zones lie above this height. An ancient example would be the late Paleozoic Appalachian Mountains chain, which was characterized by high-temperature/high-pressure metamorphics.

MID-CRETACEOUS PALEOGEOGRAPHY

Paleogeography involves more than plotting localities with lithofacies data and environmental determinations on a palinspastically restored base map. Stratigraphic patterns and the prior geologic history of a region offer valuable clues in tracing environments across areas where the record has been obscured by erosion, deep burial, or metamorphism. Accordingly, we depart from the style of the earlier portions of this paper (in which we rigorously codified tectonic elements, rock types, and topographic intervals) and present an example of the Cenomanian paleogeography of the United States and Mesoamerica (Figure 6), with a Recent map also displayed for purposes of comparison (Figure 7). It is perhaps inherent in geology that while broad generalities can be established, each area is idiosyncratic, and thus perhaps as much can be learned from the local variations as from the areas that conform to expectations.

The Cenomanian sedimentary and igneous rocks of the United States and Mesoamerica are exceptionally well known and widely distributed, and they record a western active margin, an eastern passive margin, and span the north temperate and subtropical climatic zones. The same is true of the Recent of this region, although, of course, it contrasts sharply in its much lower sea level. Despite the wealth of data available on the Cenomanian of North and Central America, a detailed paleogeographic map has never been attempted for this region. An atlas of lithofacies maps for the whole continent is available (Cook & Bally 1975), as is an integrated lithofacies/paleogeographic atlas of the Rocky Mountain states (Mallory 1972). Other general sources include symposium volumes on the Mesozoic paleogeography of both the western (Howell & McDougall 1978) and west-central states (Reynolds & Dolly 1983), as well as one on mid-Cretaceous events for many parts of the world (Reyment & Thomel 1976). These sources, together with others mentioned in the regional sections that follow, were used in compiling our map. Before describing the regional aspects of the map, however, we must first discuss correlation problems and sea-level considerations.

Correlation Problems

Accurate correlations are essential in paleogeography, and despite the fact that Cretaceous biostratigraphy is exceptionally well refined, some problems remain. In the Western Interior Seaway, these involve the time at which the sea reached various areas during the Cenomanian transgression. The *Geologic Atlas of the Rocky Mountain Region* (McGookey et al 1972, Figure 26) shows marine conditions at the beginning of the Cenomanian as restricted to central Colorado and northern states, while subsequent work indicates that the seaway was established in western Oklahoma (Kauffman et al 1977) by this time. A similar problem exists with the terrestrial and marginal marine sequences of the Atlantic seaboard. Here, early correlations based on nonmarine ostracods have been improved on by palynological studies, and we accept Valentine's (1982, p. 26) correlation of the widespread subsurface unit F with the Cenomanian, rather than with the Albian and early Cenomanian as previously thought. As supporting evidence, we note that the markedly transgressive nature of unit F has a decidedly Cenomanian signature (see discussion below).

The correlation of the paleontological and radiometric time scales also presents a problem. We accept an 8-m.y. span from 94–102 m.y. for the Cenomanian, which was derived by applying corrections for the currently accepted K/Ar decay constants (Steiger & Jager 1978) to Van Hinte's (1978) time scale. Specialists on the Cretaceous (Kauffman 1977, pp. 85–87) assume a considerably narrower range based on bentonite age determinations from the Western Interior Seaway. We point out that such radiometric measurements are rarely accurate to 2% of the total age and are therefore not useful for establishing relative lengths of geologic stages. Strict adherence by Cretaceous workers to the bentonite ages has resulted in a time scale in which stages have extremely disproportionate lengths, and the Cenomanian has been one of the victims (down to 2.5 m.y. by some estimates). To judge by the number of fossil zones, by the general number and thickness of formations, and particularly by the magnitude of the sea-level rise (see below), the Cenomanian should be regarded as of at least average Cretaceous stage duration (6 m.y.).

Correlation problems even exist for the Recent time interval. We define the Recent by depositional processes that can be seen to be occurring today, or at least occurring in historic times. Because of the extremely rapid sea-level rise following the Pleistocene, sedimentation patterns in the marine realm are in the process of readjustment. Many shelf areas are not subject to sedimentation at the present (Emery 1968), so care has been taken to select information for our map that represents current sedimentation processes.

Sea-Level Considerations

The characteristics of the Cenomanian sea-level curve must be established for both paleogeographic and correlation purposes. Kauffman (1977, pp. 85, 89) has argued effectively that Cretaceous transgressive-regressive cycles of the Western Interior Seaway reflect eustatic fluctuations, and he shows a transgression extending from the beginning of the Cenomanian into the early part of the succeeding Turonian Stage. Indeed, in New Mexico and Arizona the shoreline, as represented by the Dakota Sandstone/Mancos Shale contact, advanced about 500 km, with three minor sandstone tongues interpreted to represent stillstands of the shoreline during this transgression (Molenaar 1983, p. 212).

A rough idea of the magnitude of the sea-level rise responsible for the transgression can be derived from stratal thickness and water depth estimates published on a locality in western Oklahoma (Kauffman et al 1977). This area is at least 500 km from the thrust loading of the western margin of the Western Interior Seaway and a similar distance from the thermal subsidence of the Gulf Coast. Cenomanian strata are approximately 50 m thick, and water depth estimates are zero at the beginning and 60 m at the end of the stage. If we assume an initial 50% porosity upon deposition (Sclater & Christie 1980, p. 3731) and a 10% porosity at present, the original thickness would have been 90 m. If we correct for sediment loading, the minimum sea-level rise required to deposit this amount of sediment would have been 38 m (assuming a mantle density of 3.3 g cm^{-3}, a sediment grain density of 2.7 g cm^{-3}, and isostatic equilibrium at all times). The depression of the seafloor due to water loading would be about 31% of the depth. Therefore, a water column of 60 m at the end of the Cenomanian would imply a sea-level rise of 42 m in addition to the 38 m referred to earlier, adding up to a total eustatic rise of 80 m during this stage. Pitman (1978, p. 1392) assumes that a maximum rate of sea-level change due to alteration of the geometry of the mid-oceanic ridge system would be about 10 m $m.y.^{-1}$, and our calculated sea-level rise is exactly this figure if an 8-m.y. duration of the Cenomanian is correct.

The absolute sea level at the end of the Cenomanian can be approximated from a locality in the Mesabi Range of Minnesota. This area, like the Oklahoma site, is well away from Cenomanian tectonic effects and, in addition, is remote from the late Tertiary Great Plains uplift. The value of this area for establishing Cretaceous sea-level heights has been recognized by others (Sleep 1976, Hancock & Kauffman 1979), although a recent review of the stratigraphy of the area (Merewether 1983) allows some refinement of earlier estimates. A mid-late Cenomanian "shallow-marine or brackish-water" fauna has been described by Merewether (1983, pp. 34, 44) from an horizon presently at 390 m above sea level. The "structural relief" on the top of the Cenomanian is only 90 m over a distance of 300 km across Minnesota, and this could be largely due to compaction of the thicker western sections as well as to the original bathymetric relief. As Sleep (1976, p. 53) has pointed out, the original surface has probably been raised about 50 m from its Cretaceous level as a result of erosional unloading of surrounding Precambrian uplands. A further correction appears necessary to adjust for glacial effects. The postglacial rebound model of Walcott (1970, 1972) shows that Minnesota is located just south of the pivot marking the northern extent of the peripheral bulge, an uplifted ring around a glacier due to the flexural rigidity of the lithosphere. Free-air gravity data support this model, as north-central Minnesota has a regionally averaged positive anomaly of about 10 mgal (Bowin et al 1982), which would suggest that the area still has about 70 m to subside before reaching isostatic equilibrium (Walcott 1970, Figure 4). For a contrasting view, see Officer & Drake (1981, Figure 8), who published a map of "probable vertical movements" suggesting that Minnesota is being uplifted at a rate of 1 cm yr^{-1}. Applying our corrections, a figure of 270 m is indicated for the late Cenomanian sea-level high stand. The only other North American site useful for testing this figure lies to the east of the Mississippi Embayment in central Tennessee, where Cenomanian rocks of purported coastal origin presently lie at elevations ranging from 275 to 305 m (Marcher & Stearns 1962).

In summary, the Cenomanian seas rose approximately 80 m from an initial elevation of 190 m to 270 m above present sea level during an 8-m.y. interval. Our map (Figure 6) shows the paleogeography at the end of the stage, while the lithofacies components are summarized over the whole stage. The 50-m bathymetric contour corresponds approximately to the shoreline position at the beginning of the Cenomanian.

Western Interior Seaway

Foreland folding and thrusting along the western margin of the continent was well developed by Cretaceous times, and it served to produce and pond sediments and to load tectonically the continental crust of the Western Interior. By the Albian, the sea level was sufficiently high to result in transgression, and vast areas of the Rocky Mountain and Great Plain states were flooded. East of the thrust belt, the relief must have been extremely low, as these regions had been both tectonically quiescent and subject to deposition since Pennsylvanian time. However, the Ancestral Rockies seem to have remained a source area as late as Albian times, to judge by current directions in the braided stream deposits of the Dakota Sandstone of New Mexico (Gilbert & Asquith 1976).

Renewed transgression in the Cenomanian extended the seaway eastward across portions of Minnesota, Iowa and Kansas. A relief of over 400 m existed in the Precambrian terranes of Minnesota where Cenomanian deposits are preserved in valleys flooded at this time. Our shoreline and 200-m contours in Minnesota are based on Sloan's (1964) contours on the Precambrian bedrock surface and his assumption that this area has not been significantly warped since the Cretaceous; we have also extended this method to adjacent states. Unfortunately, the deposits representing the late Cenomanian shoreline have been stripped by erosion in states farther south. The coastline here is unlikely to have been as irregular as that in Minnesota, as the whole region is underlain by relatively uniform late Paleozoic carbonates.

The active western margin of the seaway was more complex. In southwest Wyoming, steep gradients are evident (de Chadenedes 1975), as might be expected along a thrust margin. Immediately north, however, the great Sheridan delta (Goodell 1962) seems to have prograded across the axis of subsidence to areas of west-central Wyoming. By the late Cenomanian, the sea-level rise evidently exceeded the sedimentation rate, since marine deposits overlie paralic deposits in both areas. To the south, in Utah, Arizona, and New Mexico, the effects of thrusting disappear, and the Cenomanian is represented by a very extensive, uniformly thin (~ 100 m) transgressive sheet of Dakota Sandstone and Mancos Shale (Molenaar 1983).

The area of the greatest uncertainty is the nature of the connections of the seaway with the Gulf of Mexico. Cenomanian deposits have been stripped from a large area in central Texas, although Albian shallow-marine deposits do occur. We assume from the magnitude of the Cenomanian transgression that a connection across this area would have existed. The Coahila platform of northeast Mexico seems to have been relatively stable, but to the west there evidently was a deeper water connection of the Western Interior Seaway through the rapidly subsiding Chihuahua Trough (Greenwood et al 1977, Hayes 1970, Powell 1965).

Ancestral Gulf of Mexico

The Gulf had opened as a small ocean basin in the Jurassic, and by the end of the early Cretaceous, its shelf margin was well defined by a nearly continuous reef tract (Martin & Case 1975). Portions of this reef-dammed bank in Florida and the Bahamas were drowned during the Cenomanian, resulting in the rather more dissected banks that have survived to the present (Sheridan et al 1981). The same effect was probably responsible for the extinction of the reefs in many parts of the Gulf, although few detailed paleoecological studies are available for Cenomanian sediments overlying the reef tract. An exception is the study of Mancini (1977) on the Woodbine and Eagle Fork sediments of southern Texas, which are interpreted to have been deposited in inner to middle neritic conditions. Also, along the Florida/Georgia border, dark carbonaceous silty shales of the Lower Atkinson Formation yield pelagic foraminifera (Applin & Applin 1967) and probably indicate the inception of a channel along the Suwanee basin connecting the Gulf and the Atlantic that persisted as a bathymetric feature into Tertiary times (McKinney 1984). This channel was bounded on the south by the Ocala arch, which is incompletely covered by Cenomanian clastic strata and evidently was a shoal area with some islands at this time.

The northern margin of the Gulf was complicated by local tectonic and igneous activity that began about Cenomanian time. The Sabine uplift of eastern Texas was active and helped to define a small gulf to the west where deltas formed; this area later became the petroleum reservoir rocks of the Woodbine Formation (Oliver 1971). On the southern margin of the Sabine uplift, a narrow shelf with steep gradients resulted, along which slumping occurred (Siemers 1978). The nature of the uplift is unclear to us, although igneous activity is manifest in the Monroe uplift of northeast Louisiana and the Jackson dome of west-central Mississippi (Murray 1961, pp. 116, 340) and must have been contemporaneous, to judge by the amount of tuffaceous material in the Cenomanian sediments of the area. These rocks are part of a predominantly later Cretaceous alkalic igneous province that extends along the Mississippi Embayment and parallels the inner margin of the Gulf Coastal Plain through Arkansas, Texas, and northern Mexico (Baldwin & Adams 1971). However, the Mississippi Embayment postdates the Cenomanian and was the site of the Pascola arch, an extension of the Ozark dome, at this time. Marcher & Stearns (1962) reconstructed the Cenomanian topography of this arch and inferred a narrow arm of the sea in central Tennessee.

The Cretaceous paleogeography of Mexico has been mapped in detail (Enos 1983). Here, basinal carbonates as well as rudistid-dominated reef platforms were subsequently involved in Laramide convergence and are thus available for study in surface sections. Some palinspastic restoration is shown on our map. Enos points out that "mid-Cretaceous platforms differ from Lower Cretaceous ones in that vertical accretion greatly predominated over lateral progradation, resulting in steeper forereef slopes and much higher depositional relief in the mid-Cretaceous." He concludes that a relief of 1000 m was established in some areas because great aprons of shallow water debris extend 10 to 20 km from the reef tracts. The basinal deposits consist of "dark-gray lime mudstone and wackestone with pelagic microfossils," and bedded chert is present in some areas.

The basinal deposits of the northern Gulf have been prograded by thousands of meters of Tertiary sediments and have not been penetrated by drilling. West of the Florida platform, however, they have been drilled and are described as "laminated, banded and bioturbated limestone with coarse bioclastic layers" (Schlager et al 1984). Seismic profiles indicate that they were deposited in a depth range of 1200 to 1500 m. To the southeast, in northern Cuba, Cretaceous platform, slope or scarp, and pelagic facies were thrust over the Bahamas platform in early Tertiary times (Pardo 1975). Steep slopes are implied by reef-derived talus in the slope deposits, while bedded chert occurs in the carbonates of the pelagic facies.

Atlantic Seaboard

The Atlantic margin history was similar to that of the Gulf in that rifting began in the late Triassic, drifting began in the mid-Jurassic, and thermal subsidence was well underway by the mid-Cretaceous. Like the Gulf, the entire margin up to Nova Scotia was probably reef dammed through the early Cretaceous (Austin et al 1980), with the reef tract having developed at the outer limit of stretched crust (Hutchinson et al 1982). Inundation of this reef tract during the Cenomanian transgression was evidently complete north of the Bahamas.

The position of the late Cenomanian coastline can be traced from the subsurface of the central Atlantic states to New Jersey (Sohl et al 1976, Figure 8), where it trends across the present shore to the Scotian Shelf (Sherwin 1973, Figure 15). It is well bracketed by coal deposits representing the coastal plain and shallow marine shelf deposits, and it lies about 500 m below present sea level. Assuming our sea-level datum of 270 m for the Cenomanian to be correct, one might expect that the coast would transgress the unstretched portion of the crust, including the foothills of the Appalachian Mountains. The fact that this did not happen indicates that the Appalachians have subsided in excess of 70 m since this time. We assume that the Cenomanian 200-m topographic contour would approximate the inner limit of stretched crust, and if so it would generally coincide with the present 200-m contour, except in northern New England. This contour is slightly inland of the preserved coastal plain deposits.

Igneous activity affected New England during the Mesozoic, and some authors have related this activity to the passage of the hotspot that generated the New England Seamount chain (Vogt & Tucholke 1979). This hotspot cleared the shelf margin by Cenomanian times, but late Cretaceous fission track ages from northern New England (Zimmermann et al 1975) indicate that uplift continued through this time. Accordingly, we show some highlands on our Cenomanian map.

Pacific Margin

This is the area of the map most fragmented by strike-slip faulting. (See the section on "Palinspastic Restoration.") Essentially, a once-continuous

Andean subduction accretion complex has been transported northwest from northern Mexico and the United States as part of the Pacific plate, and southeast from southern Mexico as part of the Caribbean plate. The best-preserved and most studied area is in central California, where the arc, forearc basin, and trench deposits are preserved (Ingersoll 1979). A continuous Mesozoic record of convergence is preserved here, and by mid-Cretaceous times, a submarine forearc ridge had developed and served to pond the Great Valley forearc sediments.

The arc itself is the most pervasively preserved element along the various segments, but erosion has generally stripped the surface volcanics, and only the granodiorite intrusions are left to mark its general trend. Problems associated with interpreting the various types of radiometric ages on these rocks in California and the Baja Peninsula have been summarized by Krummenacher et al (1975). The trend in the Pacific Northwest (Armstrong et al 1977) is partly obscured by Tertiary volcanism and is distorted by Basin and Range extension. The trend in Central America is east-southeast (Weyl 1980, p. 299), i.e. subparallel to the present southeast-trending arc. Our radiometric data come from these papers and from many articles in the journal *Isochron/West*.

Forearc basin sediments of Cenomanian age have been identified with confidence only in California and Oregon (Dickinson & Thayer 1978), and the forearc ridge can be recognized by radiometric ages on blueschist rocks from California and the Baja Peninsula (Suppe & Armstrong 1972). True oceanic trench deposits of this age are not known with certainty.

The Methow-Tyaughton Trough of northern Washington and adjacent British Columbia was a forearc basin in Jurassic–early Cretaceous times, but as a result of the collision of outboard terranes about the mid-Cretaceous, it became, for a time, a narrow marine inlet (Tennyson & Cole 1978). This trough is the only detected reentrant in the late Cretaceous Andean arc.

Discussion and Conclusions

The changes in paleogeography over the past 100 m.y. (Figures 6 and 7) result from the counterclockwise rotation of the North American continent, the tectonic deformation and volcanic constructs of the western and Caribbean portions of the continent, the considerably lower sea levels observed today, and the effects of erosion and sedimentation. The rotation of the continent has meant that areas east of the Mississippi are slightly higher in latitude today, while the opposite is true in western areas. The net effect is not great, and the 35°N latitude line on both maps generally divides sediments of subtropical aspect from those of the temperate zone.

The entire western portion of the continent has been remolded by the intracontinental Laramide Orogeny, followed by Basin and Range extension, and finally by strike-slip fragmentation of the western margin. These three major effects combined to produce by late Tertiary times a much more mountainous and irregular margin than in the Cenomanian, when a relatively well-integrated Andean-type margin existed. The motion of many of the faults shown on our tectonic map (Figure 1) is small in scale, but their cumulative effect is considerable, and this serves to demonstrate the need in paleogeography for palinspastic restorations. In addition, this level of structural analysis is essential in understanding the character and evolution of sedimentary basins.

The effect of sea-level change is best observed in the central portion of North America. The hypsography of the continent is profoundly different as a result of this change, with the lowland environments having gained at the expense of the shallow sea since the Cenomanian. Modification of bathymetry by sedimentation is most profound along the northwestern coast of the Gulf of Mexico, where the shelf margin has prograded about 300 km since the Cenomanian (Winker 1982) as a result of the drainage of the central part of the continent by the Mississippi and other river systems. By contrast, shallow-water carbonate areas drowned by the Cenomanian transgression or moved northward out of the subtropical belt by the rotation of the continent are today deeper as the result of continued subsidence and low sedimentation rates. Finally, erosion has played an important but poorly understood role (except for glacial scouring, which created the Great Lakes, many peat bogs, and the irregular topography of both the New England and Washington State coastal and offshore regions).

The level of detail possible in paleogeographic reconstructions, as demonstrated in this review, will provide the framework for answering a number of important geologic questions. By tracing the elevations of shorelines of the past, we can examine the long-term warping of the relatively stable cratons. Can observed variations be explained by isostatic effects, or do they instead reflect deeper mantle processes? By linking tectonic and stratigraphic approaches, we can better understand the character and evolution of sedimentary basins. Did a basin result from thermal subsidence, tectonic loading, a transtensional system, or just a sea-level rise? By mapping the elevations of past mountain chains, we can better estimate their orographic effects on atmospheric circulation patterns. How does this translate into rainfall and its influence on peat formation, or into oceanic currents and their influence on upwelling, organic productivity, and oil source rock generation? Finally, by improved paleogeographies, we can understand the context in which organic evolution took place. Did a taxon evolve in response to environmental or climatic change, to isolation in a constantly changing world, or simply to selection pressures imposed by other biotic elements? These questions are just a sampler, and they indicate the range of problems that can be addressed when accurate paleogeographic maps are prepared.

ACKNOWLEDGMENTS

The tectonic map was prepared by the authors with the help of Stephen F. Barrett, Richard M. Friedman, W. S. McKerrow, and Christopher R. Scotese at the University of Chicago, and with the advice of visitors Rob Arnott (University of Oxford), Scott Bowman (Marathon Oil), and James L. Pindell (University of Durham). The lithofacies data were assembled by the authors with the help of Richard K. Bambach, Stephen F. Barrett, Paula S. Benes, Richard M. Friedman, Aaron B. Rourke, Katharine L. Sellers, and Sammy J. Wong at the University of Chicago. Financial support for the Paleogeographic Atlas Project is provided by the following oil companies: Amoco, Arco, British Petroleum, Chevron, Cities Service, Elf Aquitaine, Exxon, Marathon, Mobil, Pennzoil, Phillips, Shell, Sohio, Sun, and Texaco. Financial support was also provided in the early stages of the project by the National Science Foundation (Grant EAR-7915133).

Literature Cited

Allmendinger, R. W., Sharp, J. W., Von Tish, D., Serpa, L., Brown, L., et al. 1983. Cenozoic and Mesozoic structure of the eastern Basin and Range province, Utah, from COCORP seismic-reflection data. *Geology* 11:532–36

Anderson, T. H., Schmidt, V. A. 1983. The evolution of Middle America and the Gulf of Mexico–Caribbean Sea region during Mesozoic time. *Geol. Soc. Am. Bull.* 94: 941–66

Applin, P. L., Applin, E. R. 1967. The Gulf Series in the subsurface in northern Florida and southern Georgia. *US Geol. Surv. Prof. Pap. 524-G.* 34 pp.

Armstrong, R. L. 1968. Sevier Orogenic Belt in Nevada and Utah. *Geol. Soc. Am. Bull.* 79:429–58

Armstrong, R. L., Suppe, J. 1973. Potassium-argon geochronometry of Mesozoic igneous rocks in Nevada, Utah and southern California. *Geol. Soc. Am. Bull.* 84:1375–92

Armstrong, R. L., Taubeneck, W. H., Hales, P. O. 1977. Rb-Sr and K-Ar geochronometry of Mesozoic granitic rocks and their Sr isotopic composition, Oregon, Washington and Idaho. *Geol. Soc. Am. Bull.* 88:397–411

Asquith, D. O. 1970. Depositional topography and major marine environments, Late Cretaceous, Wyoming. *Am. Assoc. Pet. Geol. Bull.* 54:1184–1224

Austin, J. A., Uchupi, E. 1982. Continental-oceanic crustal transition off southwest Africa. *Am. Assoc. Pet. Geol. Bull.* 66: 1328–47

Austin, J. A., Uchupi, E., Shaugnessy, D. R., Ballard, R. D. 1980. Geology of New England passive margin. *Am. Assoc. Pet. Geol. Bull.* 64:501–26

Baldwin, O. D., Adams, J. A. S. 1971. K40/Ar40 ages of the alkalic igneous rocks of the Balcones Fault trend of Texas. *Tex. J. Sci.* 22:223–31

Bally, A. W. 1981. Thoughts on the tectonics of folded belts. In *Thrust and Nappe Tectonics, Geol. Soc. London Spec. Publ. No. 9*, ed. K. R. McClay, N. J. Price, pp. 13–32

Bally, A. W., Gordy, P. L., Stewart, G. A. 1966. Structure, seismic data and orogenic evolution of southern Canadian Rockies. *Can. Pet. Geol. Bull.* 14:337–81

Beaumont, C. 1981. Foreland basins. *Geophys. J. R. Astron. Soc.* 65:291–329

Beaumont, C., Keen, C. E., Boutilier, R. 1982. On the evolution of rifted continental margins: comparison of models and observations for the Nova Scotian margin. *Geophys. J. R. Astron. Soc.* 70:667–715

Berner, R. A. 1971. *Principles of Chemical Sedimentology*. New York: McGraw-Hill. 240 pp.

Bond, G. C., Kominz, M. A. 1984. Construction of tectonic subsidence curves for early Paleozoic miogeocline, southern Canadian Rocky Mts. *Geol. Soc. Am. Bull.* 95:155–73

Bowin, C., Warsi, W., Milligan, J. 1982. Free-air gravity anomaly atlas of the world. *US Geol. Surv. Map Chart Ser. MC-46*

Boyer, S. E., Elliott, D. 1982. Thrust systems. *Am. Assoc. Pet. Geol. Bull.* 66:1196–1230

Buffler, R. T., Watkins, J. S., Shaub, F. J., Worzel, J. L. 1980. Structure and early geologic history of the deep central Gulf of Mexico Basin. In *The Origin of the Gulf of Mexico and the Early Opening of the Central North Atlantic Ocean*, ed. R. H. Pilger, pp. 3–16. Baton Rouge: La. State Univ.

Bullard, E. C., Everett, J. E., Smith, A. G. 1965. The fit of the continents around the Atlantic. *Philos. Trans. R. Soc. London Ser. A* 258:41–51

Burchfiel, B. C. 1980. Tectonics of noncollisional regimes—the modern Andes and the Mesozoic Cordilleran Orogen of the western United States. In *Continental Tectonics*, ed. Geophys. Study Comm., pp. 65–72. Washington DC: Natl. Acad. Sci.

Burchfiel, B. C., Davis, G. A. 1975. Nature and controls of Cordilleran orogenesis, western United States: extensions of an earlier synthesis. *Am. J. Sci.* 275:363–98

Cebull, S. E., Shurbet, D. H. 1980. The Ouachita Belt in the evolution of the Gulf of Mexico. In *The Origin of the Gulf of Mexico and the Early Opening of the Central North Atlantic Ocean*, ed. R. H. Pilger, pp. 17–26. Baton Rouge: La. State Univ.

Cochran, J. R. 1981. The Gulf of Aden: structure and evolution of a young ocean basin and continental margin. *J. Geophys. Res.* 86:263–87

Cochran, J. R. 1983a. Effects of finite rifting times on the development of sedimentary basins. *Earth Planet. Sci. Lett.* 66:289–302

Cochran, J. R. 1983b. A model for development of Red Sea. *Am. Assoc. Pet. Geol. Bull.* 67:41–69

Cogley, J. G. 1984. Continental margins and the extent and number of the continents. *Rev. Geophys. Space Phys.* 22:101–22

Coney, P. J. 1978. Mesozoic-Cenozoic Cordilleran plate tectonics. In *Cenozoic Tectonics and Regional Geophysics of the Western Cordillera, Geol. Soc. Am. Mem.*, ed. R. B. Smith, G. P. Eaton, 152:33–50

Coney, P. J. 1980. Cordilleran metamorphic core complexes: an overview. In *Cordilleran Metamorphic Core Complexes, Geol. Soc. Am. Mem.*, ed. M. D. Crittenden Jr., P. J. Coney, G. H. Davis, 153:7–31

Cook, F. A., Albaugh, D. S., Brown, L. D., Kaufman, S., Oliver, J. E., Hatcher, R. D. 1979. Thin-skinned tectonics in the crystalline southern Appalachians; COCORP seismic-reflection profiling of the Blue Ridge and Piedmont. *Geology* 7:563–67

Cook, T. D., Bally, A. W., eds. 1975. *Stratigraphic Atlas of North and Central America*. Princeton, NJ: Princeton Univ. Press. 272 pp.

Dahlstrom, C. D. A. 1969. The upper detachment in concentric folding. *Can. Pet. Geol. Bull.* 17:326–46

Dahlstrom, C. D. A. 1970. Structural geology in the eastern margin of the Canadian Rocky Mountains. *Can. Pet. Geol. Bull.* 18:332–406

Davis, G. A., Burchfiel, B. C. 1973. Garlock Fault: an intracontinental transform structure, southern California. *Geol. Soc. Am. Bull.* 84:1407–22

Davis, G. A., Monger, J. W. H., Burchfiel, B. C. 1978. Mesozoic construction of the Cordilleran "collage," central British Columbia to central California. See Howell & McDougall 1978, pp. 1–32

Davis, G. A., Anderson, J. L., Frost, E. G., Shackelford, T. J. 1980. Mylonitization and detachment faulting in the Whipple-Buckskin-Rawhide Mountains terrane, southeastern California and western Arizona. In *Cordilleran Metamorphic Core Complexes*, *Geol. Soc. Am. Mem.*, ed. M. D. Crittenden Jr., P. J. Coney, G. H. Davis, 153:79–129

Davis, R. A. 1983. *Depositional Systems, a Genetic Approach to Sedimentary Geology*. Englewood Cliffs, NJ: Prentice-Hall. 669 pp.

de Chadenedes, J. F. 1975. Frontier deltas of the western Green River Basin, Wyoming. In *Deep Drilling Frontiers of the Central Rocky Mountains*, ed. D. W. Bolyard, pp. 149–57. Denver: Rocky Mt. Assoc. Geol.

Dewey, J. F. 1980. Episodicity, sequence and style at convergent plate boundaries. In *The Continental Crust and Its Mineral Deposits, Geol. Assoc. Can. Spec. Pap.*, ed. D. W. Strangway, 20:553–76

Dewey, J. F. 1982. Plate tectonics and the evolution of the British Isles. *J. Geol. Soc. London* 139:371–412

Dewey, J. F., Sengör, A. M. C. 1979. Aegean and surrounding regions: complex multiplate and continuum tectonics in a convergent zone. *Geol. Soc. Am. Bull.* 90:84–92

Dickinson, W. R. 1979. Plate tectonic evolution of North Pacific Rim. In *Geodynamics of the Western Pacific*, ed. S. Uyeda, R. W. Murphy, K. Kobayashi, pp. 1–19. Tokyo: Jpn. Sci. Soc. Press

Dickinson, W. R. 1983. Cretaceous sinistral strike slip along Nacimiento Fault in coastal California. *Am. Assoc. Pet. Geol. Bull.* 67:624–45

Dickinson, W. R., Seely, D. R. 1979. Structure and stratigraphy of forearc regions. *Am. Assoc. Pet. Geol. Bull.* 63:2–31

Dickinson, W. R., Thayer, T. P. 1978. Paleogeographic and paleotectonic implications of Mesozoic stratigraphy and structure in the John Day Inlier of central Oregon. See Howell & McDougall 1978, pp. 147–61

Dietz, R. S., Holden, J. C. 1970. Reconstruction of Pangea; breakup and dispersion of continents, Permian to Present. *J. Geophys. Res.* 75:4939–56

Dingle, R. V., Siesser, W. G., Newton, A. R. 1983. *Mesozoic and Tertiary Geology of Southern Africa.* Rotterdam: Balkema. 375 pp.

Dixon, J. S. 1982. Regional structural synthesis, Wyoming salient of western overthrust belt. *Am. Assoc. Pet. Geol. Bull.* 66:1560–80

Duncan, R. A. 1982. A captured island chain in the coast range of Oregon and Washington. *J. Geophys. Res.* 87:10827–37

Emery, K. O. 1968. Relict sediments on continental shelves of the world. *Am. Assoc. Pet. Geol. Bull.* 52:445–64

Emery, K. O. 1969. The continental shelves. *Sci. Am.* 221:107–22

Emiliani, C., ed. 1983. *The Oceanic Lithosphere: The Sea, Ideas and Observations on Progress in the Study of the Seas*, Vol. 7. New York: Wiley-Interscience. 1738 pp.

Enos, P. 1983. Late Mesozoic paleogeography of Mexico. See Reynolds & Dolly 1983, pp. 133–57

Fitch, T. J. 1972. Plate convergence, transcurrent faults and internal deformation adjacent to Southeast Asia and the Western Pacific. *J. Geophys. Res.* 77: 4432–60

Friedman, G. M., Sanders, J. E. 1978. *Principles of Sedimentology*. New York: Wiley. 792 pp.

Gealey, W. K. 1980. Ophiolite obduction mechanism. *Proc. Int. Ophiolite Symp., Cyprus, 1979*, ed. A. Panayiotou, pp. 228–43. Cyprus: Rep. Cyprus Geol. Surv. Dep.

Gilbert, J. L., Asquith, G. B. 1976. Sedimentology of braided alluvial interval of Dakota Sandstone, northeastern New Mexico. *N. Mex. Bur. Mines Miner. Resour. Circ. 150*. 16 pp.

Goodell, H. G. 1962. The stratigraphy and petrology of the Frontier Formation of Wyoming. *Symp. Early Cretaceous Rocks Wyoming Adjacent Areas, 17th Ann. Field Conf.*, pp. 173–210. Casper: Wyo. Geol. Assoc.

Greenwood, E., Kottlowski, F. E., Thompson, S. 1977. Petroleum potential and stratigraphy of Pedregosa Basin: comparison with Permian and Orogrande Basins. *Am. Assoc. Pet. Geol. Bull.* 61: 1448–69

Hamilton, W. 1978. Mesozoic tectonics of the western United States. See Howell & McDougall 1978, pp. 33–70

Hamilton, W., Myers, W. B. 1966. Cenozoic tectonics of the western United States. *Rev. Geophys.* 4:509–49

Hancock, J. M., Kauffman, E. G. 1979. The great transgressions of the late Cretaceous. *J. Geol. Soc. London* 136:175–86

Hannah, J. L., Verosub, K. L. 1980. Tectonic implications of remagnetized Upper Paleozoic strata of the northern Sierra Nevada. *Geology* 8:520–24

Harding, T. P., Lowell, J. D. 1979. Structural styles, their plate-tectonic habitats and hydrocarbon traps in petroleum provinces. *Am. Assoc. Pet. Geol. Bull.* 63: 1016–58

Harland, W. B. 1971. Tectonic transpression in Caledonian Spitsbergen. *Geol. Mag.* 108:27–42

Harland, W. B., Cox, A. V., Llewellyn, P. G., Pickton, C. A. G., Smith, A. G., Walters, R. 1982. *A Geologic Time Scale*. Cambridge: Cambridge Univ. Press. 131 pp.

Hayes, P. T. 1970. Cretaceous paleogeography of southeastern Arizona and adjacent areas. *US Geol. Surv. Prof. Pap. 658-B*. 42 pp.

Henderson, L. J., Gordon, R. G., Engebretson, D. C. 1984. Mesozoic aseismic ridges on the Farallon Plate and southward migration of shallow subduction during the Laramide orogeny. *Tectonics* 3:121–32

Howell, D. G., McDougall, K. A., eds. 1978. *Mesozoic Paleogeography of the Western United States*. Los Angeles, Calif: Pac. Sect., Soc. Econ. Paleontol. Mineral. 573 pp.

Hutchinson, D. R., Grow, J. A., Klitgord, K. D., Swift, B. A. 1982. Deep structure and evolution of the Carolina Trough. In *Studies in Continental Margin Geology Memoir 34*, ed. J. S. Watkins, C. L. Drake, pp. 129–52. Tulsa, Okla: Am. Assoc. Pet. Geol.

Ingersoll, R. V. 1979. Evolution of the Late Cretaceous forearc basin, northern and central California. *Geol. Soc. Am. Bull.* 90:813–26

Jarvis, G. T., McKenzie, D. P. 1980. Sedimentary basin formation with finite extension rates. *Earth Planet. Sci. Lett.* 48: 42–52

Jones, D. L., Silberling, N. J., Hillhouse, J. 1977. Wrangellia: a displaced terrane in northwestern North America. *Can. J. Earth Sci.* 14:2565–77

Jordan, T. E. 1981. Thrust loads and foreland basin evolution, Cretaceous, western United States. *Am. Assoc. Pet. Geol. Bull.* 65:2506–20

Karig, D. E. 1974. Evolution of arc systems in the western Pacific. *Ann. Rev. Earth Planet. Sci.* 2:51–76

Kauffman, E. G. 1967. Coloradoan macroinvertebrate assemblages, central Western Interior, United States. In *Paleoenvironments of the Cretaceous Seaway —A Symposium*, ed. E. G. Kauffman, H. C. Kent, pp. 67–143. Golden: Colo. Sch. Mines

Kauffman, E. G. 1977. Second day: Upper Cretaceous cyclothems, biotas and environments, Rock Canyon Anticline, Pueblo, Colorado. *Mt. Geol.* 14:129–52

Kauffman, E. G., Hattin, D. E., Powell, J. D. 1977. Stratigraphic, paleontologic and paleoenvironmental analysis of the Upper Cretaceous rocks of Cimarron County, northwestern Oklahoma. *Geol. Soc. Am. Mem. 149*. 150 pp.

Kay, M. 1937. Stratigraphy of the Trenton Group. *Geol. Soc. Am. Bull.* 48:233–302

Kay, M. 1945. Paleogeographic and palinspastic maps. *Am. Assoc. Pet. Geol. Bull.* 29:426–50

Keen, C. E. 1981. The continental margins of eastern Canada: a review. In *Dynamics of Passive Margins, Geodyn. Ser.*, 6:45–58. Washington DC: Am. Geophys. Union

Klitgord, K. D., Popenoe, P., Schouten, H. 1984. Florida: a Jurassic transform plate boundary. *J. Geophys. Res.* 89:7753–72

Krummenacher, D., Gastil, R. G., Bushee, J., Doupont, J. 1975. K–Ar apparent ages, Peninsular Ranges Batholith, southern California and Baja California. *Geol. Soc. Am. Bull.* 86:760–68

LeFort, P. 1975. Himalayas: the collided range. Present knowledge of the continental arc. *Am. J. Sci.* 275:1–44

Le Pichon, X., Fox, P. J. 1971. Marginal offsets, fracture zones and the early opening of the North Atlantic. *J. Geophys. Res.* 76:6294–6308

Le Pichon, X., Sibuet, J. C. 1981. Passive margins: a model of formation. *J. Geophys. Res.* 86:3708–20

Livaccari, R. F., Burke, K., Sengör, A. M. C. 1981. Was the Laramide orogeny related to subduction of an oceanic plateau? *Nature* 289:276–78

Lowell, J. D. 1972. Spitsbergen Tertiary orogenic belt and the Spitsbergen fracture zone. *Geol. Soc. Am. Bull.* 83:3091–3102

Mallory, W. W., ed. 1972. *Geologic Atlas of the Rocky Mountain Region.* Denver, Colo: Rocky Mt. Assoc. Geol. 331 pp.

Mancini, E. A. 1977. Depositional environment of the Grayson Formation (Upper Cretaceous) of Texas. *Gulf Coast Assoc. Geol Soc. Trans.* 27:334–51

Mann, P., Hempton, M. R., Bradley, D. C., Burke, K. 1983. Development of pull-apart basins. *J. Geol.* 91:529–54

Marcher, M. V., Stearns, R. G. 1962. Tuscaloosa formation in Tennessee. *Geol. Soc. Am. Bull.* 73:1365–86

Martin, R. G. 1978. Northern and eastern Gulf of Mexico continental margin: stratigraphic and structural framework. In *Framework, Facies and Oil-Trapping Characteristics of the Upper Continental Margin, Am. Assoc. Pet. Geol. Stud. Geol.*, ed. A. Bouma et al, 7:21–42

Martin, R. G., Case, J. E. 1975. Geophysical studies in the Gulf of Mexico. In *The Ocean Basins and Margins, the Gulf of Mexico and the Caribbean*, ed. A. E. M. Nairn, F. G. Stehli, 3:65–106. New York: Plenum

McElhinny, M. W., Valencio, D. A., eds. 1981. *Paleoreconstruction of the Continents. Geodyn. Ser.* Vol. 2. Washington DC: Am. Geophys. Union. 192 pp.

McGookey, D. P., Haun, J. D., Hale, L. A., Goodell, H. G., McCubbin, D. G., et al. 1972. Cretaceous system. See Mallory 1972, pp. 190–228

McKenzie, D. 1972. Active tectonics of the Mediterranean region. *Geophys. J. R. Astron. Soc.* 30:109–85

McKenzie, D. 1978. Some remarks on the development of sedimentary basins. *Earth Planet. Sci. Lett.* 40:25–32

McKinney, M. L. 1984. Suwanee Channel of the Paleogene coastal plain: support for the "carbonate suppression" model of basin formation. *Geology* 12:343–45

McLaughlin, R. J., Kling, S. A., Poore, R. Z., McDougall, K., Beutner, E. C. 1982. Post–Middle Miocene accretion of Franciscan rocks, northwestern California. *Geol. Soc. Am. Bull.* 93:595–605

Merewether, E. A. 1983. Lower Upper Cretaceous strata in Minnesota and adjacent areas—time-stratigraphic correlations and structural attitudes. *US Geol. Surv. Prof. Pap. 1253-B*, pp. 27–52

Miller, E. L., Gans, P. B., Garing, J. 1983. The Snake Range Decollement: an exhumed mid-Tertiary ductile-brittle transition. *Tectonics* 2:239–63

Misch, P. 1966. Tectonic evolution of the north Cascades of Washington State. *Can. Inst. Min. Metall. Spec. Vol.* 8:101–48

Molenaar, C. M. 1983. Major depositional

cycles and regional correlations of Upper Cretaceous rocks, southern Colorado Plateau and adjacent areas. See Reynolds & Dolly 1983, pp. 201–24

Molnar, P., Atwater, T. 1978. Interarc spreading and Cordilleran tectonics as alternates related to the age of subducted oceanic lithosphere. *Earth Planet. Sci. Lett.* 41:330–40

Molnar, P., Tapponnier, P. 1975. Cenozoic tectonics of Asia: effects of a continental collision. *Science* 189:419–26

Monger, J. W. H. 1977. Upper Paleozoic rocks of the western Canadian Cordillera and their bearing on Cordilleran evolution. *Can. J. Earth Sci.* 14:1832–59

Monger, J. W. H., Price, R. A. 1979. Geodynamic evolution of the Canadian Cordillera—progress and problems. *Can. J. Earth Sci.* 16:770–91

Monger, J. W. H., Price, R. A., Tempelman-Kluit, D. J. 1982. Tectonic accretion and the origin of the two major metamorphic and plutonic welts in the Canadian Cordillera. *Geology* 10:70–75

Montadert, L., de Charpal, O., Roberts, D., Guennoc, P., Sibuet, J. C. 1979. Northeast Atlantic passive continental margins: rifting and subsidence processes. In *Deep Sea Drilling Results in the Atlantic Ocean*, ed. M. Talwani et al, pp. 154–86. Washington DC: Am. Geophys. Union

Moore, J. C., Watkins, J. S., Shipley, T. H., McMillen, K. J., Bachman, S. B., Lundberg, N. 1982. Geology and tectonic evolution of a juvenile accretionary terrane along a truncated convergent margin: synthesis of results from Leg 66 DSDP of the Deep Sea Drilling Project, southern Mexico. *Geol. Soc. Am. Bull.* 93:847–61

Murray, G. E. 1961. *Geology of the Atlantic and Gulf Coastal Province of North America.* New York: Harper. 692 pp.

Officer, C. B., Drake, C. L. 1981. Epeirogenic plate movements. *J. Geol.* 90:139–53

Oliver, W. B. 1971. Depositional systems in the Woodbine Formation (Upper Cretaceous), northeast Texas. *Univ. Tex. Bur. Econ. Geol. Rep. Invest. 73*, Austin. 28 pp.

Page, B. M., Engebretson, D. C. 1984. Correlation between the geologic record and computed plate motions for central California. *Tectonics* 3:133–55

Pardo, G. 1975. Geology of Cuba. In *The Ocean Basins and Margins, the Gulf of Mexico and the Caribbean*, ed. A. E. M. Nairn, F. G. Stehli, 3:553–615. New York: Plenum

Parrish, J. T., Curtis, R. 1982. Atmospheric circulation, upwelling and organic-rich rocks in the Mesozoic and Cenozoic Eras. *Palaeogeogr. Palaeoclimatol. Palaeoecol.* 40:31–66

Parrish, J. T., Ziegler, A. M., Scotese, C. R. 1982. Rainfall patterns and the distribution of coals and evaporites in the Mesozoic and Cenozoic. *Palaeogeogr. Palaeoclimatol. Palaeoecol.* 40:67–101

Payton, C. E., ed. 1977. *Seismic Stratigraphy—Applications to Hydrocarbon Exploration, Am. Assoc. Pet. Geol. Mem. 26.* 516 pp.

Pindell, J. L., Dewey, J. F. 1982. Permo-Triassic reconstruction of western Pangea and the evolution of the Gulf of Mexico/Caribbean region. *Tectonics* 1: 179–211

Pitman, W. C. 1978. Relationship between eustacy and stratigraphic sequences of passive margins. *Geol. Soc. Am. Bull.* 89: 1389–1403

Pitman, W. C., Talwani, M. 1972. Sea-floor spreading in the North Atlantic. *Geol. Soc. Am. Bull.* 83:619–46

Powell, J. D. 1965. Late Cretaceous platform-basin facies, northern Mexico and adjacent Texas. *Am. Assoc. Pet. Geol. Bull.* 49:511–25

Price, R. A. 1981. The Cordilleran foreland thrust and fold belt in the southern Canadian Rocky mountains. In *Thrust and Nappe Tectonics, Geol. Soc. London Spec. Publ. No. 9*, ed. K. R. McClay, N. J. Price, pp. 427–48

Proffett, J. M. 1977. Cenozoic geology of the Yerington District, Nevada, and implications for the nature and origin of Basin and Range faulting. *Geol. Soc. Am. Bull.* 88:247–66

Rabinowitz, P. D., La Brecque, J. 1979. The Mesozoic South Atlantic Ocean and evolution of its continental margins. *J. Geophys. Res.* 84:5973–6002

Rabinowitz, P. D., Coffin, M. F., Falvey, D. 1983. The separation of Madagascar and Africa. *Science* 220:67–69

Ramsay, J. G., Huber, M. I. 1983. *The Techniques of Modern Structural Geology, Vol. 1: Strain Analysis.* London: Academic. 307 pp.

Reading, H. G. 1978. *Sedimentary Environments and Facies.* New York: Elsevier. 557 pp.

Reyment, R. A., Thomel, G., eds. 1976. *Événements de la Partie Moyenne du Cretace, Uppsala 1975–Nice 1976, Ann. Mus. Hist. Nice 4.* Nice, Fr: Cent. Etud. Mediterr.

Reynolds, M. W., Dolly, E. D., eds. 1983. *Mesozoic Paleogeography of the West-Central United States, Rocky Mt. Paleogeogr. Symp. 2.* Denver, Colo: Soc. Econ. Paleontol. Mineral. 391 pp.

Royden, L., Keen, C. E. 1980. Rifting process and thermal evolution of the continental margin of eastern Canada determined from subsidence curves. *Earth Planet. Sci. Lett.* 51:343–61

Schlager, W., Buffler, R. T., Angstadt, D., Bowdler, J. L., Cotillon, P. H., et al. 1984. Deep Sea Drilling Project, Leg 77, southeastern Gulf of Mexico. *Geol. Soc. Am. Bull.* 95:226–36

Sclater, J. G., Christie, P. A. F. 1980. Continental stretching: an explanation of the post-mid-Cretaceous subsidence of the central North Sea Basin. *J. Geophys. Res.* 85:3711–39

Sclater, J. G., Parsons, B., Jaupart, C. 1981. Oceans and continents: similarities and differences in the mechanisms of heat loss. *J. Geophys. Res.* 86:11535–52

Selley, R. C. 1970. *Ancient Sedimentary Environments.* Ithaca, NY: Cornell Univ. Press. 287 pp.

Sheridan, R. E., Crosby, J. T., Bryan, G. M., Stoffa, P. L. 1981. Stratigraphy and structure of southern Blake Plateau, northern Florida Straits, and northern Bahama platform from multichannel seismic reflection data. *Am. Assoc. Pet. Geol. Bull.* 65:2571–93

Sherwin, D. F. 1973. Scotian Shelf and Grand Banks. In *The Future Petroleum Provinces of Canada*, ed. R. G. McCrossan, pp. 519–59. Calgary: Can. Soc. Pet. Geol.

Shih, T. C. 1980. Marine magnetic anomalies from the western Philippine Sea: implications for the evolution of marginal basins. In *The Tectonic and Geologic Evolution of Southeast Asian Seas and Islands*, ed. D. E. Hayes, pp. 49–75. Washington, DC: Am. Geophys. Union

Siemers, C. T. 1978. Submarine fan deposition of the Woodbine–Eagle Ford interval (Upper Cretaceous), Tyler County, Texas. *Gulf Coast Assoc. Geol. Soc. Trans.* 28:493–533

Simpson, C., Schmid, S. M. 1983. An evaluation of criteria to deduce the sense of movement in sheared rocks. *Geol. Soc. Am. Bull.* 94:1281–88

Sleep, N. H. 1976. Platform subsidence mechanisms and "eustatic" sea-level changes. *Tectonophysics* 36:45–56

Sloan, R. E. 1964. The Cretaceous System in Minnesota. *Minn. Geol. Surv. Rep. Invest.* 5. 64 pp.

Smithson, S. B., Brewer, J. A., Kaufman, S., Oliver, J. E., Hurich, C. A. 1979. Structure of the Laramide Wind River Uplift, Wyoming, from COCORP deep reflection data and from gravity data. *J. Geophys. Res.* 84:5955–72

Snavely, P. D., MacLeod, N. S., Wagner, H. C. 1968. Tholeiitic and alkalic basalts of the Eocene Siletz River Volcanics, Oregon Coast Range. *Am. J. Sci.* 266:454–81

Sohl, N. F., Smith, C. C., Christopher, R. A. 1976. Middle Cretaceous rocks of the Atlantic seaboard and eastern Gulf Coastal Plain of North America. See Reyment & Thomel 1976. 16 pp.

Srivastava, S. P. 1978. Evolution of the Labrador Sea and its bearing on the early evolution of the North Atlantic. *Geophys. J. R. Astron. Soc.* 52:313–57

Steckler, M. S., Watts, A. B. 1981. Subsidence history and tectonic evolution of Atlantic-type continental margins. In *Dynamics of Passive Margins, Geodyn. Ser.*, 6:184–96. Washington DC: Am. Geophys. Union

Steiger, R. H., Jager, E. 1978. Subcommission of geochronology: convention on the use of decay constants in geochronology and cosmochronology. In *Contributions to the Geologic Time Scale, Am. Assoc. Pet. Geol. Stud. Geol.*, ed. G. V. Cohee, M. F. Glaessner, H. D. Hedberg, 6:67–71

Stone, D. B., Panuska, B. C., Packer, D. R. 1982. Paleolatitudes versus time for southern Alaska. *J. Geophys. Res.* 87: 3697–3707

Streckeisen, A. 1979. Classification and nomenclature of volcanic rocks, lamprophyres, carbonatites, and melilitic rocks: recommendations and suggestions of the IUGS Subcommission on the Systematics of Igneous Rocks. *Geology* 7:331–35

Suppe, J. 1983. Geometry and kinematics of fault-bend folding. *Am. J. Sci.* 283:684–721

Suppe, J., Armstrong, R. L. 1972. Potassium-argon dating of Franciscan metamorphic rocks. *Am. J. Sci.* 272:217–33

Tennyson, M. E., Cole, M. R. 1978. Tectonic significance of Upper Mesozoic Methow-Pasaythen sequence, northeastern Cascade Range, Washington and British Columbia. See Howell & McDougall 1978, pp. 499–508

Thiede, J. 1982. Paleogeography and paleobathymetry: quantitative reconstructions of ocean basins. In *Tidal Friction and the Earth's Rotation II*, ed. P. Brosche, J. Sündermann, pp. 229–39. Berlin: Springer-Verlag

Valentine, P. C. 1982. Upper Cretaceous subsurface stratigraphy and structure of coastal Georgia and South Carolina. *US Geol Surv. Prof. Pap. 1222.* 33 pp.

Vance, J. A., Dungan, M. A., Blanchard, D. P., Rhodes, J. M. 1980. Tectonic setting and trace element geochemistry of Mesozoic ophiolitic rocks in western Washington. *Am. J. Sci.* 280:359–88

Van Hinte, J. E. 1978. A Cretaceous time scale. In *Contributions to the Geologic Time Scale, Am. Assoc. Pet. Geol. Stud. Geol.*, ed. G. V. Cohee, M. F. Glaessner, H. D. Hedberg, 6:269–87

Vink, G. E. 1982. Continental rifting and the

implications for plate tectonic reconstructions. *J. Geophys. Res.* 87:10677–88

Vogt, P. R., Tucholke, B. E. 1979. The New England Seamounts: testing origins. In *Initial Reports of the Deep Sea Drilling Project*, 43:847–56. Washington DC: GPO

Walcott, R. I. 1970. Isostatic response to loading of the crust in Canada. *Can. J. Earth Sci.* 7:716–34

Walcott, R. I. 1972. Past sea levels, eustacy and deformation of the Earth. *Quat. Res.* 2:1–14

Wang, H. C. 1980. Megastages in the tectonic development of Asia. *Sci. Sin.* 23:331–45

Weissel, J. K. 1981. Magnetic lineations in marginal basins of the western Pacific. *Philos. Trans. R. Soc. London Ser. A* 300: 223–47

Wernicke, B., Burchfiel, B. C. 1982. Modes of extensional tectonics. *J. Struct. Geol.* 4: 105–15

Wernicke, B., Spencer, J. E., Burchfiel, B. C., Guth, P. L. 1982. Magnitude of crustal extension in the southern Great Basin. *Geology* 10:499–502

Weyl, R. 1980. *Geology of Central America.* Berlin: Gebruder Borntraeger. 371 pp.

Wiltschko, D. V., Dorr, J. A. 1983. Timing of deformation in Overthrust Belt and Foreland of Idaho, Wyoming, and Utah. *Am. Assoc. Pet. Geol. Bull.* 67:1304–22

Winker, C. D. 1982. Cenozoic shelf margins, northwestern Gulf of Mexico. *Gulf Coast Assoc. Geol. Soc. Trans.* 32:427–48

York, D., Farquhar, R. M. 1972. *The Earth's Age and Geochronology.* Oxford: Pergamon. 178 pp.

Ziegler, A. M., Scotese, C. R., McKerrow, W. S., Johnson, M. E., Bambach, R. K. 1979. Paleozoic paleogeography. *Ann. Rev. Earth Planet. Sci.* 7:473–502

Ziegler, A. M., Barrett, S. F., Scotese, C. R. 1981. Palaeoclimate, sedimentation and continental accretion. In *The Origin and Evolution of the Earth's Continental Crust*, ed. S. Moorbath, B. F. Windley, 301:253–64. London: R. Soc. London

Ziegler, A. M., Barrett, S. F., Kazmer, C. S. 1982a. The geologic history of the upwelling zones off western South America. In *Geology of Phosphorite Deposits and Problems of Phosphorite Genesis*, pp. 38–40. Novosibirsk, SSSR: Akad. Nauk Siberskoe Otd.

Ziegler, A. M., Scotese, C. R., Barrett, S. F. 1982b. Mesozoic and Cenozoic paleogeographic maps. In *Tidal Friction and the Earth's Rotation II*, ed. P. Brosche, J. Sündermann, pp. 240–52. Berlin: Springer-Verlag

Ziegler, A. M., Hulver, M. L., Lottes, A. L., Schmachtenberg, W. F. 1984. Uniformitarianism and palaeoclimates: influences from the distribution of carbonate rocks. In *Fossils and Climate, Geol. J. Spec. Issue*, ed. P. J. Brenchley, pp. 3–25. Chichester: Wiley

Zimmermann, R. A., Reimer, G. M., Foland, K. A., Faul, H. 1975. Cretaceous fission track dates of apatites from northern New England. *Earth Planet. Sci. Lett.* 28:181–88

SUBJECT INDEX

G

H

I

J

K

L

CONTRIBUTING AUTHORS VOLUMES 1–13

CHAPTER TITLES VOLUMES 1–13

GEOPHYSICS AND PLANETARY SCIENCE

OCEANOGRAPHY, METEOROLOGY, AND PALEOCLIMATOLOGY

PALEONTOLOGY, STRATIGRAPHY, AND SEDIMENTOLOGY

TECTONOPHYSICS AND REGIONAL GEOLOGY

NEW BOOKS
FROM
ANNUAL REVIEWS INC.

NOW YOU CAN
CHARGE THEM
TO

ORDER FORM

A NONPROFIT SCIENTIFIC PUBLISHER

4139 EL CAMINO WAY • PALO ALTO, CA 94306-9981 • (415) 493-4400

Orders for Annual Reviews Inc. publications may be placed through your bookstore; subscription agent; participating professional societies; or directly from Annual Reviews Inc. by mail or telephone (paid by credit card or purchase order). Prices subject to change without notice.

Individuals: Prepayment required in U.S. funds or charged to American Express, MasterCard, or Visa.
Institutional Buyers: Please include purchase order.
Students: Special rates are available to qualified students. Refer to Annual Reviews ***Prospectus*** or contact Annual Reviews Inc. office for information.
Professional Society Members: Members whose professional societies have a contractural arrangement with Annual Reviews may order books through their society at a special discount. Check with your society for information.

Regular orders: When ordering current or back volumes, please list the volumes you wish by volume number.
Standing orders: (New volume in the series will be sent to you automatically each year upon publication. Cancellation may be made at any time.) Please indicate volume number to begin standing order.
Prepublication orders: Volumes not yet published will be shipped in month and year indicated.
California orders: Add applicable sales tax.
Postage paid (4th class bookrate /surface mail) by Annual Reviews Inc.

ANNUAL REVIEWS SERIES

		Prices Postpaid per volume USA/elsewhere	Regular Order Please send: Vol. number	Standing Order Begin with: Vol. number
Annual Review of **ANTHROPOLOGY**				
Vols. 1-10	(1972-1981)	$20.00/$21.00		
Vol. 11	(1982)	$22.00/$25.00		
Vols. 12-13	(1983-1984)	$27.00/$30.00		
Vol. 14	(avail. Oct. 1985)	$27.00/$30.00	Vol(s).______	Vol.______
Annual Review of **ASTRONOMY AND ASTROPHYSICS**				
Vols. 1-19	(1963-1981)	$20.00/$21.00		
Vol. 20	(1982)	$22.00/$25.00		
Vols. 21-22	(1983-1984)	$44.00/$47.00		
Vol. 23	(avail. Sept. 1985)	$44.00/$47.00	Vol(s).______	Vol.______
Annual Review of **BIOCHEMISTRY**				
Vols. 29-34, 36-50	(1960-1965; 1967-1981)	$21.00/$22.00		
Vol. 51	(1982)	$23.00/$26.00		
Vols. 52-53	(1983-1984)	$29.00/$32.00		
Vol. 54	(avail. July 1985)	$29.00/$32.00	Vol(s).______	Vol.______
Annual Review of **BIOPHYSICS**				
Vols. 1-10	(1972-1981)	$20.00/$21.00		
Vol. 11	(1982)	$22.00/$25.00		
Vols. 12-13	(1983-1984)	$47.00/$50.00		
Vol. 14	(avail. June 1985)	$47.00/$50.00	Vol(s).______	Vol.______
Annual Review of **CELL BIOLOGY**				
Vol. 1	(avail. Nov. 1985)	est. $27.00/$30.00	Vol. ______	Vol.______
Annual Review of **EARTH AND PLANETARY SCIENCES**				
Vols. 1-9	(1973-1981)	$20.00/$21.00		
Vol. 10	(1982)	$22.00/$25.00		
Vols. 11-12	(1983-1984)	$44.00/$47.00		
Vol. 13	(avail. May 1985)	$44.00/$47.00	Vol(s).______	Vol.______
Annual Review of **ECOLOGY AND SYSTEMATICS**				
Vols. 1-12	(1970-1981)	$20.00/$21.00		
Vol. 13	(1982)	$22.00/$25.00		
Vols. 14-15	(1983-1984)	$27.00/$30.00		
Vol. 16	(avail. Nov. 1985)	$27.00/$30.00	Vol(s).______	Vol.______

SEE ORDERING INFORMATION ON PAGE 4

		Prices Postpaid per volume USA/elsewhere	Regular Order Please send: Vol. number	Standing Order Begin with: Vol. number
Annual Review of **ENERGY**				
Vols. 1-6	(1976-1981)	$20.00/$21.00		
Vol. 7	(1982)	$22.00/$25.00		
Vols. 8-9	(1983-1984)	$56.00/$59.00		
Vol. 10	(avail. Oct. 1985)	$56.00/$59.00	Vol(s).________	Vol.________
Annual Review of **ENTOMOLOGY**				
Vols. 8-16, 18-26	(1963-1971; 1973-1981)	$20.00/$21.00		
Vol. 27	(1982)	$22.00/$25.00		
Vols. 28-29	(1983-1984)	$27.00/$30.00		
Vol. 30	(avail. Jan. 1985)	$27.00/$30.00	Vol(s).________	Vol.________
Annual Review of **FLUID MECHANICS**				
Vols. 1-5, 7-13	(1969-1973; 1975-1981)	$20.00/$21.00		
Vol. 14	(1982)	$22.00/$25.00		
Vols. 15-16	(1983-1984)	$28.00/$31.00		
Vol. 17	(avail. Jan. 1985)	$28.00/$31.00	Vol(s).________	Vol.________
Annual Review of **GENETICS**				
Vols. 1-15	(1967-1981)	$20.00/$21.00		
Vol. 16	(1982)	$22.00/$25.00		
Vols. 17-18	(1983-1984)	$27.00/$30.00		
Vol. 19	(avail. Dec. 1985)	$27.00/$30.00	Vol(s).________	Vol.________
Annual Review of **IMMUNOLOGY**				
Vols. 1-2	(1983-1984)	$27.00/$30.00		
Vol. 3	(avail. April 1985)	$27.00/$30.00	Vol(s).________	Vol.________
Annual Review of **MATERIALS SCIENCE**				
Vols. 1-11	(1971-1981)	$20.00/$21.00		
Vol. 12	(1982)	$22.00/$25.00		
Vols. 13-14	(1983-1984)	$64.00/$67.00		
Vol. 15	(avail. Aug. 1985)	$64.00/$67.00	Vol(s).________	Vol.________
Annual Review of **MEDICINE: Selected Topics in the Clinical Sciences**				
Vols. 1-3, 5-15	(1950-1952; 1954-1964)	$20.00/$21.00		
Vols. 17-32	(1966-1981)	$20.00/$21.00		
Vol. 33	(1982)	$22.00/$25.00		
Vols. 34-35	(1983-1984)	$27.00/$30.00		
Vol. 36	(avail. April 1985)	$27.00/$30.00	Vol(s).________	Vol.________
Annual Review of **MICROBIOLOGY**				
Vols. 17-35	(1963-1981)	$20.00/$21.00		
Vol. 36	(1982)	$22.00/$25.00		
Vols. 37-38	(1983-1984)	$27.00/$30.00		
Vol. 39	(avail. Oct. 1985)	$27.00/$30.00	Vol(s).________	Vol.________
Annual Review of **NEUROSCIENCE**				
Vols. 1-4	(1978-1981)	$20.00/$21.00		
Vol. 5	(1982)	$22.00/$25.00		
Vols. 6-7	(1983-1984)	$27.00/$30.00		
Vol. 8	(avail. March 1985)	$27.00/$30.00	Vol(s).________	Vol.________
Annual Review of **NUCLEAR AND PARTICLE SCIENCE**				
Vols. 12-31	(1962-1981)	$22.50/$23.50		
Vol. 32	(1982)	$25.00/$28.00		
Vols. 33-34	(1983-1984)	$30.00/$33.00		
Vol. 35	(avail. Dec. 1985)	$30.00/$33.00	Vol(s).________	Vol.________

SEE ORDERING INFORMATION ON PAGE 4

		Prices Postpaid per volume USA/elsewhere	Regular Order Please send:	Standing Order Begin with:
Annual Review of **NUTRITION**			Vol. number	Vol. number
Vol. 1	(1981)	$20.00/$21.00		
Vol. 2	(1982)	$22.00/$25.00		
Vols. 3-4	(1983-1984)	$27.00/$30.00		
Vol. 5	(avail. July 1985)	$27.00/$30.00	Vol(s).______	Vol.______
Annual Review of **PHARMACOLOGY AND TOXICOLOGY**				
Vols. 1-3, 5-21	(1961-1963; 1965-1981)	$20.00/$21.00		
Vol. 22	(1982)	$22.00/$25.00		
Vols. 23-24	(1983-1984)	$27.00/$30.00		
Vol. 25	(avail. April 1985)	$27.00/$30.00	Vol(s).______	Vol.______
Annual Review of **PHYSICAL CHEMISTRY**				
Vols. 10-21, 23-32	(1959-1970; 1972-1981)	$20.00/$21.00		
Vol. 33	(1982)	$22.00/$25.00		
Vols. 34-35	(1983-1984)	$28.00/$31.00		
Vol. 36	(avail. Nov. 1985)	$28.00/$31.00	Vol(s).______	Vol.______
Annual Review of **PHYSIOLOGY**				
Vols. 19-43	(1957-1981)	$20.00/$21.00		
Vol. 44	(1982)	$22.00/$25.00		
Vols. 45-46	(1983-1984)	$27.00/$30.00		
Vol. 47	(avail. March 1985)	$27.00/$30.00	Vol(s).______	Vol.______
Annual Review of **PHYTOPATHOLOGY**				
Vols. 2-19	(1964-1981)	$20.00/$21.00		
Vol. 20	(1982)	$22.00/$25.00		
Vols. 21-22	(1983-1984)	$27.00/$30.00		
Vol. 23	(avail. Sept. 1985)	$27.00/$30.00	Vol(s).______	Vol.______
Annual Review of **PLANT PHYSIOLOGY**				
Vols. 10, 13-32	(1959; 1962-1981)	$20.00/$21.00		
Vol. 33	(1982)	$22.00/$25.00		
Vols. 34-35	(1983-1984)	$27.00/$30.00		
Vol. 36	(avail. June 1985)	$27.00/$30.00	Vol(s).______	Vol.______
Annual Review of **PSYCHOLOGY**				
Vols. 4, 5, 8,	(1953, 1954, 1957)	$20.00/$21.00		
Vols. 10-24, 26-32	(1959-1973; 1975-1981)	$20.00/$21.00		
Vol. 33	(1982)	$22.00/$25.00		
Vols. 34-35	(1983-1984)	$27.00/$30.00		
Vol. 36	(avail. Feb. 1985)	$27.00/$30.00	Vol(s).______	Vol.______
Annual Review of **PUBLIC HEALTH**				
Vols. 1-2	(1980-1981)	$20.00/$21.00		
Vol. 3	(1982)	$22.00/$25.00		
Vols. 4-5	(1983-1984)	$27.00/$30.00		
Vol. 6	(avail. May 1985)	$27.00/$30.00	Vol(s).______	Vol.______
Annual Review of **SOCIOLOGY**				
Vols. 1-7	(1975-1981)	$20.00/$21.00		
Vol. 8	(1982)	$22.00/$25.00		
Vols. 9-10	(1983-1984)	$27.00/$30.00		
Vol. 11	(avail. Aug. 1985)	$27.00/$30.00	Vol(s).______	Vol.______

SEE ORDERING INFORMATION ON PAGE 4

SPECIAL PUBLICATIONS

		Prices Postpaid per volume USA/elsewhere	Regular Order Please send:
Annual Reviews Reprints: **Cell Membranes,** 1975-1977			
(published 1978)	Softcover	$12.00/$12.50	______ Copy(ies).
Annual Reviews Reprints: **Cell Membranes,** 1978-1980			
(published 1981)	Hardcover	$28.00/$29.00	______ Copy(ies).
Annual Reviews Reprints: **Immunology,** 1977-1979			
(published 1980)	Softcover	$12.00/$12.50	______ Copy(ies).
History of Entomology			
(published 1973)	Clothbound	$10.00/$10.50	______ Copy(ies).
Intelligence and Affectivity: Their Relationship During Child Development, by Jean Piaget			
(published 1981)	Hardcover	$8.00/$9.00	______ Copy(ies).
Some Historical and Modern Aspects of Amino Acids, Fermentations, and Nucleic Acids: Proceedings of a Symposium held in St. Louis, Missouri, June 3, 1981			
(published 1982)	Softcover	$10.00/$12.00	______ Copy(ies).
Telescopes for the 1980s			
(published 1981)	Hardcover	$27.00/$28.00	______ Copy(ies).
The Excitement and Fascination of Science, Volume 1			
(published 1965)	Clothbound	$6.50/$7.00	______ Copy(ies).
The Excitement and Fascination of Science, Volume 2			
(published 1978)	Hardcover	$12.00/$12.50	
	Softcover	$10.00/$10.50	______ Copy(ies).

TO: **ANNUAL REVIEWS INC., 4139 El Camino Way, Palo Alto, CA 94306-9981, USA • Tel. (415) 493-4400**

Please enter my order for the publications checked above. California orders, add sales tax. Prices subject to change without notice.

Institutional purchase order No. ______

Amount of remittance enclosed $ ______

Charge my account ☐ VISA ☐ MasterCard ☐ American Express

INDIVIDUALS: Prepayment required in U.S. funds or charge to bank card below. Include card number, expiration date, and signature.

Acct. No. ______

Exp. Date ______ ______
Signature

Name ______
Please print

Address ______ Zip Code ______
Please print

______ Send free copy of current *Prospectus* ☐
Area(s) of Interest